Statistical Monitoring of Complex Multivariate Processes

Statistics in Practice

Series Advisors

Human and Biological Sciences
Stephen Senn
CRP-Santé, Luxembourg

Earth and Environmental Sciences
Marian Scott
University of Glasgow, UK

Industry, Commerce and Finance
Wolfgang Jank
University of Maryland, USA

Founding Editor
Vic Barnett
Nottingham Trent University, UK

Statistics in Practice is an important international series of texts which provide detailed coverage of statistical concepts, methods and worked case studies in specific fields of investigation and study.

With sound motivation and many worked practical examples, the books show in down-to-earth terms how to select and use an appropriate range of statistical techniques in a particular practical field within each title's special topic area.

The books provide statistical support for professionals and research workers across a range of employment fields and research environments. Subject areas covered include medicine and pharmaceutics; industry, finance and commerce; public services; the earth and environmental sciences, and so on.

The books also provide support to students studying statistical courses applied to the above areas. The demand for graduates to be equipped for the work environment has led to such courses becoming increasingly prevalent at universities and colleges.

It is our aim to present judiciously chosen and well-written workbooks to meet everyday practical needs. Feedback of views from readers will be most valuable to monitor the success of this aim.

A complete list of titles in this series appears at the end of the volume.

Statistical Monitoring of Complex Multivariate Processes

With Applications in Industrial Process Control

Uwe Kruger and Lei Xie

Institute of Cyber-Systems and Control, State Key Laboratory of Industrial Control Technology, Zhejiang University, China

WILEY

A John Wiley & Sons, Ltd., Publication

This edition first published 2012

Registered office
John Wiley & Sons Ltd, The Atrium, Southern Gate, Chichester, West Sussex, PO19 8SQ, United Kingdom

For details of our global editorial offices, for customer services and for information about how to apply for permission to reuse the copyright material in this book please see our website at www.wiley.com.

Library of Congress Cataloging-in-Publication Data

Kruger, Uwe, Dr.
Advances in statistical monitoring of complex multivariate processes : with applications in industrial process control / Uwe Kruger and Lei Xie.
p. cm.
Includes bibliographical references and index.
ISBN 978-0-470-02819-3 (hardback)
1. Multivariate analysis. I. Xie, Lei. II. Title.
QA278.K725 2012
519.5′35 – dc23

2012016445

A catalogue record for this book is available from the British Library.

ISBN: 978-0-470-02819-3

Set in 10/12pt Times by Laserwords Private Limited, Chennai, India
Printed and bound in Singapore by Markono Print Media Pte Ltd

Dedicated to Dr. Xun Wang, my wife, and Melanie Kruger, my daughter, whose abundant support has made this book possible.

Uwe Kruger

Without the dedication and support of my wife, Jingjing Sha, this book would never have been possible.

Lei Xie

Contents

Preface

This book provides a timely reference text for academics, undergraduate and graduate students, and practitioners alike in the area of process monitoring and safety, as well as product quality assurance using multivariate statistics. The rapid evolution of this research area over the past 20 years is mainly driven by significant advances in computer horsepower and the ever growing demand from industry to effectively and efficiently monitor production processes. As an example, Nimmo (1995) outlined that the US-based petrochemical industry could save an estimated \$10 bn annually if abnormal conditions could be detected, diagnosed and appropriately dealt with. Moreover, the demand from the oil and gas industry, other chemical engineering and general manufacturing industries is also a result of ever tighter government legislation on emissions and increased safety standards of their products.

The wide range of applications of multivariate statistics for process monitoring, safety and product quality is of considerable interest to the readership in chemical, mechanical, manufacturing, electrical and electronic, industrial and other related engineering and science disciplines. This research text serves as a reference for introductory and advanced courses on process safety, process monitoring and product quality assurance, total quality management of complex technical systems and is a supplementary text for courses on applied statistics and process systems engineering. As a textbook and reference, this book pays particular attention to a balanced presentation between the required theory and the industrial exploitation of statistical-based process monitoring, safety and quality assurance.

To cater for the different audiences with their partially conflicting demands, the scope of the book is twofold. The main thrust lies on outlining the relevant and important fundamental concept of multivariate statistical process control or, in short, MSPC and to demonstrate the working of this technology using recorded data from complex process systems. This addresses the needs for the more how-does-it-work and what-does-it-do oriented readership of this book, which includes undergraduate students, industrial practitioners and industrially oriented researchers. The second pillar is the theoretical analysis of the underlying MSPC component technology, which is important for the more research-oriented audience including graduate students and academicians.

The twofold coverage of the material results from the research background of both authors, which is centered on academic research in process monitoring, safety, product quality assurance and general process systems engineering, and their participation in numerous industrial R&D projects, including consultancy concerning the application of MSPC and the development of commercial software packages. As this book carefully outlines and discusses, the main advantage of the MSPC technology is its simplicity and reliance on recorded data and some *a priori* knowledge regarding the operation of the process system. On the other hand, this simplicity comes at the expense of stringent assumptions, including that the process is stationary and time-invariant, and that the process variables follow a Gaussian distribution.

With this in mind and based on academic and industrial R&D experience, the authors are convinced that MSPC technology has the potential to play an important role in commercial applications of process monitoring, safety and product quality assurance. This view is also supported by the arrival of software that entered the value-added market for commercially available packages, which includes AspenMultivariate™, Wonderware, SIMCA-P (to name but a few), consultancy companies, such as Perceptive Engineering Ltd., Eigenvector Research Inc. and statistical data analysis software, e.g. STATISTICA, SAS®.

The first thrust of MSPC work for monitoring complex process systems emerged in the late 1980 and the early 1990s and lays out a statistically sound concept under these basic assumptions. It is important to note, however, that if a process 'unfortunately forgets' to meet the above assumptions, the corresponding monitoring charts may produce false alarms or the sensitivity in detecting minor upsets is compromised. From the end of the 1990s until now, research work that has enhanced the core MSPC methodology has removed some of these stringent assumptions. This, in turn, allows the enhanced MSPC technology to be applicable in a more practically relevant environment.

Besides the required theoretical foundation of the MSPC methodology, this book also includes a detailed discussion of these advances, including (i) the monitoring of time-variant process systems, where the mean and variance of the recorded variables, and the relationship between and among these sets, change over time, (ii) the development and application of more practically relevant data structures for the underlying MSPC monitoring models and (iii) the development of a different construction of monitoring statistics and charts which significantly improves their sensitivity in detecting incipient fault conditions.

This book ideally supplements the good number of research texts available on multivariate statistics, statistical process control, process safety and product quality assurance. In particular, the research text brings together the theory of MSPC with industrial applications to demonstrate its usefulness. In particular, the mix of theory and practice in this area is rare; (exceptions include Mason and Young (2001)). Moreover, good and solid reference that address the theory as well as the application of component technology are rarely written for the industrial practitioner whose experience is pivotal in any process monitoring, safety and product quality assurance application.

To comprehend the content of this book, the readership is expected to possess basic knowledge of calculus including differentiation, integration and matrix computation. For the application study, a basic understanding of principles in physics and chemistry is helpful in following the analysis of the application studies and particularly the diagnosis of the recorded fault conditions. To enhance the understanding of the presented material and to improve the learning experience, each chapter presenting theoretical material, except the last two, includes a tutorial session which contains questions and homework-style projects. The questions assist with the familiarization of the covered material and the projects help the reader to understand the underlying principles through experimenting and discovering the facts and findings presented in this book either through self-study reports or team-based project reports. The calculations can be carried out using standard computational software, for example Matlab®.

Acknowledgements

This book would not have been possible without encouragement, dedicated help and constructive comments from a large number of people. We would like to thank SupCon Software Co Ltd., Hangzhou, P.R. China, for providing access to the data sets used in Chapters 7 and 8. We particularly thank Dr. Yong Gu and Mr. Yanhui Zhang for technical advice regarding these data sets and for interpreting associated results. Our thanks also extend to Dr. Jian Chu, Dr. Hongye Su and Dr. Shuqing Wang for facilitating numerous research visits by Dr. Uwe Kruger to the Institute of Cyber Systems and Control, Zhejiang University, P.R. China, from 2006 onwards.

Dr. Xie is grateful for financial support from the National Science Foundation of China (Grant No. 60904039, 61134007) and the Fundamental Research Funds for the Central Universities. Furthmore, Dr. Kruger would like to acknowledge the financial support of the Ministry Of Education Program of Introducing Talents of Discipline (111 Project, Grant No. B07031).

With regards to the recorded data from the chemical reaction process in Chapter 4, Dr. Uwe Kruger is grateful to Dr. Keith Smith for advice on how to interpret the process data, and ICI Polymer Chemicals for providing access to the operating data used and for the permission to present associated results in Kruger *et al* (2001). Dr. Kruger would also like to thank Mr. Steve Robinson for providing helpful advice in analyzing and interpreting the process data of the distillation process and is grateful to BP Amoco Scotland for providing access to the operating data used in Chapter 5 and for the permission to present associated results in Wang *et al* (2003). We wish to acknowledge the contribution of Dr. Randall C. McFarlane, who introduced the mechanistic simulator of the fluid catalytic cracking unit and for offering helpful advice regarding the generation of realistic operating scenarios for the application study in Chapter 7.

Dr. Uwe Kruger is indebted to Dr. David J. Sandoz for the mentoring and the care as adviser for his doctorate degree at the University of Manchester and for introducing the area of industrial process control and monitoring during his attachment to the Control Technology Center Ltd. and Predictive Control Ltd. between 1996 and 2000. Dr. Sandoz's leadership and vision has always been a source of inspiration and a reference for technology transfer, improving existing methods and for generating conceptual ideas. In addition, Dr. Kruger would like to acknowledge the mentoring as well as the helpful and constructive advice by Dr. George W. Irwin during his employment at Queen's University Belfast between 2001 and 2007. Dr. Irwin's leadership of the Intelligent Systems and Control Group contributed in large parts to the research work in Chapter 7. During that time, Dr. Kruger's research activities on process monitoring, process safety and quality assurance were financially supported by DuPont (UK) Ltd., Invest Northern Ireland, the Engineering and Physical Science Research Council, the European Social Fund, the Department of Education and Learning, the Center for the Theory and Application of Catalysis and the Virtual Engineering Center. From 2007 to 2012, Dr. Kruger acknowledges financial support from The Petroleum Institute to continue the development of industrially relevant techniques for process monitoring, process safety and product quality assurance. Dr. Kruger would particularly like to acknowledge the helpful assistance by Dr. Cornelis Peters and Dr. Ali Almansoori of the Chemical Engineering Program and the advice by Dr. Jaap Geluk of the Department of Mathematics regarding the central limit theorem.

We are also in debt to many graduate students, colleagues and friends for their encouragement, helpful suggestions and invaluable contributions in generating the research work in this book. As it is difficult to provide an inclusive list of all contributors to our work, we would like to mention in particular the academic colleagues Dr. Xun Wang, Dr. Qian Chen, Dr. Tim Littler, Dr. Barry Lennox, Dr. Günter Wozny, Dr. Sebstian Engell, Dr. Yiqi Zhou, Dr. Enrique Luis Lima, Dr. José Carlos Pinto and Dr. Zhihuan Song. The following former postdoctoral researchers and graduate students strongly contributed to the work in this book: Dr. Zhiqiang Ge, Dr. David Antory, Dr. Dirk Lieftucht, Dr. Yan Zhou, Dr. Xueqin Liu, Dr. Thiago Feital and Dr. Udo Schubert. The authors also want to acknowledge the contribution by the graduate students Mr. Omar AlJaberi, Ms. Zhe Li and Mr. Gui Chen. Dr. Uwe Kruger finally wishes to thank Mr. Marcel Meronk for his support in relation to the application studies in Chapters 4 and 5 and for his friendship and encouragement.

Finally, the authors would like to thank the Wiley team and, in particular, Mr. Richard Davies, Miss. Heather Kay and Mrs. Susan Barclay for their patience, invaluable support and encouragement for drafting and completing this book.

Abbreviations

CLT	Central Limit Theorem
flops	Number of floating point operations
i.d.	Independently distributed
i.i.d.	Identically and independently distributed
LCL	Lower Confidence Limit
LV	Latent Variable
MLPCA	Maximum Likelihood Principal Component Analysis
MLPLS	Maximum Likelihood Partial Least Squares
MRPLS	Maximum Redundancy Partial Least Squares
MSPC	Multivariate Statistical Process Control
MWPCA	Moving Window Principal Component Analysis
OLS	Ordinary Least Squares
PCA	Principal Component Analysis
PDF	Probability Density Function
PLS	Partial Least Squares
RPCA	Recursive Principal Component Analysis
RPLS	Recursive Partial Least Squares
SPC	Statistical Process Control
SVD	Singular Value Decomposition
UCL	Upper Control Limit
w.r.t	with respect to

Symbols

$\widehat{\ }$	Estimated variable
$\sim$	Follows a specific distribution function, e.g. $z \sim F(\cdot)$
$\mathring{\ }$	Stochastic component of variable
$\bar{\ }$	Mean value of variable
lim	Limit operator, e.g. $\lim\limits_{x \to \infty} \frac{1}{x} \to 0$
$\otimes$	Kronecker product of two matrices
i	Imaginary unit, $\mathrm{i} = \sqrt{-1}$
$\mathbf{a}$	Vector $\mathbf{a} \in \mathbb{R}^N$
$\mathbf{A}$	Matrix $\mathbf{A} \in \mathbb{R}^{N \times M}$
$\mathbf{I}$	Identity matrix
$\mathrm{diag}\left\{ a_1 \cdots a_N \right\}$	Diagonal matrix $\mathbf{A} = \begin{bmatrix} a_1 & \cdots & 0 \\ \vdots & \ddots & \vdots \\ 0 & \cdots & a_N \end{bmatrix}$
$\lvert\mathbf{A}\rvert$	Determinant of $\mathbf{A}$, $N = M$
$\lVert\mathbf{a}\rVert$	Length of $\mathbf{a}$, $\lVert\mathbf{a}\rVert = \sqrt{\sum\limits_{i=1}^{N} a_i^2}$
$\lVert\mathbf{A}\rVert$	Frobenius norm of $\mathbf{A}$, $\lVert\mathbf{A}\rVert = \sqrt{\sum\limits_{i=1}^{N}\sum\limits_{j=1}^{M} a_{ij}^2}$
$\mathbf{A}^{\dagger}$	Generalized inverse of a matrix, $\mathbf{A}^{\dagger}\mathbf{A} = \mathbf{I}$ if $N > M$ or $\mathbf{A}\mathbf{A}^{\dagger} = \mathbf{I}$ if $N < M$
$\mathbf{A}^{\perp}$	Orthogonal complement of column or row space of $\mathbf{A}$, $\mathbf{A}^{\perp}\mathbf{A} = \mathbf{0}$ if $N > M$ or $\mathbf{A}\mathbf{A}^{\perp} = \mathbf{0}$ if $N < M$
$\mathrm{trace}\left\{\mathbf{A}\right\}$	Trace of $\mathbf{A}$, $\mathrm{trace}\left\{\mathbf{A}\right\} = \sum\limits_{i=1}^{N} a_{ii}$, $N = M$
$E\left\{\cdot\right\}$	Expectation operator, e.g. $E\{z\} = \bar{z}$
$\mathrm{vec}\left\{\mathbf{A}\right\}$	Vector representation of matrix, $\mathrm{vec}\left\{\mathbf{A}\right\} = \left(\mathbf{a}_1^T \cdots \mathbf{a}_M^T\right)^T \in \mathbb{R}^{N \cdot M}$

Nomenclature

$\mathfrak{A}$, $\mathfrak{B}$, c, d, ϑ, ϱ	Constants or correction terms
$\mathcal{C}_{z_0 z_0}$	Correlation matrix for variable set z_0
χ^2, χ^2_α	Cumulative chi-squared distribution function and critical value for significance α
$\mathfrak{C}$	Projection matrix, $\mathfrak{C} = \mathbf{P}\mathbf{P}^T$ (PCA model)
δ_{ij}	Kronecker delta, $\delta_{ij} = 1$ for $i = j$ and 0 for $i \neq j$
$\mathbf{e}$, $\mathbf{f}$, $\mathbf{E}$, $\mathbf{F}$	Residual vectors and matrices for input and output variables (PLS/MRPLS model)
$\mathfrak{e}$, $\mathfrak{f}$	Error vector for input and variables (PLS/MRPLS model)
$f(\cdot)$, $F(\cdot)$	Probability density and cumulative distribution function
$\mathcal{F}$, $\mathcal{F}_\alpha(m_1, m_2)$	Cumulative distribution and critical value of an F distribution with m_1 and m_2 degrees of freedom and a significance α
$\mathbf{g}$, $\mathbf{G}$	Residual vector and matrix (PCA model)
$\mathfrak{g}$	Error vector for process variables (PCA model)
$\Gamma(m)$	Gamma function, $m > 0$
H_0,H_1	Null and alternative hypothesis
i, j	Variable indices, e.g. z_i or x_j
k, l	Sampling indices, e.g. $z(k)$ is the kth sample of z
$\mathfrak{I}$	Index set
K, $\mathfrak{K}$ and $\mathcal{K}$	Number of reference samples, application delay for adapted MSPC models and length of moving window
$\mathbf{\Lambda}$, λ_i	Diagonal matrix of eigenvalues, ith eigenvalue
n	Number of retained pairs of latent variables (PCA and PLS/MRPLS models)
n_x, n_y and n_z	Number of input and output variables (PLS/MRPLS model), and number of process variables (PCA model)
$\mathcal{N}\left\{\bar{\mathbf{z}}, \mathbf{S}_{zz}\right\}$	Cumulative Gaussian distribution function of mean $\bar{\mathbf{z}}$ and covariance matrix $\mathbf{S}_{zz}$
$\mathbf{\Omega}$, $\boldsymbol{\omega}$	General projection matrix and vector
$\mathbf{p}$, $\mathbf{P}$	Loading vector and matrix for process variables (PCA model) and input variable sets (PLS/MRPLS model)
$\mathbf{q}$, $\acute{\mathbf{q}}$, $\mathbf{Q}$, $\acute{\mathbf{Q}}$	Weight and loading vectors and matrices for output variables (PLS/MRPLS model)

Q, Q_α	Residual Q statistic and its control limit for significance α
r_{ij}	Correlation coefficient between ith and jth variable
$\mathbf{r}$, $\mathbf{w}$, $\mathbf{R}$, $\mathbf{W}$	Loading vectors and matrices for input variables (PLS/MRPLS model)
σ, $\mathbf{\Sigma}$	Standard deviation and diagonal matrix storing standard deviations
$\mathbf{S}_{ss} = \mathfrak{T}\mathfrak{L}\mathfrak{T}^T$	Covariance matrix of source variables and its eigendecomposition
$\mathbf{t}$, $\mathbf{T}$	Score vector and matrix for data vector of PCA model and input vector of PLS/MRPLS model
T^2, T_α^2	Hotelling's T^2 statistic and control limit for significance α
$\mathbf{u}$, $\mathbf{U}$	Score vector and matrix for output vector (PLS model)
$\mathbf{\Upsilon}$	Fault subspace
$\mathbf{w}$, $\mathfrak{w}$, $\mathbf{W}$	Weight vectors and matrix for input variables (PLS/MRPLS model)
$\mathbf{x}$, $\mathbf{X}$	Input vector and matrix (PLS/MRPLS model)
Ξ and ($\mathfrak{P}$, $\mathfrak{Q}$)	Parameter matrices (PCA and PLS/MRPLS models)
$\mathbf{y}$, $\mathbf{Y}$	Output vector and matrix (PLS/MRPLS model)
$\mathbf{z}$, $\mathbf{Z}$	Data vector and matrix (PCA model)
$\mathfrak{z}$	Parameter vector determining confidence interval for covariance matrix (Chapter 7)

Metric conversions

1lb/s	0.453 kb/s
1scf/s	$0.0283 m^3/s$
1psi/psia	6894.75732Pa
Degrees in K	(80 × degrees in F − 32)/1.8 + 273.15
1 ICFM	$1.699 m^3/h$
degrees in C	(80 × degrees in F − 32)/1.8

Introduction

Performance assessment and quality control of complex industrial process systems are of ever increasing importance in the chemical and general manufacturing industries as well as the building and construction industry (Gosselin and Ruel 2007; Marcon *et al.* 2005; Miletic *et al.* 2004; Nimmo 1995). Besides other reasons, the main drivers of this trend are: the ever more stringent legislation based on process safety, emissions and environmental pollution (ecological awareness); an increase in global competition; and the desire of companies to present a green image of their production processes and products.

Associated tasks entail the on-line monitoring of production facilities, individual processing units and systems (products) in civil, mechanical, automotive, electrical and electronic engineering. Examples of such systems include the automotive and the aerospace industries for monitoring operating conditions and emissions of internal combustion and jet engines; buildings for monitoring the energy consumption and heat loss; and bridges for monitoring stress, strain and temperature levels and hence assess elastic deformation.

To address the need for rigorous process monitoring, the level of instrumentation of processing units and general engineering systems, along with the accuracy of the sensor readings, have consequently increased over the past few decades. The information that is routinely collected and stored, for example in distributed control systems for chemical production facilities and the engine management system for internal combustion engines, is then benchmarked against conditions that are characterized as normal and/or optimal.

The data records therefore typically include a significant number of process variables that are frequently sampled. This, in turn, creates huge amounts of process data, which must be analyzed online or archived for subsequent analysis. Examples are reported for:

- the chemical industry (Al-Ghazzawi and Lennox 2008; MacGregor *et al.* 1991; Piovoso and Kosanovich 1992; Simoglou *et al.* 2000; Wang *et al.* 2003);

- the general manufacturing industry (Kenney *et al.* 2002; Lane *et al.* 2003; Martin *et al.* 2002; Monostori and Prohaszka 1993; Qin *et al.* 2006);

- internal combustion engines (Gérard *et al.* 2007; Howlett *et al.* 1999; Kwon *et al.* 1987; McDowell *et al.* 2008; Wang *et al.* 2008);

- aircraft systems (Abbott and Person 1991; Boller 2000; Jaw 2005; Jaw and Mattingly 2008; Tumer and Bajwa 1999); and
- civil engineering systems (Akbari *et al.* 2005; Doebling *et al.* 1996; Ko and Ni 2005; Pfafferott *et al.* 2004; Westergren *et al.* 1999).

For the chemical and manufacturing industries, the size of the data records and the ever increasing complexity of such systems have caused efficient process monitoring by plant operators to become a difficult task. This complexity stems from increasing levels of process optimization and intensification, which gives rise to operating conditions that are at the limits of operational constraints and which yield complex dynamic behavior (Schmidt-Traub and Górak 2006). A consequence of these trends is a reduced safety margin if the process shows some degree of abnormality, for example caused by a fault (Schuler 2006).

Examples for monitoring technical systems include internal combustion engines and gearbox systems. Process monitoring of internal combustion engines relates to tackling increasing levels of pollution caused by the emissions of an ever growing number of registered vehicles and has resulted in the introduction of the first on-board-diagnostic (OBD) system in the United States in 1988, and in Europe (EURO1) in 1992. The requirement for more advanced monitoring systems culminated in the introduction of OBDII (1994), EURO2 (1997) and EURO3 (2000) legislation. This trend has the aim of continuously decreasing emissions and is supported through further regulations, which relate to the introduction of OBDIII (considered since 2000), EURO4 (2006) and EURO5 (2009) systems.

Current and future regulations demand strict monitoring of engine performance at certain intervals under steady-state operating conditions. This task entails the diagnosis of any fault condition that could potentially cause the emissions to violate legislated values at the earliest opportunity. With respect to this development, a prediction by Powers and Nicastri (1999) indicated that the integration of model-based control systems and design techniques have the potential to produce safer, more comfortable and manoeuvrable vehicles. According to Kiencke and Nielsen (2000), there are a total of three main objectives that automotive control systems have to adhere to: (i) maintaining efficiency and low fuel consumption, (ii) producing low emissions to protect the environment and (iii) ensuring safety. Additional benefits of condition monitoring are improved reliability and economic operation (Isermann and Ballé 1997) through early fault detection.

For gearbox systems, the early detection of incipient fault conditions is of fundamental importance for their operation. Gearboxes can be found in aerospace, civil and general mechanical systems. The consequences of not being able to detect such faults at early stages can, for example, include reduced productivity in manufacturing processes, reduced efficiency of engines, equipment damage or even failure. Early detection of such faults can therefore provide significant improvements in the reduction of operational and maintenance costs, system down-time, and lead to increased levels of safety, which is of ever growing importance. An incipiently developing fault in a mechanical system usually affects certain parameters, such as vibration, noise and temperature. The analysis of these *external variables* therefore allows the monitoring of internal components, such as gears, which are usually inaccessible without the dismantling of the system. It is consequently essential to extract relevant information from the recorded signals with the aim of detecting any irregularities that could be caused by such faults.

The research community has utilized a number of different approaches to monitor complex technical systems. These include model-based approaches (Ding 2008; Frank *et al.* 2000; Isermann 2006; Simani *et al.* 2002; Venkatasubramanian *et al.* 2003) that address a wide spectrum of application areas, signal-based approaches (Bardou and Sidahmed 1994; Chen *et al.* 1995; Hu *et al.* 2003; Kim and Parlos 2003) which are mainly applied to mechanical systems, rule-based techniques (Iserman 1993; Kramer and Palowitch 1987; Shin and Lee 1995; Upadhyaya *et al.* 2003) and more recently knowledge-based techniques (Lehane *et al.* 1998; Ming *et al.* 1998; Qing and Zhihan 2004; Shing and Chee 2004) that blend heuristic knowledge into monitoring application. Such techniques have shown their potential whenever cost-benefit economics have justified the required effort in developing applications.

Given the characteristics of modern production and other technical systems, however, such complex technical processes may present a large number of recorded variables that are affected by a few common trends, which may render these techniques difficult to implement in practice. Moreover, such processes often operate under steady-state operation conditions that may or may not be predefined. To some extent, this also applies to automotive systems as routine technical inspections, for example once per year, usually include emission tests that are carried out at a reference steady state operation condition of the engine.

Underlying trends are, for example, resulting from known or unknown disturbances, interactions of the control system with the technical system, and minor operator interventions. This produces the often observed high degree of correlated among the recorded process variables that mainly describe common trends or common cause variation. The sampled data has therefore embedded within it information for revealing the current state of process operation. The difficult issue here is to extract this information from the data and to present it in a way that can be easily interpreted.

Based on the early work on quality control and monitoring (Hotelling 1947; Jackson 1959, 1980; Jackson and Morris 1956, 1957; Jackson and Mudholkar 1979), several research articles around the 1990s proposed a multivariate extension to statistical process control Kresta *et al.* (1989, 1991) MacGregor *et al.* (1991) Wise *et al.* (1989b, 1991) to generate a statistical fingerprint of a technical system based on recorded reference data. Methods that are related to this extension are collectively referred to as multivariate statistical process control or MSPC. The application of MSPC predominantly focussed on the chemical industry (Kosanovich and Piovoso 1991; Morud 1996; Nomikos and MacGregor 1994; Piovoso and Kosanovich 1992; Piovoso *et al.* 1991) but was later extended to general manufacturing areas (Bissessur *et al.* 1999; 2000; Lane *et al.* 2003; Martin *et al.* 2002; Wikström *et al.* 1998).

Including this earlier work, the last two decades have seen the development and application of MSPC gaining substantial interest in academe and industry alike. The recipe for the considerable interest in MSPC lies in its simplicity and adaptability for developing monitoring applications, particularly for larger numbers of recorded variables. In fact, MSPC relies on relatively few assumptions and only requires routinely collected operating data from the process to be monitored. The first of four parts of this book outlines and describes these assumptions, and is divided into a motivation for MSPC, a description of the main MSPC modeling methods and the underlying data structures, and the construction of charts to carry out on-line monitoring.

For monitoring processes in the chemical industry, the research community has proposed two different MSPC approaches. The first one relates to processes that produce a specific product on a continuous basis, i.e. they convert a constant stream of inputs into a constant stream of outputs and are referred to as a *continuous processes*. Typical examples of continuous processes can be found in the petrochemical industry. The second approach has been designed to monitor processes that convert a discontinuous feed into the required product over a longer period of time. More precisely, and different from a continuous process, this type of process receives a feed that remains in the reactor over a significantly longer period of time before the actual production process is completed. Examples of the second type of process can be found in the pharmaceutical industry and such processes are referred to as *batch processes*. This book focuses on continuous processes to provide a wide coverage of processes in different industries. References that discuss the monitoring of batch processes include Chen and Liu (2004), Lennox *et al.* (2001), Nomikos and MacGregor (1994, 1995), van Sprang *et al.* (2002) to name only a few.

The second part of this book then presents two application studies of a chemical reaction process and a distillation process. Both applications demonstrate the ease of utilizing MSPC for process monitoring and detecting as well as diagnosing abnormal process behavior. The detection is essentially a boolean decision whether current process behavior still matches the statistical fingerprint describing behavior that is deemed normal and/or optimal. If it matches, the process is *in-statistical-control* and if it does not the process is *out-of-statistical-control*. The diagnosis of abnormal events entails the identification and analysis of potential root causes that have led to the anomalous behavior. In other words, it assesses why the current plant behavior deviates from that manifested in the statistical fingerprint, constructed from a historic data record, that characterizes normal process behavior. The second part of this book also demonstrates that the groundwork on MSPC in the early to mid 1990s may rely on oversimplified assumptions that may not represent true process behavior.

The aim of the third part is then to show advances in MSPC which the research literature has proposed over the past decade in order to overcome some of the pitfalls of this earlier work. These advances include:

- improved data structures for MSPC monitoring models;

- the removal of the assumption that the stochastic process variables have a constant mean and variance, and the variable interrelationships are constant over time; and
- a fresh look at constructing MSPC monitoring charts, resulting in the introduction of a new paradigm which significantly improves the sensitivity of the monitoring scheme in detecting incipient fault conditions.

In order to demonstrate the practical usefulness of these improvements, the application studies of the chemical reactor and the distillation processes in the second part of this book are revisited. In addition, the benefits of the adaptive MSPC scheme is also shown using recorded data from a furnace process and the enhanced monitoring scheme is applied to recorded data from gearbox systems.

Finally, the fourth part of this book presents a detailed treatment of the core MSPC modeling methods, including their objective functions, and their statistical and geometric properties. The analysis also includes the discussion of computational issues in order to obtain data models efficiently.

PART I

FUNDAMENTALS OF MULTIVARIATE STATISTICAL PROCESS CONTROL

1

Motivation for multivariate statistical process control

This first chapter outlines the basic principles of multivariate statistical process control. For the reader unfamiliar with statistical-based process monitoring, a brief revision of statistical process control (SPC) and its application to industrial process monitoring are provided in Section 1.1.

The required extension to MSPC to address data correlation is then motivated in Section 1.2. This section also highlights the need to extract relevant information from a large dimensional data space, that is the space in which the variation of recorded variables is described. The extracted information is described in a reduced dimensional data space that is a subspace of the original data space.

To help readers unfamiliar with MSPC technology, Section 1.3 offers a tutorial session, which includes a number of questions, small calculations/examples and projects to help familiarization with the subject and to enhance the learning outcomes. The answers to these questions can be found in this chapter. Project 2 to 4 require some self study and result in a detailed understanding on how to interpret SPC monitoring charts for detecting incipient fault conditions.

1.1 Summary of statistical process control

Statistical process control has been introduced into general manufacturing industry for monitoring process performance and product quality, and to observe the general process variation, exhibited in a few key process variables. Although this indicates that SPC is a process monitoring tool, the reference to *control* (in control engineering often referred to as describing and analyzing the feedback or feed-forward controller/process interaction), is associated with product or, more

Statistical Monitoring of Complex Multivariate Processes: With Applications in Industrial Process Control, First Edition. Uwe Kruger and Lei Xie.

precisely, process improvement. In other words, the *control* objective here is to reduce process variation and to increase process reliability and product quality. One could argue that the controller function is performed by process operators or, if a more fundamental interaction with the process is required, a task force of experienced plant personnel together with plant managers. The next two subsections give a brief historical review of its development and outline the principles of SPC charts. The discussion of SPC in this section only represents a brief summary for the reader unfamiliar with this subject. A more in-depth and detailed treatment of SPC is available in references Burr (2005); Montgomery (2005); Oakland (2008); Smith (2003); Thompson and Koronacki (2002).

1.1.1 Roots and evolution of statistical process control

The principles of SPC as a system monitoring tool were laid out by Dr. Walter A. Shewhart during the later stages of his employment at the Inspection Engineering Department of the Western Electric Company between 1918 and 1924 and from 1925 until his retirement in 1956 at the Bell Telephone Laboratories. Shewhart summarized his early work on statistical control of industrial production processes in his book (Shewhart, 1931). He then extended this work which eventually led to the applications of SPC to the measurement processes of science and stressed the importance of operational definitions of basic quantities in science, industry and commerce (Shewhart, 1939). In particular, the latter book has had a profound impact upon statistical methods for research in behavioral, biological and physical sciences, as well as general engineering.

The second pillar of SPC can be attributed to Dr. Vilfredo Pareto, who first worked as a civil engineer after graduation in 1870. Pareto became a lecturer at the University of Florence, Italy from 1886, and from 1893 at the University of Lausanne, Switzerland. He postulated that many system failures are a result of relatively few causes. It is interesting to note that these pioneering contributions culminated in two different streams of SPC, where Shewhart's work can be seen as *observing a system*, whilst Pareto's work serves as a *root cause analysis* if the observed system behaves abnormally. Attributing the control aspect (*root cause analysis*) of SPC to the 'Pareto Maxim' implies that system improvement requires skilled personnel that are able to find and correct the causes of 'Pareto glitches', those being abnormal events that can be detected through the use of SPC charts (*observing the system*).

The work by Shewhart drew the attention of the physicists Dr. W. Edwards Deming and Dr. Raymond T. Birge. In support of the principles advocated by Shewart's early work, they published a landmark article on measurement errors in science in 1934 (Deming and Birge 1934). Predominantly Deming is credited, and to a lesser extend Shewhart, for introducing SPC as a tool to improved productivity in wartime production during World War II in the United States, although the often proclaimed success of the increased productivity during that time is contested, for example Thompson and Koronacki (2002, p5). Whilst the influence of SPC faded substantially after World War II in the United States, Deming became an 'ambassador' of Shewhart's SPC principles in Japan from the mid 1950s. Appointed by the United States Department of the Army, Deming taught engineers, managers including top management, and scholars in SPC and concepts of quality. The quality and reliability of Japanese products, such as cars and electronic devices, are predominantly attributed to the rigorous transfer of these SPC principles and the introduction of Taguchi methods, pioneered by Dr. Genichi Taguchi (Taguchi 1986), at all production levels including management.

SPC has been embedded as a cornerstone in a wider quality context, that emerged in the 1980s under the buzzword *total quality management* or TQM. This philosophy involves the entire organization, beginning from the supply chain management to the product life cycle. The key concept of 'total quality' was developed by the founding fathers of today's quality management, Dr. Armand V. Feigenbaum (Feigenbaum 1951), Mr. Philip B. Crosby (Crosby 1979), Dr. Kaoru Ishikawa (Ishikawa 1985) and Dr. Joseph M. Juran (Juran and Godfrey 2000). The application of SPC nowadays includes concepts such as Six Sigma, which involves DMAIC (Define, Measure, Analyze, Improve and Control), QFD (Quality Function Deployment) and FMEA (Failure Modes and Effect Analysis) (Brussee, 2004). A comprehensive timeline for the development and application of quality methods is presented in Section 1.2 in Montgomery (2005).

1.1.2 Principles of statistical process control

The key measurements discretely taken from manufacturing processes do not generally describe constant values that are equal to the required and predefined set points. In fact, if the process operates at a steady state condition, then these set points remain constant over time. The recorded variables associated with product quality are of a stochastic nature and describe a random variation around their set point values in an ideal case.

1.1.2.1 Mean and variance of a random variable

The notion of an ideal case implies that the expectation of a set of discrete samples for a particular key variable converges to the desired set point. The expectation, or 'average', of a key variable, further referred to as a process variable z, is described as follows

$$E\left\{z\right\} = \bar{z} \tag{1.1}$$

where $E\left\{\cdot\right\}$ is the expectation operator. The 'average' is the *mean value*, or *mean*, of z, $\widehat{\bar{z}}$, which is given by

$$\lim_{K\to\infty} \widehat{\bar{z}} = \lim_{K\to\infty} \frac{z(1) + z(2) + \cdots + z(K)}{K} = \lim_{K\to\infty} \tfrac{1}{K} \sum_{k=1}^{K} z(k) \to \bar{z}. \tag{1.2}$$

In the above equation, the index k represents time and denotes the order when the specific sample (quality measurement) was taken. Equation (1.2) shows that

$\widehat{\bar{z}} \rightarrow \bar{z}$ as $K \rightarrow \infty$. For large values of K, however, we can assume that $\widehat{\bar{z}} \approx \bar{z}$ and small K values may lead to significant differences between $\widehat{\bar{z}}$ and $\bar{z}$. The latter situation, that is, small sample sizes, may present difficulties if no set point is given for a specific process variable and the average therefore needs to be estimated. A detailed discussion of this is given in Section 6.4.

So far, the mean of a process variable is assumed to be equal to a predefined set point $\bar{z}$ and the recorded samples describe a stochastic variation around this set point. The following data model can therefore be assumed to describe the samples

$$z = z_0 + \bar{z}. \tag{1.3}$$

The stochastic variation is described by the stochastic variable z_0 and can be captured by an upper bound and a lower bound or *the control limits* which can be estimated from a reference set of the process variable. Besides a constant mean, the second main assumption for SPC charts is a constant *variance* of the process variable

$$E\left\{(z-\bar{z})^2\right\} = E\left\{z^2\right\} - \bar{z}^2 = \lim_{K\rightarrow\infty} \frac{1}{K-1}\sum_{k=1}^{K}\left(z(k) - \widehat{\bar{z}}\right)^2 = \sigma^2 \tag{1.4}$$

where σ is defined as the *standard deviation* and σ^2 as the *variance* of the stochastic process variable. This parameter is a measure for the spread or the variability that a recorded process variable exhibits. It is important to note that the control limits depend on the variance of the recorded process variable.

For a sample size K the estimate $\widehat{\bar{z}}$ may accordingly depart from $\bar{z}$ and Equation (1.4) is, therefore, an estimate of the variance σ^2, $\widehat{\sigma}^2$. It is also important to note that the denominator $K-1$ is required in (1.4) instead of K since one degree of freedom has been used for determining the estimate of the mean value, $\widehat{\bar{z}}$.

1.1.2.2 Probability density function of a random variable

Besides a constant mean and variance of the process variable, the third main assumption for SPC charts is that the recorded variable follows a Gaussian distribution. The distribution function of a random variable is discussed later and depends on the *probability density function* or PDF. Equation (1.5) shows the PDF of the Gaussian distribution

$$f(z) = \frac{1}{\sqrt{2\sigma^2\pi}} e^{-\frac{(z-\bar{z})^2}{2\sigma^2}}. \tag{1.5}$$

Figure 1.1 shows the Gaussian density function for $\bar{z} = 0$ and various values of σ. In this figure the abscissa refers to values of z and the ordinate represents the 'likelihood of occurrence' of a specific value of z. It follows from Figure 1.1 that the smaller σ the narrower the Gaussian density function becomes and vice versa. In other words, the variation of the variable depends on the parameter σ.

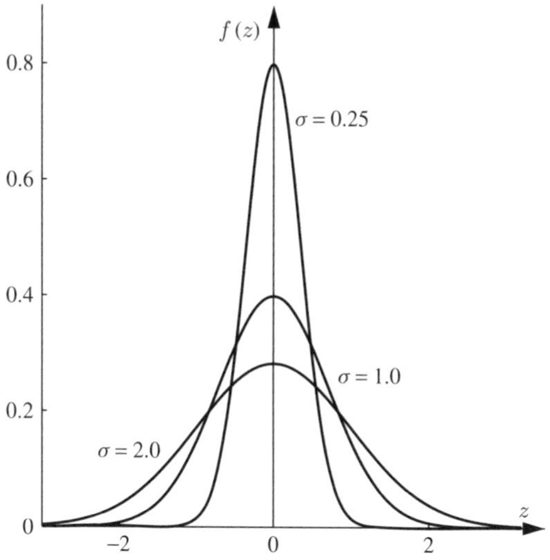

Figure 1.1 Gaussian density function for $\bar{z} = 0$ *and* $\sigma = 0.25$*,* $\sigma = 1.0$ *and* $\sigma = 2.0$.

It should also be noted that the value of $\pm\sigma$ represents the point of inflection on the curve $f\,(z)$ and the maximum of this function is at $z = \bar{z}$, i.e. this value has the highest chance of occurring. Traditionally, a stochastic variable that follows a Gaussian distribution is abbreviated by $z \sim \mathcal{N}\left\{\bar{z}, \sigma^2\right\}$.

By closer inspection of (1.4) and Figure 1.1, it follows that the variation (spread) of the variables covers the entire range of real numbers, from minus to plus infinity, since likelihood values for very small or large values are nonzero. However, the likelihood of large absolute values is very small indeed, which implies that most values for the recorded variable are centered in a narrow band around $\bar{z}$. This is graphically illustrated in Figure 1.2, which depicts a total of 20 samples and the probability density function $f(z)$ describing the likelihood of

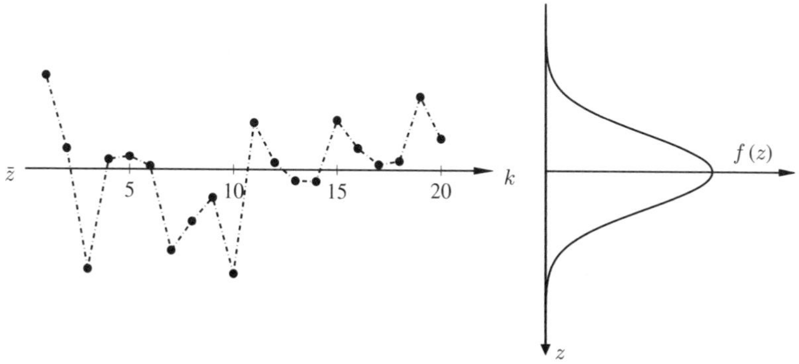

Figure 1.2 Random Gaussian distributed samples of mean $\bar{z}$ *and variance* σ.

occurrence for each sample. This figure shows that large departures from $\bar{z}$ can occur, e.g. samples 1, 3 and 10, but that most of samples center closely around $\bar{z}$.

1.1.2.3 Cumulative distribution function of a random variable

We could therefore conclude that the *probability* of z values that are far away from $\bar{z}$ is small. In other words, we can simplify the task of monitoring the process variable by defining an upper and a lower boundary that includes the vast majority of possible cases and excludes those cases that have relatively small likelihood of occurrence. Knowing that the integral over the entire range of the probability density function is equal to 1.0, the *probability* is therefore a measure for defining these upper and lower boundaries. For the symmetric Gaussian probability density function, the probability within the range bounded by $\bar{z} - z_{\alpha/2}$ and $\bar{z} + z_{\alpha/2}$ is defined as

$$F\left(\bar{z} - z_{\alpha/2}, \bar{z} + z_{\alpha/2}\right) = \int_{\bar{z}-z_{\alpha/2}}^{\bar{z}+z_{\alpha/2}} f(z)\,\mathrm{d}z = 1 - \alpha. \tag{1.6}$$

Here, $\bar{z} - z_{\alpha/2}$ and $\bar{z} + z_{\alpha/2}$ defines the size of this range that is centered at $\bar{z}$, $0 \leq F(\cdot) \leq 1.0$, $F(\cdot)$ is the *cumulative distribution function* and α is the *significance*, that is the percentage, $\alpha \cdot 100\%$, of samples that could fall outside the range between the upper and lower boundary but still belong to the probability density function $f(\cdot)$. Given that the Gaussian PDF is symmetric, the chance that a sample has an 'extreme' value falling in the left or the right tail end is $\alpha/2$. The general definition of the Gaussian *cumulative distribution function* $F(a, b)$ is as follows

$$F(a, b) = \Pr\{a \leq z \leq b\} = \frac{1}{\sqrt{2\pi\sigma^2}} \int_a^b e^{-\frac{(z-\bar{z})^2}{2\sigma^2}}\,\mathrm{d}z, \tag{1.7}$$

where $\Pr\{\cdot\}$ is defined as the *probability* that z assumes values that are within the interval $[a, b]$.

1.1.2.4 Shewhart charts and categorization of process behavior

Assuming that $\bar{z} = 0$ and $\sigma = 1.0$, the probability of $1 - \alpha = 0.95$ and $1 - \alpha = 0.99$ yield ranges between $z_{\alpha/2} = \pm 1.96$ and $z_{\alpha/2} = \pm 2.58$. This implies that 5% and 1% of recorded values can be outside this 'normal' range by chance alone, respectively. Figure 1.3 gives an example of this for $\bar{z} = 10.0$, $\sigma = 1.0$ and $\alpha = 0.01$. This implies that the upper boundary or *upper control limit*, UCL, and the lower boundary or *lower control limit*, LCL, are equal to $10 + 2.58 = 12.58$ and $10 - 2.58 = 7.42$, respectively. Figure 1.3 includes a total of 100 samples taken from a Gaussian distribution and highlights that one sample, sample number 90, is outside the 'normal' region. Out of 100 recorded samples, this is 1% and in line with the way the *control limits*, that is, the

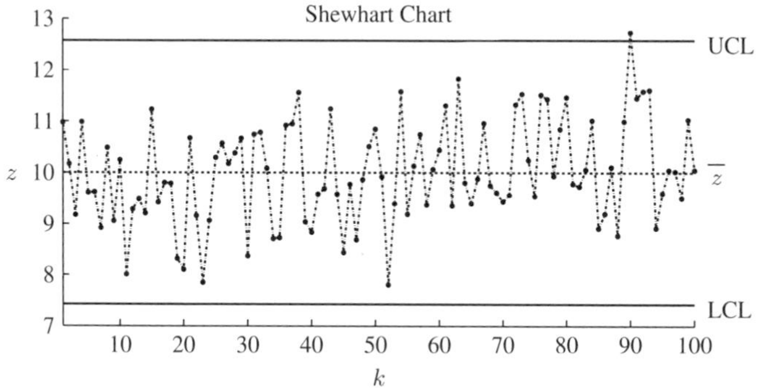

Figure 1.3 Schematic diagram showing statistical process control chart.

upper and lower control limits, have been determined. Loosely speaking, 1% of samples might violate the control limits by chance alone.

From the point of an interpretation of the SPC chart in Figure 1.3, which is defined as a *Shewhart chart*, samples that fall between the UCL and the LCL categorize in-statistical-control behavior of the recorded process variable and samples that are outside this region are indicative of an out-of-statistical-control situation. As discussed above, however, it is possible that $\alpha \cdot 100\%$ of samples fall outside the control limits by chance alone. This is further elaborated in Subsection 1.1.3.

1.1.2.5 Trends in mean and variance of random variables

Statistically, for quality related considerations a process is out-of-statistical-control if at least one of the following six conditions is met:

1. one point is outside the control limits;
2. two out of three consecutive points are two standard deviations above/below the set point;
3. four out of five consecutive points are one standard deviation above/below one standard deviation;
4. seven points in a row are all above/below the set point;
5. ten out of eleven points in a row are all above/below the set point; and
6. seven points in a row are all increasing/decreasing.

The process that is said to be an in-statistical-control process if none of the above hypotheses are accepted. Such a process is often referred to as a *stable* process or a process that does not present a *trend*. Conversely, if at least one of the above conditions is met the process has a trend that manifest itself in changes of the mean and/or variance of the recorded random variable. This, in turn, requires a detailed and careful inspection in order to identify the root cause of this trend. In essence, the assumption of a stable process is that a recorded quality variable follows a Gaussian distribution that has a constant mean and variance over time.

1.1.2.6 Control limits vs. specification limits

Up until now, the discussion has focussed on the process itself. This discussion has led to the definition of the control limits for process variables that follow a Gaussian distribution function and have a constant mean value, or set point, and variances have been obtained. More precisely, rejecting all of the above six hypotheses implies that the process is in-statistical control or *stable* and does not describe any trend. For SPC, it is of fundamental importance that the control limits of the key process variable(s) are inside the *specification limits* for the product. The specification limits are production tolerances that are defined by the customer and must be met. If the upper and lower control limits are within the range defined by the *upper and lower specification limits*, or *USL* and *LSL*, a stable process produces items that are, by default, within the specification limits. Figure 1.4 shows the relationship between the specification limits, the control limits and the set point of a process variable z for which 20 consecutively recorded samples are available.

1.1.2.7 Types of processes

Using the definition of the specification and control limits, a process can be categorized into a total of four distinct types:

1. an ideal process;
2. a promising process;
3. a treacherous process; and
4. a turbulent process.

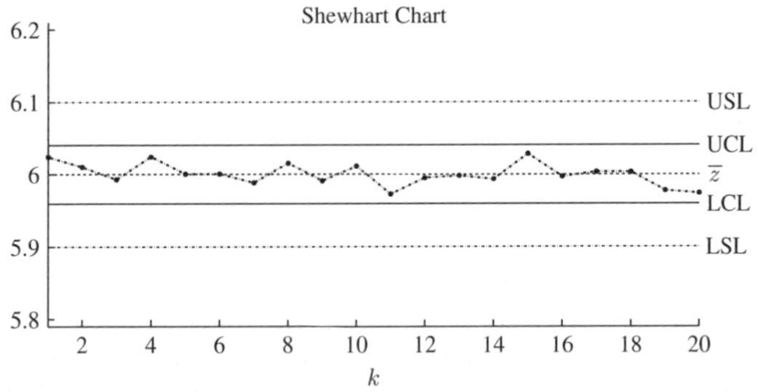

Figure 1.4 Upper and lower specification limit as well as upper and lower control limits and set point value for key variable z.

The process shown in Figure 1.4 is an *ideal process*, where the product is almost always within the specification limits. An *ideal process* is therefore a *stable process*, since the mean and variance of the key product variable z is time invariant. A *promising process* is a stable process but the control limits are outside the region defined by the specification limits. The promising process has the potential to produce a significant amount of off-spec product.

The *treacherous process* is an unstable process, as the mean and/or variance of z varies over time. For this process, the absolute difference of the control limits is assumed to be smaller than the absolute difference of the specification limits. Similar to a promising process, a treacherous process has the potential to produce significant off-spec product although this is based on a change in mean/variance of z. Finally, a *turbulent process* is an unstable process for which the absolute difference of the control limits is larger than the absolute difference of the specification limits. The turbulent process therefore often produces off-spec products.

1.1.2.8 Determination of control limits

It is common practice for SPC applications to determine the control limits z_α as a product of σ, for example the range for the UCL and LCL are $\pm\sigma$, $\pm 2\sigma$ etc. Typical are *three sigma* and *six sigma* regions. It is interesting to note that the control limits that represent three sigma capture 99.73% of cases, which appears to describe almost all possible cases. It is important to note, however, that if a product is composed of say 50 items each of which has been produced within a UCL and LCL that correspond to $\pm 3\sigma$, then the probability that any of the products does not conform to the required specification is $1 - (1 - \alpha)^{50} = 1 - 0.9973^{50} = 1 - 0.8736 = 0.1664$, which is 16.64% and not 0.27%. It is common practice in such circumstances to determine UCL and LCL with respect to $\pm 6\sigma$, that is $\alpha = 1 - 0.999999998$, for which the same calculation yields that the probability that one product does not conform to the required specification reduces to 0.01 parts per million.

1.1.2.9 Common cause vs. special cause variation

Another concept that is of importance is the analysis as to what is causing the variation of the process variable z. Whilst this can be regarded as a process specific entity, two distinct sources have been proposed to describe this variation, the *common cause variation* and the *special cause variation*. The properties of common cause variation are that it arises all the time and is relatively small in magnitude. As an example for common cause variation, consider two screws that are produced in the same shift and selected randomly. These screws are not identical although the differences in thread thickness, screw length etc. are relatively small. The differences in these key variables must not be a result of an assignable cause. Moreover, the variation in thread length and total screw length must be process specific and cannot be removed. An attempt to reduce common cause variation is often regarded as tampering and may, in fact, lead to an increase in the variance of the recorded process variable(s). A special cause variation on the other hand, has an assignable cause, e.g. the introduction of disturbances, a process fault, a grade change or a transition between two operating regions. This variation is usually rare but may be relatively large in magnitude.

1.1.2.10 Advances in designing statistical process control charts

Finally, improvements for Shewhart type charts have been proposed in the research literature for detecting incipient shifts in $\bar{z}$ (that is $\widehat{\bar{z}}$ departs from $\bar{z}$ over time), and for dealing with cases where the samples distribution function slightly departs from a Gaussian distribution. This has led to the introduction of cumulative sum or *CUSUM* charts (Hawkins 1993; Hawkins and Olwell 1998) and exponentially weighted moving average or EWMA charts (Hunter 1986; Lucas and Saccucci 1990).

Next, Subsection 1.1.3 summarizes the statistically important concept of hypothesis testing. This test is fundamental in evaluating the current state of the process, that is, to determine whether the process is in-statistical-control or out-of-statistical-control. Moreover, the next subsection also introduces errors associated with this test.

1.1.3 Hypothesis testing, Type I and II errors

To motivate the underlying meaning of a hypothesis test in an SPC context, Figure 1.5 describes the two scenarios introduced in the preceding discussion. The upper graph in this figure exemplifies an in-statistical-control situation, since:

- the recorded samples, $z(k)$, are drawn from the distribution described by $f_0(z)$; and
- the *confidence region*, describing the range limited by the upper and lower control limits of this process, has been calculated by Equation 1.6 using $f_0(z)$

Hence, the recorded samples fall inside the confidence region with a significance of α. The following statement provides a formal description of this situation.

$$H_0 \quad : \quad \text{The process is in-statistical-control.}$$

In mathematical statistics, such a statement is defined as a *hypothesis* and referred to as H_0. As Figure 1.5 highlights, a hypothesis is a statement concerning the probability distribution of a random variable and therefore its population parameters, for example the mean and variance of the Gaussian distribution function. Consequently, the hypothesis H_0 that the process is in-statistical-control can be tested by determining whether newly recorded samples fall within the confidence

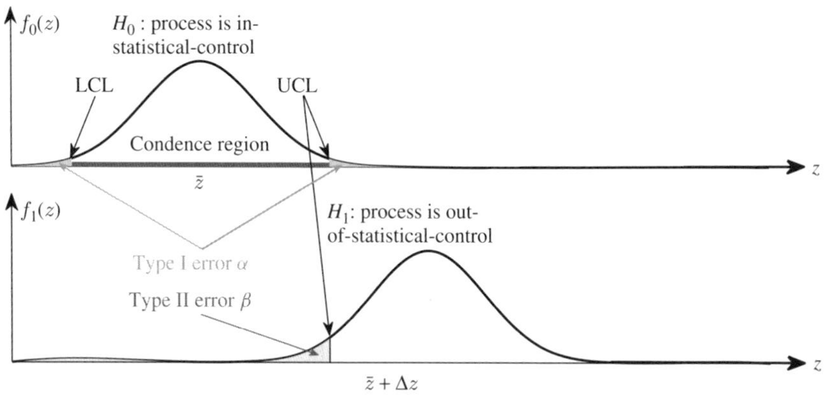

Figure 1.5 Graphical illustration of Type I and II errors in an SPC context.

region. If this is the case then the *hypothesis* that the process is in-statistical-control is accepted.

For any hypothesis testing problem, the hypothesis H_0 is defined as the *null hypothesis* and is accompanied by the *alternative hypothesis* H_1. In relation to the terminology introduced in the previous subsection, the statement governing the alternative hypothesis is as follows:

$$H_1 \quad : \quad \text{The process is out-of-statistical-control.}$$

The lower plot in Figure 1.5 gives an example of an out-of-statistical-control situation by a shift in the mean of z from $\bar{z}$ to $\bar{z} + \Delta z$. In general, if the null hypothesis is rejected the alternative hypothesis is accepted. This implies that if the newly recorded samples fall outside the *confidence region* the alternative hypothesis is accepted and the process is out-of-statistical-control. It should be noted that detecting an out-of-statistical-control situation, which is indicative of abnormal process behavior, is important but does not address the subsequent question as to what has caused this behavior. In fact, the diagnosis of anomalous process behavior can be considerably more challenging than detecting this event (Jackson 2003).

It is also important to note that testing the null hypothesis relies on probabilistic information, as it is related to the significance level α. If we assume a significance of 0.01, 99% of samples are expected to fall within the confidence region on average. In other words, this test is prone to mistakes and a sample that has an extreme value is likely to be outside the confidence region although it still follows $f_0(z)$. According to the discussion above, however, this sample must be considered to be associated with the alternative hypothesis H_1. This error is referred to as a Type I error.

Definition 1.1.1 *A* ***Type I error*** *arises if* H_0 *is rejected while, in fact, it must be accepted. The probability of Type I error is defined as*

$$F_0\left(\text{ rejecting } H_0 | H_0 \text{ is true }\right) = \alpha = \int_{-\infty}^{LCL} f_0(z)\,\mathrm{d}z + \int_{UCL}^{-\infty} f_0(z)\,\mathrm{d}z,$$

where $f_0(\cdot)$ *is the PDF of* z.

Figure 1.5 also illustrates a second error that is associated with the hypothesis testing. Defining the PDF corresponding to the shift in mean of z from $\bar{z}$ to $\bar{z} + \Delta z$ by $f_1(\cdot)$, it is possible that a recorded sample belongs to $f_1(\cdot)$ but its value is with the control limits. This scenario is defined as a Type II error.

Definition 1.1.2 *A* ***Type II error*** *arises if* H_0 *is accepted, while, in fact, it must be rejected. In the context of the scenario described in the lower plot in Figure 1.5, the probability of a Type II error is defined as follows*

$$F_1\left(\text{failing to reject } H_0 | H_1 \text{ is true}\right) = \beta = \int_{-\infty}^{UCL} f_1(z)\,\mathrm{d}z.$$

The Type I error is equal to the significance level α for determining the upper and lower control limits. However, the probability of a Type II error β is not a constant and, according to the lower plot in Figure 1.5, depends on the size of Δz. It should be noted that the statement 'failing to reject H_0' does not necessarily mean that there is a high probability that H_0 is true but simply implies that a Type II error can be significant if the magnitude of the fault condition is small or incipient. Subsection 8.7.3 presents a detailed examination of detecting incipient fault conditions.

From SPC charts, it is desirable to minimize both Type I and II errors. However, Figure 1.5 highlights that decreasing α produces an increase in β and vice versa. One could argue that selecting α could depend on what abnormal conditions are expected. For SPC, however, the Type I error is usually considered to be more serious, since rejecting H_0 although it is in fact true implies that a false alarm has been raised. If the online monitoring scheme produces numerous such false alarms, the confidence of process operators in the SPC/MSPC technology would be negatively affected. This argument suggests smaller α values. In support of this, the discussion on determining control limits in the previous subsection also advocates smaller α values.

The preceding discussion in this section has focussed on charting individual key variables. The next section addressed the problem of correlation among key process variables and motivates the need for a multivariate extension of the SPC framework.

1.2 Why multivariate statistical process control

The previous section has shown how a recorded process variable that follows a Gaussian distribution can be charted and how to determine whether the process is an *ideal process*. The use of Shewhart charts, however, relies on analyzing individual key variables of the process in order to analyze the current product quality and to assess the current state of the process operation. Despite the widespread success of the SPC methodology, it is important to note that correlation between process variables can substantially increase the number of Type II errors. If the null hypothesis is accepted, although it must be rejected, yields that the process is assumed to be in-statistical-control although it is, in fact, out-of-statistical-control. The consequence is that a large Type II error may render abnormal process behavior difficult to detect.

Before describing the effect of correlation between a set of process variables, it is imperative to define variable correlation. In here, it is strictly related to the correlation coefficients between a set of variables. For the ith and the jth process variable, which have the variances of $E\left\{z_{0i}^2\right\} = \sigma_i^2$ and $E\left\{z_{0j}^2\right\} = \sigma_j^2$ and the covariance $E\left\{z_{0i}z_{0j}\right\} = \sigma_{ij}^2$, the correlation coefficient r_{ij} is defined as

$$r_{ij} = \frac{E\left\{z_{0i}z_{0j}\right\}}{\sqrt{E\left\{z_{0i}^2\right\}E\left\{z_{0j}^2\right\}}} = \frac{\sigma_{ij}^2}{\sigma_i\sigma_j} \qquad -1 \le r_{ij} \le 1. \tag{1.8}$$

The above equation also shows the following and well known relationship between the variable variances, their covariance and the correlation coefficient

$$E\left\{z_{0i}z_{0j}\right\} = r_{ij}\sqrt{E\left\{z_{0i}^2\right\}E\left\{z_{0j}^2\right\}}. \tag{1.9}$$

Equation (1.9) outlines that a large covariance between two variables arises if (i) the correlation coefficient is large and (ii) their variances are large. Moreover, if the variances of both variables are 1, the correlation coefficient reduces to the covariance.

To discuss the correlation issue, the next three subsections present examples that involve two Gaussian distributed variables, $z_1 = z_{0_1} + \bar{z}_1$ and $z_2 = z_{0_2} + \bar{z}_2$ that have a mean of $\bar{z}_1$ and $\bar{z}_2$ and a variance of σ_1^2 and σ_2^2, respectively. Furthermore, the upper and lower control limits for these variables are given by UCL_1 and LCL_1 for z_1 and UCL_2 and LCL_2 for z_2. The presented examples describe the following three different cases:

1. no correlation between z_{0_1} and z_{0_2};
2. perfect correlation between z_{0_1} and z_{0_2}; and
3. a high degree of correlation between z_{0_1} and z_{0_2}.

Cases 1 and 2 imply that the correlation coefficient between z_{0_1} and z_{0_2}, is zero and one, respectively. The third case describes a large absolute correlation coefficient.

1.2.1 Statistically uncorrelated variables

Figure 1.6 gives an example of two process variables that have a correlation coefficient of zero. Both process variables can, of course, be plotted with a time base in individual Shewhart charts. In Figure 1.6 the horizontal and vertical plot represents the Shewhart charts for process variables z_1 and z_2, respectively. Individually, each of the process variables show that the process is in-statistical-control.

Projecting the samples of the individual charts into the central box between both charts yields a *scatter diagram*. The *scatter points* marked by '+' are the intercept of the projections associated with the same sample index, e.g. $z_1(k)$ and $z_2(k)$ represent the kth point in the scatter diagram. The confidence region for the scatter diagram can be obtained from the joint PDF. Defining $f_1(\cdot)$ and $f_2(\cdot)$ as the PDF of z_1 and z_2, respectively, and given that $r_{12} = 0$ the joint PDF $f(\cdot)$ is equal to[1]

$$f\left(z_1, z_2\right) = f_1\left(z_1\right) f_2\left(z_2\right) = \frac{1}{\sqrt{2\pi\sigma_1^2}} \frac{1}{\sqrt{2\pi\sigma_2^2}} e^{-\frac{(z_1-\bar{z}_1)^2}{2\sigma_1^2}} e^{-\frac{(z_2-\bar{z}_2)^2}{2\sigma_2^2}}$$

$$f\left(z_1, z_2\right) = \frac{1}{2\pi\sigma_1\sigma_2} e^{-\frac{(z_1-\bar{z}_1)^2}{2\sigma_1^2} - \frac{(z_2-\bar{z}_2)^2}{2\sigma_2^2}}. \tag{1.10}$$

The joint PDF in (1.10) can also be written in matrix-vector form

$$f\left(z_1, z_2\right) = \frac{1}{2\pi\left|\mathbf{S}_{z_0z_0}\right|^{1/2}} e^{-\frac{1}{2}\begin{pmatrix} z_1 - \bar{z}_1 & z_2 - \bar{z}_2 \end{pmatrix}\mathbf{S}_{z_0z_0}^{-1}\begin{pmatrix} z_1 - \bar{z}_1 \\ z_2 - \bar{z}_2 \end{pmatrix}} \tag{1.11}$$

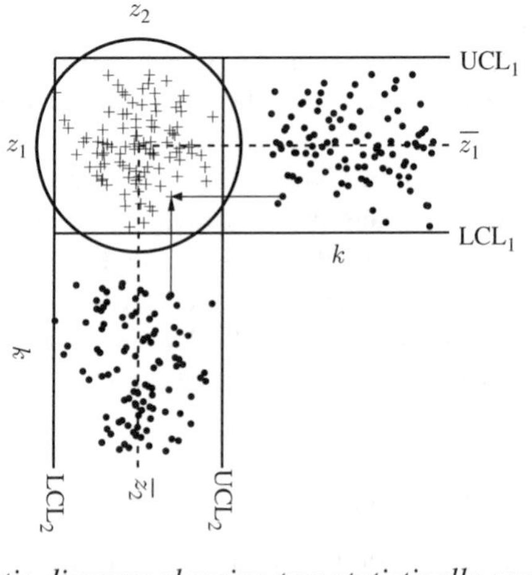

Figure 1.6 Schematic diagram showing two statistically uncorrelated variables.

[1] It follows from a correlation coefficient of zero that the covariance is also zero. This, in turn, implies that two Gaussian distributed variables are statistically independent.

where

$$\mathbf{S}_{z_0 z_0} = \begin{bmatrix} \sigma_1^2 & 0 \\ 0 & \sigma_2^2 \end{bmatrix} \quad (1.12)$$

is the covariance matrix of z_{0_1} and z_{0_2} and $|\cdot|$ is the determinant of a matrix.

In a similar fashion to the covariance matrix, the correlation between a set of n_z variables can be described by the correlation matrix. Using (1.8) for $i = j$, the diagonal elements of this matrix are equal to one. Moreover, the non-diagonal elements possess values between $-1 \le r_{ij} \le 1$. The concept of correlation is also important to assess time-based trends within the process variables. However, the application of MSPC assumes that the process variables do not possess such trends.

Based on (1.11), a confidence region can be obtained as follows. Intercept a plane located close and parallel to the $z_1 - z_2$ plane with $f\left(z_1, z_2\right)$. The integral over the interception area hugging the joint PDF is equal to $1 - \alpha$. The contour describing this interception is defined as the *control ellipse* and represents the confidence region. It should be noted that if the variance of both variables are identical, the control ellipse reduces to a circle, which is the case described in Figure 1.6.[2] Subsection 1.2.3 shows how to construct a control ellipse.

One could naively draw a 'rectangular' confidence region that is bounded by the upper and lower control limits of the individual Shewhart charts. Since the individual samples are all inside the upper and lower control limits for both charts, the scatter points must fall within this 'rectangle'. By directly comparing the 'rectangle' with the control ellipse in Figure 1.6, it can be seen both areas are comparable in size and that the scatter points fall within both.

The four corner areas of the rectangle that do not overlap with the circular region are small. Statistically, however, the circular region is the correct one, as it is based on the joined PDF. The comparison between the 'rectangle' and the circle, however, shows that the difference between them is negligible. Thus, the individual and joint analysis of both process variables yield an in-statistical-control situation.

1.2.2 Perfectly correlated variables

In this second case, the two variables z_{0_1} and z_{0_2} have a correlation coefficient of -1. According to (1.8), this implies that the covariance σ_{12}^2 is equal to

$$E\left\{z_{0_1} z_{0_2}\right\} = r_{12}\sqrt{E\left\{z_{0_1}^2\right\} E\left\{z_{0_2}^2\right\}} \quad \Rightarrow \quad \sigma_{12}^2 = -\sigma_1\sigma_2. \quad (1.13)$$

For identical variances, $\sigma_1^2 = \sigma_2^2 = \sigma^2$, it follows that

$$\sigma_{12}^2 = -\sigma^2. \quad (1.14)$$

[2] Two uncorrelated Gaussian distributed variables that have the same variance describe *independently and identically distributed* or *i.i.d.* sequences.

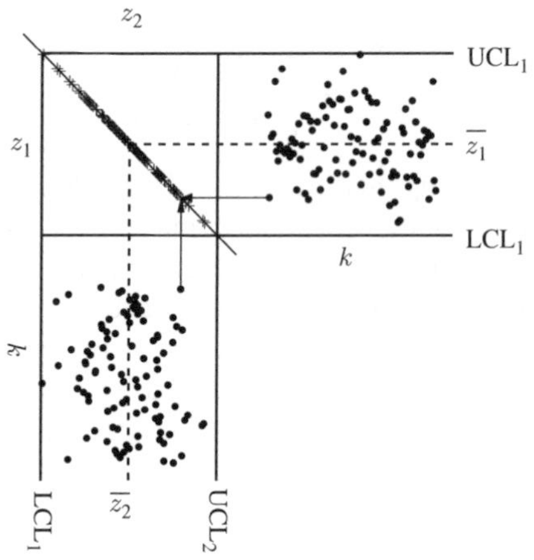

Figure 1.7 Schematic diagram showing two perfectly correlated variables.

In other words, $z_{0_1} = -z_{0_2}$. For unequal variances, both signals are equal up to the scaling factor $z_{0_1} = a z_{0_2} \Rightarrow a = -{}^{\sigma_1}/_{\sigma_2}$. Figure 1.7 shows the Shewhart charts for z_1 and z_2 which are plotted horizontally and vertically, respectively. Given that both variables have the same variance, both of them produce identical absolute values for each sample. This, however, implies that the projections of each sample fall onto a line that has an angle of 135° and 45° to the abscissas of the Shewhart chart for variable z_1 and z_2, respectively.

The 2D circular confidence region if z_{0_1} and z_{0_2} are statistically uncorrelated therefore reduces to a 1D line if they are perfectly correlated. Moreover, the joint PDF $f\left(z_1, z_2\right)$ in this case is equal to

$$f\left(z_1, z_2\right) = \frac{1}{2\pi} \frac{1}{\left|\sigma^2 \left[\begin{smallmatrix} 1 & -1 \\ -1 & 1 \end{smallmatrix}\right]\right|^{1/2}} e^{-\frac{1}{2\sigma^2}\left(\begin{matrix} z_{0_1} & z_{0_2} \end{matrix}\right)\left[\begin{matrix} 1 & -1 \\ -1 & 1 \end{matrix}\right]^{-1}\left(\begin{matrix} z_{0_1} \\ z_{0_2} \end{matrix}\right)} \tag{1.15}$$

if the variables have equal variance. Equation (1.15), however, presents two problems:

- the determinant of $\mathbf{S}_{z_0 z_0} = \sigma^2 \left[\begin{smallmatrix} 1 & -1 \\ -1 & 1 \end{smallmatrix}\right]$ is equal to zero; and
- the inverse of $\mathbf{S}_{z_0 z_0}$ therefore does not exist.

This results from the fact that the rank of $\mathbf{S}_{z_0 z_0}$ is equal to one.

To determine the joint PDF, Figure 1.8 'summarizes' the scatter diagram of Figure 1.7 by assuming that the control limits for z_1 and z_2 are $\bar{z} \pm 3\sigma$ and $\bar{z}_1 = \bar{z}_2 = \bar{z}$ for simplicity. The catheti of the right triangle depicted in Figure 1.8 are the ordinates of both Shewhart charts and the hypotenuse is the semimajor

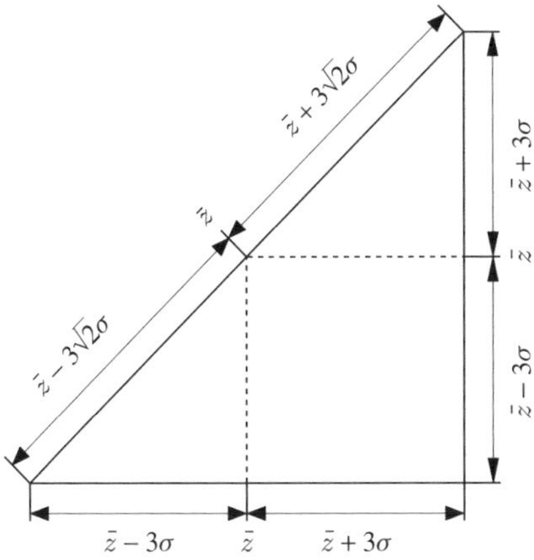

Figure 1.8 Geometric interpretation of the scatter diagram in Figure 1.7.

of the 'control ellipse'. The length of both catheti is 6σ and the length of the hypotenuse is $6\sqrt{(2)}\sigma$, accordingly.

As the projections of the recorded samples fall onto the hypotenuse, the control limits for the projected points are the endpoints of the hypotenuse. Defining the two identical variables z_1 and z_2 by z the projected samples of z follow a Gaussian distribution and are scaled by $2^{1/2}$. Next, defining the projected samples of z onto the hypotenuse as t, the 'joint' PDF of $z_1 = z_2 = z$ reduces to the PDF of t

$$f\left(z_1, z_2\right) = f\left(t\right) = \frac{1}{\sqrt{2\pi 2\sigma^2}} e^{-\frac{1}{4}\frac{(t-\bar{z})^2}{\sigma^2}}. \tag{1.16}$$

One could argue that only one variable needs to be monitored. An inspection of Figure 1.7, however, yields that the joint analysis is a sensitive mechanism for detecting abnormal process behavior. If the process is in-statistical-control, the sample projections fall onto the hypotenuse. If not, the process is out-of-statistical-control, even if each sample is within the control limits of the individual Shewhart charts. Although a perfect correlation is a theoretical assumption, this extreme case highlights one important reason for conducting SPC on the basis of a multivariate rather than a univariate analysis of the individual variables.

1.2.3 Highly correlated variables

The last two subsections presented two extreme cases for the correlation between two variables. The third case examined here relates to a high degree of correlation between z_1 and z_2, which is an often observed phenomenon between the recorded process variables, particulary for large-scale systems. For example, temperature and pressure readings, flow rate measurements and concentrations or other product quality measures frequently possess similar patterns. Using a correlation coefficient of -0.95 and unity variance for z_1 and z_2 yields the following covariance matrix

$$\mathbf{S}_{z_0 z_0} = \begin{bmatrix} 1.00 & -0.95 \\ -0.95 & 1.00 \end{bmatrix}. \tag{1.17}$$

Figure 1.9 shows, as before, the two Shewhart charts and the scatter diagram, which represents the projected samples of each variable. Given that z_1 and z_2 are assumed to be normally distributed, the joint PDF is given by

$$f\left(z_1, z_2\right) = \frac{1}{2\pi \left|\mathbf{S}_{z_0 z_0}\right|^{1/2}} e^{-\frac{1}{2}\left(z_1 - \bar{z}_1 \quad z_2 - \bar{z}_2 \right)\mathbf{S}_{z_0 z_0}^{-1}\begin{pmatrix} z_1 - \bar{z}_1 \\ z_2 - \bar{z}_2 \end{pmatrix}}. \tag{1.18}$$

The control ellipse can be obtained by intercepting a plane that is parallel to the $z_1 - z_2$ plane to the surface of the joint PDF, such that the integral evaluated within the interception area is equal to $1 - \alpha$. Different to two uncorrelated variables of equal variance, this procedure yields an ellipse. Comparing Figures 1.6 and 1.9 highlights that both axes of the circle are parallel to the abscissas of both Shewhart charts for uncorrelated variables, whilst the semimajor of the control ellipse for highly correlated variables has an angle to both abscissas.

1.2.3.1 Size and orientation of control ellipse for correlation matrix $\mathcal{C}_{z_0 z_0}$

The following study presents a lucid examination of the relationship between the angle of the semimajor and the abscissa, and the correlation coefficient between z_1 and z_2. This study assumes that the variance of both Gaussian variables is 1 and that they have a mean of 0. Defining the correlation coefficient between z_1

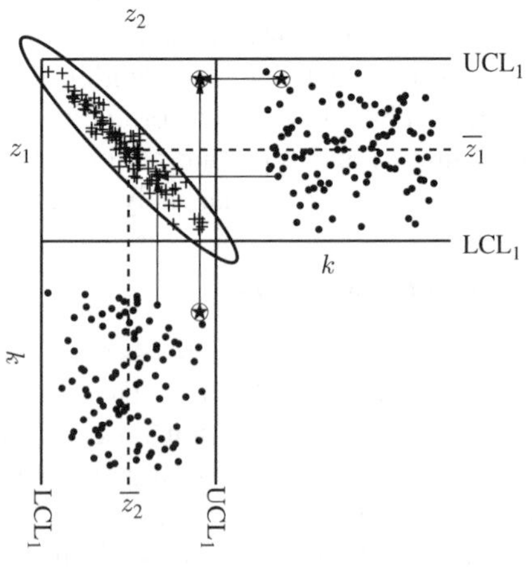

Figure 1.9 Schematic diagram showing two highly correlated variables.

and z_2 by r_{12}, produces the following covariance/correlation matrix

$$\mathbf{S}_{z_0z_0} = \mathcal{C}_{z_0z_0} = \begin{bmatrix} 1 & r_{12} \\ r_{12} & 1 \end{bmatrix}. \tag{1.19}$$

It follows from (1.9) that the correlation coefficient is equal to the covariance, since $\sigma_1^2 = \sigma_2^2 = 1$. As discussed in Jackson (2003), the orientation of the semimajor and semiminor of the control ellipse are given by the eigenvectors associated with the largest and the smallest eigenvalue of the $\mathbf{S}_{z_0z_0}$, respectively,

$$\mathcal{C}_{z_0z_0} = \begin{bmatrix} \mathbf{p}_1 & \mathbf{p}_2 \end{bmatrix} \begin{bmatrix} \lambda_1 & 0 \\ 0 & \lambda_2 \end{bmatrix} \begin{bmatrix} \mathbf{p}_1^T \\ \mathbf{p}_2^T \end{bmatrix}. \tag{1.20}$$

In the above equation, $\mathbf{p}_1$ and $\mathbf{p}_2$ are the eigenvectors associated with the eigenvalues λ_1 and λ_2, respectively, and $\lambda_1 > \lambda_2$.

Eigenvalues of $\mathcal{C}_{z_0z_0}$. For a correlation coefficient ranging from -1 to 1, Figure 1.10 shows how the eigenvalues λ_1 and λ_2 depend on the absolute value of the correlation coefficient r_{12}, i.e. $\lambda_1 (r_{12}) = 1 + |r_{12}|$ and $\lambda_2 (r_{12}) = 1 - |r_{12}|$. This analysis also includes the two extreme cases discussed in both previous subsections. For $r_{12} = 0$, both eigenvalues are equal to 1. On the other hand, if $r_{12} = -1$, the larger eigenvalue is equal to 2 and the other one is 0. The eigenvalues represent the variance of the sample projections on the semimajor (larger eigenvalue) and the semiminor (smaller eigenvalue).

For $r_{12} = 0$ the variances of the projected samples onto both axis of the ellipse are identical, which explains why the control ellipse reduces to a circle if $\sigma_1^2 = \sigma_2^2$. For $r_{12} = 1$, however, there is no semiminor since all of the projected samples fall onto the hypotenuse of Figure 1.8. Consequently, λ_2 is equal to zero (no variance) and the scaling factor between the projections of z, t, and $z = z_1 = z_2$ is equal to $\sqrt{2}$ since $E\left\{(t - \bar{z})^2\right\} = E\left\{\left(\sqrt{2}z - \bar{z}\right)^2\right\} = 2$. The introduced variable t describes the distance of the projected point measured from the center of the hypotenuse that is represented by the interception of both abscissas.

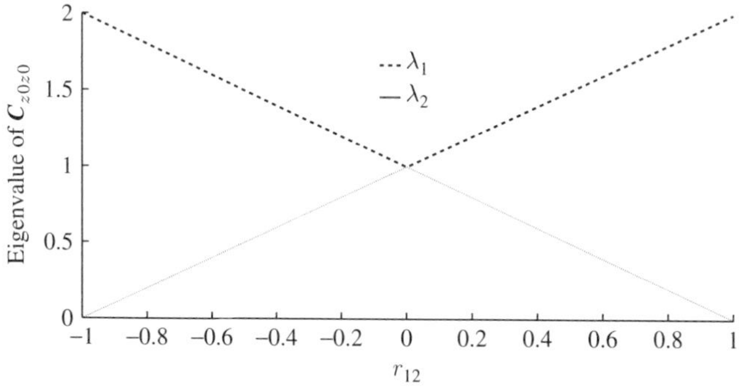

Figure 1.10 Eigenvalues of $\mathbf{S}_{z_0z_0}$, λ_1 and λ_2, vs. correlation coefficient r_{12}.

Orientation of semimajor of control ellipse. The second issue is the orientation of the semimajor relative to the abscissas of both Shewhart charts, which is determined by the direction of $\mathbf{p}_1^T = \begin{pmatrix} p_{11} & p_{21} \end{pmatrix}$. The angle of the semimajor and semiminor is given by $\arctan(p_{21}/p_{11}) \times 180/\pi$ and $\arctan(p_{22}/p_{12}) \times 180/\pi$ relative to the z_2 axis. This yields the following angles for the ellipse axes:

$$135^\circ \text{ if } -1 \le r_{12} < 0,\ 0 \text{ if } r_{12} = 0 \text{ and } 45^\circ \text{ if } 0 < r_{12} \le 1 \text{ (semimajor)} \tag{1.21}$$

and

$$45^\circ \text{ if } -1 \le r_{12} < 0,\ 0 \text{ if } r_{12} = 0 \text{ and } 135^\circ \text{ if } 0 < r_{12} \le 1 \text{ (semiminor)}, \tag{1.22}$$

respectively.

1.2.3.2 Size and orientation of control ellipse for covariance matrix $\mathbf{S}_{z_0z_0}$

In a general case, $E\left\{\left(z_1 - \bar{z}_1\right)^2\right\} = \sigma_1^2 \neq E\left\{\left(z_2 - \bar{z}_2\right)^2\right\} = \sigma_2^2$, the covariance matrix of z_1 and z_2, is

$$\mathbf{S}_{z_0z_0} = E\left\{\begin{pmatrix} z_1 - \bar{z}_1 \\ z_2 - \bar{z}_2 \end{pmatrix}\begin{pmatrix} z_1 - \bar{z}_1 & z_2 - \bar{z}_2 \end{pmatrix}\right\}$$

$$\mathbf{S}_{z_0z_0} = \begin{bmatrix} \sigma_1^2 & r_{12}\sigma_1\sigma_2 \\ r_{12}\sigma_1\sigma_2 & \sigma_2^2 \end{bmatrix} = \sigma_2^2\begin{bmatrix} \sigma_1^2/\sigma_2^2 & r_{12}\sigma_1/\sigma_2 \\ r_{12}\sigma_1/\sigma_2 & 1 \end{bmatrix}. \tag{1.23}$$

Fixing r_{12} to, say 0.8, and taking into account that the eigenvectors do not change if this matrix is multiplied by a scalar factor allows examining the effect of σ_1^2/σ_2^2 upon the orientation of the eigenvectors. More generally, varying this parameter within the interval $\sigma_1^2/\sigma_2^2 \in \left[\ 0 \quad \infty\ \right)$ and the correlation coefficient $r_{12} \in \left[\ 0 \quad 1\ \right]$ as well as defining $\sigma_2^2 = 1$ allows examination of:

- the angle between $\mathbf{p}_1$ and the abscissa; and
- the values of both eigenvalues of $\mathbf{S}_{z_0z_0}$.

Eigenvalues of $\mathbf{S}_{z_0z_0}$. The left plot in Figure 1.11 shows the resultant parametric curves for both eigenvalues vs. σ_1/σ_2. It is interesting to note that small ratios yield eigenvalues that are close to one for λ_1 and zero λ_2. This is no surprise given that the variance of z_2 was selected to be one, whilst that of z_1 is close to zero. In other words, the variance of z_2 predominantly contributes to the joint PDF.

Given that the length of the semimajor and semiminor is proportional to the eigenvalues λ_1 and λ_2, respectively, the ellipse becomes narrower as the ratio σ_1/σ_2 decreases. In the extreme case of $\sigma_1/\sigma_2 \rightarrow 0$ the control ellipse reduces to a line. On the other hand, larger ratios produce larger values for λ_1 and values

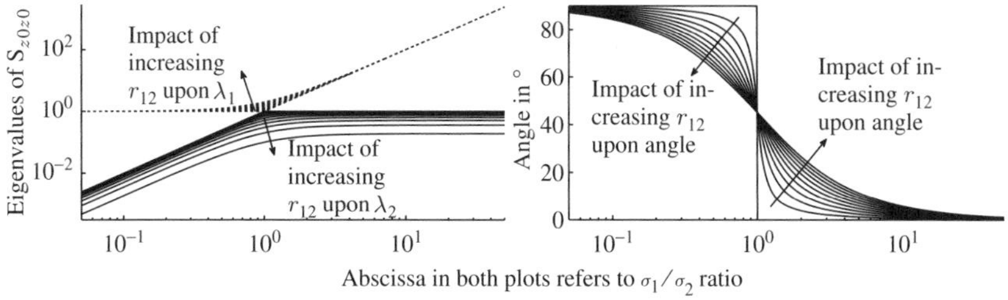

Figure 1.11 Eigenvalues of $\mathbf{S}_{z_0z_0}$ *(left plot) and angle of eigenvector associated with larger eigenvalue (right plot) vs.* σ_1/σ_2 *and the parameter* r_{12}.

below 1 for λ_2. If no correlation between z_1 and z_2 exists, that is $r_{12} \to 0$, the eigenvalue λ_2 converges to one for large σ_1/σ_2 ratios. However, the left plot in Figure 1.11 highlights that λ_2 reduces in value and eventually converges to zero if there is a perfect correlation between both variables. For $\sigma_1^2 = \sigma_2^2$, this plot also includes both extreme cases discussed in the previous two subsections. For $\sigma_1/\sigma_2 \to 1$, letting $r_{12} \to 0$ then both eigenvalues are equal to one and $r_{12} = 1$, $\lambda_1 = 2$ and $\lambda_2 = 0$.

Orientation of semimajor of control ellipse. The right plot in Figure 1.11 shows how the angle between the semimajor and the abscissa of the Shewhart chart for z_1 changes with σ_1/σ_2 and r_{12}. For cases $\sigma_1 \ll \sigma_2$ this angle asymptotically converges to 90°. Together with the fact that the eigenvalues in this case are $\lambda_1 = 1$ and $\lambda_2 = 0$ the control ellipse reduces to a line that is parallel to the abscissa of the Shewhart chart for z_1 and orthogonal to that of the Shewhart chart for z_2.

Larger ratios of σ_1/σ_2 produce angles that asymptotically converge to 0°. Given that λ_1 converges to infinity and λ_2 between zero and one, depending upon the correlation coefficient, the resultant control ellipse is narrow with an infinitely long semimajor that is orthogonal to the abscissa of the Shewhart chart for z_1. If $r_{12} \to 1$, the ellipse reduces to a line.

The case of $r_{12} \to 0$ is interesting, as it represents the asymptotes of the parametric curves. If $0 \le \sigma_1/\sigma_2 < 1$ the semimajor has an angle of 90°, whilst for values in the range of $1 < \sigma_1/\sigma_2 < \infty$, the angle becomes zero. For $\sigma_1^2 = \sigma_2^2$, the control ellipse becomes a circle and a semimajor therefore does not exist.

1.2.3.3 Construction of control ellipse

What has not been discussed thus far is how to construct the control ellipse. The analysis above, however, pointed out that the orientation of this ellipse depends on the eigenvectors. The direction of the semimajor and semiminor is defined by the direction of the eigenvectors associated with the larger and the smaller eigenvalues, respectively. The exact length of the semimajor, a, and semiminor, b,

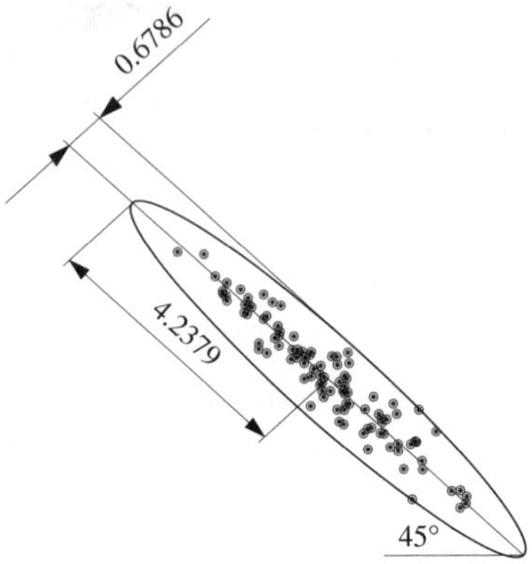

Figure 1.12 Control ellipse for $\mathbf{z} \sim \mathcal{N}\left\{\bar{\mathbf{z}}, \mathbf{S}_{z_0z_0}\right\}$, *where* $\mathbf{S}_{z_0z_0} = \mathcal{C}_{z_0z_0}$ *is the covariance/correlation matrix in (1.17), with a significance of 0.01.*

depends on the eigenvalues of the covariance matrix

$$a = \sqrt{T_\alpha^2 \cdot \lambda_1} \qquad b = \sqrt{T_\alpha^2 \cdot \lambda_2} \tag{1.24}$$

where T_α^2 is defined by

$$T_\alpha^2 = \chi_\alpha^2(2) \tag{1.25}$$

and $\chi_\alpha^2(2)$ is the critical value of a χ^2 distribution with two degrees of freedom and a significance α, for example selected to be 0.05 and 0.01.

Applying (1.24) and (1.25) to the covariance matrix in (1.17) for $\alpha = 0.01$ yields $T_{0.01}^2 = 9.2103$, implying that $a = \sqrt{9.2103 \times 1.95} = 4.2379$ and $b = \sqrt{9.2103 \times 0.05} = 0.6786$. As z_1 and z_2 have an equal variance of 1, the angle between the semimajor and the abscissa of the Shewhart chart for z_2 is 45° as discussed in Equation (1.22). Figure 1.12 shows this control ellipse along with a total of 100 samples of z_1 and z_2. Jackson (1980) introduced an alternative construction

$$\frac{1}{1-r_{12}^2}\left(\frac{\left(z_1-\bar{z}_1\right)^2}{\sigma_1^2}+\frac{\left(z_2-\bar{z}_2\right)^2}{\sigma_2^2}-\frac{2r_{12}\left(z_1-\bar{z}_1\right)\left(z_2-\bar{z}_2\right)}{\sigma_1\sigma_2}\right)=T_\alpha^2. \tag{1.26}$$

Based on the preceding discussion, the next subsection addresses the question laid out at the beginning of this section: *why multivariate statistical process control?*

1.2.4 Type I and II errors and dimension reduction

For the extreme case analyzed in Subsection 1.2.2, it follows that the projections of two perfectly correlated variables fall onto a 1D line and any departure from this line confirms that r_{12} is no longer equal to 1 for the 'violating' samples.

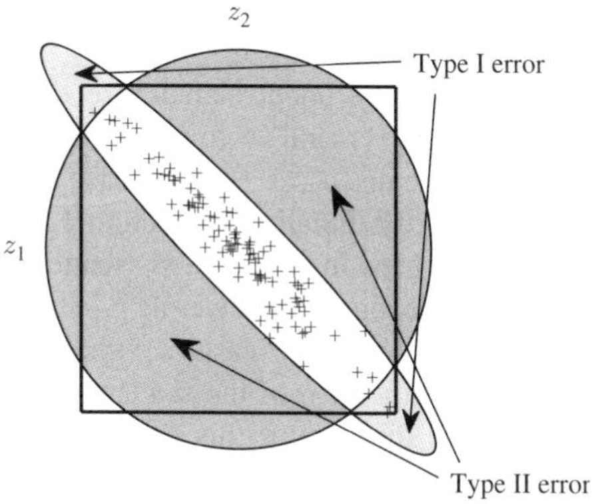

Figure 1.13 Graphical illustration of Type I and II errors for correlated variable sets.

Moreover, inspecting Figure 1.9, describing high correlation between two variables with respect to the sample represented by the asterisk yields that this sample shows an in-statistical-control situation, since it is within the control limits of both variables. However, if this sample is analyzed with respect to the multivariate control ellipse it lies considerably outside the normal operating region and hence, describes an out-of-statistical-control situation.

The joint analysis therefore suggests that this sample is indicative of an out-of-statistical-control situation. Comparing this to the case where the two variables are uncorrelated, Figure 1.6 outlines that such a situation can only theoretically arise and is restricted to the small corners of the naive 'rectangular control region'. Following the introduction of hypothesis testing in Subsection 1.1.3, accepting the null hypothesis although it must be rejected constitutes a Type II error. Figure 1.13 shows graphically that the Type II error can be very substantial.

The slightly darker shaded area in this figure is proportional to the Type II error. More precisely, this area represents the difference between the circular and the elliptical confidence regions describing the uncorrelated case and the highly correlated case, respectively. It is interesting to note that correlation can also give rise to Type I errors if larger absolute values for $z_1 - z_2$ arise. The brightly shaded areas in Figure 1.13 give a graphical account of the Type I error, which implies that the alternative hypothesis H_1, i.e. the process is out-of-statistical-control, is accepted although the null hypothesis must be accepted.

Besides the impact of correlation upon the hypothesis testing, particularly the associated Type II errors, Figure 1.9 highlights another important aspect. Projecting the samples of z_1 and z_2 onto the semimajor of the control ellipse describes most of the variance that is encapsulated within both variables. In contrast, the remaining variance that cannot be described by projecting the samples onto the semimajor is often very small in comparison and depends on r_{12}. More precisely, the ratio of the larger over the smaller eigenvalue of $\mathbf{S}_{z_0z_0}$ is a measure to compute how much the projection of the recorded samples onto the semimajor contribute to the variance within both variables.

A ratio equal to 1 describes the uncorrelated case discussed in Subsection 1.2.1. This ratio increases with r_{12} and asymptotically describes the case of $r_{12} = 1$ which Subsection 1.2.2 discusses. In this case, the variable $t = \sqrt{2}z$, $z_1 = z_2 = z$ describes both exactly. Finally, Subsection 1.2.3 discuss large ratios of λ_1/λ_2 representing large r_{12} values. In this case, the scatter diagram for z_1 and z_2 produces a control ellipse that becomes narrower as r_{12} increases and vice versa.

In analogy to the perfectly correlated case, a variable t can be introduced that represents the orthogonal projection of the scatter point onto the semimajor. In other words, t describes the distance of this projected point from the origin, which is the interception of the abscissas of both Shewhart charts. The variable t consequently captures most of the variance of z_1 and z_2. The next chapter introduces data models that are based on approximating the recorded process variables by defining a set of such t-variables. The number of these t-variables is smaller than the number of recorded process variables.

1.3 Tutorial session

Question 1: What is the main motivation for using the multivariate extension of statistical process control? Discuss the principles of statistical process control and the disadvantage of analyzing a set of recorded process variables separately to monitoring process performance and product quality.

Question 2: Explain how a Type I and a Type II error affect the monitoring of a process variable and the detection of an abnormal operating condition.

Question 3: With respect to Figure 1.13, use the area of overlap between the control ellipse and the naive rectangular confidence region, approximate the Type I and II error for using the naive rectangular confidence region for various correlation coefficients, $0 \leq r_{12} \leq 1$.

Question 4: Using a numerical integration, for example the *quad2d* and *dblquad* commands in Matlab, determine the correct Type I and II error in Question 2.

Question 5: Following Questions 2 and 3, determine and plot an empirical relationship between the Type II error and correlation coefficient for two variables.

Project 1: Simulate 1000 samples from a Gaussian distributed random variable such that the first 500 samples have a mean of zero, the last 500 samples have a mean of 0.25 and the variance of each sample is 1. Determine a Shewhart chart based on the assumption that the process mean is zero and comment on the detectability of the shift in mean from 0 to 0.25. Next, vary the shift in mean and comment on the detectability of the mean.

Project 2: Construct a CUSUM chart and repeat the experiments in Project 1 for various window lengths. Comment upon the detectability and the average run length of the CUSUM chart depending on the window length. Empirically estimate the PDF of the CUSUM samples and comment on the relationship between the distribution function and the window length with respect to the central limit theorem. Is a CUSUM chart designed to detect small changes in the variable variance? Tip: carry out a Monte Carlo simulation and examine the asymptotic definition of a CUSUM sample.

Project 3: Develop EWMA charts and repeat the experiments in Project 1 for various weighting parameters. Comment upon the detectability and the average run length depending on the weighting parameter. Is an EWMA chart designed to detect small changes in the variable variance? Tip: examine the asymptotic PDF of the EWMA samples for a change in variance.

Project 4: Based on the analysis in Projects 2 and 3, study the literature and propose ways on how to detect small changes in the variable variance. Is it possible to construct *hypothetical* cases where a shift in mean and a simultaneous reduction in variance remains undetected? Suggest ways to detect such hypothetical changes.

2

Multivariate data modeling methods

The last chapter has introduced the principles of SPC and motivated the required multivariate extension to prevent excessive Type II errors if the recorded process variables are highly correlated. The aim of this chapter is to present different methods that generate a set of t-variables that are defined as *score variables*. Under the assumption that the process variables follow a multivariate Gaussian distribution, these score variables are statistically independent to circumvent increased levels of Type II errors. According to Figures 1.7 and 1.8, the generation of these score variables relies on projecting the recorded samples onto predefined directions in order to extract as much information from the recorded process variables as possible.

The data reduction techniques, introduced in the literature, are firmly based on the principle of establishing sets of *latent variables* that capture significant and important variation that is encapsulated within the recorded data. The score variables form part of these latent variable sets. For process monitoring, the variation that the latent variable sets extract from the recorded process variables is of fundamental importance for assessing product quality, process safety and, more generally, whether the process is in-statistical-control. These aspects are of ever growing importance to avert risks to the environment and to minimize pollution.

Data analysis and reduction techniques can be divided into single-block and dual-block techniques. The most notable single-block techniques include:

- Principal Component Analysis (Pearson 1901);

Statistical Monitoring of Complex Multivariate Processes: With Applications in Industrial Process Control, First Edition. Uwe Kruger and Lei Xie.

- Linear or Fisher's Discriminant Analysis (Duda and Hart 1973); and
- Independent Component Analysis (Hyvärinen *et al.* 2001).

Dual-block techniques, on the other hand, divide the recorded data sets into one block of predictor or cause variables and one block of response or effect variables and include:

- Canonical Correlation Analysis (Hotelling 1935; Hotelling 1936);
- Reduced Rank Regression (Anderson 1951);
- Partial Least Squares (Wold 1966a,b); and
- Maximum Redundancy (van den Wollenberg 1977),

among others. These listed single- and dual-block techniques are collectively referred as *latent variable* techniques.

From this list of techniques, the focus in the research literature has been placed on variance/covariance-based techniques as most appropriate for process monitoring applications. This has been argued on the basis of capturing the process variation, that is, encapsulated in the variance among and the covariance between the recorded process variables. These techniques are Principal Component Analysis (PCA) and Partial Least Squares (PLS), which are discussed and applied in this chapter and described and analyzed in Part IV of this book.

It should be noted that the research community has also developed latent variable techniques for multiple variable blocks, referred to as multi-block methods (MacGregor *et al.* 1994; Wangen and Kowalski 1989). These methods, however, can be reduced to single-block PCA or dual-block PLS models, for example discussed in Qin *et al.* (2001), Wang *et al.* (2003), Westerhuis *et al.* (1998). The methods used in this book are therefore limited to PCA and PLS.

As the focus for presenting MSPC technology in this chapter is based on its exploitation as a statistically based process monitoring tool, details of PCA and PLS are given using an introduction of the underlying data model, a geometric analysis and by presenting simple simulation examples in Sections 2.1 and 2.2, respectively. This allows a repetition of the results in order to gain familiarization with both techniques. A detailed statistical analysis of both techniques are given in Chapters 9 and 10.

Section 2.3 presents an extension of the PLS algorithm after analyzing that PCA and PLS fail to produce a latent variable data representation for a more general data structure. The validity of the general data structure is demonstrated by an application study of a distillation process in Part II of this book, which also includes an application study involving the applications of PCA. Section 2.4 then introduces methods for determining the number of the latent variable sets for each method. To enhance the learning outcomes, this chapter concludes with a tutorial session including short questions and calculations as well as homework type projects in Section 2.5.

2.1 Principal component analysis

This section introduces PCA using a geometrical analysis. Chapter 9 provides a more comprehensive treatment of PCA, including its properties, and further information may also be taken from the research literature, for example references Anderson (2003); Jolliffe (1986); Mardia *et al.* (1979); Wold *et al.* (1987). For a set of highly correlated process variables, PCA allows reducing the number of variables to be monitored by defining a significantly reduced set of latent variables, referred to as principal components, that describe the important process variation that is encapsulated within the recorded process variables.

2.1.1 Assumptions for underlying data structure

According to Figure 1.9, the important process variation can be described by projecting the two variables onto the semimajor of the control ellipse. This is further illustrated in Figure 2.1, which shows that the two correlated variables can be approximated with a high degree of accuracy by their projection onto the semimajor of the control ellipse. It can be seen further that the variance of the error of approximating both process variables using their projection onto the semimajor is relatively small compared to the variance of both process variables.

This analysis therefore suggests utilizing the following data structure for the two process variables

$$\begin{pmatrix} z_1 \\ z_2 \end{pmatrix} = \begin{pmatrix} z_{0_1} \\ z_{0_2} \end{pmatrix} + \begin{pmatrix} \bar{z}_1 \\ \bar{z}_2 \end{pmatrix} + \begin{pmatrix} \mathfrak{g}_1 \\ \mathfrak{g}_2 \end{pmatrix} = \begin{pmatrix} \xi_1 \\ \xi_2 \end{pmatrix} s + \begin{pmatrix} \bar{z}_1 \\ \bar{z}_2 \end{pmatrix} + \begin{pmatrix} \mathfrak{g}_1 \\ \mathfrak{g}_2 \end{pmatrix} \tag{2.1}$$

where $\begin{pmatrix} z_{0_1} & z_{0_2} \end{pmatrix} + \begin{pmatrix} \bar{z}_1 & \bar{z}_2 \end{pmatrix}$ are the approximated values of the original process variables z_1 and z_2. In analogy to Figure 2.1, the vector $\boldsymbol{\xi}^T = \begin{pmatrix} \xi_1 & \xi_2 \end{pmatrix}$ describes the orientation of the semimajor of the control ellipse.

With this in mind, approximating the samples of z_1 and z_2 relies on projecting the scatter points onto the semimajor. If the length of $\boldsymbol{\xi}$ is 1, the approximation is equal to $\boldsymbol{\xi}s$[1], which the proof of Lemma 2.1.1 highlights. With respect to (2.1), the variable s is defined as the source signal, whilst $\mathfrak{g}_1$ and $\mathfrak{g}_2$ are error variables.

On the basis of the two-variable example above, the following general data model can be assumed for $n_z \geq 2$ recorded process variables

$$\mathbf{z} = \Xi \mathbf{s} + \bar{\mathbf{z}} + \mathfrak{g} = \mathbf{z}_s + \bar{\mathbf{z}} + \mathfrak{g} = \mathbf{z}_t + \mathfrak{g} = \mathbf{z}_0 + \bar{\mathbf{z}}. \tag{2.2}$$

Here, $\mathbf{z} \in \mathbb{R}^{n_z}$ is a vector of measured variables, $\Xi \in \mathbb{R}^{n_z \times n}$ is a parameter matrix of rank $n < n_z$, $\mathbf{s} \in \mathbb{R}^n$ is a vector of *source* variables representing the *common cause variation* of the process, $\mathbf{z}_s \in \mathbb{R}^{n_z}$ describes the stochastic variation of the

[1] The variable s describes the distance between the projection of the sample $\mathbf{z}_0^T = \begin{pmatrix} z_1 & z_2 \end{pmatrix}$ onto the semimajor from the origin of the control ellipse which, according to Figure 1.9, is given by the interception the abscissas of both Shewhart charts.

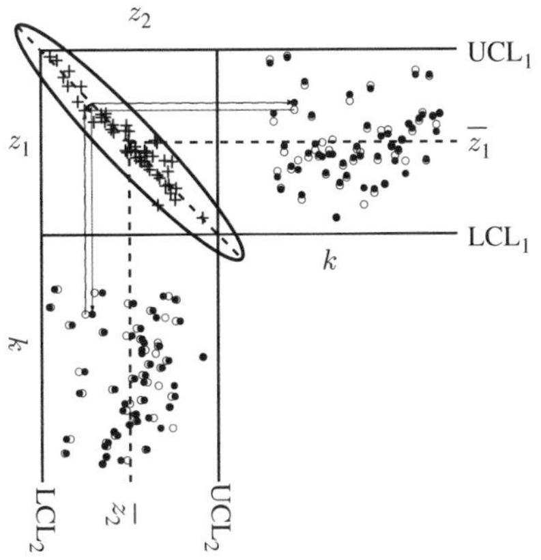

Figure 2.1 *Schematic diagram of reconstructing two process variables by their projection onto the semimajor.*

process driven by common cause variation which is centered around the mean vector $\bar{\mathbf{z}} \in \mathbb{R}^{n_z}$, $\mathbf{g} \in \mathbb{R}^{n_z}$ is an error vector, $\mathbf{z}_t \in \mathbb{R}^{n_z}$ is the approximation of $\mathbf{z}$ using common cause variation $\Xi\mathbf{s} + \bar{\mathbf{z}}$, and $\mathbf{z}_0 \in \mathbb{R}^{n_z}$ represents the stochastic variation of the recorded variables $\Xi\mathbf{s} + \mathbf{g}$.

It should be noted that the subscript $_t$ symbolically implies that $\Xi\mathbf{s} + \bar{\mathbf{z}}$ is the *true* representation of the variable interrelationships, whilst the error vector $\mathbf{g}$ represents measurement uncertainty and the impact of unmeasured and naturally occurring stochastic disturbances. With respect to SPC, unmeasured deterministic disturbances or stochastic disturbances of a large magnitude describe *special cause variation* that lead to a change of the mean vector $\bar{\mathbf{z}}$ and/or changes in the covariance matrix $\mathbf{S}_{z_0 z_0}$.

The space spanned by the linearly independent column vectors in Ξ is defined as the *model subspace*, which is an n-dimensional subspace of the original n_z-dimensional data space. The data model in (2.2) gives rise to the construction of a second subspace that is orthogonal to the model subspace and referred to as the *residual subspace*. The residual subspace is complementary to the model subspace and of dimension $n_z - n$.

With respect to Figure 2.1, the semimajor and semiminor are the model subspace and the residual subspace, respectively. It is important to note these spaces only describe the stochastic component of the data vector $\mathbf{z}$, which is $\mathbf{z}_0 = \mathbf{z}_s + \mathbf{g} = \mathbf{z} - \bar{\mathbf{z}}$. Otherwise, both subspaces do not include the element $\mathbf{0}$ unless $\bar{\mathbf{z}} = \mathbf{0}$ and are, by definition, not subspaces.

Assumptions imposed on the data model in (2.2), describing highly correlated process variables, include:

- that each vector $\mathbf{z}$, $\mathbf{z}_0$, $\mathbf{s}$, and $\mathbf{g}$, stores random variables that follow Gaussian distributions; and
- that each of these vectors do not possess any time-based correlation.

The second assumption implies that the vectors $\mathbf{s}$ and $\mathbf{g}$ have the following properties:

- $E\left\{\mathbf{s}(k)\,\mathbf{s}^T(l)\right\} = \delta_{kl}\mathbf{S}_{ss}$;
- $E\left\{\mathbf{g}\mathbf{g}^T\right\} = \sigma_g^2\mathbf{I}$;
- $E\left\{\mathbf{g}(k)\,\mathbf{g}^T(l)\right\} = \delta_{kl}\mathbf{S}_{gg}$; and
- $E\left\{\mathbf{s}(k)\,\mathbf{g}^T(l)\right\} = \mathbf{0}$.

Here, k and l are sample instances, δ_{kl} is the Kroneker delta, that is 0 for all $k \neq l$ and 1 if $k = l$, and $\mathbf{S}_{ss} \in \mathbb{R}^{n \times n}$ and $\mathbf{S}_{gg} \in \mathbb{R}^{n_z \times n_z}$ are covariance matrices for $\mathbf{s}$ and $\mathbf{g}$, respectively. Table 2.1 shows the mean and covariance matrices for each vector in (2.2). The condition that $E\left\{\mathbf{s}(k)\,\mathbf{g}^T(l)\right\} = \mathbf{0}$ implies that $\mathbf{s}$ and $\mathbf{g}$ are statistically independent.

It should be noted that the assumption of $E\left\{\mathbf{g}\mathbf{g}^T\right\} = \sigma_g^2\mathbf{I}$ is imposed for convenience. Under this condition, the eigendecomposition of $\mathbf{S}_{z_0 z_0}$ provides a consistent estimation of the model subspace spanned by the column vectors of Ξ if the number of recorded samples goes to infinity. This, however, is a side issue as the main aim of this subsection is to introduce the working of PCA as a MSPC tool. Section 6.1 shows how to consistently estimate the model subspace if this assumption is relaxed, that is $\mathbf{S}_{gg}$ is no longer a diagonal matrix storing equal diagonal elements.

Prior to the analysis of how PCA reduces the number of variables, let us reconsider the perfect correlation situation discussed in Subsection 1.2.2. This situation arises if the error vector $\mathbf{g}$ in (2.2) is set to zero. In this case, it is possible to determine the source variable set, $\mathbf{s}$, directly from the process variables $\mathbf{z}$ if the column vectors of Ξ are orthonormal, i.e. mutually orthogonal and of unit length.

Lemma 2.1.1 *If the column vectors of Ξ are mutually orthonormal, the source variables, $\mathbf{s}$, are equal to the orthogonal projection of the stochastic*

Table 2.1 Mean vector and covariance matrices of stochastic vectors in Equation (2.2).

Vector	Mean vector	Covariance matrix
$\mathbf{s}$	$\mathbf{0}$	$\mathbf{S}_{ss}$
$\mathbf{z}_s$	$\mathbf{0}$	$\mathbf{S}_{z_s z_s} = \Xi\mathbf{S}_{ss}\Xi^T$
$\mathbf{z}_t = \mathbf{z}_s + \bar{\mathbf{z}}$	$\bar{\mathbf{z}}$	$\mathbf{S}_{z_t z_t} = \mathbf{S}_{z_s z_s}$
$\mathbf{g}$	$\mathbf{0}$	$\mathbf{S}_{gg} = \sigma_g^2\mathbf{I}$
$\mathbf{z}_0 = \mathbf{z}_s + \mathbf{g}$	$\mathbf{0}$	$\mathbf{S}_{z_0 z_0} = \mathbf{S}_{z_s z_s} + \mathbf{S}_{gg}$
$\mathbf{z}$	$\bar{\mathbf{z}}$	$\mathbf{S}_{zz} = \mathbf{S}_{z_s z_s} + \mathbf{S}_{gg}$

component of the measured vector, $\mathbf{z}_0 = \mathbf{z} - \bar{\mathbf{z}}$*, onto* $\Xi = \begin{bmatrix} \boldsymbol{\xi}_1 & \boldsymbol{\xi}_2 & \cdots & \boldsymbol{\xi}_n \end{bmatrix}$*, that is* $s_1 = \mathbf{z}_0^T \boldsymbol{\xi}_1$, $s_2 = \mathbf{z}_0^T \boldsymbol{\xi}_2$, $\ldots$, $s_n = \mathbf{z}_0^T \boldsymbol{\xi}_n$ *in the error free case, i.e.* $\mathbf{g} = \mathbf{0}$.

Proof. If the column vectors of Ξ are orthonormal, the matrix product $\Xi^T \Xi$ is equal to the identity matrix. Consequently, if $\mathbf{z}_0 = \Xi \mathbf{s}$, the source signals can be extracted by $\Xi^T \mathbf{z}_0 = \Xi^T \Xi \mathbf{s} = \mathbf{s}$.

On the other hand, if the column vectors of Ξ are mutually orthonormal but the error vector is no longer assumed to be zero, the source signals can be approximated by $\Xi^T \mathbf{z}_0$, which follows from

$$\mathbf{z}_0 = \Xi \mathbf{s} + \mathbf{g} \Rightarrow \Xi^T \mathbf{z}_0 = \mathbf{s} + \Xi^T \mathbf{g} \approx \mathbf{s} = \widehat{\mathbf{s}}. \tag{2.3}$$

The variance of $\boldsymbol{\xi}_i^T \mathbf{s}$, however, must be assumed to be larger than that of g_i, i.e. $E\left\{\left(\boldsymbol{\xi}_i^T \mathbf{s}\right)^2\right\} \gg \sigma_g^2$ for all $1 \le i \le n_z$, to guarantee an accurate estimation of $\mathbf{s}$.

2.1.2 Geometric analysis of data structure

The geometric analysis in Figure 2.2 confirms the result in (2.3), since

$$\cos\left(\varphi_{(\mathbf{z}_0, \boldsymbol{\xi}_i)}\right) = \frac{\mathbf{z}_0^T \boldsymbol{\xi}_i}{\left\|\mathbf{z}_0\right\| \left\|\boldsymbol{\xi}_i\right\|}, \tag{2.4}$$

where $\varphi_{(\mathbf{z}_0, \boldsymbol{\xi}_i)}$ is the angle between $\mathbf{z}_0$ and $\boldsymbol{\xi}_i$. Given that $\left\|\boldsymbol{\xi}_i\right\| = 1$, reformulating (2.3) yields

$$\cos\left(\varphi_{(\mathbf{z}_0, \boldsymbol{\xi}_i)}\right) \left\|\mathbf{z}_0\right\| = \mathbf{z}_0^T \boldsymbol{\xi}_i = \widehat{s}_i. \tag{2.5}$$

The projection of a sample onto the column vectors of Ξ is given by

$$\widehat{\mathbf{z}} = \sum_{i=1}^{n} \boldsymbol{\xi}_i \widehat{s}_i + \bar{\mathbf{z}} = \sum_{i=1}^{n} \boldsymbol{\xi}_i \boldsymbol{\xi}_i^T \mathbf{z}_0 + \bar{\mathbf{z}} = \widehat{\mathbf{z}}_0 + \bar{\mathbf{z}}. \tag{2.6}$$

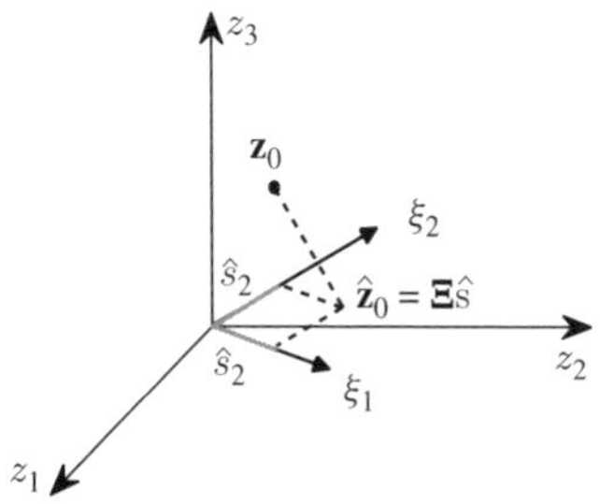

Figure 2.2 Orthogonal projection of $\mathbf{z}_0$ *onto orthonormal column vector of* Ξ.

The estimation of $\mathbf{s}$, however, does not reduce to the simple projection shown in (2.4) and (2.5) if the column vectors of Ξ are not mutually orthonormal. To address this, PCA determines n_z orthonormal *loading vectors* such that n of them span the same column space as Ξ, which are stored as column vectors in the matrix $\mathbf{P} \in \mathbb{R}^{n_z \times n}$. The remaining $n_z - n$ loading vectors are stored in the matrix $\mathbf{P}_d \in \mathbb{R}^{n_z \times (n_z - n)}$. These two matrices have the following orthogonality properties

$$\begin{aligned} \mathbf{P}^T\mathbf{P} &= \mathbf{I} & \mathbf{P}^T\mathbf{P}_d &= \mathbf{0} \\ \mathbf{P}_d^T\mathbf{P}_d &= \mathbf{I} & \mathbf{P}_d^T\mathbf{P} &= \mathbf{0} \\ \mathbf{P}_d^T\Xi &= \mathbf{0} & \mathbf{P}^T\Xi &= \mathcal{X} \in \mathbb{R}^{n \times n}. \end{aligned} \tag{2.7}$$

The loading vectors are eigenvectors of $\mathbf{S}_{z_0z_0}$ and the above orthogonality properties give rise to the calculation of the following orthogonal projections

$$\begin{aligned} \mathbf{t} &= \mathbf{P}^T\mathbf{z}_0 = \mathcal{X}\mathbf{s} + \mathbf{P}^T\mathfrak{g} \\ \mathbf{t}_d &= \mathbf{P}_d^T\mathbf{z}_0 = \mathbf{P}_d^T\mathfrak{g} \\ \mathbf{t} &\approx \mathcal{X}\mathbf{s} = \widehat{\mathbf{t}}. \end{aligned} \tag{2.8}$$

The ith element stored in $\mathbf{t}$ represents the coordinate describing the orthogonal projection of $\mathbf{z}_0$ onto the ith column vector in $\mathbf{P}$. Note that the column space of $\mathbf{P}$ is identical to the column space of Ξ. Moreover, the column vectors of $\mathbf{P}$ and $\mathbf{P}_d$ are *base vectors* spanning the model subspace and the residual subspace, respectively.

Given that the column vectors stored in $\mathbf{P}_d$ are orthogonal to those in $\mathbf{P}$, they are also orthogonal to those in Ξ. Consequently, $\mathbf{P}_d^T\Xi = \mathbf{0}$. In this regard, the jth element of $\mathbf{t}_d$ is equal to the coordinate describing the orthogonal projection of $\mathbf{z}_0$ onto the jth column vector in $\mathbf{P}_d$. In other words, the elements in $\mathbf{t}$ are the coordinates describing the orthogonal projection of $\mathbf{z}_0$ onto the model subspace and the elements in $\mathbf{t}_d$ are the coordinates describing the orthogonal projection of $\mathbf{z}_0$ onto the residual subspace. This follows from the geometric analysis in Figure 2.2.

On the basis of the preceding discussion, Figure 2.3 shows an extension of the simple 2-variable example to a 3-variable one, where two common cause 'source' variables describe the variation of 3 process variables. This implies that the dimensions of the model and residual subspaces are 2 and 1, respectively.

2.1.3 A simulation example

Using the geometric analysis in Figure 2.3, this example shows how to obtain an estimate of the model subspace $\Xi = \begin{bmatrix} \boldsymbol{\xi}_1 & \boldsymbol{\xi}_2 \end{bmatrix}$ and the residual subspace, defined by the cross product of $\boldsymbol{\xi}_1$ and $\boldsymbol{\xi}_2$. The data model for this example is

$$\begin{pmatrix} z_{0_1} \\ z_{0_2} \\ z_{0_3} \end{pmatrix} = \begin{bmatrix} 0.2 & -0.5 \\ 0.8 & 0.4 \\ -0.3 & -0.7 \end{bmatrix} \begin{pmatrix} s_1 \\ s_2 \end{pmatrix} + \begin{pmatrix} \mathfrak{g}_1 \\ \mathfrak{g}_2 \\ \mathfrak{g}_3 \end{pmatrix}, \tag{2.9}$$

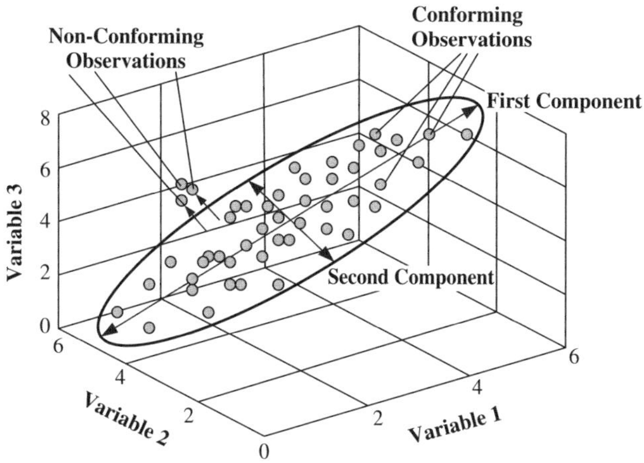

Figure 2.3 Schematic diagram of showing the PCA model subspace and its complementary residual subspace for 3 process variables.

which has a mean vector of zero. The elements in **s** follow a Gaussian distribution

$$\mathbf{s} \sim \mathcal{N}\left\{\mathbf{0}, \mathbf{S}_{ss}\right\} \qquad \mathbf{S}_{ss} = \begin{bmatrix} 1 & -0.3 \\ -0.3 & 1 \end{bmatrix}. \tag{2.10}$$

The error vector $\mathbf{g}$ contains random variables that follow a Gaussian distribution too

$$\mathbf{g} \sim \mathcal{N}\left\{\mathbf{0}, \mathbf{S}_{gg}\right\} \qquad \mathbf{S}_{gg} = 0.05\mathbf{I}. \tag{2.11}$$

From this process, a total of $K = 100$ samples, $\mathbf{z}_0(1), \ldots, \mathbf{z}_0(k), \ldots, \mathbf{z}_0(100)$ are simulated. Figure 2.4 shows time-based plots for each of the 3 process variables. PCA analyzes the stochastic variation encapsulated within this reference set, which leads to the determination of the model subspace, spanned by the column vectors of $\boldsymbol{\Xi}$, and the complementary residuals subspace. Chapter 9 highlights that this involves the data covariance matrix, which must be estimated from the recorded data

$$\widehat{\mathbf{S}}_{z_0 z_0} = \frac{1}{K}\sum_{k=1}^{K}\mathbf{z}_0(k)\,\mathbf{z}_0^T(k) = \begin{bmatrix} 0.3794 & 0.0563 & 0.2858 \\ 0.0563 & 0.6628 & -0.2931 \\ 0.2858 & -0.2931 & 0.4781 \end{bmatrix}. \tag{2.12}$$

For a nonzero mean vector, it must be estimated from the available samples first

$$\widehat{\bar{\mathbf{z}}} = \frac{1}{K}\sum_{k=1}^{K}\mathbf{z}(k), \tag{2.13}$$

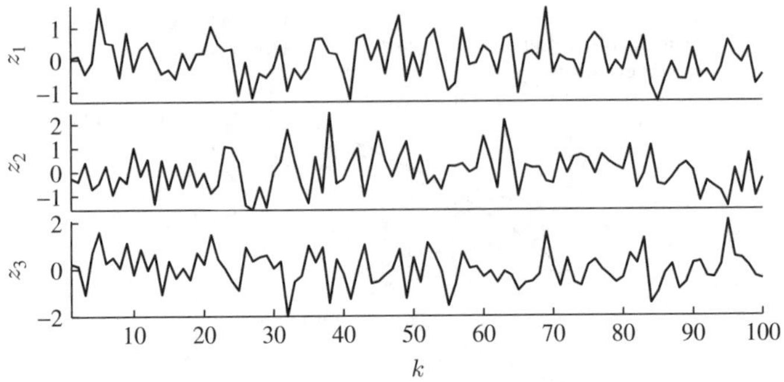

Figure 2.4 Time-based plot of simulated process variables.

which yields the following estimation of the data covariance matrix

$$\widehat{\mathbf{S}}_{z_0z_0} = \frac{1}{K-1}\sum_{k=1}^{K}\left(\mathbf{z}(k) - \widehat{\bar{\mathbf{z}}}\right)\left(\mathbf{z}(k) - \widehat{\bar{\mathbf{z}}}\right)^T. \tag{2.14}$$

The estimation of the data covariance matrix from the recorded reference data is followed by determining its eigendecomposition

$$\widehat{\mathbf{S}}_{z_0z_0} = \widehat{\mathbf{P}}\widehat{\mathbf{\Lambda}}\widehat{\mathbf{P}}^T \tag{2.15}$$

which produces the following estimates for the eigenvector and eigenvalue matrices

$$\widehat{\mathbf{P}} = \begin{bmatrix} -0.2763 & -0.7201 & -0.6365 \\ 0.7035 & -0.6028 & 0.3766 \\ -0.6548 & -0.3437 & 0.6731 \end{bmatrix} \tag{2.16}$$

and

$$\widehat{\mathbf{\Lambda}} = \begin{bmatrix} 0.9135 & 0 & 0 \\ 0 & 0.5629 & 0 \\ 0 & 0 & 0.0439 \end{bmatrix}, \tag{2.17}$$

respectively.

Given that $\mathbf{\Xi}$, $\mathbf{S}_{ss}$ and $\mathbf{S}_{gg}$ are known, the covariance matrix for the recorded variables can be determined as shown in Table 2.1

$$E\left\{\mathbf{z}_0\mathbf{z}_0^T\right\} = \mathbf{\Xi}\mathbf{S}_{ss}\mathbf{\Xi}^T + \mathbf{S}_{gg} = \begin{bmatrix} 0.4000 & 0.0560 & 0.2870 \\ 0.0560 & 0.6580 & -0.3160 \\ 0.2870 & -0.3160 & 0.5040 \end{bmatrix}. \tag{2.18}$$

Subsection 6.1 points out that $\widehat{\mathbf{S}}_{z_0z_0}$ asymptotically converges to $\mathbf{S}_{z_0z_0}$. To examine how accurate the PCA model has been estimated from $K = 100$ samples, the eigendecomposition of $\widehat{\mathbf{S}}_{z_0 z_0}$ can be compared with that of $\mathbf{S}_{z_0 z_0}$

$$\mathbf{P} = \begin{bmatrix} -0.2836 & -0.7338 & -0.6173 \\ 0.6833 & -0.6063 & 0.4068 \\ -0.6728 & -0.3064 & 0.6734 \end{bmatrix}$$

$$\mathbf{\Lambda} = \begin{bmatrix} 0.9459 & 0 & 0 \\ 0 & 0.5661 & 0 \\ 0 & 0 & 0.0500 \end{bmatrix}. \qquad (2.19)$$

The departures of the estimated eigenvalues are:

- $|\frac{0.9135-0.9459}{0.9459}| \cdot 100\% = 3.43\%$;
- $|\frac{0.5829-0.5661}{0.5661}| \cdot 100\% = 0.57\%$; and
- $|\frac{0.0439-0.0500}{0.0500}| \cdot 100\% = 12.2\%$.

To determine the accuracy of the estimated model subspace, we can compare the normal vector of the actual model subspace with the estimated one. The one for the model subspace is proportional to the cross product, denoted here by the symbol $\times$, of the two column vectors of $\mathbf{\Xi}$

$$\mathbf{n} = \frac{\boldsymbol{\xi}_1 \times \boldsymbol{\xi}_2}{\left\| \boldsymbol{\xi}_1 \times \boldsymbol{\xi}_2 \right\|} = \begin{pmatrix} -0.6173 \\ 0.4068 \\ 0.6734 \end{pmatrix}. \qquad (2.20)$$

As the simulated process has two normally distributed source signals, the two principal components associated with the two largest eigenvalues must, accordingly, be associated with the model subspace, whilst the third one represents the complementary residual subspace, spanned by the third eigenvector. This is based on the fact that the eigenvectors are mutually orthonormal, as shown in Chapter 9. The last column of the matrix $\widehat{\mathbf{P}}$ stores the third eigenvector and the scalar product of this vector with $\mathbf{n}$ yields the minimum angle between the true and estimated residual subspace

$$\cos\left(\varphi_{(\mathbf{n},\widehat{\mathbf{p}}_3)}\right) = \mathbf{n}^T \widehat{\mathbf{p}}_3 = 0.9994 \qquad \arccos(0.9994) \cdot \frac{180^\circ}{\pi} = 2.0543^\circ. \qquad (2.21)$$

Equation (2.21) shows that the estimated model subspace is rotated by just over 2° relative to the actual one. In contrast, the one determined from $\mathbf{S}_{z_0 z_0}$, as expected, is equal to $\mathbf{n}$.

Figure 2.2 shows that storing the 100 samples consecutively as row vectors in the matrix $\mathbf{Z}_0^T = \left[\, \mathbf{z}_0(1) \quad \cdots \quad \mathbf{z}_0(k) \quad \cdots \quad \mathbf{z}_0(K) \,\right]$ allows determining the orthogonal projection of these samples onto the estimated model subspace as follows

$$\left[\, \widehat{\mathbf{t}}_1 \quad \widehat{\mathbf{t}}_2 \,\right] = \mathbf{Z}_0 \left[\, \widehat{\mathbf{p}}_1 \quad \widehat{\mathbf{p}}_2 \,\right] \qquad (2.22)$$

where $\widehat{\mathbf{t}}_1$ and $\widehat{\mathbf{t}}_2$ store the coordinates that determine the location of samples when projected orthogonally onto $\widehat{\mathbf{p}}_1$ and $\widehat{\mathbf{p}}_2$, respectively.

It should be noted that even if the column vectors of Ξ are orthonormal they may be different to the eigenvectors of $\mathbf{S}_{z_0 z_0}$. This is because PCA determines the *principal directions* such that the orthogonal projection of $\mathbf{z}_0$ produces a maximum variance for each of them. More precisely, $E\left\{t_1^2\right\} \geq E\left\{t_2^2\right\} \geq \cdots \geq E\left\{t_n^2\right\}$, which is equal to $E\left\{\mathbf{p}_1^T \mathbf{S}_{z_0 z_0} \mathbf{p}_1\right\} \geq E\left\{\mathbf{p}_2^T \mathbf{S}_{z_0 z_0} \mathbf{p}_2\right\} \geq \cdots \geq E\left\{\mathbf{p}_n^T \mathbf{S}_{z_0 z_0} \mathbf{p}_n\right\}$, and follows from the analysis of PCA in Chapter 9. These expectations, however, are equal to the eigenvalues of $\mathbf{S}_{z_0 z_0}$, which, accordingly, represent the variances of the projections, i.e. the t-scores or *principal components* such that $\lambda_1 \geq \lambda_2 \geq \cdots \geq \lambda_n$.

Another aspect that this book discusses is the use of scatter diagrams for the loading vectors. Figure 1.9 shows a scatter diagram for two highly correlated variables. Moreover, Subsection 3.1.1 introduces scatter diagrams and the construction of the control ellipse, or ellipsoid if the dimension exceeds 2, for the score variables or principal components. Scatter diagrams for the loading vectors, on the other hand, plot the elements of the pairs or triples of loading vectors, for example the ith and the jth loading vector. This allows identifying groups of variables that have a similar covariance structure. An example and a detailed discussion of this is available in Kaspar and Ray (1992). The application studies in Chapters 4 and 5 also present a brief analysis of the variable interrelationships for recorded data sets from a chemical reaction and a distillation process, respectively.

2.2 Partial least squares

As in the previous section, the presentation of PLS relies on a geometric analysis. Chapter 10 provides a more detailed analysis of the PLS algorithm, including its properties and further information is available from the research literature, for example (de Jong 1993; de Jong *et al.* 2001; Geladi and Kowalski 1986; Höskuldsson 1988; Lohmoeller 1989; ter Braak and de Jong 1998). In contrast to PCA, PLS relies on the analysis of two variable sets that represent the process input and output variable sets shown in Figure 2.5. Alteratively, these variable sets are also referred to as the *predictor* and *response*, the *cause* and *effect*, the *independent* and *dependent* or the *regressor* and *regressand* variables sets in the literature. For simplicity, this book adopts the notation *input* and *output* variable sets to denote $\mathbf{x} \in \mathbb{R}^{n_x}$ as the input and $\mathbf{y} \in \mathbb{R}^{n_y}$ as the output variable sets. These sets span separate data spaces denoted as the input and output spaces, which Figure 2.5 graphically illustrates.

Between these variables sets, there is the following linear parametric relationship

$$\mathbf{y}_0 = \mathcal{B}^T \mathbf{x}_0 + \mathfrak{f} = \mathbf{y}_s + \mathfrak{f} \tag{2.23}$$

where $\mathbf{x}_0$ and $\mathbf{y}_0$ are zero mean random vectors that follow a Gaussian distribution. Similar to (2.2), the recorded variables are defined by $\mathbf{x} = \mathbf{x}_0 + \bar{\mathbf{x}}$ and $\mathbf{y} = \mathbf{y}_0 + \bar{\mathbf{y}}$

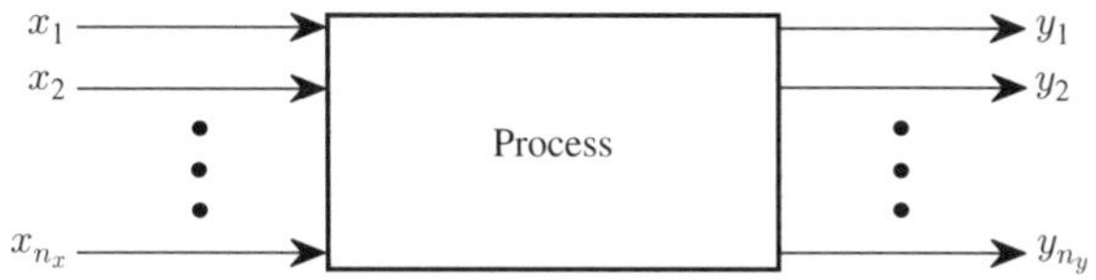

Figure 2.5 Division of the process variables into input and output variables.

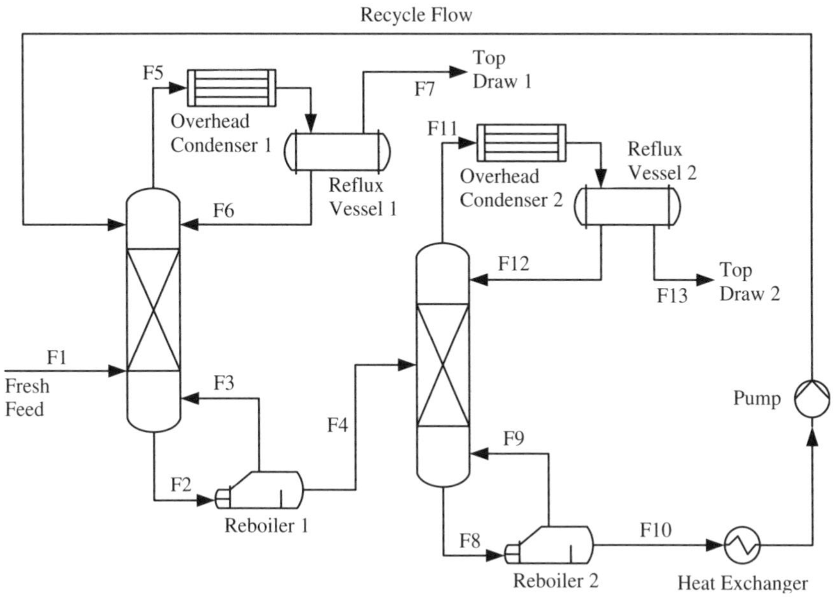

Figure 2.6 Schematic diagram of a distillation unit.

with $\bar{\mathbf{x}}$ and $\bar{\mathbf{y}}$ being mean vectors. The matrix $\mathcal{B} \in \mathbb{R}^{n_x \times n_y}$ is a parameter matrix describing the linear relationships between $\mathbf{x}_0$ and the uncorrupted output variables $\mathbf{y}_s = \mathcal{B}^T \mathbf{x}_0$, and $\mathfrak{f}$ is an error vector, representing measurement uncertainty for the output variables or the impact of unmeasured disturbances for example.

The error vector $\mathfrak{f}$ is also assumed to follow a zero mean Gaussian distribution and is statistically independent of the input vector $\mathbf{x}_0$, implying that $E\left\{\mathbf{x}_0 \mathfrak{f}^T\right\} = \mathbf{0}$. Moreover, the covariance matrices for $\mathbf{x}_0$, $\mathbf{y}_s$ and $\mathfrak{f}$ are $\mathbf{S}_{x_0x_0}$, $\mathcal{B}^T \mathbf{S}_{x_0x_0} \mathcal{B}$ and $\mathbf{S}_{\mathfrak{f}\mathfrak{f}}$, respectively. To denote the parametric matrix $\mathcal{B}$ by its transpose relates to the identification of this matrix from recorded samples of $\mathbf{x}$ and $\mathbf{y}$ which are stored as row vectors in data matrices. This is discussed further in Chapter 10.

2.2.1 Assumptions for underlying data structure

With respect to the preceding discussion, the recorded variables are highly correlated. Separating them into the mean centered input and output variable sets implies that the individual sets are also highly correlated. According to (2.23), there is also considerable correlation between the input and output variables:

- as the uncorrupted output variables are a linear combination of the input variables; and
- the assumption that $E\left\{\left(\mathfrak{b}_i^T \mathbf{x}_0\right)^2\right\} \gg E\left\{\mathfrak{f}_i^2\right\}$ for all $1 \leq i \leq n_x$, where $\mathfrak{b}_i$ is the ith column vector of $\mathcal{B}$.

To illustrate the correlation issue in more detail, consider the distillation process in Figure 2.6. The output variables of this process are mainly tray temperature, pressure and differential pressure measurements inside the columns, and concentrations (if measured). These variables follow common cause variation, for example introduced by variations of the fresh feed and its composition as well as the temperatures and flow rate of the input streams into the reboilers and overhead condensers. Other sources that introduce variation are, among others, unmeasured disturbances, changes in ambient temperature and pressure, and operator interventions. Through controller feedback, the variations of the output variables will propagate back to the input variables, which could include flow rates, temperatures of the heating/cooling streams entering and leaving the reboilers and overhead condensers. The degree of correlation within both variable sets suggests the following data structure for the input and output variables

$$\mathbf{x}_0 = \sum_{i=1}^{n} \mathfrak{p}_i s_i + \mathbf{e}_{n+1} = \mathfrak{P}\mathbf{s} + \mathbf{e}_{n+1} \qquad \mathbf{y}_0 = \sum_{i=1}^{n} \mathfrak{q}_i s_i + \mathbf{f}_{n+1} = \mathfrak{Q}\mathbf{s} + \mathbf{f}_{n+1}. \tag{2.24}$$

Here, $\mathfrak{P} = \begin{bmatrix} \mathfrak{p}_1 & \cdots & \mathfrak{p}_n \end{bmatrix} \in \mathbb{R}^{n_x \times n}$ and $\mathfrak{Q} = \begin{bmatrix} \mathfrak{q}_1 & \cdots & \mathfrak{q}_n \end{bmatrix} \in \mathbb{R}^{n_y \times n}$ are parameter matrices, $\mathbf{e}_{n+1} \in \mathbb{R}^{n_x}$ and $\mathbf{f}_{n+1} \in \mathbb{R}^{n_y}$ are the residual vectors of the input and output sets, respectively, which describe a negligible contribution for predicting the output set. The vector $\mathbf{s}$ stores the source signals describing common cause variation of the input and output sets. Recall that $\mathfrak{f}$ is the error vector associated with the output variables and $\mathbf{f}_{n+1} \to \mathfrak{f}$ under the assumptions (i) that the covariance matrix of the input variables has full rank n_x, (ii) that $n = n_x$ and (iii) that the number of samples for identifying the PLS model in (2.24) $K \to \infty$.

The source and error signals are assumed to be statistically independent of each other and follow a zero mean Gaussian distribution

$$\begin{aligned} E\left\{\begin{pmatrix} \mathbf{s} \\ \mathfrak{f} \end{pmatrix}\right\} &= \mathbf{0} \\ E\left\{\begin{pmatrix} \mathbf{s}(k) \\ \mathfrak{f}(k) \end{pmatrix}\begin{pmatrix} \mathbf{s}^T(l) & \mathfrak{f}^T(l) \end{pmatrix}\right\} &= \delta_{kl}\begin{bmatrix} \mathbf{S}_{ss} & \mathbf{0} \\ \mathbf{0} & \mathbf{S}_{\mathfrak{f}\mathfrak{f}} \end{bmatrix} \\ \mathbf{s}(k) \sim \mathcal{N}\left\{\mathbf{0}, \mathbf{S}_{ss}\right\} \qquad & \mathfrak{f}(k) \sim \mathcal{N}\left\{\mathbf{0}, \mathbf{S}_{\mathfrak{f}\mathfrak{f}}\right\}. \end{aligned} \tag{2.25}$$

Moreover, the residual vectors $\mathbf{e}$ and $\mathbf{f}$ are also assumed to follow zero mean Gaussian distributions with covariance matrices $\mathbf{S}_{ee}$ and $\mathbf{S}_{ff}$, respectively.

The residual vectors, however, are generally not statistically independent, i.e. $E\left\{\mathbf{e}\mathbf{f}^T\right\} \neq \mathbf{0}$. Subsection 2.3.2 discusses the independence of the error vectors in more detail. Asymptotically, if $n = n_x$ and $K \to \infty$, however, $\mathbf{S}_{ff} \to \mathbf{S}_{\mathfrak{f}\mathfrak{f}}$ and $\mathbf{S}_{ee} \to \mathbf{0}$.

By comparing the causal data model for PLS with that of the non-causal PCA one in (2.2), it should be noted that there are similarities. The parameter matrix Ξ for the PCA data model becomes $\mathfrak{P}$ and $\mathfrak{Q}$ to describe the influence of the source variables upon the input and output variables, respectively. Moreover, the error variable $\mathbf{g}$ for the PCA data structure becomes $\mathbf{e}$ and $\mathbf{f}$ for the input and output variable sets, respectively. For PCA, however, if the number of source signals is assumed to be $n = n_z$, the variable set $\mathbf{z}_0$ can be described by $\mathbf{Pt}$. This follows from the fact that the covariance matrix of $\mathbf{z}_0$ is equal to its eigendecomposition for $n = n_z$, as shown in (2.15) for $\widehat{\mathbf{S}}_{z_0 z_0}$. With regards to PLS, however, this property is only maintained for the input variable set $\mathbf{x}_0$, as $\mathbf{e} \to \mathbf{0}$ for $n \to n_x$. In contrast, as $n \to n_x$ the error vector $\mathbf{f} \to \mathfrak{f} \neq \mathbf{0}$.

Using the terminology for training artificial neural networks in an MSPC context, assuming that the variable sets $\mathbf{z}_0$ and $\mathbf{x}_0$ are identical PCA is an unsupervised learning algorithm for determining latent variable sets. In contrast, PLS is a supervised learning algorithm, which incorporates the parametric relationship relationship $\mathbf{y}_s = \mathcal{B}^T \mathbf{x}_0$ into the extraction of sets of latent variables. Although this comparison appears hypothetical, this is a practically relevant case. An example is if the output variable set $\mathbf{y}_0$ consists of concentration measurements that represent quality variables which are not recorded with the same frequency as the variable set $\mathbf{x}_0$. In this case, only the $\mathbf{z}_0 = \mathbf{x}_0$ is available for on-line process monitoring.

2.2.2 Deflation procedure for estimating data models

PLS computes sequences of linear combinations of the input and output variables to determine sets of latent variables that describe common cause variation. The first set of latent variables includes

$$t_1 = \mathbf{x}_0^T \mathbf{w}_1 \qquad u_1 = \mathbf{y}_0^T \mathbf{q}_1, \tag{2.26}$$

where $\mathbf{w}_1$ and $\mathbf{q}_1$ are *weight vectors* of unit length that determine a set of linear combinations of $\mathbf{x}_0$ and $\mathbf{y}_0$, respectively, and yield the *score variables* t_1 and u_1. Geometrically, the linear combinations result in the orthogonal projections of the data vectors $\mathbf{x}_0$ and $\mathbf{y}_0$ onto the directions defined by $\mathbf{w}_1$ and $\mathbf{q}_1$, respectively. This follows from the fact that $\mathbf{x}_0^T \mathbf{w}_1$ and $\mathbf{y}_0^T \mathbf{q}_1$ are scalar products

$$\left\|\mathbf{x}_0\right\| \left\|\mathbf{w}_1\right\| \cos\left(\varphi_{(\mathbf{x}_0, \mathbf{w}_1)}\right) = \mathbf{x}_0^T \mathbf{w}_1 \Rightarrow t_1 = \left\|\mathbf{x}_0\right\| \cos\left(\varphi_{(\mathbf{x}_0, \mathbf{w}_1)}\right) \tag{2.27}$$

and

$$\left\|\mathbf{y}_0\right\| \left\|\mathbf{q}_1\right\| \cos\left(\varphi_{(\mathbf{y}_0, \mathbf{q}_1)}\right) = \mathbf{y}_0^T \mathbf{q}_1 \Rightarrow u_1 = \left\|\mathbf{y}_0\right\| \cos\left(\varphi_{(\mathbf{y}_0, \mathbf{q}_1)}\right) \tag{2.28}$$

where $\varphi_{(\mathbf{x}_0,\mathbf{w}_1)}$ and $\varphi_{(\mathbf{y}_0,\mathbf{q}_1)}$ are the angles between the vector pairs $\mathbf{x}_0$ and $\mathbf{w}_1$, and $\mathbf{y}_0$ and $\mathbf{q}_1$, respectively. Consequently, the score variables t_1 and u_1 describe the minimum distance between the origin of the coordinate system and the orthogonal projection of $\mathbf{x}_0$ and $\mathbf{y}_0$ onto $\mathbf{w}_1$ and $\mathbf{q}_1$, respectively. The weight vectors are determined to maximize the covariance between t_1 and u_1.

Chapter 10 gives a detailed account of the PLS objective functions for computing the weight vectors. After determining the score variables, the t-score variable is utilized to predict the input and output variables. For this, PLS computes a set of *loading vectors*, leading to the following prediction of both variable sets

$$\widehat{\mathbf{x}}_0 = t_1\mathbf{p}_1 \qquad \widehat{\mathbf{y}}_0 = t_1\acute{\mathbf{q}}_1. \tag{2.29}$$

Here, $\mathbf{p}_1$ and $\acute{\mathbf{q}}_1$ are the loading vectors for the input and output variables, respectively. As before, the notation $\widehat{\cdot}$ represents the prediction or estimation of a variable. Chapter 10, again, shows the objective function for determining the loading vectors. The aim of this introductory section on PLS is to outline its working and how to apply it.

It should be noted, however, that the weight and the loading vector of the output variables, $\mathbf{q}_1$ and $\acute{\mathbf{q}}_1$, are equal up to a scalar factor. The two weight vectors, $\mathbf{w}_1$ and $\mathbf{q}_1$, the two loading vectors, $\mathbf{p}_1$ and $\acute{\mathbf{q}}_1$, and the two score variables, t_1 and u_1 are referred to as the first set of latent variables (LVs). For computing further sets, the PLS algorithm carries out a *deflation procedure*, which subtracts the contribution of previously computed LVs from the input and output variables. After computing the first set of LVs, the deflation procedure yields

$$\mathbf{e}_2 = \mathbf{x}_0 - t_1\mathbf{p}_1 \qquad \mathbf{f}_2 = \mathbf{y}_0 - t_1\acute{\mathbf{q}}_1 \tag{2.30}$$

where $\mathbf{e}_2$ and $\mathbf{f}_2$ are residual vectors that represent variation of the input and output variable sets which can be exploited by the second set of LVs, comprising of the weight vectors $\mathbf{w}_2$ and $\mathbf{q}_2$, the loading vectors $\mathbf{p}_2$ and $\acute{\mathbf{q}}_2$ and the score variables t_2 and u_2. Applying the deflation procedure again yields

$$\mathbf{e}_3 = \mathbf{e}_2 - t_2\mathbf{p}_2 \qquad \mathbf{f}_3 = \mathbf{f}_2 - t_2\acute{\mathbf{q}}_2. \tag{2.31}$$

Defining the original data vectors $\mathbf{x}_0$ and $\mathbf{y}_0$ as $\mathbf{e}_1$ and $\mathbf{f}_1$, the general formulation of the PLS deflation procedure becomes

$$\mathbf{e}_{i+1} = \mathbf{e}_i - t_i\mathbf{p}_i \qquad \mathbf{f}_{i+1} = \mathbf{f}_i - t_i\acute{\mathbf{q}}_i \tag{2.32}$$

and the ith pair of LVs include the weight vectors $\mathbf{w}_i$ and $\mathbf{q}_i$, the loading vectors $\mathbf{p}_i$ and $\acute{\mathbf{q}}_i$ and the score variables t_i and u_i.

Compared to the data structure in (2.24), the objective of the PLS modeling procedure is to:

- estimate the column space of parameter matrices $\mathfrak{P}$ and $\mathfrak{Q}$; and
- extract the variation of the source variable set $\mathbf{s}$.

From the n sets of LVs, the p- and q́-loading vectors, stored in separate matrices

$$\mathbf{P} = \begin{bmatrix} \mathbf{p}_1 & \mathbf{p}_2 & \cdots & \mathbf{p}_n \end{bmatrix} \qquad \acute{\mathbf{Q}} = \begin{bmatrix} \acute{\mathbf{q}}_1 & \acute{\mathbf{q}}_2 & \cdots & \acute{\mathbf{q}}_n \end{bmatrix}, \tag{2.33}$$

are an estimate for the column space of $\mathfrak{P}$ and $\mathfrak{Q}$. The t-score variables

$$\mathbf{t} = \begin{pmatrix} t_1 & t_3 & \cdots & t_n \end{pmatrix}^T \tag{2.34}$$

represent the variation of the source variables.

2.2.3 A simulation example

To demonstrate the working of PLS, an application study of data from a simulated process is now presented. According to (2.23), the process includes three input and two output variables and the following parameter matrix

$$\mathcal{B} = \begin{bmatrix} 0.3412 & 0.5341 & 0.7271 \\ 0.3093 & 0.8385 & 0.5681 \end{bmatrix}. \tag{2.35}$$

The input variable set follows a zero mean Gaussian distribution with a covariance

$$\mathbf{S}_{x_0 x_0} = \begin{bmatrix} 1 & 0.8 & 0.9 \\ 0.8 & 1 & 0.5 \\ 0.9 & 0.5 & 1 \end{bmatrix}. \tag{2.36}$$

The error variable set, $\mathfrak{f}$ follows a zero mean Gaussian distribution describing i.i.d. sequences $\mathfrak{f} \sim \mathcal{N}\{\mathbf{0}, 0.05\mathbf{I}\}$. Figure 2.7 shows a total of 100 samples, that were simulated from this process, and produced the following covariance matrices

$$\widehat{\mathbf{S}}_{y_0 y_0} = \begin{bmatrix} 2.1318 & 2.2293 \\ 2.2293 & 2.5130 \end{bmatrix} \quad \widehat{\mathbf{S}}_{x_0 y_0} = \begin{bmatrix} 1.4292 & 1.5304 \\ 1.1756 & 1.4205 \\ 1.3118 & 1.2877 \end{bmatrix}. \tag{2.37}$$

Equations 2.38 and 2.39 show how to compute the cross-covariance matrix

$$\widehat{\mathbf{S}}_{x_0 y_0} = \tfrac{1}{K} \sum_{k=1}^{K} \mathbf{x}(k)\, \mathbf{y}^T(k) \tag{2.38}$$

or

$$\widehat{\mathbf{S}}_{x_0 y_0} = \tfrac{1}{K-1} \sum_{k=1}^{K} \left(\mathbf{x}(k) - \widehat{\bar{\mathbf{x}}}\right) \left(\mathbf{y}(k) - \widehat{\bar{\mathbf{y}}}\right)^T. \tag{2.39}$$

If $\bar{\mathbf{x}}$ and $\bar{\mathbf{y}}$ are equal to zero, the estimation of the covariance and cross-covariance matrices requires the use of (2.13) and (2.38). If this is not the case for at least one of the two variable sets, use (2.14) and (2.38) to estimate them.

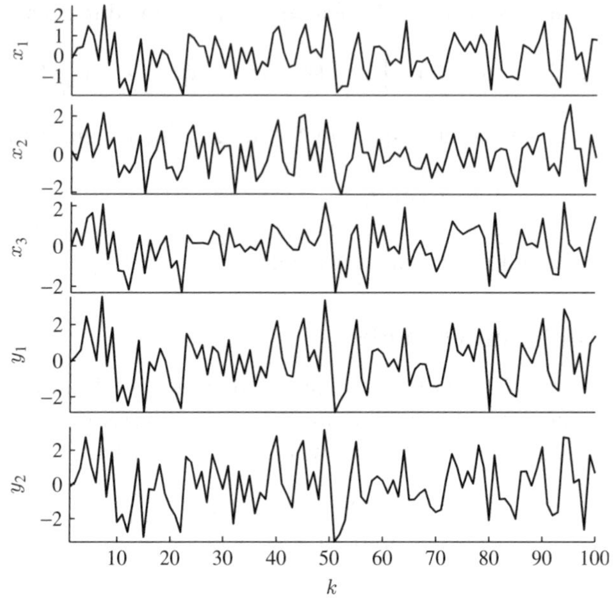

Figure 2.7 Simulated samples of input and output variables.

Knowing that $\mathfrak{f}$ is statistically independent of $\mathbf{x}_0$, (2.23) shows that these covariance matrices $\mathbf{S}_{x_0 y_0}$ and $\mathbf{S}_{y_0 y_0}$ are equal to

$$E\left\{\mathbf{x}_0\mathbf{y}_0^T\right\} = E\left\{\mathbf{x}_0\mathbf{x}_0^T\boldsymbol{\mathcal{B}} + \mathbf{x}_0\mathfrak{f}^T\right\} = E\left\{\mathbf{x}_0\mathbf{x}_0^T\right\}\boldsymbol{\mathcal{B}} = \mathbf{S}_{x_0x_0}\boldsymbol{\mathcal{B}} \tag{2.40}$$

and

$$\begin{aligned} E\left\{\mathbf{y}_0\mathbf{y}_0^T\right\} &= E\left\{\boldsymbol{\mathcal{B}}^T\mathbf{x}_0\mathbf{x}_0^T\boldsymbol{\mathcal{B}} + \boldsymbol{\mathcal{B}}^T\mathbf{x}_0\mathfrak{f}^T + \mathfrak{f}\mathbf{x}_0^T\boldsymbol{\mathcal{B}} + \mathfrak{f}\mathfrak{f}^T\right\} \\ E\left\{\mathbf{y}_0\mathbf{y}_0^T\right\} &= \boldsymbol{\mathcal{B}}^T E\left\{\mathbf{x}_0\mathbf{x}_0^T\right\}\boldsymbol{\mathcal{B}} + E\left\{\mathfrak{f}\mathfrak{f}^T\right\} = \boldsymbol{\mathcal{B}}^T\mathbf{S}_{x_0x_0}\boldsymbol{\mathcal{B}} + \mathbf{S}_{\mathfrak{f}\mathfrak{f}}, \end{aligned} \tag{2.41}$$

respectively. Inserting $\mathbf{S}_{x_0x_0}$, $\mathbf{S}_{\mathfrak{f}\mathfrak{f}}$ and $\boldsymbol{\mathcal{B}}$, defined in (2.35) and (2.36), into (2.40) and (2.41) yields

$$\mathbf{S}_{y_0y_0} = \begin{bmatrix} 2.0583 & 2.1625 \\ 2.1625 & 2.3308 \end{bmatrix} \qquad \mathbf{S}_{x_0y_0} = \begin{bmatrix} 1.4237 & 1.4926 \\ 1.1715 & 1.3709 \\ 1.3022 & 1.2666 \end{bmatrix}. \tag{2.42}$$

Comparing the true matrices with their estimates shows a close agreement.

Using the estimated matrices $\widehat{\mathbf{S}}_{x_0x_0}$ and $\widehat{\mathbf{S}}_{x_0y_0}$, a PLS model is determined next. The preceding discussion has outlined that a PLS model relies on the calculation of weight vectors of length 1. The projection of the input and output variables onto these weight vectors then produces the score variables. To complete the computation of one set of latent variables, the final step is to determine the loading vectors and the application of the deflation procedure to the input and output variables.

Figure 2.8 illustrates the working of the iterative PLS approach to the input and output data shown in Figure 2.7. The left and right column of plots present the results for the individual sets of latent variables, respectively. The top, middle and bottom rows of plots summarize the results of the first, the second and the third sets of latent variables, respectively. The first set of latent variables are

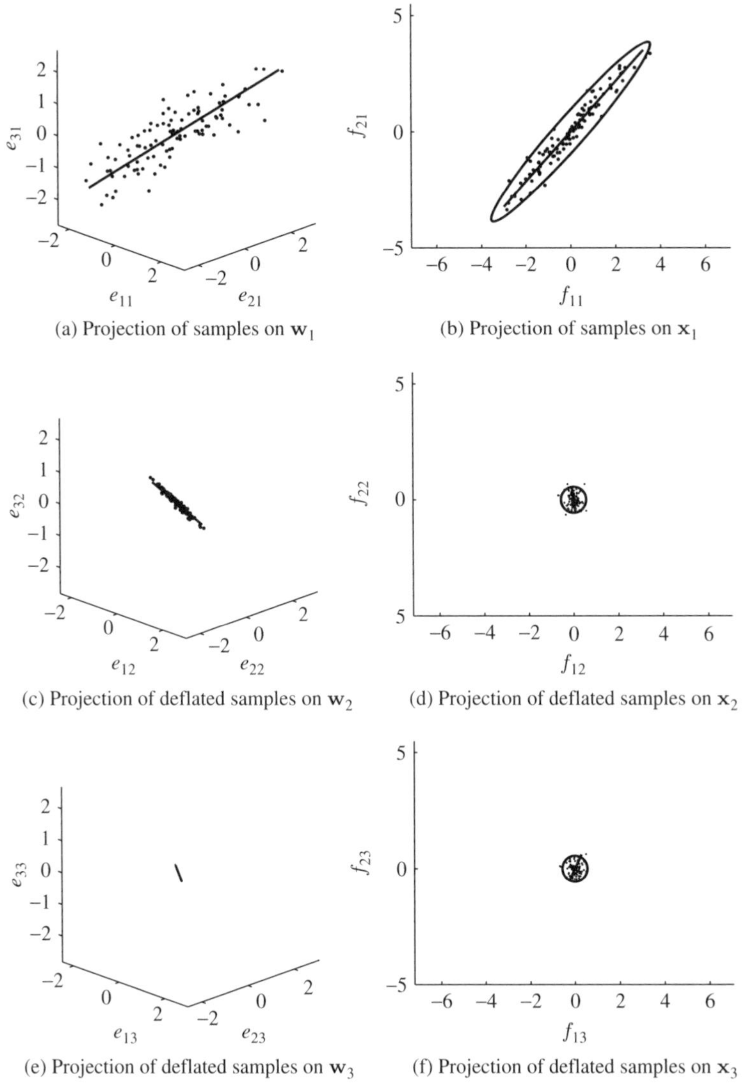

Figure 2.8 Graphical illustration of the sample projections in the input and output spaces for determining the first, second and third set of latent variables.

computed from the original input and output variable sets and the first two plots at the top show the samples and the computed direction of the weight vectors.

The control ellipses in the right plots are for the two output variables. The depicted samples in the middle and lower rows of plots represent the samples after the first and second deflation procedure has been carried out. It is interesting to note that after applying the first deflation procedure to the output variables, there is little variation left in this variable set, noticeable by the small control ellipse constructed on the basis of the covariance matrix of $\mathbf{S}_{\mathfrak{ff}} = \text{diag}\left\{ \; 0.05 \quad 0.05 \; \right\}$. The deflation procedure also reduces the remaining variation of the input variables when comparing the top left with the middle left plot.

The third and final set of LVs is determined from the input and output variable sets after deflating the first and second sets of LVs. Comparing the plots in the bottom row with those in the middle of Figure 2.8 suggests that there is hardly any reduction in the remaining variance of the output variables but a further reduction in variation of the input variables. The analysis in Chapter 10 shows that after deflating the third set of latent variables from the input and output variables, the residuals of the input variable set is zero and the residuals of the output variables are identical to those of applying a regression model obtained from the ordinary least squares (OLS). Asymptotically, the residuals $\mathbf{f}$ converge to $\mathfrak{f}$ as $K \to \infty$.

Equation 2.43 lists the estimates for the w- and q-weight, the p- and q́-loading matrices and the maximum covariance values for the t- and u-score variables

$$\begin{aligned}
\widehat{\mathbf{W}} &= \begin{bmatrix} 0.6272 & -0.1005 & -0.7723 \\ 0.5515 & 0.7576 & 0.3492 \\ 0.5500 & -0.6450 & 0.5306 \end{bmatrix} \\
\widehat{\mathbf{Q}} &= \begin{bmatrix} 0.6788 & -0.3722 & 0.8072 \\ 0.7343 & 0.9282 & 0.5903 \end{bmatrix} \\
\widehat{\mathbf{P}} &= \begin{bmatrix} 0.6310 & -0.0889 & -0.7723 \\ 0.5361 & 0.7523 & 0.3492 \\ 0.5611 & -0.6530 & 0.5306 \end{bmatrix} \qquad (2.43) \\
\widehat{\acute{\mathbf{Q}}} &= \begin{bmatrix} 0.9147 & -0.1101 & 0.4601 \\ 0.9894 & 0.2747 & 0.3365 \end{bmatrix} \\
\widehat{\mathbf{J}}^T &= \begin{pmatrix} 3.3385 & 0.1496 & 0.0091 \end{pmatrix}.
\end{aligned}$$

Using the true covariance matrices, it is possible to compare the accuracy of the estimated ones. It follows from the analysis in Chapter 10 that each LV in one set can be computed either from the w- or the q-weight vector. It is therefore sufficient to determine the departure of the estimated w-weight vectors. The estimation error of the other LVs can be computed from the estimation error of the covariance matrices and the w-weight vector. For example, the estimation error for the q-weight vector is

$$\mathbf{q} = \frac{\mathbf{S}_{x_0 y_0}^T \mathbf{w}}{\|\mathbf{S}_{x_0 y_0}^T \mathbf{w}\|} \qquad \widehat{\mathbf{q}} = \frac{\widehat{\mathbf{S}}_{x_0 y_0}^T \widehat{\mathbf{w}}}{\|\widehat{\mathbf{S}}_{x_0 y_0}^T \widehat{\mathbf{w}}\|}$$

$$\Delta\mathbf{q} = \frac{\mathbf{S}_{x_0 y_0}^T \mathbf{w}}{\|\mathbf{S}_{x_0 y_0}^T \mathbf{w}\|} - \frac{\widehat{\mathbf{S}}_{x_0 y_0}^T \widehat{\mathbf{w}}}{\|\widehat{\mathbf{S}}_{x_0 y_0}^T \widehat{\mathbf{w}}\|} = \frac{\mathbf{S}_{x_0 y_0}^T \mathbf{w}}{\|\mathbf{S}_{x_0 y_0}^T \mathbf{w}\|_2} - \frac{\left[\mathbf{S}_{x_0 y_0}^T - \Delta\mathbf{S}_{x_0 y_0}^T\right](\mathbf{w} - \Delta\mathbf{w})}{\|\widehat{\mathbf{S}}_{x_0 y_0}^T \widehat{\mathbf{w}}\|}$$

$$\Delta\mathbf{q} \approx \frac{\Delta\mathbf{S}_{x_0 y_0}^T \mathbf{w} + \mathbf{S}_{x_0 y_0}^T \Delta\mathbf{w}}{\|\mathbf{S}_{x_0 y_0}^T \mathbf{w}\|}. \tag{2.44}$$

It is assumed here that $\mathbf{S}_{x_0 y_0} = \widehat{\mathbf{S}}_{x_0 y_0} + \Delta\mathbf{S}_{x_0 y_0}$, $\|\mathbf{S}_{x_0 y_0}^T \mathbf{w}\|_2 \approx \|\widehat{\mathbf{S}}_{x_0 y_0}^T \widehat{\mathbf{w}}\|_2$ and $\|\Delta\mathbf{S}_{x_0 y_0} \Delta\mathbf{w}\|_2 \ll \|\mathbf{S}_{x_0 y_0} \mathbf{w}\|_2$. The true w-weight matrix is equal to

$$\mathbf{W} = \begin{bmatrix} 0.6279 & -0.1061 & -0.7710 \\ 0.5482 & 0.7635 & 0.3414 \\ 0.5525 & -0.6371 & 0.5375 \end{bmatrix}. \tag{2.45}$$

Since the w-weight vectors are of unit length, the angles between the estimated and true ones can directly be obtained using the scalar product $\varphi_{(\mathbf{w}_k, \widehat{\mathbf{w}}_k)} = \mathbf{w}_k^T \widehat{\mathbf{w}}_k \cdot \frac{180}{\pi}$ and are 0.2374°, 0.6501° and 0.6057° for the first, second and third vectors, respectively. The covariances of the first, the second and the third pair of score variables, obtained from the true covariance matrices, are 3.2829, 0.1296 and 0.0075 respectively, and close to the estimated ones stored in the vector $\hat{\mathbf{J}}$ in (2.43). The estimation error for the w-weight vectors are around 0.25° for the first and around 0.65° for the second and third ones and is therefore small. The estimation accuracy, however, increases with the number of recorded samples. After inspecting the estimation accuracy, a very important practical aspect, namely how to interpret the results obtained, is given next.

So far, the analysis of the resultant PLS regression model has been made from Figure 2.8 by eye, for example, noticing that the number of samples outside the control ellipse describing the error vector $\mathfrak{f}$. A sound statistically-based conclusions, however, requires a more detailed investigation. For example, such an analysis helps in determining how many sets of latent variables need to be retained in the PLS model and how many sets can be discarded. One possibility to assess this is the analysis of the residual variance, given in Table 2.2.

Table 2.2 Variance reduction of PLS model to $\mathbf{x}_0$ and $\mathbf{y}_0$.

LV Set	Input Variables $\mathbf{x}_0$ (s_{e_i})	Output Variables $\mathbf{y}_0$ (s_{f_i})
1	17.3808%	3.1522%
2	0.5325%	2.1992%
3	0.0000%	2.0875%

The percentage values describe the cumulative variance remaining.

Equation (2.46) introduces a measure for the residual variance of both variable sets, s_{e_i} and s_{f_i}, after deflating the previously computed $i-1$ LVs

$$s_{e_i} = \frac{\text{trace}\left\{E\left\{\mathbf{e}_i\mathbf{e}_i^T\right\}\right\}}{\text{trace}\left\{E\left\{\mathbf{x}_0\mathbf{x}_0^T\right\}\right\}} \cdot 100\% \quad s_{f_i} = \frac{\text{trace}\left\{E\left\{\mathbf{f}_i\mathbf{f}_i^T\right\}\right\}}{\text{trace}\left\{E\left\{\mathbf{y}_0\mathbf{y}_0^T\right\}\right\}} \cdot 100\% \tag{2.46}$$

where trace$\{\cdot\}$ is the sum of the diagonal elements of a squared matrix,

$$E\left\{\mathbf{e}_i\mathbf{e}_i^T\right\} = E\left\{\mathbf{x}_0\mathbf{x}_0^T\right\} - \sum_{j=1}^{i-1}\mathbf{p}_j E\left\{t_j^2\right\}\mathbf{p}_j^T = \mathbf{S}_{x_0x_0} - \mathbf{P}E\left\{\mathbf{t}\mathbf{t}^T\right\}\mathbf{P}^T \tag{2.47}$$

and

$$E\left\{\mathbf{f}_i\mathbf{f}_i^T\right\} = E\left\{\mathbf{y}_0\mathbf{y}_0^T\right\} - \sum_{j=1}^{i-1}\acute{\mathbf{q}}_j E\left\{t_j^2\right\}\acute{\mathbf{q}}_j^T = \mathbf{S}_{y_0y_0} - \acute{\mathbf{Q}}E\left\{\mathbf{t}\mathbf{t}^T\right\}\acute{\mathbf{Q}}^T. \tag{2.48}$$

The assumption that the process variables are normally distributed implies that the t-score variables $\mathbf{t} = \begin{pmatrix} t_1 & t_2 & \cdots & t_k \end{pmatrix}$ are statistically independent, which the analysis in Chapter 10 yields. Hence, $E\left\{\mathbf{t}\mathbf{t}^T\right\}$ reduces to a diagonal matrix.

Summarizing the results in Table 2.2, the first set of LVs contribute to a relative reduction in variance of 82.6192% for the input and 96.8478% for the output variable set. For the second set of LVs, a further relative reduction of 16.8483% can be noticed for the input variable set, whilst the reduction for the output variables only amounts to 0.9530%. Finally, the third set of LVs only contribute marginally to the input and output variables by 0.5225% and 0.1117%, which is negligible.

The analysis in Table 2.2 therefore confirms the visual inspection of Figure 2.8. Given that PLS aims to determine a covariance representation of $\mathbf{x}_0$ and $\mathbf{y}_0$ using a reduced set of linear combinations of these sets, a parsimonious selection is to retain the first set of LVs and discard the second and third sets as insignificant contributors.

The final analysis of the PLS model relates to the accuracy of the estimated parameter matrix, $\mathcal{B}$. Table 2.2 shows that $\mathbf{x}_0$ is completely exhausted after deflating 3 sets of LVs. Furthermore, the theoretical value for s_{f_3} can be obtained

$$s_{f_3} = \frac{0.05 + 0.05}{2.0583 + 2.3308} \cdot 100\% = 2.2784\%. \tag{2.49}$$

As stated in the preceding discussion, the estimated regression matrix, including all three sets of LVs, is equivalent to that obtained using the OLS approach. Equation (2.50) shows this matrix from the simulated 100 samples

$$\widehat{\mathcal{B}} = \widehat{\mathbf{W}}\left[\widehat{\mathbf{P}}^T\widehat{\mathbf{W}}\right]^{-1}\widehat{\acute{\mathbf{Q}}}^T = \begin{bmatrix} 0.2281 & 0.3365 \\ 0.5864 & 0.8785 \\ 0.8133 & 0.5457 \end{bmatrix}. \tag{2.50}$$

Comparing the estimated parameter matrix with the true one, shown in (2.35), it should be noted that particularly the first column of $\widehat{\mathcal{B}}$ departs from $\mathcal{B}$, whilst the second column provides a considerably closer estimate. Larger mismatches between the estimated and true parameter matrix can arise if:

- there is substantial correlation among the input variables (Wold *et al.* 1984); and
- the number of observations is 'small' compared to the number of variables (Ljung 1999; Söderström and Stoica 1994).

By inspecting the $\mathbf{S}_{x_0 x_0}$ in (2.36), non-diagonal elements of 0.9 and 0.8 show indeed a high degree of correlation between the input variables. Subsection 6.2.1 presents a further and more detailed discussion of the issue of parameter identification. The issue related to the accuracy of the PLS model is also a subject in the Tutorial Session of this chapter and further reading material covering the aspect of model accuracy is given in Höskuldsson (1988, 1996).

2.3 Maximum redundancy partial least squares

This section examines the legitimate question of why do we need both, the single-block PCA and the dual-block PLS methods for process monitoring. A more precise formulation of this question is: what can the separation of the recorded variable set to produce a dual-block approach offer that a single-block technique cannot? To address this issue, the first subsection extends the data models describing *common cause variation* in (2.2) and (2.24). Subsection 2.3.2 then shows that PCA and PLS cannot identify this generic data structure correctly. Finally, Subsection 2.3.3 introduces a different formulation of the PLS objective function that enables the identification of this generic data structure, and Subsection 2.3.4 presents a simulation example to demonstrate the working of this revised PLS algorithm.

2.3.1 Assumptions for underlying data structure

The preceding discussion in this chapter has outlined that PCA is a single-block technique that analyzes a set of variables. According to (2.2), this variable set is a linear combination of a smaller set of source signals that represent common cause variation. For each process variable, a statistically independent error variable is then superimposed to the contribution from the source signals.

On the other hand, PLS is a dual-block technique for which the recorded variables are divided into an input and an output set. Figure 2.6 shows that this division may not be straightforward. Whilst the fresh feed (stream F1) is easily identified as an input and top draw 1 (stream F7) and top draw 2 (stream F14) are outputs, how can the remaining streams (flow rates), temperature variables, pressure measurements, differential pressures or concentrations (if measured online) be divided?

An approach that the literature has proposed is selecting the variables describing the product quality as the outputs and utilizing the remaining ones as 'inputs'. This arrangement separates the variables between a set of *cause variables* that describe, or predict, the variation of the output or *effect variables*. A question that one can justifiably ask is why do we need PLS if PCA is able to analyze a single-block arrangement of these variables, which is conceptually simpler? In addition to that, the division into input and output variables may not be straightforward either.

The need for a dual-block technique becomes clear by revisiting Figure 2.6. The concentrations (the quality variables $\mathbf{y}_0$), are influenced by changes affecting the energy balance within the distillation towers. Such changes manifest themselves in the recorded temperatures and pressures for example. On the other hand, there are also variables that relate to the operation of reboilers 1 and 2, overhead condensers 1 and 2, both reflux vessels, the heat exchanger and the pump that do not affect the quality variables. The variation in these variables, however, may be important to monitor the operation of the individual units and consequently cannot be ignored.

A model to describe the above scenario is an extension of (2.24)

$$\mathbf{y}_0 = \sum_{i=1}^{n} \mathfrak{q}_i s_i + \mathfrak{f} = \mathfrak{Q}\mathbf{s} + \mathfrak{f} \qquad \mathbf{x}_0 = \sum_{i=1}^{n} \mathfrak{p}_i s_i + \sum_{j=1}^{m} \mathfrak{p}'_j s'_j = \mathfrak{P}\mathbf{s} + \mathfrak{P}'\mathbf{s}'$$

$$\mathbf{y} = \mathbf{y}_0 + \bar{\mathbf{y}} \qquad \mathbf{x} = \mathbf{x}_0 + \bar{\mathbf{x}} \tag{2.51}$$

where $\mathbf{s}$ represents common cause variation in both variable sets and $\mathbf{s}'$ describes variation among the input or cause variables that is uncorrelated to the output variables and hence, uninformative for predicting them. The next subsection examines whether PCA and PLS can identify the data structure in (2.51).

2.3.2 Source signal estimation

The model estimation w.r.t. (2.51) is separately discussed for PCA/PLS.

2.3.2.1 Model identification using PCA

The advantage of a dual block method over a single block approach, when applied to the above data structure, is best demonstrated by reformulating (2.51)

$$\mathbf{z}_0 = \begin{pmatrix} \mathbf{y}_0 \\ \mathbf{x}_0 \end{pmatrix} = \begin{bmatrix} \mathfrak{Q} \\ \mathfrak{P} \end{bmatrix} s + \begin{bmatrix} \mathbf{0} \\ \mathfrak{P}' \end{bmatrix} s' + \begin{pmatrix} \mathfrak{f} \\ \mathbf{0} \end{pmatrix}. \tag{2.52}$$

Now, applying PCA to the data structure in (2.52) yields the following estimate for the source signals and residuals

$$\mathbf{P}^T \mathbf{z}_0 = \mathbf{P}^T \begin{bmatrix} \mathfrak{Q} \\ \mathfrak{P} \end{bmatrix} s + \mathbf{P}^T \left(\begin{bmatrix} \mathbf{0} \\ \mathfrak{P}' \end{bmatrix} s' + \begin{pmatrix} \mathfrak{f} \\ \mathbf{0} \end{pmatrix} \right) = \mathbf{t} \tag{2.53}$$

and

$$\mathbf{P}_d^T \mathbf{z}_0 = \mathbf{P}_d^T \left(\begin{bmatrix} \mathbf{0} \\ \mathfrak{P}' \end{bmatrix} \mathbf{s}' + \begin{pmatrix} \mathfrak{f} \\ \mathbf{0} \end{pmatrix} \right) = \mathbf{t}_d, \tag{2.54}$$

respectively. Here, $\mathbf{P}$ and $\mathbf{P}_d$ store the first n and the remaining $n_z - n$ eigenvectors of the data covariance matrix $\mathbf{S}_{z_0 z_0}$, respectively, where

$$\mathbf{S}_{z_0 z_0} = \underbrace{\begin{bmatrix} \mathfrak{Q}\mathbf{S}_{ss}\mathfrak{Q}^T & \mathfrak{Q}\mathbf{S}_{ss}\mathfrak{P}^T \\ \mathfrak{P}\mathbf{S}_{ss}\mathfrak{Q}^T & \mathfrak{P}\mathbf{S}_{ss}\mathfrak{P}^T \end{bmatrix}}_{\text{common cause variation}} + \underbrace{\begin{bmatrix} \mathbf{S}_{\mathfrak{f}\mathfrak{f}} & \mathbf{0} \\ \mathbf{0} & \mathfrak{P}'\mathbf{S}_{s's'}\mathfrak{P}'^T \end{bmatrix}}_{\text{remaining variation}}. \tag{2.55}$$

Note that above covariance matrix is divided into a part that represents common cause variation and a second part that describes the common cause variation that only affects input variables and the error term for the output variables. Assuming that the model subspace, spanned by the eigenvectors stored in $\mathbf{P}$ is consistently estimated,[2] the elements in $\mathbf{t}_d$ are linear combinations of $\mathfrak{P}'\mathbf{s}' + \mathfrak{f}$. Consequently, it may not be possible to extract and independently monitor $\mathfrak{P}'\mathbf{s}'$ using PCA.

Moreover, the covariance matrix $\mathfrak{P}'\mathbf{S}_{s's'}\mathfrak{P}'^T$ is not known *a priori* and may have significantly larger entries compared to the error covariance matrix $\mathbf{S}_{\mathfrak{f}\mathfrak{f}}$. It is also possible that $\mathfrak{P}'\mathbf{S}_{s's'}\mathfrak{P}'^T$ is the dominant contribution of the joint variable set $\mathbf{z}_0$. Both aspects render the estimation of the column space $\Xi^T = \begin{bmatrix} \mathfrak{Q}^T & \mathfrak{P}^T \end{bmatrix}$ using PCA a difficult task, given that the error covariance matrix is not of the form $\mathbf{S}_{gg} = \sigma_g^2 \mathbf{I}$. More precisely, Subsection 6.1.1 discusses how to estimate the error covariance matrix and the model subspace simultaneously using maximum likelihood PCA.

Based on this simultaneous estimate, the source signals contribution $\mathfrak{P}'\mathbf{s}'$ must be considered as additional error variables that:

- may have a considerably larger variance and covariance values compared to those of $\mathbf{S}_{\mathfrak{f}\mathfrak{f}}$; and
- the rank of the covariance matrix $\mathfrak{P}'\mathbf{S}_{s's'}\mathfrak{P}'^T$ is $n_x - n$ and not n_x.

The assumption for estimating the error covariance matrix, however, is that it is a full rank matrix. Hence, PCA is (i) unable to separate the source signals of the input variables into those commonly shared by the input and output variables, and the remaining ones that are only encapsulated in the input variables and (ii) unable to identify the data structure using a maximum likelihood implementation.

2.3.2.2 Model Identification Using PLS

Different from PCA, PLS extracts t-score variables from the input variables. It is therefore tempting to pre-conclude that PLS extracts common cause variation

[2] The assumptions for this are discussed in Subsection 6.1.1.

by determining the n t-score variables that discard the *non-predictive* variation in $\mathfrak{P}'\mathbf{s}'$. The fact that the cross-covariance matrix $\mathbf{S}_{x_0y_0} = \mathfrak{P}\mathbf{S}_{ss}\mathfrak{Q}^T$ does not represent the signal contributions $\mathfrak{P}'\mathbf{s}'$ and $\mathfrak{f}$ reinforces this assumption.

A more detailed analysis, however, yields that this is not the case. Equation 2.56 reexamines the construction of the weight vectors assuming that $\mathbf{q}$ is predetermined

$$\mathbf{w} = \arg\max_{\mathbf{w}} E\left\{\mathbf{w}^T\mathbf{x}_0\mathbf{y}_0^T\mathbf{q}\right\} - \tfrac{1}{2}\lambda\left(\mathbf{w}^T\mathbf{w} - 1\right). \tag{2.56}$$

The score variables are linear combination of $\mathbf{x}_0$ and $\mathbf{y}_0$, which implies that

$$t = \mathbf{s}^T\mathfrak{P}^T\mathbf{w} + \mathbf{s}'^T\mathfrak{P}'^T\mathbf{w} \qquad u = \mathbf{s}^T\mathfrak{Q}^T\mathbf{q} + \mathfrak{f}^T\mathbf{q}. \tag{2.57}$$

Equation 2.57 dictates the condition for separating $\mathbf{s}$ and $\mathbf{s}'$ is $\mathfrak{P}'^T\mathbf{w} = \mathbf{0}$. Applying (1.8) to reformulate the covariance of the pair of score variables yields

$$E\{tu\} = r_{tu}\sqrt{E\left\{\left(\mathbf{s}^T\mathfrak{P}^T\mathbf{w}\right)^2 + \left(\mathbf{s}'^T\mathfrak{P}'^T\mathbf{w}\right)^2\right\}E\left\{\left(\mathbf{s}^T\mathfrak{Q}^T\mathbf{q}\right)^2 + \left(\mathfrak{f}^T\mathbf{q}\right)^2\right\}} \tag{2.58}$$

where r_{tu} is the correlation coefficient between the score variables. If $\mathfrak{P}'^T\mathbf{w} = \mathbf{0}$, it follows from (2.58) that

$$E\{tu\} = r_{tu}\sqrt{E\left\{\left(\mathbf{s}^T\mathfrak{P}^T\mathbf{w}\right)^2\right\}E\left\{\left(\mathbf{s}^T\mathfrak{Q}^T\mathbf{q}\right)^2 + \left(\mathfrak{f}^T\mathbf{q}\right)^2\right\}} \tag{2.59}$$

and hence, the t-score variable does not include the *non-predictive* contribution $\mathfrak{P}'\mathbf{s}'$. This, however, generally cannot be assumed. It therefore follows that PCA and PLS cannot estimate a model that discriminates between:

- the common cause variation of the input and output variables;
- the non-predictive variation encapsulated in the input variables only; and
- the error variables corrupting the outputs.

The next subsection develops an alternative PLS formulation that extracts the common cause variation and discriminates between the three different types of variation.

2.3.3 Geometric analysis of data structure

The detailed examination of (2.56) to (2.58) yields that PLS effectively does not produce score variables that are related to model accuracy. This follows from the fact that the covariance criterion can be expressed by the product of the correlation coefficient times the square root of the variance products of the score variable. A larger variance for any of the score variables at the expense of a smaller correlation coefficient may, consequently, still produce a larger covariance. Model accuracy in the score space, however, is related to the correlation coefficient. The larger the correlation coefficient between two variables the more they have in common and hence, the more accurately one of these variables can predict the other.

Preventing PLS from incorporating $\mathbf{P}'\mathbf{s}'$ into the calculation of the t-score variables requires, therefore, a fresh look at its objective function. As outlined above, the key lies in determining weight vectors based on an objective function that relates to model accuracy rather than covariance. Starting with the following data structure

$$\begin{gathered}\mathbf{y}_0 = \mathcal{B}^T\mathbf{x}_0 + \mathfrak{f} \\ E\left\{\begin{pmatrix} \mathbf{x}_0 \\ \mathfrak{f} \end{pmatrix}\right\} = \mathbf{0} \\ E\left\{\begin{pmatrix} \mathbf{x}_0 \\ \mathfrak{f} \end{pmatrix}\begin{pmatrix} \mathbf{x}_0^T & \mathfrak{f}^T \end{pmatrix}\right\} = \begin{bmatrix} \mathbf{S}_{x_0x_0} & \mathbf{0} \\ \mathbf{0} & \mathrm{diag}\left\{ \sigma_{\mathfrak{f}_1}^2 \ \cdots \ \sigma_{\mathfrak{f}_{n_y}}^2 \right\} \end{bmatrix}\end{gathered} \tag{2.60}$$

for which the best linear unbiased estimator is the OLS solution (Henderson 1975)

$$\mathcal{B}^T = \mathbf{S}_{y_0x_0}\mathbf{S}_{x_0x_0}^{-1}. \tag{2.61}$$

Using (2.60) and (2.61) gives rise to reformulate $\mathbf{S}_{y_0y_0}$ as follows

$$\mathbf{S}_{y_0y_0} = E\left\{\mathbf{y}_0\mathbf{y}_0^T\right\} = \mathbf{S}_{y_0x_0}\mathbf{S}_{x_0x_0}^{-1}\mathbf{S}_{x_0y_0} + \mathbf{S}_{\mathfrak{ff}}, \tag{2.62}$$

where $\mathbf{S}_{\mathfrak{ff}} = \mathrm{diag}\left\{ \sigma_{\mathfrak{f}_1}^2 \ \cdots \ \sigma_{\mathfrak{f}_{n_y}}^2 \right\}$. It follows from (2.60) that the only contribution to $\mathbf{S}_{y_0y_0}$ that can be predicted by the linear model is $\mathbf{S}_{\widehat{y}_0\widehat{y}_0} = \mathbf{S}_{y_0x_0}\mathbf{S}_{x_0x_0}^{-1}\mathbf{S}_{x_0y_0}$, since $\mathbf{S}_{y_0y_0} = \mathbf{S}_{\widehat{y}_0\widehat{y}_0} + \mathbf{S}_{\mathfrak{ff}}$. In a similar fashion to PCA, it is possible to determine a direction vector to maximize the following objective function

$$\mathbf{q} = \arg\max_{\mathbf{q}} E\left\{\mathbf{q}^T\widehat{\mathbf{y}}_0\widehat{\mathbf{y}}_0^T\mathbf{q}\right\} - \lambda\left(\mathbf{q}^T\mathbf{q} - 1\right), \tag{2.63}$$

where $\widehat{\mathbf{y}}_0 = \mathcal{B}^T\mathbf{x}_0$. The optimal solution for (2.63) is

$$\mathbf{S}_{y_0x_0}\mathbf{S}_{x_0x_0}^{-1}\mathbf{S}_{x_0y_0}\mathbf{q} - \lambda\mathbf{q} = \mathbf{0}. \tag{2.64}$$

The eigenvalue λ is the variance of the orthogonal projection of $\widehat{\mathbf{y}}_0$ onto $\mathbf{q}$. The solution to (2.64) is the eigenvector associated with the largest eigenvalue of $\mathbf{S}_{y_0x_0}\mathbf{S}_{x_0x_0}^{-1}\mathbf{S}_{x_0y_0}$. The eigenvector associated with the second largest eigenvalue captures the second largest contribution and so on.

Whilst this allows to extract weight vectors for $\mathbf{y}_0$, how to determine weight vectors for $\mathbf{x}_0$ to predict the u-score variable $\mathbf{q}^T\widehat{\mathbf{y}}_0$ as accurately as possible? By revisiting (2.57) and (2.58) it follows that the correlation coefficient r_{tu} must yield a maximum to achieve this

$$\frac{\mathbf{q}^T\widehat{\mathbf{y}}_0}{\sqrt{E\left\{\left(\mathbf{q}^T\widehat{\mathbf{y}}_0\right)^2\right\}}} = r_{tu}\frac{\mathbf{w}^T\mathbf{x}_0}{\sqrt{E\left\{\left(\mathbf{w}^T\mathbf{x}_0\right)^2\right\}}} \tag{2.65}$$

where $u = \mathbf{q}^T\widehat{\mathbf{y}}_0$ and $t = \mathbf{w}^T\mathbf{x}_0$. By incorporating the constraint $\mathbf{w}^T\mathbf{S}_{x_0x_0}\mathbf{w} - 1 = 0$, setting the variance of $\mathbf{w}^T\mathbf{x}_0$ to be 1, $t \sim \mathcal{N}\{0, 1\}$ and (2.65) becomes

$$\mathbf{w} = \arg\max_{\mathbf{w}} E\left\{\frac{\mathbf{q}^T}{\sqrt{\lambda}}\widehat{\mathbf{y}}_0\mathbf{x}_0^T\mathbf{w}\right\} - \tfrac{1}{2}\mu\left(\mathbf{w}^T\mathbf{S}_{x_0x_0}\mathbf{w} - 1\right). \tag{2.66}$$

The fact that $\lambda = E\left\{\left(\mathbf{q}^T\mathbf{y}_0\right)^2\right\}$ follows from

- $\mathbf{q}^T\mathbf{S}_{y_0x_0}\mathbf{S}_{x_0x_0}^{-1}\mathbf{S}_{x_0y_0}\mathbf{q} = \lambda$,
- $\mathbf{S}_{\widehat{y}_0\widehat{y}_0} = \mathbf{S}_{y_0x_0}\mathbf{S}_{x_0x_0}^{-1}\mathbf{S}_{x_0y_0}$, and
- $\mathbf{q}^T\widehat{\mathbf{y}}_0 \sim \mathcal{N}\{0, \lambda\}$,

so that ${}^{\mathbf{q}^T\mathbf{y}_0}/_{\sqrt{\lambda}} \sim \mathcal{N}\{0, 1\}$. The objective function in (2.66) therefore maximizes the correlation coefficient, $r_{tu} = \left(\mathbf{q}^T\widehat{\mathbf{y}}_0\right)\left(\mathbf{w}^T\mathbf{x}_0\right)/\sqrt{\lambda} = ({}^{u}/_{\sqrt{\lambda}})\,t$, and has the following solution

$$\mathbf{S}_{x_0y_0}\mathbf{q} - \mu\sqrt{\lambda}\mathbf{S}_{x_0x_0}\mathbf{w} = \mathbf{0} \qquad \mathbf{w} = \frac{\mathbf{S}_{x_0x_0}^{-1}\mathbf{S}_{x_0y_0}\mathbf{q}}{\mu\sqrt{\lambda}}, \tag{2.67}$$

where the Lagrangian multiplier, μ, satisfies the constraint $\mathbf{w}^T\mathbf{S}_{x_0x_0}\mathbf{w} = 1$. Next, (2.63) and (2.66) can be combined to produce the objective function

$$J = \mathbf{w}^T\mathbf{S}_{x_0y_0}\mathbf{q} - \tfrac{1}{2}\lambda\left(\mathbf{w}^T\mathbf{S}_{x_0x_0}\mathbf{w} - 1\right) - \tfrac{1}{2}\lambda\left(\mathbf{q}^T\mathbf{q} - 1\right), \tag{2.68}$$

which has the following solution for $\mathbf{w}$ and $\mathbf{q}$

$$\frac{\partial J}{\partial \mathbf{w}} = \mathbf{S}_{x_0y_0}\mathbf{q} - \lambda\mathbf{S}_{x_0x_0}\mathbf{w} = \mathbf{0} \qquad \frac{\partial J}{\partial \mathbf{q}} = \mathbf{S}_{y_0x_0}\mathbf{w} - \lambda\mathbf{q} = \mathbf{0} \tag{2.69}$$

and hence

$$\mathbf{S}_{x_0x_0}^{-1}\mathbf{S}_{x_0y_0}\mathbf{S}_{y_0x_0}\mathbf{w} - \lambda^2\mathbf{w} = \mathbf{0} \qquad \mathbf{S}_{y_0x_0}\mathbf{S}_{x_0x_0}^{-1}\mathbf{S}_{x_0y_0}\mathbf{q} - \lambda^2\mathbf{q} = \mathbf{0}. \tag{2.70}$$

That both Lagrangian multiples have the same value follows from

$$\mathbf{w}^T\frac{\partial J}{\partial \mathbf{w}} = \lambda \qquad \mathbf{q}^T\frac{\partial J}{\partial \mathbf{q}} = \lambda. \tag{2.71}$$

This solution relates to a nonsymmetric index of redundancy, introduced by Stewart and Love (1968) to describe the amount of predicted variance, and was first developed by van den Wollenberg (1977). Moreover, ten Berge (1985) showed that van den Wollenberg's *maximum redundancy* analysis represents a special case of Fortier's simultaneous linear prediction (Fortier 1966). The objective of the work in Fortier (1966) is to determine a linear combination of a set of predictors (inputs) that has a maximum predictability for all predictants (outputs) simultaneously.

The next step is to apply the standard PLS deflation procedure to determine subsequent sets of LVs. According to the data model in (2.51), only the contribution $\mathbf{Ps}$ in $\mathbf{x}_0$ is predictive for $\mathbf{y}_0$. By default, the solution of the objective function in (2.68) must discard the contribution $\mathbf{P}'\mathbf{s}'$. The next question is how many sets of latent variables can be determined by solving (2.68) and carrying out the PLS deflation procedure? The answer to this lies in the cross covariance matrix $\mathbf{S}_{x_0 y_0}$ as it only describes the common cause variation, that is, $\mathbf{S}_{x_0 y_0} = \mathbf{P}\mathbf{S}_{ss}\mathbf{Q}^T$.

The loading vectors $\mathbf{p}_i$ and $\acute{\mathbf{q}}_i$ can now be computed by

$$\mathbf{p}_k = \mathbf{S}_{ee}^{(i)}\mathbf{w}_i \qquad \acute{\mathbf{q}}_i = \mathbf{S}_{fe}^{(i)}\mathbf{w}_i. \tag{2.72}$$

Utilizing (2.72), the deflation of the covariance matrix is

$$\mathbf{S}_{ee}^{(i+1)} = \mathbf{S}_{ee}^{(i)} - E\left\{\mathbf{p}_i t_i^2 \mathbf{p}_i^T\right\} = \mathbf{S}_{ee}^{(i)} - \mathbf{p}_i\mathbf{p}_i^T \tag{2.73}$$

and similarly for the cross-covariance matrix

$$\mathbf{S}_{ef}^{(i+1)} = \mathbf{S}_{ef}^{(i)} - \mathbf{p}_i\acute{\mathbf{q}}_i^T. \tag{2.74}$$

If the cross-covariance matrix is exhausted, there is no further common cause variation in the input variable set. One criterion for testing this, or a stopping rule according to the next section, would be to determine the Frobenius norm of the cross-covariance matrix after applying the ith deflation procedure

$$\left\|\mathbf{S}_{ef}^{(i+1)}\right\|^2 = \sum_{j_1=1}^{n_x}\sum_{j_2=1}^{n_y}\left(s_{ef_{(j_1 j_2)}}^{(i+1)}\right)^2 \geq 0. \tag{2.75}$$

If (2.75) is larger than zero, obtain the $(i+1)$th pair of weight vectors, $\mathbf{w}_{i+1}$ and $\mathbf{q}_{i+1}$, by solving (2.70). On the other hand, if (2.75) is zero, the common cause variation has been extracted from the input variables.

It is important to note that (2.70) presents an upper limit for determining the maximum number of weight vector pairs. Assuming that $n_y \leq n_x$, the rank of the matrix products $\mathbf{S}_{x_0x_0}^{-1}\mathbf{S}_{x_0y_0}\mathbf{S}_{y_0x_0}$ and $\mathbf{S}_{y_0x_0}\mathbf{S}_{x_0x_0}^{-1}\mathbf{S}_{x_0y_0}$ is n_y. This follows from the fact that the rank of $\mathbf{S}_{x_0y_0}$ is equal to n_y. If $n \leq \min\left(n_y, n_x\right)$, alternative stopping rules are discussed in Subsection 2.4.2. After extracting the common cause variation from $\mathbf{x}_0$, the objective function in (2.68) can be replaced by

$$\begin{pmatrix}\mathbf{w}_i\\ \mathbf{q}_i\end{pmatrix} = \arg\max_{\mathbf{w},\mathbf{q}} \mathbf{w}^T\mathbf{S}_{ef}^{(i)}\mathbf{q} - \tfrac{1}{2}\lambda\left(\mathbf{w}^T\mathbf{w}-1\right) - \tfrac{1}{2}\lambda\left(\mathbf{q}^T\mathbf{q}-1\right), \tag{2.76}$$

which is the PLS one. Table 2.3 shows the steps of this *maximum redundancy* PLS or *MRPLS* algorithm. This algorithm is an extension of the NIPALS algorithm for PLS, for example discussed in Geladi and Kowalski (1986), and incorporates the constraint objective function in (2.68). This implies that the actual data matrices $\mathbf{X}_0$ and $\mathbf{Y}_0$, storing a total of K samples of $\mathbf{x}_0$ and $\mathbf{y}_0$ in a consecutive order as row vectors, are utilized instead of $\mathbf{S}_{x_0x_0}$ and $\mathbf{S}_{x_0y_0}$.

Table 2.3 Algorithm for maximum redundancy PLS.

Step	Description	Equation
1	Initiate iteration	$n = 1,\ i = 1,\ \mathbf{F}^{(1)} = \mathbf{Y}_0$
2	Set up ${}_0\widehat{\mathbf{u}}_i$	${}_0\widehat{\mathbf{u}}_i = \mathbf{F}^{(i)}\ (:, 1)$
3	Determine auxiliary vector	$\widehat{\mathfrak{w}}_i = \mathbf{X}_0^T\,{}_0\widehat{\mathbf{u}}_i / \left\Vert \mathbf{X}_0^T\,{}_0\widehat{\mathbf{u}}_i \right\Vert$
4	Calculate w-weight vector	if $i = n$ $\widehat{\mathbf{w}}_i = \left[\mathbf{X}_0^T\mathbf{X}_0\right]^{-1}\widehat{\mathfrak{w}}_i / \sqrt{\widehat{\mathfrak{w}}_i^T\left[\mathbf{X}_0^T\mathbf{X}_0\right]^{-1}\widehat{\mathfrak{w}}_i}$ else $\widehat{\mathbf{w}}_i = \widehat{\mathfrak{w}}_i$
5	Determine r-weight vector	if $i = n$ $\widehat{\mathbf{r}}_i = \widehat{\mathbf{w}}_i$ else $\widehat{\mathbf{r}}_i = \widehat{\mathbf{w}}_i - \sum_{j=1}^{i-1} \widehat{\mathbf{p}}_j^T\widehat{\mathbf{w}}_i\widehat{\mathbf{r}}_j$
6	Compute t-score vector	$\widehat{\mathbf{t}}_i = \mathbf{X}_0\widehat{\mathbf{r}}_i$
7	Determine q-weight vector	$\widehat{\mathbf{q}}_i = \mathbf{F}^{(i)^T}\widehat{\mathbf{t}}_i / \left\Vert \mathbf{F}^{(i)^T}\widehat{\mathbf{t}}_i \right\Vert$
8	Calculate u-score vector	${}_1\widehat{\mathbf{u}}_i = \mathbf{F}^{(i)}\widehat{\mathbf{q}}_i$
9	Check for convergence	if $\left\Vert {}_1\widehat{\mathbf{u}}_i - {}_0\widehat{\mathbf{u}}_i \right\Vert > \epsilon$ set ${}_0\widehat{\mathbf{u}}_i = {}_1\widehat{\mathbf{u}}_i$ and go to Step 3 else set $\widehat{\mathbf{u}}_i = {}_1\widehat{\mathbf{u}}_i$ and go to Step 10
10	Determine p-loading vector	if $i = n : \widehat{\mathbf{p}}_i = \mathbf{X}_0^T\widehat{\mathbf{t}}_i$ else : $\widehat{\mathbf{p}}_i = \mathbf{X}_0^T\widehat{\mathbf{t}}_i / \widehat{\mathbf{t}}_i^T\widehat{\mathbf{t}}_i$
11	Determine q́-loading vector	if $i = n : \widehat{\acute{\mathbf{q}}}_i = \mathbf{F}^{(i)^T}\widehat{\mathbf{t}}_i$ else : $\widehat{\acute{\mathbf{q}}}_i = \mathbf{F}^{(i)^T}\widehat{\mathbf{t}}_i / \widehat{\mathbf{t}}_i^T\widehat{\mathbf{t}}_i$
12	Deflate output data matrix	$\mathbf{F}^{(i+1)} = \mathbf{F}^{(i)} - \widehat{\mathbf{t}}_i\widehat{\acute{\mathbf{q}}}_i^T$
13	Check whether there is still significant variation remaining in $\mathbf{X}_0^T\mathbf{F}^{(i+1)}$	if so $i = i + 1,\ n = n + 1$ and go to Step 3 if not $i = i + 1$, go to Step 14
14	Check whether $i = n_x$	if so then terminate else go to Step 2

The preceding discussion in this subsection has assumed the availability of $\mathbf{S}_{x_0x_0}$ and $\mathbf{S}_{x_0y_0}$, which has been for the convenience and simplicity of the presentation. Removing this assumption, the MRPLS algorithm relies on the data matrices $\mathbf{X}_0$ and $\mathbf{Y}_0$. The covariance and cross-covariance matrices can then be estimated, implying that the weight, score and loading vectors are estimates too.

That the MRPLS algorithm in Table 2.3 produces the optimal solution of the objective function in (2.68) follows from the iterative procedure described in Steps 3 to 8 in Table 2.3. With respect to Equation (2.70), the optimal solutions for $\widehat{\mathbf{w}}_k$ and $\widehat{\mathbf{q}}_k$ are the dominant eigenvectors[3] of the positive semi-definite matrices

$$\left[\widehat{\mathbf{S}}_{ee}^{(i)}\right]^{\dagger}\widehat{\mathbf{S}}_{ef}^{(i)}\widehat{\mathbf{S}}_{fe}^{(i)} \quad \text{and} \quad \widehat{\mathbf{S}}_{fe}^{(i)}\left[\widehat{\mathbf{S}}_{ee}^{(i)}\right]^{\dagger}\widehat{\mathbf{S}}_{ef}^{(i)}, \tag{2.77}$$

respectively. Substituting Step 5 into Step 6 yields

$$\widehat{\mathbf{q}}_i \propto \mathbf{F}^{(i)^T}\mathbf{X}_0\widehat{\mathbf{w}}_i. \tag{2.78}$$

Now, substituting consecutively Step 4 and Step 3 into (2.78) gives rise to

$${}_0\widehat{\mathbf{q}}_i \propto \mathbf{F}^{(i)^T}\mathbf{X}_0\left[\mathbf{X}_0^T\mathbf{X}_0\right]^{-1}\mathbf{X}_0^T\left({}_0\widehat{\mathbf{u}}_i\right). \tag{2.79}$$

Finally, substituting Step 8 into (2.79)

$$\widehat{\lambda}_i^2\left({}_1\widehat{\mathbf{q}}_i\right) = \mathbf{F}^{(i)^T}\mathbf{X}_0\left[\mathbf{X}_0^T\mathbf{X}_0\right]^{-1}\mathbf{X}_0^T\mathbf{F}^{(i)}\left({}_0\widehat{\mathbf{q}}_i\right) \tag{2.80}$$

confirms that the iteration procedure in Table 2.3 yields the dominant eigenvector of

$$\mathbf{F}^{(i)^T}\mathbf{X}_0\left[\mathbf{X}_0^T\mathbf{X}_0\right]^{-1}\mathbf{X}_0^T\mathbf{F}^{(i)} = (K-1)\,\widehat{\mathbf{S}}_{fe}^{(i)}\left[\widehat{\mathbf{S}}_{ee}^{(i)}\right]^{\dagger}\widehat{\mathbf{S}}_{ef}^{(i)} \tag{2.81}$$

as the q-weight vector. The equality in (2.81) is discussed in Chapter 10, Lemma 10.5.3 and Theorem 10.5.7. In fact, the iteration procedure of the MRPLS algorithm represents the iterative Power method for determining the dominant eigenvector of a symmetric positive semi-definite matrix (Golub and van Loan 1996). The dominant eigenvalue of $\mathbf{F}^{(i)^T}\mathbf{X}_0\left[\mathbf{X}_0^T\mathbf{X}_0\right]^{-1}\mathbf{X}_0^T\mathbf{F}^{(i)}$ is $K-1$ times the dominant eigenvalue of $\widehat{\mathbf{S}}_{fe}^{(i)}\left[\widehat{\mathbf{S}}_{ee}^{(i)}\right]^{\dagger}\widehat{\mathbf{S}}_{ef}^{(i)}$. Now, substituting Step 3 into Step 4 gives rise to

$${}_0\widehat{\mathbf{w}}_i \propto \left[\mathbf{X}_0^T\mathbf{X}_0\right]^{-1}\mathbf{X}_0^T\left({}_0\widehat{\mathbf{u}}_i\right). \tag{2.82}$$

Next, consecutively substituting Steps 8, 7, 6 and then 5 into Equation (2.82) yields

$$\widehat{\lambda}_i^2\left({}_1\widehat{\mathbf{w}}_i\right) = \left[\mathbf{X}_0^T\mathbf{X}_0\right]^{-1}\mathbf{X}_0^T\mathbf{F}^{(i)}\mathbf{F}^{(i)^T}\mathbf{X}_0\left({}_0\widehat{\mathbf{w}}_i\right). \tag{2.83}$$

[3] A dominant eigenvector is the eigenvector associated with the largest eigenvalue of a symmetric positive semi-definite matrix under the assumption that this eigenvalue is not a multiple eigenvalue of that matrix.

Hence, the iteration procedure of the MRPLS algorithm in Table 2.3 computes the optimal solution of the MRPLS objective function.

It should also be noted that, different from the PLS algorithm, the MRPLS algorithm produces an auxiliary vector $\mathfrak{w}_i$. This vector is, in fact, the w-weight vector for PLS. Furthermore, the w-weight vector for MRPLS is the product of $\mathfrak{w}_i$ and the inverse of $\mathbf{S}_{x_0x_0}$ or $\mathbf{X}_0^T\mathbf{X}_0$ when using the data matrices.

The algorithm presented in Table 2.3 relies on the fact that only the output data matrix needs to be deflated. Hence, the length constraint for the w-weight vector $\mathbf{w}_i^T\mathbf{E}^{(i)^T}\mathbf{E}^{(i)}\mathbf{w}_i - 1$ is equivalent to $\mathbf{w}_i^T\mathbf{X}_0^T\mathbf{X}_0\mathbf{w}_i - 1$. It is important to note that deflating the output data matrix for the PLS algorithm requires the introduction of r-weight vectors, which is proven in Chapter 10, together with the geometric property that the w-weight vectors are mutually orthogonal to the p-loading vectors. Hence, MRPLS does not require the introduction of r-weight vectors.

Another important aspect that needs to be considered here relates to the deflated cross-covariance matrix. Equation (2.75) outlines that the Frobenius norm of $\mathbf{S}_{ef}^{(i)}$ is larger than or equal to zero. For a finite data set, the squared elements of $\mathbf{S}_{ef}^{(n)}$ may not be zero if the cross-covariance matrix is estimated. Hence, the PLS algorithm is able to obtain further latent variables to exhaust the input variable set. It is important to note, however, that the elements of $\mathbf{S}_{ef}^{(n)}$ asymptotically converge to zero

$$\lim_{K\to\infty} \widehat{\mathbf{S}}_{x_0y_0} - \sum_{i=1}^{n} \widehat{\mathbf{p}}_i\widehat{\mathbf{q}}_i^T \to \mathbf{0}. \tag{2.84}$$

This presents the following problem for a subsequent application of PLS

$$\begin{aligned}\begin{pmatrix}\widehat{\mathbf{w}}_i\\ \widehat{\mathbf{q}}_i\end{pmatrix} &= \arg\max_{\mathbf{w},\mathbf{q}} \mathbf{w}^T\left[\lim_{K\to\infty}\widehat{\mathbf{S}}_{x_0y_0} - \sum_{i=1}^{n}\widehat{\mathbf{p}}_i\widehat{\mathbf{q}}_i^T\right]\mathbf{q}\\ &\quad - \tfrac{1}{2}\lambda\left(\mathbf{w}^T\mathbf{w}-1\right) - \tfrac{1}{2}\lambda\left(\mathbf{q}^T\mathbf{q}-1\right)\\ \begin{pmatrix}\widehat{\mathbf{w}}_i\\ \widehat{\mathbf{q}}_i\end{pmatrix} &= \arg\max_{\mathbf{w},\mathbf{q}} \mathbf{w}^T\mathbf{0}\mathbf{q} - \tfrac{1}{2}\lambda\left(\mathbf{w}^T\mathbf{w}-1\right) - \tfrac{1}{2}\lambda\left(\mathbf{q}^T\mathbf{q}-1\right)\end{aligned} \tag{2.85}$$

which yields an infinite number of solutions for $\widehat{\mathbf{w}}_i$ and $\widehat{\mathbf{q}}_i$. In this case, it is possible to apply PCA to the deflated input data matrix in order to generate a set of $n_x - n$ t-score variables that are statistically independent of the t-score variables obtained from the MRPLS algorithm.

2.3.4 A simulation example

This example demonstrates the shortcomings of PLS and highlights that MRPLS can separately extract the common cause variation that affects the input and output variables and the remaining variation of the input variables that is not predictive to the output variables. The simulation example relies on the data model introduced in (2.51), where the parameter matrices $\mathfrak{P}$, $\mathfrak{P}'$ and $\mathfrak{Q}$ were populated by random values obtained from a Gaussian distribution of zero mean and variance 1.

The number of input and output variables is 10 and 6, respectively. Moreover, these variable sets are influenced by a total of 4 source variables describing common cause variation. The remaining variation of the input variables is simulated by a total of 6 stochastic variables. The dimensions of the parameter matrices are, consequently, $\mathfrak{P} \in \mathbb{R}^{10\times4}$, $\mathfrak{P}' \in \mathbb{R}^{10\times6}$ and $\mathfrak{Q} \in \mathbb{R}^{6\times4}$. Equations (2.86) to (2.88) show the elements determined for each parameter matrix.

$$\mathfrak{Q} = \begin{bmatrix} 0.174 & 0.742 & -0.149 & 0.024 \\ -0.486 & 0.243 & -0.313 & 0.449 \\ 0.405 & -0.470 & -0.002 & -0.212 \\ 0.230 & -0.997 & 0.562 & 0.381 \\ 0.268 & 0.685 & 0.424 & 0.242 \\ 0.810 & -0.005 & 0.208 & 0.431 \end{bmatrix} \tag{2.86}$$

$$\mathfrak{P} = \begin{bmatrix} -0.899 & -0.871 & -0.559 & 0.765 \\ 0.083 & 0.122 & -0.638 & -0.011 \\ 0.041 & -0.719 & -0.857 & -0.774 \\ -0.138 & 0.473 & 0.451 & -0.484 \\ 0.668 & 0.596 & -0.719 & 0.036 \\ 0.363 & 0.801 & 0.507 & -0.580 \\ 0.538 & -0.644 & 0.719 & -0.220 \\ -0.868 & 0.000 & -0.880 & -0.478 \\ 0.458 & 0.170 & -0.256 & 0.277 \\ 0.600 & 0.662 & 0.324 & -0.721 \end{bmatrix} \tag{2.87}$$

$$\mathfrak{P}' = \begin{bmatrix} 0.565 & -0.370 & -0.279 & -0.618 & 0.219 & -0.890 \\ 0.447 & -0.819 & -0.427 & -0.297 & -0.844 & -0.935 \\ -0.910 & 0.116 & 0.593 & -0.768 & 0.807 & -0.391 \\ -0.106 & -0.556 & 0.986 & 0.753 & -0.535 & -0.536 \\ 0.132 & -0.217 & 0.715 & -0.921 & 0.693 & 0.715 \\ -0.234 & 0.547 & 0.775 & 0.448 & -0.289 & -0.650 \\ 0.005 & 0.035 & 0.929 & 0.224 & -0.230 & -0.442 \\ 0.854 & 0.341 & 0.500 & 0.388 & -0.814 & -0.844 \\ 0.987 & 0.843 & -0.613 & -0.963 & 0.807 & -0.537 \\ 0.976 & -0.587 & -0.504 & 0.861 & -0.588 & 0.648 \end{bmatrix}. \tag{2.88}$$

The common cause variation $\mathbf{s} \in \mathbb{R}^4$ as well as the uninformative variation in the input variables for predicting the outputs, $\mathbf{s}' \in \mathbb{R}^6$, were Gaussian distributed i.i.d. sets of zero mean and unity covariance matrices, that is, $\mathbf{s} \sim \mathcal{N}\{\mathbf{0}, \mathbf{I}\}$ and $\mathbf{s}' \sim \mathcal{N}\{\mathbf{0}, \mathbf{I}\}$. Both source signals were statistically independent of each other, that is, $\mathbf{S}_{ss'} = \mathbf{0}$. Finally, the error variables, $\mathfrak{f} \in \mathbb{R}^6$, were statistically independent of the source signals, that is, $\mathbf{S}_{\mathfrak{f},s,s'} = \mathbf{0}$, and followed a zero mean Gaussian distribution. The variance of the error variables was also randomly selected between 0.01 and 0.06: $\sigma_1^2 = 0.0276$, $\sigma_2^2 = 0.0472$, $\sigma_3^2 = 0.0275$, $\sigma_4^2 = 0.0340$, $\sigma_5^2 = 0.0343$ and $\sigma_6^2 = 0.0274$.

To contrast MRPLS with PLS, a total of 5000 samples were simulated and analyzed using both techniques. The estimated covariance matrices for the source signals which are encapsulated in the input and output variable sets, **s**, the second set of source signals that is not predictive for the output variables, **s**′, and the error signals $\mathfrak{f}$, are listed in (2.89) to (2.91).

$$\widehat{\mathbf{S}}_{ss} = \begin{bmatrix} 0.989 & 0.009 & 0.008 & -0.010 \\ 0.009 & 1.008 & 0.005 & 0.014 \\ 0.008 & 0.005 & 1.004 & 0.010 \\ -0.010 & 0.014 & 0.010 & 1.027 \end{bmatrix} \tag{2.89}$$

$$\widehat{\mathbf{S}}_{s's'} = \begin{bmatrix} 1.014 & -0.011 & -0.005 & -0.010 & -0.007 & -0.017 \\ -0.011 & 1.036 & 0.004 & 0.006 & 0.010 & -0.025 \\ -0.005 & 0.004 & 0.987 & 0.010 & 0.008 & 0.018 \\ -0.010 & 0.006 & 0.010 & 1.039 & -0.004 & 0.006 \\ -0.007 & 0.010 & 0.008 & -0.004 & 1.005 & -0.011 \\ -0.017 & -0.025 & 0.018 & 0.006 & -0.011 & 0.993 \end{bmatrix} \tag{2.90}$$

$$\widehat{\mathbf{S}}_{\mathfrak{ff}} = \begin{bmatrix} 0.028 & 0.001 & -0.001 & 0.001 & 0.001 & 0.000 \\ 0.001 & 0.048 & 0.000 & 0.001 & -0.000 & 0.000 \\ -0.001 & 0.000 & 0.028 & 0.001 & -0.001 & -0.000 \\ 0.001 & 0.001 & 0.001 & 0.035 & 0.001 & 0.001 \\ 0.001 & -0.000 & -0.001 & 0.001 & 0.035 & -0.001 \\ 0.000 & 0.000 & -0.000 & 0.001 & -0.001 & 0.028 \end{bmatrix}. \tag{2.91}$$

Comparing the estimates of $\mathbf{S}_{ss}$, $\mathbf{S}_{s's'}$ and $\mathbf{S}_{\mathfrak{ff}}$ signals with the true covariance matrices shows a close agreement. This was expected given that 5000 is a relatively large number of simulated samples. Next, (2.92) to (2.94) show the estimates of $\mathbf{S}_{x_0x_0}$, $\mathbf{S}_{x_0y_0}$ and $\mathbf{S}_{y_0y_0}$.

$$\widehat{\mathbf{S}}_{x_0x_0} = \begin{bmatrix} 4.319 & 1.709 & 0.684 & -1.169 & \cdots & -2.154 \\ 1.709 & 3.156 & -0.449 & 0.442 & \cdots & 0.756 \\ 0.684 & -0.449 & 4.507 & -0.559 & \cdots & -2.805 \\ -1.169 & 0.442 & -0.559 & 3.057 & \cdots & 1.097 \\ -0.678 & -0.473 & 1.470 & -0.804 & \cdots & -0.305 \\ -2.128 & -0.423 & -0.124 & 2.103 & \cdots & 0.503 \\ -0.583 & -0.394 & 0.401 & 1.462 & \cdots & -0.130 \\ 1.457 & 1.724 & -0.006 & 1.303 & \cdots & 0.226 \\ 1.519 & 0.364 & 0.326 & -2.307 & \cdots & -0.840 \\ -2.154 & 0.756 & -2.805 & 1.097 & \cdots & 4.483 \end{bmatrix} \tag{2.92}$$

$$\widehat{\mathbf{S}}_{x_0 y_0} = \begin{bmatrix} -0.701 & 0.753 & -0.118 & 0.669 & -0.875 & -0.512 \\ 0.190 & 0.191 & -0.021 & -0.436 & -0.167 & -0.083 \\ -0.427 & -0.296 & 0.527 & -0.054 & -1.050 & -0.507 \\ 0.223 & -0.185 & -0.158 & -0.398 & 0.330 & -0.230 \\ 0.679 & 0.065 & -0.032 & -0.858 & 0.301 & 0.403 \\ 0.569 & -0.410 & -0.104 & -0.651 & 0.723 & 0.152 \\ -0.505 & -0.752 & 0.568 & 1.092 & -0.054 & 0.479 \\ -0.032 & 0.477 & -0.225 & -0.853 & -0.734 & -1.088 \\ 0.282 & 0.035 & 0.020 & -0.128 & 0.253 & 0.448 \\ 0.512 & -0.551 & 0.095 & -0.608 & 0.550 & 0.240 \end{bmatrix} \tag{2.93}$$

$$\widehat{\mathbf{S}}_{y_0 y_0} = \begin{bmatrix} 0.634 & 0.162 & -0.291 & -0.777 & 0.502 & 0.121 \\ 0.162 & 0.650 & -0.405 & -0.357 & 0.018 & -0.253 \\ -0.291 & -0.405 & 0.455 & 0.476 & -0.276 & 0.218 \\ -0.777 & -0.357 & 0.476 & 1.541 & -0.291 & 0.461 \\ 0.502 & 0.018 & -0.276 & -0.291 & 0.828 & 0.418 \\ 0.121 & -0.253 & 0.218 & 0.461 & 0.418 & 0.906 \end{bmatrix}. \tag{2.94}$$

Equations (2.96) to (2.98) show the actual matrices. With respect to the data model in (2.51), using $\mathfrak{P}$, $\mathfrak{P}'$ and $\mathfrak{Q}$, given in (2.86) to (2.88), $\mathbf{S}_{ss} = \mathbf{I}$, $\mathbf{S}_{s's'} = \mathbf{I}$ and $\mathbf{S}_{\mathfrak{f}\mathfrak{f}}$, the covariance matrices $\mathbf{S}_{x_0 x_0}$ and $\mathbf{S}_{y_0 y_0}$ allows computing the true covariance and cross-covariance matrices

$$\begin{aligned} \mathbf{S}_{x_0 x_0} &= \mathfrak{P}\mathbf{S}_{ss}\mathfrak{P}^T + \mathfrak{P}'\mathbf{S}_{s's'}\mathfrak{P}'^T \\ \mathbf{S}_{x_0 y_0} &= \mathfrak{P}\mathbf{S}_{ss}\mathfrak{Q}^T \\ \mathbf{S}_{y_0 y_0} &= \mathfrak{Q}\mathbf{S}_{ss}\mathfrak{Q}^T + \mathbf{S}_{\mathfrak{f}\mathfrak{f}}. \end{aligned} \tag{2.95}$$

A direct comparison between the estimated matrices in (2.89) to (2.91) and the actual ones in (2.96) to (2.98) yields an accurate and very close estimation of the elements of $\mathbf{S}_{y_0 y_0}$ and $\mathbf{S}_{x_0 y_0}$. However, slightly larger departures can be noticed for the estimation of the elements in $\mathbf{S}_{x_0 x_0}$. This can be explained by the fact that the asymptotic dimension of $\mathbf{S}_{x_0 y_0}$ is 4 and the source signals have a much more profound impact upon $\mathbf{S}_{y_0 y_0}$ than $\mathfrak{f}$. With this in mind, the last two eigenvalues of $\mathbf{S}_{y_0 y_0}$ are expected to be significantly smaller than the first four, which describe the impact of the source variables. In contrast, there are a total of 10 source signals, including 4 that the input and output variables share in common and an additional 6 source signals that are not describing the variation of the output variables. Hence, the estimation accuracy of the 10-dimensional covariance matrix of the input variables is less than the smaller dimensional covariance matrix of the input and the cross-covariance matrix of the input and output variables.

$$
\mathbf{S}_{x_0x_0} = \begin{bmatrix}
4.220 & 1.674 & 0.752 & -1.146 & \cdots & -2.178 \\
1.674 & 3.158 & -0.372 & 0.480 & \cdots & 0.698 \\
0.752 & -0.372 & 4.439 & -0.542 & \cdots & -2.816 \\
-1.146 & 0.480 & -0.542 & 3.115 & \cdots & 1.069 \\
-0.650 & -0.461 & 1.453 & -0.787 & \cdots & -0.305 \\
-2.063 & -0.355 & -0.134 & 2.164 & \cdots & 0.526 \\
-0.558 & -0.373 & 0.405 & 1.476 & \cdots & -0.153 \\
1.457 & 1.745 & 0.022 & 1.346 & \cdots & 0.186 \\
1.464 & 0.339 & 0.339 & -2.280 & \cdots & -0.771 \\
-2.178 & 0.698 & -2.816 & 1.069 & \cdots & 4.482
\end{bmatrix} \tag{2.96}
$$

$$
\mathbf{S}_{x_0y_0} = \begin{bmatrix}
-0.701 & 0.745 & -0.116 & 0.639 & -0.889 & -0.510 \\
0.200 & 0.185 & -0.020 & -0.465 & -0.167 & -0.071 \\
-0.417 & -0.273 & 0.521 & -0.049 & -1.032 & -0.474 \\
0.248 & -0.177 & -0.176 & -0.435 & 0.361 & -0.229 \\
0.667 & 0.061 & -0.016 & -0.831 & 0.291 & 0.404 \\
0.568 & -0.401 & -0.107 & -0.652 & 0.721 & 0.146 \\
-0.497 & -0.742 & 0.565 & 1.086 & -0.046 & 0.493 \\
-0.031 & 0.483 & -0.248 & -0.876 & -0.721 & -1.092 \\
0.250 & 0.023 & 0.047 & -0.102 & 0.197 & 0.436 \\
0.530 & -0.556 & 0.084 & -0.615 & 0.578 & 0.240
\end{bmatrix} \tag{2.97}
$$

$$
\mathbf{S}_{y_0y_0} = \begin{bmatrix}
0.632 & 0.153 & -0.283 & -0.775 & 0.498 & 0.117 \\
0.153 & 0.642 & -0.405 & -0.359 & 0.012 & -0.267 \\
-0.283 & -0.405 & 0.457 & 0.480 & -0.266 & 0.238 \\
-0.775 & -0.359 & 0.480 & 1.542 & -0.291 & 0.472 \\
0.498 & 0.012 & -0.266 & -0.291 & 0.814 & 0.406 \\
0.117 & -0.267 & 0.238 & 0.472 & 0.406 & 0.912
\end{bmatrix}. \tag{2.98}
$$

To verify the problem for PLS in identifying a model that relies on the underlying data structure in (2.51), the following matrix product shows that the w-weight vectors, obtained by PLS, are not orthogonal to the column vectors of

$\mathfrak{P}'$. According to (2.58), however, this is a condition for separating $\mathbf{s}$ from $\mathbf{s}'$.

$$\widehat{\mathbf{W}}^T \mathfrak{P}' = \begin{bmatrix} -0.358 & -0.031 & -0.128 & -0.608 & 0.292 & -0.852 \\ 0.564 & -0.048 & -0.058 & -0.432 & -0.241 & -0.670 \\ 0.803 & 0.192 & -0.980 & -1.394 & 0.728 & -0.367 \\ 1.097 & -0.258 & 0.135 & -0.042 & -0.572 & -0.843 \\ 0.639 & -0.417 & 0.199 & 1.333 & -1.540 & -1.008 \\ 1.053 & -0.197 & -1.522 & 0.491 & -0.320 & 1.041 \\ -0.160 & -0.900 & 0.516 & -0.100 & -0.327 & 0.704 \\ 0.390 & 1.093 & 0.822 & -0.056 & 0.388 & 0.024 \\ 0.314 & -0.568 & 0.374 & -0.128 & 0.493 & 0.196 \\ 0.005 & -0.002 & 0.000 & 0.009 & 0.012 & -0.003 \end{bmatrix}. \tag{2.99}$$

Carrying out the same analysis by replacing the w-weight matrix computed by PLS with that determined by MRPLS, only marginal elements remain with values below 10^{-4}. This can be confirmed by analyzing the estimated cross-covariance matrix between $\mathbf{s}'$ and $\mathbf{t}$, that is, the 4 t-score variables extracted by MRPLS

$$\widehat{\mathbf{S}}_{ts'} = \begin{bmatrix} -0.001 & 0.010 & 0.007 & -0.006 & -0.004 & 0.009 \\ 0.002 & 0.024 & -0.001 & -0.025 & -0.008 & 0.006 \\ -0.006 & -0.004 & -0.005 & -0.019 & -0.011 & -0.016 \\ 0.015 & -0.005 & -0.003 & 0.008 & 0.010 & -0.008 \end{bmatrix}. \tag{2.100}$$

In contrast, the estimated cross-covariance matrix between $\mathbf{t}$ and $\mathbf{s}$ is equal to

$$\widehat{\mathbf{S}}_{ts} = \begin{bmatrix} 0.989 & 0.009 & 0.008 & -0.010 \\ 0.009 & 1.008 & 0.005 & 0.014 \\ 0.008 & 0.005 & 1.004 & 0.010 \\ -0.010 & 0.014 & 0.010 & 1.027 \end{bmatrix}. \tag{2.101}$$

That $\widehat{\mathbf{S}}_{ts}$ is close to an identity matrix is a coincidence and relates to the fact that the covariance matrices of the original source signals and the extracted t-score variables are equal to the identity matrix. In general, the extracted t-score variable set is asymptotically equal to $\mathbf{s}$ up to a similarity transformation, that is, $\mathbf{t} = \mathbf{S}_{ss}^{-1/2}\mathbf{s}$.

Finally, Figure 2.9 compares the impact of the extracted LVs by PLS and MRPLS upon the deflation of the covariance and cross-covariance matrices. The presented analysis relies on the squared Frobenius norm of the deflated matrices

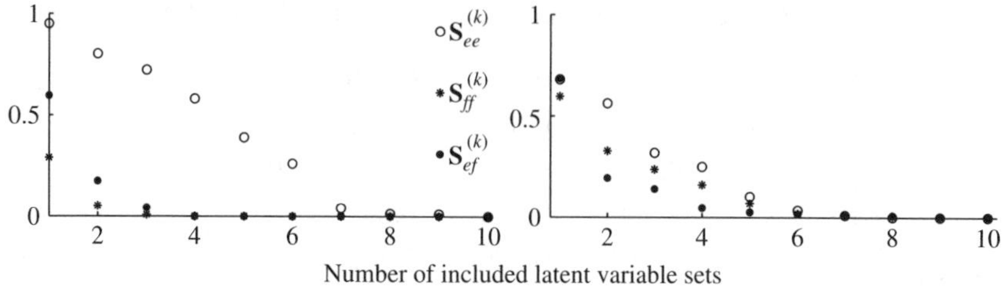

Figure 2.9 *Deflation of* $\mathbf{S}_{x_0x_0}$, $\mathbf{S}_{y_0y_0}$ *and* $\mathbf{S}_{x_0y_0}$ *using extracted latent variable sets (left plot → MRPLS model; right plot → PLS model).*

over the squared Frobenius norm of the original matrices

$$\frac{\left\|\widehat{\mathbf{S}}^{(i)}_{ee}\right\|^2}{\left\|\widehat{\mathbf{S}}_{x_0x_0}\right\|^2} = \frac{\left\|\widehat{\mathbf{S}}_{x_0x_0} - \sum_{j=1}^{i-1}\widehat{\mathbf{p}}_j E\left\{\widehat{t_j^2}\right\}\widehat{\mathbf{p}}_j^T\right\|^2}{\left\|\widehat{\mathbf{S}}_{x_0x_0}\right\|^2}, \tag{2.102}$$

$$\frac{\left\|\widehat{\mathbf{S}}^{(i)}_{ff}\right\|^2}{\left\|\widehat{\mathbf{S}}_{y_0y_0}\right\|^2} = \frac{\left\|\widehat{\mathbf{S}}_{y_0y_0} - \sum_{j=1}^{i-1}\widehat{\acute{\mathbf{q}}}_j E\left\{\widehat{t_j^2}\right\}\widehat{\acute{\mathbf{q}}}_j^T\right\|^2}{\left\|\widehat{\mathbf{S}}_{y_0y_0}\right\|^2} \tag{2.103}$$

and

$$\frac{\left\|\widehat{\mathbf{S}}^{(i)}_{ef}\right\|^2}{\left\|\widehat{\mathbf{S}}_{x_0y_0}\right\|^2} = \frac{\left\|\widehat{\mathbf{S}}_{x_0y_0} - \sum_{j=1}^{i-1}\widehat{\mathbf{p}}_j E\left\{\widehat{t_j^2}\right\}\widehat{\acute{\mathbf{q}}}_j^T\right\|^2}{\left\|\widehat{\mathbf{S}}_{x_0y_0}\right\|^2}. \tag{2.104}$$

Comparing both plots in Figure 2.9 yields that MRPLS rapidly deflates $\widehat{\mathbf{S}}_{y_0y_0}$. The retention of only one set of LVs produces a value below 0.3 for (2.101) and retaining a second LV set reduces this value to 0.05. In contrast, PLS deflates $\mathbf{S}_{x_0x_0}$ more rapidly than MRPLS. The retention of only three sets of LVs yields values of 0.33 and 0.72 for PLS and MRPLS, respectively. Increasing this number to six retained LV sets produces values of 0.04 and 0.28 for PLS and MRPLS, respectively. Comparing the impact of the deflation with regards to (2.104) yields a favorable performance of the MRPLS algorithm. For each number of retained LV sets, MRPLS yields a smaller value that is close to zero for $i = 4$. In sharp contrast, even for seven or eight sets of retained LVs, PLS does not completely deflate $\widehat{\mathbf{S}}^{(i)}_{ef}$.

On the basis of the analysis above, particularly the result presented in (2.99), however, this is expected and confirms that PLS is generally not capable of extracting common cause variation that is encapsulated in the input and output variables in an efficient manner unless the weight vectors are constrained to be orthogonal to the column space of $\mathfrak{P}'$. As this matrix is generally not known *a priori*, such a constraint cannot be incorporated into the PLS routine.

2.4 Estimating the number of source signals

This section discusses the important issue of how to estimate the number of sets of latent components describing common cause variation and, accordingly, the number of source signals. This number, n, is of fundamental importance for the following reasons. If too few latent components are retained, variation that is to be attributed to the source signals is partially encapsulated in the residuals of the PCA/PLS models. On the other hand, the retention of too many latent components produces a model subspace that may capture a significant portion of the error vector(s). In the latter case, the monitoring statistics, summarized in the next chapter, describe common cause variation that is corrupted by a stronger influence of the error vector $\mathbf{\mathfrak{g}}$ (PCA) or variation of the input variables that is not significantly predictive for the output variables, that is, $\mathbf{e}_{n+1}$ or $\mathbf{s}'$ for small data sets (PLS/MRPLS).

An estimation of n that is too small or too large will affect the sensitivity in detecting and diagnosing special cause variation that negatively influences product quality and/or the general state of the process operation. Hence, abnormal events may consequently not be detected nor correctly diagnosed. The question, therefore, is when to stop retaining more sets of LVs in PCA/PLS monitoring models. This question has been addressed in the literature by developing *stopping rules*. The stopping rules for PCA, proposed in the research literature, are summarized in Subsection 2.4.1, followed by those of PLS in Subsection 2.4.2. For the subsequent discussion of stopping rules, $\mathfrak{n}$ denotes the inclusion of $\mathfrak{n}$ sets of LVs that are currently being evaluated and as before, n is the selected number of sets.

2.4.1 Stopping rules for PCA models

The literature has introduced and discussed numerous approaches for determining how many principal components should be included, or, in relation to (2.2), how many source signals the process has. Jackson (2003) and Valle *et al.* (1999) provide surveys and comparisons of various stopping rules for PCA models. The aim of this subsection is to bring together the most important stopping rules, which can be divided into (i) information theoretic criterion, (ii) eigenvalue-based criterion and (iii) cross-validation-based criterion. An additional criterion that is difficult to attribute to one of these three categories is (iv) the Velicer's partial correlation procedure. These four different approaches are now separately discussed below.

2.4.1.1 Information-based criteria

These include the Akaike's Information Criterion (AIC) (Akaike 1974) and the Minimum Description Length (MDL) (Rissanen 1978; Schwarz 1978). Both criteria rely on the utilization of (2.2) under the assumption that $\bar{\mathbf{z}} = \mathbf{0}$ and $E\left\{\mathbf{g}\mathbf{g}^T\right\} = \sigma_g^2\mathbf{I}$. The covariance structure of the stochastic signal component is therefore

$$\mathbf{S}_{z_0z_0} = \boldsymbol{\Xi}\mathbf{S}_{ss}\boldsymbol{\Xi}^T + \sigma_g^2\mathbf{I} \tag{2.105}$$

with $\boldsymbol{\Xi}\mathbf{S}_{ss}\boldsymbol{\Xi}^T$ being of rank n and the discarded eigenvalues of $\mathbf{S}_{z_0z_0}$, λ_{n+1}, λ_{n+2}, $\ldots$, λ_{n_z} are equal to σ_g^2. The eigendecomposition of the $\mathbf{S}_{z_0z_0}$ allows a consistent estimation of $\mathbf{S}_{z_sz_s} = \boldsymbol{\Xi}\mathbf{S}_{ss}\boldsymbol{\Xi}^T$ and $\mathbf{S}_{gg} = \sigma_g^2\mathbf{I}$

$$\mathbf{S}_{z_0z_0} = \sum_{i=1}^{n_z}\lambda_i\mathbf{p}_i\mathbf{p}_i^T = \sum_{i=1}^{n}\left(\lambda_i - \sigma_g^2\right)\mathbf{p}_i\mathbf{p}_i^T + \sigma_g^2\sum_{i=1}^{n_z}\mathbf{p}_i\mathbf{p}_i^T. \tag{2.106}$$

Given that the eigenvectors are mutually orthonormal, the above equation reduces to

$$\mathbf{S}_{z_0z_0} = \underbrace{\sum_{i=1}^{n}\left(\lambda_i - \sigma_g^2\right)\mathbf{p}_i\mathbf{p}_i^T}_{=\boldsymbol{\Xi}\mathbf{S}_{ss}\boldsymbol{\Xi}^T} + \sigma_g^2\mathbf{I}. \tag{2.107}$$

The next step involves the utilization of the following parameter vector

$$\mathcal{P} = \begin{pmatrix} \lambda_1 & \lambda_2 & \cdots & \lambda_n & \sigma_g^2 & \mathbf{p}_1^T & \mathbf{p}_2^T & \cdots & \mathbf{p}_n^T \end{pmatrix}^T \tag{2.108}$$

which allows the construction of the following maximum likelihood function[4]

$$J\left(\mathbf{z}(1), \cdots, \mathbf{z}(K)\,|\mathcal{P}\right) = \prod_{k=1}^{K}\frac{1}{(2\pi)^{n_z/2}\left|\mathbf{S}_{z_0z_0}\right|^{1/2}}\exp\left(-\tfrac{1}{2}\mathbf{z}_0(k)\mathbf{S}_{z_0z_0}^{-1}\mathbf{z}_0(k)\right). \tag{2.109}$$

Wax and Kailath (1985) rewrote the above equation to be a log-likelihood function

$$J^*\left(\mathbf{z}_0(1), \cdots, \mathbf{z}_0(K)\,|\mathcal{P}\right) = K\log\left(\left|\mathbf{S}_{z_0z_0}\right|\right) + \mathrm{trace}\left(\mathbf{S}_{z_0z_0}\widehat{\mathbf{S}}_{z_0z_0}\right) \tag{2.110}$$

where $\widehat{\mathbf{S}}_{z_0z_0}$ is the estimate of $\mathbf{S}_{z_0z_0}$. The maximum likelihood estimate for $\mathcal{P}$ maximizes (2.110). Anderson (1963) showed that these estimates are

$$\begin{aligned} \widehat{\lambda}_i &= l_i \qquad i = 1, \cdots, n \\ \widehat{\sigma}_g^2 &= \frac{1}{n_z - n}\sum_{i=n+1}^{n_z} l_i \\ \widehat{\mathbf{p}}_i &= \mathbf{l}_i \qquad i = 1, \cdots, n. \end{aligned} \tag{2.111}$$

[4] Information about parameter estimation using maximum likelihood is given in Subsection 6.1.3.

Here, l_i and $\mathbf{l}_i$ are the eigenvalue and the eigenvector of $\widehat{\mathbf{S}}_{z_0z_0}$. Wax and Kailath (1985) highlighted that substituting these estimates into (2.110) yields

$$J^*\left(\mathbf{z}(1),\cdots,\mathbf{z}(K)\,|\mathcal{P}\right) = \log\left(\frac{\prod_{i=n+1}^{n_z}\widehat{\lambda}_i^{1/(n_Z-n)}}{\frac{1}{n_z-n}\sum_{i=n+1}^{n_z}\widehat{\lambda}_i}\right)^{(n_z-n)K}. \tag{2.112}$$

The AIC and MDL objective functions include the above term but rely on different terms to penalize model complexity. The objective functions for AIC and MDL are

$$\mathrm{AIC}(\mathfrak{n}) = -2\log\left(\frac{\prod_{i=\mathfrak{n}+1}^{n_z}\widehat{\lambda}_i^{1/(n_Z-\mathfrak{n})}}{\frac{1}{n_z-\mathfrak{n}}\sum_{i=\mathfrak{n}+1}^{n_z}\widehat{\lambda}_i}\right)^{(n_z-\mathfrak{n})K} + 2\mathfrak{n}\left(2n_z-\mathfrak{n}\right) \tag{2.113}$$

and

$$\mathrm{MDL}(\mathfrak{n}) = -\log\left(\frac{\prod_{i=\mathfrak{n}+1}^{n_z}\widehat{\lambda}_i^{1/(n_z-\mathfrak{n})}}{\frac{1}{n_z-\mathfrak{n}}\sum_{i=\mathfrak{n}+1}^{n_z}\widehat{\lambda}_i}\right)^{(n_z-\mathfrak{n})K} + \tfrac{1}{2}\mathfrak{n}\left(2n_z-\mathfrak{n}\right)\log\left(K\right), \tag{2.114}$$

respectively. Here, $\mathfrak{n}$ is the number of principal components $1 \leq \mathfrak{n} \leq n_z - 1$. The selected number of principal components, $\mathfrak{n} = n$, is the minimum of the AIC($\mathfrak{n}$) or MDL($\mathfrak{n}$) objective functions, depending which one is used. Wax and Kailath (1985) pointed out that the MDL objective function provides a consistent estimation of n, whilst the AIC one is inconsistent and tends, asymptotically, to overestimate n.

2.4.1.2 Eigenvalue-based criteria

Eigenvalue-based stopping rules include the cumulative percentage variance, the SCREE test, the residual percentage variance, the eigenvector-one-rule and other methods that derive from those.

Cumulative percentage variance or CPV test. This is the simplest and perhaps most intuitive eigenvalue-based test and determines the ratio of the sum of the first $\mathfrak{n}$ estimated eigenvalues over the sum of all estimated eigenvalues

$$\mathrm{CPV}(\mathfrak{n}) = \left(\frac{\sum_{i=1}^{\mathfrak{n}}\widehat{\lambda}_i}{\sum_{i=1}^{n_z}\widehat{\lambda}_i}\right)100\%. \tag{2.115}$$

The CPV criterion relies on the fact that the sum of the squared variables of $\mathbf{z}_0$ is equal to the sum of squared values of the score variables. This follows from

$$\mathbf{z}_0 = \begin{pmatrix} z_{0_1} \\ z_{0_2} \\ \vdots \\ z_{0_{n_z}} \end{pmatrix} = \begin{bmatrix} \widehat{p}_{11} & \widehat{p}_{12} & \cdots & \widehat{p}_{1n_z} \\ \widehat{p}_{21} & \widehat{p}_{22} & \cdots & \widehat{p}_{2n_z} \\ \vdots & \vdots & \ddots & \vdots \\ \widehat{p}_{n_z 1} & \widehat{p}_{n_z 1} & \cdots & \widehat{p}_{n_z n_z} \end{bmatrix} \begin{pmatrix} \widehat{t}_1 \\ \widehat{t}_2 \\ \vdots \\ \widehat{t}_{n_z} \end{pmatrix}, \tag{2.116}$$

and yields the relationship between the jth process and the n_z score variables

$$z_{0_j} = \sum_{i=1}^{n_z} \widehat{p}_{ji} \widehat{t}_i. \tag{2.117}$$

The squared value of z_{0_j} then becomes

$$z_{0_j}^2 = \sum_{i=1}^{n_z} \widehat{p}_{ji}^2 \widehat{t}_i^2 + 2 \sum_{h=1}^{n_z-1} \sum_{i=h+1}^{n_z} \left(\widehat{p}_{jh}\widehat{t}_h\right)\left(\widehat{p}_{ji}\widehat{t}_i\right), \tag{2.118}$$

producing the following sum over the complete variable set, $z_{0_1}, \cdots, z_{0_{n_z}}$,

$$\sum_{m=1}^{n_z} z_{0_m}^2 = \sum_{m=1}^{n_z} \left(\sum_{i=1}^{n_z} \widehat{p}_{mi}^2 \widehat{t}_i^2 + 2 \sum_{h=1}^{n_z-1} \sum_{i=h+1}^{n_z} \left(\widehat{p}_{mh}\widehat{t}_h\right)\left(\widehat{p}_{mi}\widehat{t}_i\right) \right). \tag{2.119}$$

As the score variables do not include the index m, rewriting the above sum yields

$$\sum_{m=1}^{n_z} z_{0_m}^2 = \sum_{i=1}^{n_z} \widehat{t}_i^2 \underbrace{\sum_{m=1}^{n_z} \widehat{p}_{mi}^2}_{\|\widehat{\mathbf{p}}_i\|^2=1} + 2 \sum_{h=1}^{n_z-1} \sum_{i=h+1}^{n_z} \widehat{t}_h \widehat{t}_i \underbrace{\sum_{m=1}^{n_z} \widehat{p}_{mh} \widehat{p}_{mi}}_{\widehat{\mathbf{p}}_h^T \widehat{\mathbf{p}}_i = 0, h \neq i}. \tag{2.120}$$

Hence, (2.118) reduces to

$$\sum_{j=1}^{n_z} z_{0_j}^2 = \sum_{j=1}^{n_z} \widehat{t}_j^2. \tag{2.121}$$

Finally, taking the expectation of (2.121) yields

$$E\left\{\sum_{j=1}^{n_z} z_{0_j}^2\right\} = \sum_{j=1}^{n_z} E\left\{z_{0_j}^2\right\} = \sum_{j=1}^{n_z} E\left\{\widehat{t}_j^2\right\} = \sum_{j=1}^{n_z} \widehat{\lambda}_j. \tag{2.122}$$

Equation (2.122) implies that the sum of the variances of the recorded process variables is equal to the sum of the eigenvalues of the data covariance matrix. Moreover, the variance of the ith score variable is equal to the ith eigenvalue of the data covariance matrix. This is analyzed and discussed in more detail in Chapter 9.

The denominator of the CPV criterion is therefore the sum of the variances of the process variables and the numerator is the variance contribution of the retained components to this sum. Hence, the larger n the closer the CPV criterion becomes to 100%. A threshold, for example 95% or 99%, can be selected and n is the number for which Equation (2.115) exceeds this threshold. Despite the simplicity of the CPV criterion, the selection of the threshold is often viewed as arbitrary and subjective, for example (Valle *et al.* 1999). Smaller threshold suggests including fewer components and a less accurate recovery of $\mathbf{z}_0$ and a larger threshold increases n. The threshold is therefore a tradeoff between parsimony and accuracy in recovering $\mathbf{z}_0$.

SCREE test. This test plots the eigenvalues of the $\widehat{\mathbf{S}}_{z_0z_0}$ against their number in descending order, which is referred to as a SCREE plot. Cattell (1966) highlighted that SCREE plots often show that the first few eigenvalues decrease sharply in value whilst most of the remaining ones align along a line that slowly decreases and further suggested to retain the first few sharply decreasing eigenvalues and the first one of the second set of slowly decreasing eigenvalues. If more than one such *elbow* emerges, Jackson (2003) pointed out that the first of these breaks determines the number of retained principal components. Conditions under which a larger number of principal components should be retained if the SCREE plot produces multiple elbows are discussed in Box *et al.* (1973); Cattell and Vogelmann (1977).

Residual percentage variance or RPV test. Similar to the CPV test, the RPV test determines n from the last few eigenvalues (Cattell 1966; Rozett and Petersen 1975)

$$\mathrm{RPV}(\mathrm{n}) = \left(\frac{\sum_{i=\mathrm{n}+1}^{n_z} \widehat{\lambda}_j}{\sum_{i=1}^{n_z} \widehat{\lambda}_i} \right) 100\%. \tag{2.123}$$

Average-eigenvalue test. Kaiser (1960) proposed an extension of the SCREE test that relies on the property that the trace of the covariance/correlation matrix is equal to the sum of the eigenvalues, which follows from the relationship in (2.117) to (2.122). Using (2.122), the average eigenvalue, $\bar{\lambda}$, can be directly calculated from the trace of the data covariance/correlation matrix

$$\bar{\lambda} = \frac{1}{n_z} \sum_{i=1}^{n_z} \widehat{\sigma}_i^2. \tag{2.124}$$

This rule suggests that eigenvalues that are larger or equal to $\bar{\lambda}$ should be associates with the source signals and those below $\bar{\lambda}$ corresponding to the error vector. If $\widehat{\mathcal{C}}_{z_0z_0}$ is used instead of $\widehat{\mathbf{S}}_{z_0z_0}$ the average eigenvalue is 1, as all of the diagonal elements are 1. With the use of $\widehat{\mathcal{C}}_{z_0z_0}$ this rule is referred to as the *eigenvalue-one-rule*.

Alternative methods. Jolliffe (1972, 1973) conducted a critical review of the average-eigenvalue rule and concluded that the threshold for selecting the number of retained components may be too high. Based on a number of simulation examples, a recommendation in these references was to discard components that correspond to eigenvalues up 70% of the average eigenvalue.

To automate the SCREE test, Horn (1965) proposed the utilization of a second data set that includes the same number of samples and variables. This second data set, however, should include statistically uncorrelated Gaussian variables, so that the covariance matrix reduces to a diagonal matrix. The eigenvalues of both covariance matrices are then plotted in a single SCREE plot where the interception determines the cutoff point for separating retained from discarded components.

The use of the correlation matrix, that is, the identify matrix, reduces this method to the *eigenvalue-one-rule*. Farmer (1971) proposed a similar approach to that in (Horn 1965) using logarithmic SCREE plots. Procedures that rely on the incorporation of a second artificially generated data set are also referred to as *parallel analysis*. Other techniques that utilize the eigenvalues include the indicator function, the embedded error function (Malinowski 1977) and the broken stick model (Jolliffe 1986).

2.4.1.3 Cross-validation-based criteria

Cross-validation relies on the residuals $\mathbf{g}_{n+1} = \left[\mathbf{I} - \sum_{i=1}^{n} \widehat{\mathbf{p}}_i \widehat{\mathbf{p}}_i^T\right] \mathbf{z}_0$ and was first proposed by Mosteller and Wallace (1963) and further discussed in Allen (1974) and Stone (1974) among others. The main principle behind cross-validation is:

1. remove some of the samples from the reference data set;
2. construct a PCA model from the remaining samples;
3. apply this PCA model to the removed samples; and
4. remove a different set of samples from the reference set and continue with Step 2 until a preselected number of disjunct sets have been removed.

Figure 2.10 illustrates the structured cross-validation approach, which segments the reference data set equally into groups. The first group is used to test the PCA model constructed from the remaining groups, then the second group is used etc.

Stone (1974) argued on theoretical grounds that the number of groups should be equal to the number of observations, which leads to an excessive computation. Geisser (1974) showed that using fewer groups is sufficient. This view is also echoed in Wold (1978). The research literature has proposed a number of performance indices, including the R and W statistics. A different cross-validation

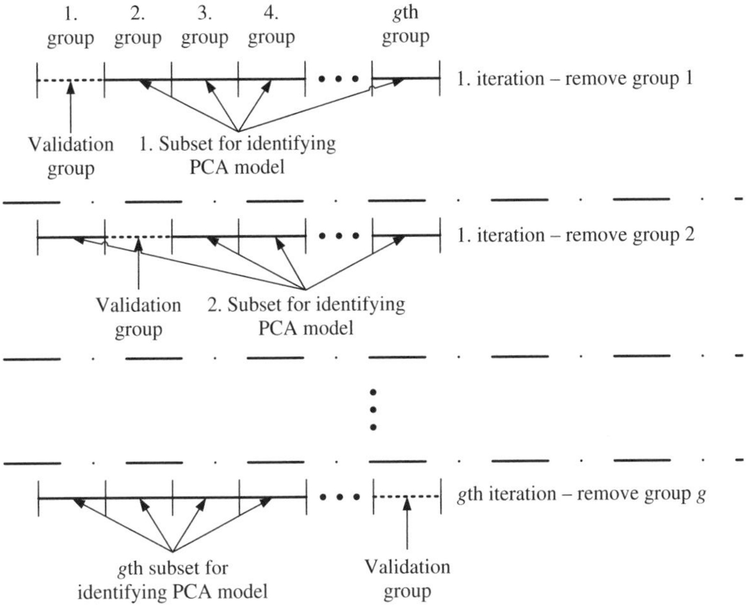

Figure 2.10 Schematic representation of a structured cross-validation approach.

approach that omits variables rather than observations was proposed in Qin and Dunia (2000).

Cross-validation based on the R statistic. For the ith group, Wold (1978) suggested using the ratio of the PRESS statistic (PREdiction Sum of Squares) over the RSS statistic (Residual Sum of Squares)

$$R_i\,(\mathrm{n}) = \frac{\mathrm{PRESS}_i\,(\mathrm{n})}{\mathrm{RSS}_i\,(\mathrm{n}-1)}, \tag{2.125}$$

where

$$\begin{aligned} \mathrm{PRESS}_i\,(\mathrm{n}) &= \frac{1}{\widetilde{K}_i n_z} \sum_{k=1}^{\widetilde{K}_i} \mathbf{g}_{\mathrm{n}+1}^T\,(k)\,\mathbf{g}_{\mathrm{n}+1}\,(k) \\ \mathrm{RSS}_i\,(\mathrm{n}-1) &= \sum_{k=1}^{\widetilde{K}_i} \mathbf{g}_{\mathrm{n}}^T\,(k)\,\mathbf{g}_{\mathrm{n}}\,(k) \end{aligned} \tag{2.126}$$

and

$$\begin{aligned} \mathbf{g}_{\mathrm{n}+1}\,(k) &= \left[\mathbf{I} - \sum_{i=1}^{\mathrm{n}} \widehat{\mathbf{p}}_i \widehat{\mathbf{p}}_i^T\right] \mathbf{z}_0\,(k) \\ \mathbf{g}_{\mathrm{n}}\,(k) &= \left[\mathbf{I} - \sum_{i=1}^{\mathrm{n}-1} \widehat{\mathbf{p}}_i \widehat{\mathbf{p}}_i^T\right] \mathbf{z}_0\,(k) \end{aligned} \tag{2.127}$$

with $\widetilde{K}_i$ being the number of samples in the ith group. The sum over the g groups is

$$R(\mathfrak{n}) = \sum_{i=1}^{g} R_i(\mathfrak{n}) = \sum_{i=1}^{g} \frac{\text{PRESS}_i(\mathfrak{n})}{\text{RSS}_i(\mathfrak{n}-1)}. \tag{2.128}$$

If $R(\mathfrak{n})$ is below one then increase $\mathfrak{n}$ to $\mathfrak{n}+1$, since the inclusion of the $\mathfrak{n}$th component increases the prediction accuracy relative to the $(\mathfrak{n}-1)$th one. In contrast, if $R(\mathfrak{n})$ exceeds one than this new component does not reduce the prediction accuracy. This stopping rule is often referred to as the R ratio or R statistic and the number of source signals is equal to the first n R ratios below one, that is, $R(1), \ldots, R(n) < 1$.

Cross-validation based on the W statistic. Eastment and Krzanowski (1982) proposed an alternative criterion, defined as the W statistic, that involves the PRESS statistics for PCA models that include $\mathfrak{n}-1$ and $\mathfrak{n}$ retained components

$$W(\mathfrak{n}) = \frac{(\text{PRESS}(\mathfrak{n}-1) - \text{PRESS}(\mathfrak{n}))\, D_r}{\text{PRESS}(\mathfrak{n})\, D_{\mathfrak{n}}} \tag{2.129}$$

where D_r and $D_{\mathfrak{n}}$ are the number of degrees of freedom that remain after determining the $\mathfrak{n}$th component and for constructing the $\mathfrak{n}$th component, respectively,

$$D_r = n_z(K-1) - \sum_{j=1}^{\mathfrak{n}} K + n_z - 2j \qquad D_{\mathfrak{n}} = K + n_z - 2\mathfrak{n}. \tag{2.130}$$

Components that have a $W(\mathfrak{n})$ value larger than 1 should be included in the PCA model. Eastment and Krzanowski (1982) suggested that the optimum number of source signals is the last one for which $W(\mathfrak{n}) > 1$. A discussion of these cross-validatory stopping rules in Wold *et al.* (1987), page 49, highlighted that they work well and the use of a proper algorithm does not render them too computationally expensive and concluded that cross-validation is slightly conservative yielding too few rather than too many components. The discussion, however, deemed this as an advantage as it circumvents an over-interpretation of the encapsulated information.

Variance of the reconstruction error (VRE). A different approach to those by Wold (1978) and Eastment and Krzanowski (1982) is discussed in Qin and Dunia (2000). Instead of leaving portions of the reference data out, this technique omits the samples of one variable and reconstructs it by the remaining $n_z - 1$ ones. Evaluating the accuracy of this reconstruction by PCA models for different numbers of source signals, each variable is removed and reconstructed by the remaining ones. This produces a total of n_z contributions to the VRE performance index.

Using the eigendecomposition $\mathbf{S}_{z_0 z_0} = \mathbf{P}\boldsymbol{\Lambda}\mathbf{P}^T$ and defining $\mathfrak{C}^{(\mathrm{n})} = \sum_{j=1}^{\mathrm{n}} \mathbf{p}_j \mathbf{p}_j^T$, the projection of $\mathbf{z}_0$ onto the model subspace $\widehat{\mathbf{z}}_0 = \mathfrak{C}^{(\mathrm{n})}\mathbf{z}_0$ is for the ith element

$$\widehat{z}_{0_i} = \sum_{j=1}^{n_z} \mathfrak{c}_{ij}^{(\mathrm{n})} z_{0_j} \tag{2.131}$$

where $\mathfrak{c}_{ij}^{(\mathrm{n})}$ is the element of $\mathfrak{C}^{(\mathrm{n})}$ stored in the ith row and the jth column. Replacing the variable z_{0_i} by $\widehat{z}_{0_i}$ the above equation becomes

$$\left(1 - \mathfrak{c}_{ii}^{(\mathrm{n})}\right)\widehat{z}_{0_i} \approx \sum_{j=1 \neq i}^{n_z} \mathfrak{c}_{ij}^{(\mathrm{n})} z_{0_j}. \tag{2.132}$$

The reconstruction of z_{0_j}, $\widetilde{z}_{0_{i_\mathrm{n}}}$, is therefore

$$\widetilde{z}_{0_i} = \sum_{j=1 \neq i}^{n_z} \frac{\mathfrak{c}_{ij}^{(\mathrm{n})}}{1 - \mathfrak{c}_{ii}^{(\mathrm{n})}} z_{0_j}. \tag{2.133}$$

A more detailed treatment of variable reconstruction is given in Section 3.2. Equation (2.133) outlines that the number of retained components can vary from 1 to $n_z - 1$. For $\mathrm{n} = n_z$, $\mathfrak{C} = \mathbf{I}$ and the denominator becomes zero. The use of (2.133) gives rise to the following reconstruction error

$$z_{0_i} - \widetilde{z}_{0_{i_\mathrm{n}}} = \frac{1}{1 - \mathfrak{c}_{ii}^{(\mathrm{n})}} \left(-\sum_{j=1}^{i-1} \mathfrak{c}_{ij}^{(\mathrm{n})} z_{0_j} + \left(1 - \mathfrak{c}_{ii}^{(\mathrm{n})}\right) z_{0_i} - \sum_{j=i+1}^{n_z} \mathfrak{c}_{ij}^{(\mathrm{n})} z_{0_j} \right). \tag{2.134}$$

Next, abbreviating $z_{0_i} - \widetilde{z}_{0_i} = \widetilde{g}_i$ and rewriting (2.134) yields

$$\widetilde{g}_i^{(\mathrm{n})} = -\frac{\begin{pmatrix} \mathfrak{c}_{i1}^{(\mathrm{n})} & \cdots & \mathfrak{c}_{i,i-1}^{(\mathrm{n})} & \mathfrak{c}_{ii}^{(\mathrm{n})} - 1 & \mathfrak{c}_{i,i+1}^{(\mathrm{n})} & \mathfrak{c}_{i,n_z}^{(\mathrm{n})} \end{pmatrix}}{1 - \mathfrak{c}_{ii}^{(\mathrm{n})}} \begin{pmatrix} z_{0_1} \\ \vdots \\ z_{0_{i-1}} \\ z_{0_i} \\ z_{0_{i+1}} \\ \vdots \\ z_{0_{n_z}} \end{pmatrix}. \tag{2.135}$$

Noting that $-\begin{pmatrix} \mathfrak{c}_{i1}^{(\mathrm{n})} & \cdots & \mathfrak{c}_{i,i-1}^{(\mathrm{n})} & \mathfrak{c}_{ii}^{(\mathrm{n})} - 1 & \mathfrak{c}_{i,i+1}^{(\mathrm{n})} & \mathfrak{c}_{i,n_z}^{(\mathrm{n})} \end{pmatrix}^T$ is equal to the ith row or column of the symmetric matrix $\mathbf{I} - \mathfrak{C}^{(\mathrm{n})}$, this vector is also equal to

$$-\begin{pmatrix} \mathfrak{c}_{i1}^{(\mathrm{n})} & \cdots & \mathfrak{c}_{i,i-1}^{(\mathrm{n})} & \mathfrak{c}_{ii}^{(\mathrm{n})} - 1 & \mathfrak{c}_{i,i+1}^{(\mathrm{n})} & \mathfrak{c}_{i,n_z}^{(\mathrm{n})} \end{pmatrix} = \boldsymbol{\upsilon}_i^T \left[\mathbf{I} - \mathfrak{C}^{(\mathrm{n})}\right] \tag{2.136}$$

where $\boldsymbol{\upsilon}_i$ is the ith Euclidean vector whose ith element is 1, whilst any other element is 0. Equation (2.135) can therefore be expressed as follows

$$\widetilde{g}_i^{(\mathfrak{n})} = \frac{\boldsymbol{\upsilon}_i^T \left[\mathbf{I} - \mathfrak{C}^{(\mathfrak{n})}\right] \mathbf{z}_0}{1 - \mathfrak{c}_{ii}^{(\mathfrak{n})}}. \tag{2.137}$$

Expressing $1 - \mathfrak{c}_{ii}^{(\mathfrak{n})}$ as a function of $\mathbf{I} - \mathfrak{C}^{(\mathfrak{n})}$ and $\boldsymbol{\upsilon}_i$

$$\begin{aligned} 1 - \mathfrak{c}_{ii}^{(\mathfrak{n})} &= \underbrace{-\begin{pmatrix} \mathfrak{c}_{i1}^{(\mathfrak{n})} & \cdots & \mathfrak{c}_{i,i-1}^{(\mathfrak{n})} & \mathfrak{c}_{ii}^{(\mathfrak{n})} - 1 & \mathfrak{c}_{i,i+1}^{(\mathfrak{n})} & \cdots & \mathfrak{c}_{i,n_z}^{(\mathfrak{n})} \end{pmatrix}}_{=\boldsymbol{\upsilon}_i^T\left[\mathbf{I}-\mathfrak{C}^{(\mathfrak{n})}\right]} \boldsymbol{\upsilon}_i \\ 1 - \mathfrak{c}_{ii}^{(\mathfrak{n})} &= \boldsymbol{\upsilon}_i^T \left[\mathbf{I} - \mathfrak{C}^{(\mathfrak{n})}\right] \boldsymbol{\upsilon}_i, \end{aligned} \tag{2.138}$$

the variance of the reconstruction error for the ith variable becomes

$$E\left\{\left(\widetilde{g}_i^{(\mathfrak{n})}\right)^2\right\} = \frac{\boldsymbol{\upsilon}_i^T \left[\mathbf{I} - \mathfrak{C}^{(\mathfrak{n})}\right] E\left\{\mathbf{z}_0 \mathbf{z}_0^T\right\} \left[\mathbf{I} - \mathfrak{C}^{(\mathfrak{n})}\right] \boldsymbol{\upsilon}_i}{\boldsymbol{\upsilon}_i^T \left[\mathbf{I} - \mathfrak{C}^{(\mathfrak{n})}\right] \boldsymbol{\upsilon}_i}. \tag{2.139}$$

Since $E\left\{\mathbf{z}_0 \mathbf{z}_0^T\right\} = \mathbf{S}_{z_0 z_0}$, the above equation reduces to

$$E\left\{\left(\widetilde{g}_i^{(\mathfrak{n})}\right)^2\right\} = \frac{\boldsymbol{\upsilon}_i^T \left[\mathbf{I} - \mathfrak{C}^{(\mathfrak{n})}\right] \mathbf{S}_{z_0 z_0} \left[\mathbf{I} - \mathfrak{C}^{(\mathfrak{n})}\right] \boldsymbol{\upsilon}_i}{\boldsymbol{\upsilon}_i^T \left[\mathbf{I} - \mathfrak{C}^{(\mathfrak{n})}\right] \boldsymbol{\upsilon}_i}. \tag{2.140}$$

Finally, defining $\mathfrak{S}_{z_0 z_0} = \left[\mathbf{I} - \mathfrak{C}^{(\mathfrak{n})}\right] \mathbf{S}_{z_0 z_0} \left[\mathbf{I} - \mathfrak{C}^{(\mathfrak{n})}\right]$, the VRE criteria is given by

$$\mathrm{VRE}\left(\mathfrak{n}\right) = \sum_{i=1}^{n_z} \frac{\boldsymbol{\upsilon}_i^T \mathfrak{S}_{z_0 z_0} \boldsymbol{\upsilon}_i}{\left(\boldsymbol{\upsilon}_i^T \left[\mathbf{I} - \mathfrak{C}^{(\mathfrak{n})}\right] \boldsymbol{\upsilon}_i\right) \left(\boldsymbol{\upsilon}_i^T \mathbf{S}_{z_0 z_0} \boldsymbol{\upsilon}_i\right)} \tag{2.141}$$

where $\boldsymbol{\upsilon}_i^T \mathbf{S}_{z_0 z_0} \boldsymbol{\upsilon}_i$ is the variance of the ith process variable. Valle *et al.* (1999) pointed out that the scaling of the reconstruction error is necessary, as variables that have a larger variance produce, by default, larger reconstruction errors and may have a dominant influence upon the calculation of the VRE performance index. The value of $\mathfrak{n}$ that yields a minimum for the VRE performance index is then n.

A detailed discussion in Qin and Dunia (2000) showed that (2.141) yields a minimum, which is related to the fact that, according to (2.2), the data space is separated into a model subspace and a complementary residual subspace. Moreover, Valle *et al.* (1999) (i) proved that the VRE approach gives a consistent estimation of the number of source signals under the assumptions that the error vector $\mathbf{g}(i)$ contains Gaussian i.i.d. sequences and (ii) postulated that it also gives a consistent estimation of the number of source signals if the error vector contains Gaussian sequences that have slightly different variances.

Valle *et al.* (1999) argued that the VRE method is to be preferred over cross-validatory methods for consistently estimating the number of source signals and in terms of computational costs. By directly comparing various stopping rules, including VRE, AIC, MDL, CPV, RPV and cross-validation based on the R-statistics, Valle *et al.* (1999) showed that the VRE method performed favorably. Despite the conceptual ease and computational efficiency of the VRE stopping rule, however, Subsection 6.1.1 shows that the above postulate, differences in the error variances, may not yield a consistent estimate for n, which is also demonstrated in Feital *et al.* (2010).

2.4.1.4 Velicer's Partial Correlation Criterion (VPC)

Velicer (1976) proposed this technique, which carries out a scaled deflation of the covariance matrix each time $\mathfrak{n}$ is increased by 1. Starting with the deflation of the covariance matrix

$$\mathbf{M}^{(\mathfrak{n}+1)} = \mathbf{S}_{z_0 z_0} - \sum_{i=1}^{\mathfrak{n}} \lambda_i \mathbf{p}_i \mathbf{p}_i^T, \tag{2.142}$$

the scaling process for $\mathbf{M}^{(\mathfrak{n}+1)}$ involves the diagonal matrix $\mathfrak{D}$

$$\widetilde{\mathbf{M}}^{(\mathfrak{n}+1)} = \mathfrak{D}^{-1/2} \mathbf{M}^{(\mathfrak{n}+1)} \mathfrak{D}^{-1/2} \tag{2.143}$$

where

$$\mathfrak{D} = \begin{bmatrix} m_{11}^{(\mathfrak{n}+1)} & 0 & \cdots & 0 \\ 0 & m_{22}^{(\mathfrak{n}+1)} & \cdots & 0 \\ \vdots & \vdots & \ddots & \vdots \\ 0 & 0 & \cdots & m_{n_z n_z}^{(\mathfrak{n}+1)} \end{bmatrix} \tag{2.144}$$

and $m_{11}^{(\mathfrak{n}+1)}, m_{22}^{(\mathfrak{n}+1)}, \cdots, m_{n_z n_z}^{(\mathfrak{n}+1)}$ are the diagonal elements of $\mathbf{M}^{(\mathfrak{n}+1)}$. The VPC criterion relies on the sum of the non-diagonal elements of $\widetilde{\mathbf{M}}^{(\mathfrak{n}+1)}$

$$\mathrm{VPC}(\mathfrak{n}) = \frac{2}{n_z\left(n_z - 1\right)} \sum_{i=1}^{n_z} \sum_{j=1 \neq i}^{n_z} \left(\widetilde{m}_{ij}^{(\mathfrak{n}+1)}\right)^2. \tag{2.145}$$

In fact, $\mathbf{M}^{(\mathfrak{n}+1)}$ is a correlation matrix. Hence, $\widetilde{m}_{ij}^{(\mathfrak{n}+1)}$ are correlation coefficients for $i \neq j$ and the VPC performance index is larger than zero within the range of $\mathfrak{n} = 1$ and $\mathfrak{n} = n_z - 1$. The estimation of n is given by the minimum of VPC($\mathfrak{n}$). The underlying assumption for the VPC method is that the VPC curve decreases in value for an additional source variable if the average remaining covariance reduces faster than the remaining variance of the variable set. In contrast, an increase indicates that an additional source signal explains more variance than covariance.

2.4.2 Stopping rules for PLS models

The literature has proposed a number of different stopping rules, which include:

- analysis of variance approaches;
- cross validation criteria based on the accuracy of the PLS model in predicting the output variables;
- bootstrapping criteria; and
- the H-principle.

These different methods are now discussed separately.

2.4.2.1 Analysis of variance criteria

An analysis of variance can be carried out for the u-score or output variables (Jackson 2003). More practical and intuitive, however, is the use of the residuals of $\mathbf{y}_0$. In a similar fashion to the SCREE test for PCA, the Frobenius norm of the residual matrix

$$\mathbf{F}^{(\mathrm{n}+1)} = \mathbf{Y}_0 - \sum_{i=1}^{\mathrm{n}} \widehat{\mathbf{t}}_i \widehat{\mathbf{q}}_i^T \tag{2.146}$$

can be plotted vs. n. The norm $\|\mathbf{F}^{(\mathrm{n}+1)}\|^2$ often shows an initial sharp decrease when retaining the first few sets of LVs and then slowly decays as additional sets are included. Like a SCREE plot, n can be estimated from the intercept between a tangent that represents the first (few) sharp decrease(s) and a parallel line to the abscissa of value, $\|\mathbf{F}^{(n_z+1)}\|^2$. Section 2.2 presents an example of using $\|\mathbf{F}^{(\mathrm{n}+1)}\|^2$ to determine the number of source signals (Table 2.2). This example, however, divides $\|\mathbf{F}^{(\mathrm{n}+1)}\|^2$ by $\|\mathbf{Y}_0\|^2$ and uses a percentage figure.

2.4.2.2 Cross-validation criterion

Lindberg *et al.* (1983) proposed a cross-validatory stopping rule that relies on the principle outlined in Figure 2.10. Segmenting the reference data into g groups, each group is once omitted for the identification of a PLS model. The prediction of the output variables is then assessed on the basis of the group that is left out. The performance index for the gth group is the PRESS statistic

$$\mathrm{PRESS}_g(\mathrm{n}) = \frac{1}{n_x \widetilde{K}_g} \sum_{k=1}^{\widetilde{K}_g} \mathbf{f}_{\mathrm{n}+1}^T(k)\mathbf{f}_{\mathrm{n}+1}(k) \tag{2.147}$$

for which

$$\mathbf{f}_{\mathrm{n}+1}(k) = \mathbf{y}_0(k) - \sum_{i=1}^{\mathrm{n}} \widehat{t}_i(k)\widehat{\mathbf{q}}_i. \tag{2.148}$$

Including each of the G groups, the PRESS statistic finally becomes

$$\mathrm{PRESS}(\mathrm{n}) = \sum_{g=1}^{G} \mathrm{PRESS}_g(\mathrm{n}). \tag{2.149}$$

If minimum of the resulting curve for PRESS(n) exists, then this is the selected number of source signals. If this curve. however, decreases monotonically without showing a clear minimum n can be selected by the intercepting the tangent that describes the first (few) steep decrease(s) and the parallel line to the abscissa. Published work on cross-validation include Qin (1998) and Rännar *et al.* (1995). The latter work, however, discusses cases where there are considerably more variables than samples.

2.4.2.3 Bootstrapping criterion

As the analysis of variance relies on the user to select n and cross-validation may be computationally expensive, bootstrapping can be proposed as an alternative. Bootstrapping, in this context, relates to recent work on entropy-based independence tests (Dionisio and Mendes 2006; Wu *et al.* 2009), which tests the hypothesis that two variables are independent. Scaling the nth pair of score variables to unit variance, that is, $\widetilde{t}_\mathrm{n} = t_\mathrm{n}/\sigma_{t_\mathrm{n}}$ and $\widetilde{u}_\mathrm{n} = u_\mathrm{n}/\sigma_{u_\mathrm{n}}$, the corresponding hypothesis is as follows

$$\begin{array}{ll} H_0 : \widetilde{t}_\mathrm{n} \text{ and } \widetilde{u}_\mathrm{n} \text{ are independent, such that } f\left(\widetilde{t}_\mathrm{n}, \widetilde{u}_\mathrm{n}\right) = f\left(\widetilde{t}_\mathrm{n}\right) f\left(\widetilde{u}_\mathrm{n}\right) \\ H_1 : \widetilde{t}_\mathrm{n} \text{ and } \widetilde{u}_\mathrm{n} \text{ are dependent, such that } f\left(\widetilde{t}_\mathrm{n}, \widetilde{u}_\mathrm{n}\right) \neq f\left(\widetilde{t}_\mathrm{n}\right) f\left(\widetilde{u}_\mathrm{n}\right). \end{array} \tag{2.150}$$

The above hypothesis can alternatively be expressed as

$$\begin{aligned} H_0 &: \mathbf{w}_\mathrm{n}^T \mathbf{S}_{ef}^{(\mathrm{n})} \mathbf{q}_\mathrm{n} = 0 \\ H_1 &: \mathbf{w}_\mathrm{n}^T \mathbf{S}_{ef}^{(\mathrm{n})} \mathbf{q}_\mathrm{n} > 0. \end{aligned} \tag{2.151}$$

Here, $\mathbf{S}_{ef}^{(\mathrm{n}+1)} = \mathbf{S}_{ef}^{(\mathrm{n})} - \mathbf{p}_\mathrm{n}\sigma_{t_\mathrm{n}}^2\mathbf{q}_\mathrm{n}^T$, $\mathbf{S}_{ef}^{(1)} = \mathbf{S}_{x_0y_0}$, $\mathbf{p}_\mathrm{n}$ and $\mathbf{q}_\mathrm{n}$ are the nth p-weight and q-loading vectors, respectively, and $\sigma_{t_\mathrm{n}}^2$ is the covariance of t_n and u_n. Chapter 10 provides a description of PLS including the relationship used in (2.151).

To test the null hypothesis, Granger *et al.* (2004) proposed the following procedure. Defining two independent Gaussian distributed white noise sequences of zero mean and variance 1, θ and ϕ, the critical value for testing the null hypothesis can be obtained from a smoothed bootstrap procedure (Efron and Tibshirani 1993) for the upper α-percentile using $\widetilde{K} \gg K$ samples of θ and ϕ, with K being the number of elements of $\widetilde{t}_\mathrm{n}$ and $\widetilde{u}_\mathrm{n}$.

1. Select randomly b bootstrapped sets containing K samples of θ and ϕ with replacement, defined here as

$$\mathbf{D}_i = \begin{bmatrix} \theta_i(1) & \theta_i(2) & \cdots & \theta_i(K) \\ \phi_i(1) & \phi_i(2) & \cdots & \phi_i(K) \end{bmatrix}^T \in \mathbb{R}^{K\times 2}, 1 \leq i \leq b.$$

2. Compute the product for each pair of bootstrapped samples $(\theta_i(k)\phi_i(k))$, that is, the product of the row elements in $\mathbf{D}_i$, and store the products in vectors $\mathbf{d}_i^T = \begin{pmatrix} \theta_i(1)\phi_i(1) & \theta_i(2)\phi_i(2) & \cdots & \theta_i(K)\phi_i(K) \end{pmatrix}$.
3. Work out the absolute estimates of the expectation of each set stored in the b vectors $\mathbf{d}_i$, i.e. $\bar{d}_i = {}^1/_K \sum_{k=1}^{K} \left|\theta_i(k)\,\phi_i(k)\right|$ and arrange these absolute estimates in the vector $\boldsymbol{\eth}^T = \begin{pmatrix} \bar{d}_1 & \bar{d}_2 & \cdots & \bar{d}_b \end{pmatrix}$ in ascending order.
4. Determine the critical value, $\eth_\alpha$, as the upper α percentile of $\boldsymbol{\eth}$.

The null hypothesis, H_0, is accepted if

$$\widetilde{t}_{\mathrm{n}}\widetilde{u}_{\mathrm{n}} = \frac{\mathbf{w}_{\mathrm{n}}^T \mathbf{S}_{ef}^{(\mathrm{n})} \mathbf{q}_{\mathrm{n}}}{\sqrt{\left(\mathbf{w}_{\mathrm{n}}^T \mathbf{S}_{ee}^{(\mathrm{n})} \mathbf{w}_{\mathrm{n}}\right)\left(\mathbf{q}_{\mathrm{n}}^T \mathbf{S}_{ff}^{(\mathrm{n})} \mathbf{q}_{\mathrm{n}}\right)}} \le \eth_\alpha, \tag{2.152}$$

and rejected if $\widetilde{t}_{\mathrm{n}}\widetilde{u}_{\mathrm{n}} > \eth_\alpha$. In (2.152), $\mathbf{S}_{ee}^{(\mathrm{n}+1)} = \mathbf{S}_{ee}^{(\mathrm{n})} - \mathbf{p}_{\mathrm{n}}\sigma_{t_{\mathrm{n}}}^2 \mathbf{p}_{\mathrm{n}}^T$, $\mathbf{S}_{ff}^{(\mathrm{n}+1)} = \mathbf{S}_{ff}^{(\mathrm{n})} - \acute{\mathbf{q}}_{\mathrm{n}}\sigma_{t_{\mathrm{n}}}^2 \acute{\mathbf{q}}_{\mathrm{n}}^T$, where $\sigma_{t_{\mathrm{n}}}^2$ is the variance of the nth t-score variable. Moreover, $\mathbf{S}_{ee}^{(1)} = \mathbf{S}_{v_0 v_0}$ and $\mathbf{S}_{ff}^{(1)} = \mathbf{S}_{y_0 y_0}$. The computation of $\mathbf{S}_{ee}^{(\mathrm{n})}$ and $\mathbf{S}_{ff}^{(\mathrm{n})}$ follows from the deflation procedure that is discussed and analyzed in Chapter 10.

It should be noted that rejection of H_0 results in accepting H_1, which implies that the nth pair of latent variables need to be included in the PLS model and requires the calculation of the $(\mathrm{n}+1)$th pair of latent variables after carrying out the deflation procedure. On the other hand, accepting the null hypothesis sets $n = \mathrm{n}$.

Table 2.4 lists estimated confidence limits for significance levels of:

- $\alpha = 5\%, 1\%, 0.5\%, 0.1\%, 0.05\%$ and 0.01%

and different sample sizes:

- $K = 100, 200, 500, 1000, 2000, 5000$ and 10000.

The entries in Table 2.4 are averaged values over a total of 10 runs for each combination of sample size and significance level. For a reference set containing 5000

Table 2.4 Confidence limits for various sample sizes, K and significance levels α.

K/α	5%	1%	0.5%	0.1%	0.05%
100	0.1781	0.2515	*0.2613*	*0.2653*	*0.2531*
200	0.1341	0.1691	0.1855	*0.1797*	*0.1890*
500	0.0800	0.1181	0.1403	0.1546	*0.1536*
1000	0.0596	0.0841	0.0883	0.0984	*0.0951*
2000	0.0401	0.0513	0.0593	0.0686	0.0707
5000	0.0229	0.0385	0.0433	0.0452	0.0512
10000	0.0222	0.0260	0.0309	0.0318	0.0343

samples, for example, the confidence limit for the smoothed bootstrap approach was selected to be $\alpha = 0.5\%$, i.e. 0.0433.

2.4.2.4 H-Principle

Finally, Höskuldsson (1994) showed an alternative approach for deriving the PCA and PLS objective functions, which is inspired by the Heisenberg uncertainty inequality and referred to as the H-principle. This objective function is a product of a goodness of fit and a precision criterion. More precisely, Höskuldsson (1994, 1995, 2008) showed that the PCA and PLS objective functions can be derived from the H-principle including an estimation of n. A more detailed discussion regarding the H-principle, however, is beyond the scope of this book.

2.5 Tutorial Session

Question 1: Compare the maximum number of sets of LVs that can be obtained by applying the PLS, the maximum redundancy and the CCA (Chapter 10) objective function if the covariance and cross-covariance matrices are known and of arbitrary dimension. Why can PLS exhaust the input variable set irrespective of the number of input and output variables, whilst maximum redundancy and CCA cannot?

Question 2: Following from Question 1, why does MSPC rely on the use of variance and covariance-based methods, i.e. PCA and PLS, for providing a data model for the recorded variable set(s)?

Question 3: Assuming that $\mathbf{z}_0 = \mathbf{x}_0$, why can PCA and PLS be seen as unsupervised and supervised learning algorithms, respectively?

Question 4: Why is it beneficial to rely on statistically independent score variables, which PCA and PLS extract from the data and input variable set, respectively, instead of the original variable sets?

Question 5: Explain the difference between the PLS and MRPLS objective function. What do the extracted score variables explain in both cases?

Project 1: With respect to the simulation example in Subsection 2.1.3, use a Monte-Carlo simulation and vary the number of reference samples, and analyze this impact on the accuracy of estimating the data covariance matrix and its eigendecomposition.

Project 2: Carry out a Monte-Carlo simulation to estimate the elements of the regression matrix (simulation example in Subsection 2.2.3) by varying the number of reference samples and the number of retained sets of LVs and comment upon your findings. Contrast your results with the simulation example in Subsection 6.2.2.

Project 3: Develop a deflation-based method for CCA to extract the common cause variation encapsulated in the input and output variables with respect to the data structure in (2.51). Use the simulation example in Subsection 2.3.4 and compare the performance of the developed CCA method with that of MRPLS with particular focus on the predictability of the output variables.

Project 4: Generate a data model with respect to (2.2) that includes a total of $n_z = 20$ process variables, a varying number of source signals $1 \leq n < 19$, a varying error variance σ_g^2, a varying number of reference samples K and apply each of the stopping rules in Section 2.4.1 to estimate n. Comment and explain the results. Which method is most successful in correctly estimating n?

3

Process monitoring charts

The aim of this chapter is:

- to design monitoring charts on the basis of the extracted LV sets and the residuals;
- to show how to utilize these charts for evaluating the performance of the process and for assessing product quality on-line; and
- to outline how to diagnose behavior that is identified as abnormal by these monitoring charts.

For monitoring a complex process on-line, the set of score and residual variables give rise to the construction of a *statistical fingerprint* of the process. This fingerprint serves as a benchmark for assessing whether the process is in-statistical control or out-of-statistical-control. Based on Chapters 1 and 2, the construction of this fingerprint relies on the following assumptions for identifying PCA/PLS data models:

- the error vectors associated with the PCA/PLS data models follow a zero mean Gaussian distribution that is described by full rank covariance matrices;
- the score variables, describing common cause variation of the process, follow a zero mean Gaussian distribution that is described by a full rank covariance matrix;
- for any recorded process variable, the variance contribution of the source signals (common cause variation) is significantly larger than the variance contribution of the corresponding error signal;

Statistical Monitoring of Complex Multivariate Processes: With Applications in Industrial Process Control, First Edition. Uwe Kruger and Lei Xie.

- the number of source variables is smaller than the number of recorded process (PCA) or input variables (PLS);
- recorded variable sets have constant mean and covariance matrices over time;
- the process is a representation of the data models in either (2.2), (2.24) or (2.51);
- none of the process variables possess any autocorrelation; and
- the cross-correlation function of any pair of process variables is zero for two different instances of time, as described in Subsection 2.1.1.

Part III of this book presents extensions of conventional MSPC which allow relaxing the above assumptions, particularly the assumption of Gaussian distributed source signals and time-invariant (steady state) process behavior.

The statistical fingerprint includes scatter diagrams, which Section 1.2 briefly touched upon, and non-negative squared statistics involving the t-score variables and the residuals of the PCA and PLS models. For the construction of monitoring models, this chapter assumes the availability of the data covariance and cross-covariance matrices. In this regard, the weight, loading and score variables do not need to be estimated from a reference data set. This simplifies the presentation of the equations derived, as the hat notation is not required.

Section 3.1 introduces the tools for constructing the statistical fingerprint for on-line process monitoring and detecting abnormal process conditions that are indicative of a fault condition. Fault conditions could range from simple sensor or actuator faults to complex process faults. Section 3.2 then summarizes tools for diagnosing abnormal conditions to assist experienced plant personnel in narrowing down potential root causes. Such causes could include, among many other possible scenarios, open bypass lines, a deteriorating performance of a heat exchanger, a tray or a pump, a pressure drop in a feed stream, a change in the input composition of input feeds, abnormal variation in the temperature of input or feed streams, deterioration of a catalyst, partial of complete blockage of pipes and valve stiction.

The diagnosis offered in Section 3.2 identifies to what extent a recorded variable is affected by an abnormal event. This section also shows how to extract time-based signatures for process variables if the effect of a fault condition deteriorates the performance of the process over time. Section 3.3 finally presents (i) a geometric analysis of the PCA and PLS projections to demonstrate that fault diagnosis based on the projection of a single sample along predefined directions may lead to erroneous and incorrect diagnosis results in the presence of complex fault conditions and (ii) discusses how to overcome this issue. A tutorial session concerning the material covered in this chapter is given in Section 3.4.

3.1 Fault detection

Following from the discussion in Chapter 2, PCA and PLS extract latent information in form of latent score variables and residuals from the recorded variables. According to the data models for PCA in (2.2) and (2.6) and PLS in (2.23), (2.24) and (2.51), the t-score variables describe common cause variation that is introduced by the source vector **s**. Given that the number of t-score variables is typically significantly smaller than the number of recorded variables, MSPC allows process monitoring on the basis of a reduced set of score variables rather than relying on charting a larger number of recorded process variables.

With respect to the assumptions made for the data structures for PCA and PLS in Chapter 2, the variation described by the t-score variables recovers the variation of the source variables. Hence, the variation encapsulated in the t-score variables recovers significant information from recorded variables, whilst the elements in the error vector have an insignificant variance contribution to the process variable set. Another fundamental advantage of the t-score variables is that they are statistically independent, which follows from the analysis of PCA and PLS in Chapters 9 and 10, respectively.

The t-score variables can be plotted in scatter diagrams for which the confidence regions are the control ellipses discussed in Subsection 1.2.3. For a time-based analysis, MSPC relies on nonnegative quadratics that includes Hotelling's T^2 statistics and residual-based *squared prediction error* statistics, referred to here as Q statistics. Scatter diagrams are not time-based but allow the monitoring of pairs or triples of combinations of t-score variables. In contrast, the time-based Hotelling's T^2 statistics present an overall measure of the variation within the process.

The next two subsections provide a detailed discussion of scatter diagrams and the Hotelling's T^2 statistic. It is important to note that MRPLS may generate two Hotelling's T^2 statistics, one for the common cause variation in the predictor and response variable sets and one for variation that is only manifested in the input variables and is not predictive for the output variables. This is discussed in more detail in the next paragraph and Subsection 3.1.2.

The mismatch between the recorded variables and what the t-score variables, or source variables, can recover from the original variables are model residuals. Depending on the variance of the discarded t-score variables (PLS) or the variance of the t′-scores (MRPLS), these score variables may be used to construct a Hotelling's T^2 or a residual Q statistic. Whilst Hotelling's T^2 statistics present a measure that relates to the source signals, or significant variation to recover the input variables, the residual Q statistic is a measure that relates to the model residuals.

Loosely speaking, a Q statistic is a measure of how well the reduced dimensional data representation in (2.2), (2.24) or (2.51) describe the recorded data. Figure 1.7 presents an illustration of perfect correlation, where the sample projections fall onto the line describing the relationship between both variables. In this extreme case, the residual vector is of course zero, as the values of both variables can be recovered without an error from the projection of the associated sample.

If, however, the projection of a sequence of samples do not fall onto this line an error for the recovery of the original variables has occurred, which is indicative of abnormal process behavior. For a high degree of correlation, Figure 1.9 shows that the recovered values of each sample using the projection of the samples onto the semimajor of the control ellipse are close to the recorded values. The perception of 'close' can be statistically described by the residual variables, its variances and the control limit of the residual based monitoring statistic.

3.1.1 Scatter diagrams

Figures 1.6, 1.7 and 1.9 show that the shape of the scatter diagrams relate to the correlation between two variables. Different from the 2D scatter diagrams, extensions to 3D scatter diagrams are possible, although it is difficult to graphically display a 3D control sphere. For PCA and PLS, the t-score variables, $\mathbf{t}_{PCA} = \mathbf{P}^T\mathbf{z}_0$ for PCA and $\mathbf{t}_{PLS} = \mathbf{R}^T\mathbf{x}_0$ for PLS are uncorrelated and have the following covariance matrices

$$E\left\{\mathbf{t}_{PCA}\mathbf{t}_{PCA}^T\right\} = \begin{bmatrix} \lambda_1 & 0 & \cdots & 0 \\ 0 & \lambda_2 & \cdots & 0 \\ \vdots & \vdots & \ddots & \vdots \\ 0 & 0 & \cdots & \lambda_n \end{bmatrix} = \mathbf{\Lambda} \tag{3.1}$$

and

$$E\left\{\mathbf{t}_{PLS}\mathbf{t}_{PLS}^T\right\} = \begin{bmatrix} \sigma_{t_1}^2 & 0 & \cdots & 0 \\ 0 & \sigma_{t_2}^2 & \cdots & 0 \\ \vdots & \vdots & \ddots & \vdots \\ 0 & 0 & \cdots & \sigma_{t_n}^2 \end{bmatrix} = \mathbf{S}_{tt}. \tag{3.2}$$

The construction of the weight matrix $\mathbf{R}$ is shown in Chapter 10. Under the assumption that the exact data covariance matrix is known *a priori*, the control ellipse for $i \neq j$ has the following mathematical description

$$\frac{t_i^2}{a_i^2} + \frac{t_j^2}{a_j^2} = T_\alpha^2 \tag{3.3}$$

where $T_\alpha^2 = \chi_\alpha^2(2)$ is the critical value of a χ^2 distribution with two degrees of freedom and a significance of α. The length of both axes depends on the variance of the ith and jth t-score variable, denoted by $a_i = \sqrt{\lambda_i}$ and $a_j = \sqrt{\lambda_j}$ for PCA, and $a_i = \sigma_{t_i}^2$ and $a_j = \sigma_{t_j}^2$ for PLS. These variances correspond to the diagonal elements of the matrices in (3.1) and (3.2), noting that $1 \leq i, j \leq n$.

It is straightforward to generate a control ellipse for any combination of score variables for $n > 2$. This, however, raises the following question: how can such scatter plots be adequately depicted? A naive solution would be to extend the 2D concept into an nD concept, where the 2D control ellipse becomes an nD-ellipsoid

$$\sum_{i=1}^{n} \frac{t_i^2}{a_i^2} = T_\alpha^2 \tag{3.4}$$

where $T_\alpha^2 = \chi_\alpha^2(n)$ is the critical value of a χ^2 distribution with n degrees of freedom. While it is still possible to depict a control ellipsoid that encompasses the orthogonal projections of the data points onto the $n = 3$-dimensional model subspace, however, this is not the case for $n > 3$. A pragmatic solution could be to display pairs of score variables, an example of which is given in Chapter 5.

It is important to note that (3.3) and (3.4) only hold true if the exact covariance matrix of the recorded process variables is known (Tracey *et al.* 1992). If the covariance matrix must be estimated from the reference data, as shown in Sections 2.1 and 2.2, the approximation by a χ^2-distribution may be inaccurate if few reference samples, K, are available. In this practically important case, (3.4) follows an F-distribution under the assumption that the covariance matrix of the score variables has been estimated independently from the score variables. For a detailed discussion of this, refer to Theorem 5.2.2. in Anderson (2003). The critical value T_α^2 in this case is given by (MacGregor and Kourti 1995; Tracey *et al.* 1992)

$$T_\alpha^2 = \frac{n\left(K^2 - 1\right)}{K\left(K - n\right)} \mathcal{F}_\alpha\left(n, K - n\right) \tag{3.5}$$

where $\mathcal{F}_\alpha(n, K-n)$ is the critical value of an F-distribution for n and $K - n$ degrees of freedom, and a significance of α. It should be noted that the value of $\frac{n\left(K^2-1\right)}{K(K-n)} \mathcal{F}_\alpha(n, K-n)$ converges to $\chi_\alpha^2(n)$ as $K \to \infty$ (Tracey *et al.* 1992) and if the variable mean is known *a priori*, T_α^2 becomes (Jackson 1980)

$$T_\alpha^2 = \frac{n\left(K - 1\right)}{K - n} \mathcal{F}_\alpha\left(n, K - n\right). \tag{3.6}$$

3.1.2 Non-negative quadratic monitoring statistics

Non-negative quadratic statistics could be interpreted as a kinetic energy measure that condenses the variation of a set of n score variables or the model residuals into single values. The reference to non-negative quadratics was proposed by Box (1954) and implies that it relies on the sum of squared values of a given set of stochastic variables. For PCA, the t-score variables and the residual variables can be used for such statistics. In the case of PLS, however, a total of three univariate statistics can be established, one that relates to the t-score variables and two further that correspond to the residuals of the output and the remaining variation of the input variables.

The next two paragraphs present the definition of non-negative quadratics for the t-score variables and detail the construction of the residual-based ones for PCA and PLS. For the reminder of this book, the loading matrix for PCA and PLS are denoted as **P** and only contain the first n column vectors, that is, the ones referring to common-cause variation. For PCA and PLS, this matrix has n_z and n, and n_x and n columns and rows, respectively. The discarded loading vectors are stored in a second matrix, defined as $\mathbf{P}_d$. Moreover, the computed score vector, $\mathbf{t} = \mathbf{P}^T \mathbf{z}_0$ for PCA and $\mathbf{t} = \mathbf{R}^T \mathbf{x}_0$ for PLS is of dimension n.

3.1.2.1 PCA monitoring models

The PCA data model includes the estimation of:

- the model subspace;
- the residual subspace;
- the error covariance matrix;
- the variance of the orthogonal projection of the samples onto the loading vectors; and
- the control ellipse/ellipsoid.

The model and residual subspaces are spanned by the column vectors of **P** and $\mathbf{P}_d$, respectively. Sections 6.1 and 9.1 provide a detailed analysis of PCA, where these geometric aspects are analyzed in more detail.

According to (2.2), the number of source signals determines the dimension of the model subspace. The projection of the samples onto the model subspace therefore yields the source variables that are corrupted by the error variables, which the relationship in (2.8) shows. Moreover, the mismatch between the data vector $\mathbf{z}_0$ and the orthogonal projection of $\mathbf{z}_0$ onto the model subspace, $\mathbf{g}$, does not include any information of the source signals, which follows from

$$
\begin{aligned}
\mathbf{g} &= \mathbf{z}_0 - \widehat{\mathbf{z}}_0 = \mathbf{z}_0 - \mathbf{P}\mathbf{t} = \mathbf{z}_0 - \mathbf{P}\mathbf{P}^T\mathbf{z}_0 = \left[\mathbf{I} - \mathbf{P}\mathbf{P}^T\right]\mathbf{z}_0 \\
\mathbf{g} &= \mathbf{P}_d\mathbf{P}_d^T\left(\boldsymbol{\Xi}\mathbf{s} + \mathbf{g}\right) = \mathbf{P}_d\mathbf{t}_d = \mathbf{P}_d\mathbf{P}_d^T\mathbf{g}.
\end{aligned}
\tag{3.7}
$$

The above relationship relies on the fact that $\mathbf{P}_d^T\boldsymbol{\Xi} = \mathbf{0}$, which (2.7) outlines, and the eigenvectors of the data covariance matrix are mutually orthonormal. The score vector **t**, approximating the variation of the source vector **s**, and the residual vector **g** give rise to the construction of two non-negative squared monitoring statistics, the Hotelling's T^2 and Q statistics that are introduced below.

Hotelling's T^2 statistic. The univariate statistic for the t-score variables, $\mathbf{t} = \mathbf{P}^T \mathbf{z}_0$ is defined as follows

$$T^2 = \mathbf{t}^T \mathbf{\Lambda}^{-1} \mathbf{t} = \mathbf{z}_0^T \mathbf{P} \mathbf{\Lambda}^{-1} \mathbf{P}^T \mathbf{z}_0. \tag{3.8}$$

The matrix $\mathbf{\Lambda}$ includes the largest n eigenvalues of $\mathbf{S}_{z_0 z_0}$ as diagonal elements. For a significance α, the control limit for the above statistic, T_α^2 is equal to $\chi^2(n)$ if the covariance matrix of the recorded process variables is known. If this is not the case the control limit can be obtained as shown in (3.5) or (3.6). The above non-negative quadratic is also referred to as a Hotelling's T^2 statistic. The null hypothesis for testing whether the process is in-statistical-control, H_0, is as follows

$$H_0 \quad : \quad T^2 \leq T_\alpha^2 \tag{3.9}$$

and the hypothesis H_0 is rejected if

$$H_0 \quad : \quad T^2 > T_\alpha^2. \tag{3.10}$$

The alternative hypothesis H_1, the process is out-of-statistical-control, is accepted if H_0 is rejected.

Assuming that the fault condition, representing the alternative hypothesis H_1, describes a bias of the mth sensors, denoted by $\mathbf{z}_f = \mathbf{z}_0 + \Delta\mathbf{z}$ where the mth element of $\Delta\mathbf{z}$ is nonzero and the remaining entries are zeros, the score variables become $\mathbf{t}_f = \mathbf{t} + \Delta\mathbf{t}$. This yields the following impact upon the Hotelling's T^2 statistic, denoted here by T_f^2 where the subscript f refer to the fault condition

$$T_f^2 = (\mathbf{t} + \Delta\mathbf{t})^T \mathbf{\Lambda}^{-1} (\mathbf{t} + \Delta\mathbf{t}) = \mathbf{t}^T \mathbf{\Lambda}^{-1} \mathbf{t} + 2\mathbf{t}^T \mathbf{\Lambda}^{-1} \Delta\mathbf{t} + \Delta\mathbf{t}^T \mathbf{\Lambda}^{-1} \Delta\mathbf{t}. \tag{3.11}$$

The above equation uses $\mathbf{t}_f = \mathbf{P}^T \left(\mathbf{z}_0 + \Delta\mathbf{z}\right) = \mathbf{t} + \Delta\mathbf{t}$. The alternative hypothesis, H_1, is therefore accepted if

$$H_1 \quad : \quad T_f^2 > T_\alpha^2 \tag{3.12}$$

and rejected if $T_f^2 \leq T_\alpha^2$. A more detailed analysis of the individual terms in (3.11) $T_1^2 = \mathbf{t}^T \mathbf{\Lambda}^{-1} \mathbf{t}$, $T_2^2 = 2\mathbf{t}^T \mathbf{\Lambda}^{-1} \Delta\mathbf{t}$ and $T_3^2 = \Delta\mathbf{t}^T \mathbf{\Lambda}^{-1} \Delta\mathbf{t}$ yields that

- $T_1^2 \leq T_\alpha^2$;
- $E\left\{T_2^2\right\} = 0$;
- $E\left\{\left(T_2^2\right)^2\right\} = 4\Delta\mathbf{t}^T \mathbf{\Lambda}^{-1} E\left\{\mathbf{t}\mathbf{t}^T\right\} \mathbf{\Lambda}^{-1} \Delta\mathbf{t} = 4\Delta\mathbf{t}^T \mathbf{\Lambda}^{-1} \Delta\mathbf{t}$;
- $T_2^2 \sim \mathcal{N}\left\{0, 4\Delta\mathbf{t}^T \mathbf{\Lambda}^{-1} \Delta\mathbf{t}\right\}$; and
- $T_3^2 > 0$.

If the term T_2^2 is hypothetically set to zero, $T_f^2 = T_1^2 + T_3^2$. The larger the fault magnitude, Δz_m the more the original T^2 statistic is shifted, which follows from

$$T_3^2 = \Delta z_m^2 \begin{pmatrix} p_{m1} & p_{m2} & \cdots & p_{mn} \end{pmatrix} \begin{bmatrix} 1/\lambda_1 & 0 & \cdots & 0 \\ 0 & 1/\lambda_2 & \cdots & 0 \\ \vdots & \vdots & \ddots & \vdots \\ 0 & 0 & \cdots & 1/\lambda_n \end{bmatrix} \begin{pmatrix} p_{m1} \\ p_{m2} \\ \vdots \\ p_{mn} \end{pmatrix}, \tag{3.13}$$

which is equal to

$$T_3^2 = \Delta z_m^2 \sum_{i=1}^{n} \frac{p_{mi}^2}{\lambda_i}. \tag{3.14}$$

The impact of the term T_2^2 upon T_f^2 is interesting since it represents a Gaussian distributed contribution. This, in turn, implies that the PDF describing T_f^2 is not only a shift of T^2 by T_3^2 but it has also a different shape.

Figure 3.1 presents the PDFs that describe the T^2 and T_f^2 and illustrate the impact of Type I and II errors for the hypothesis testing. It follows from Subsections 1.1.3 and 1.2.4 that a Type I error is a rejecting of the null hypothesis although it is true and a Type II error is the acceptance of the null hypothesis although it is false. Figure 3.1 shows that the significance level for the Type II, β, depends on the exact PDF for a fault condition, which, even the simple

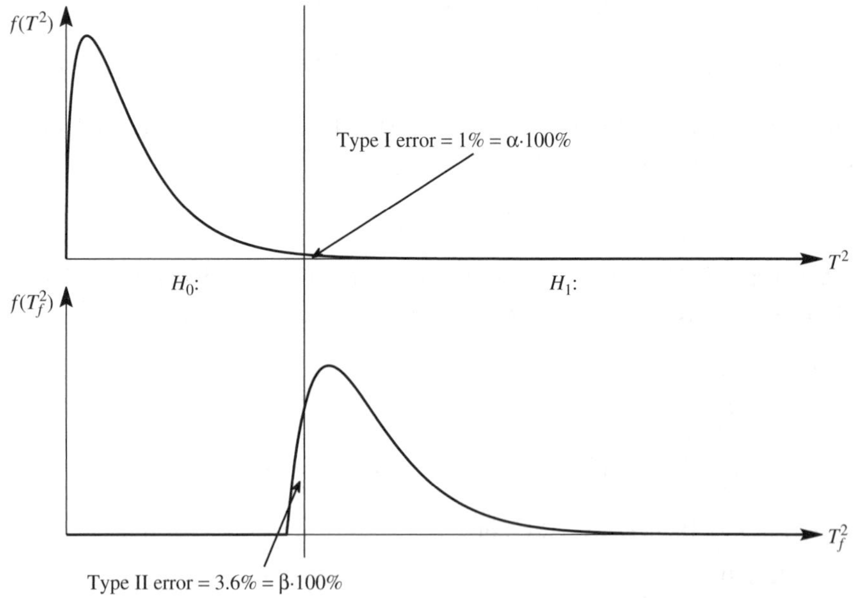

Figure 3.1 Illustration of Type I and II errors for testing null hypothesis.

sensor fault, is difficult to determine. The preceding discussion, however, highlights that *the larger the magnitude of the fault condition the more the PDF will be shifted* and hence, the smaller β becomes. In other words, incipient fault conditions are more difficult to detect than faults that have a profound impact upon the process.

Q statistic. The second non-negative quadratic statistic relates to the PCA model residuals $\mathbf{g} = \mathbf{z}_0 - \mathbf{Pt} = \left[\mathbf{I} - \mathbf{PP}^T\right]\mathbf{z}_0$ and is given by

$$Q = \mathbf{g}^T\mathbf{g} = \mathbf{z}_0^T\left[\mathbf{I} - \mathbf{PP}^T\right]\mathbf{z}_0. \tag{3.15}$$

The control limit for the Q statistic is difficult to obtain although the above sum appears to be the sum of squared values. More precisely, Subsection 3.3.1 highlights that the PCA residuals are linearly dependent and are therefore not statistically independent. Approximate distributions for such quadratic forms were derived in Box (1954) and Jackson and Mudholkar (1979). Appendix B in Nomikos and MacGregor (1995) yielded that both approximations are close. Using the method by Jackson and Mudholkar (1979) the control limit for the Q statistic is as follows

$$Q_\alpha = \theta_1\left(\frac{c_\alpha\sqrt{2\theta_2 h_0^2}}{\theta_1} + \frac{\theta_2 h_0\left(h_0 - 1\right)}{\theta_1^2} + 1\right)^{\frac{1}{h_0}}, \tag{3.16}$$

where $\theta_1 = \sum_{i=n+1}^{n_z} \lambda_i$, $\theta_2 = \sum_{i=n+1}^{n_z} \lambda_i^2$, $\theta_3 = \sum_{i=n+1}^{n_z} \lambda_i^3$, $h_0 = 1 - \frac{2\theta_1\theta_3}{3\theta_2}$ and the variable c_α is the normal deviate evaluated for the significance α. Defining the matrix product $\mathbf{PP}^T = \mathfrak{C}$, (3.15) can be rewritten as follows

$$Q = \sum_{i=1}^{n_z}\left(1 - \mathfrak{c}_{ii}\right) z_{0_i}^2 - \sum_{i=1}^{n_z} z_{0_i}\left(\sum_{j=1\neq i}^{n_z} \mathfrak{c}_{ji} z_{0_j}\right) \tag{3.17}$$

where $\mathfrak{c}_{ji} = \sum_{k=1}^{n} p_{kj} p_{ki}$ is the element of $\mathfrak{C}$ stored in the jth row and the ith column. Given that $\mathfrak{C}$ is symmetric, i.e. $\mathfrak{c}_{ji} = \mathfrak{c}_{ij}$, (3.17) becomes

$$Q = \sum_{i=1}^{n_z}\left(1 - \mathfrak{c}_{ii}\right) z_{0_i}^2 - 2\sum_{i=1}^{n_z-1} z_{0_i} \sum_{j=i+1}^{n_z} \mathfrak{c}_{ij} z_{0_j}. \tag{3.18}$$

A modified version of the Q statistic in (3.15) was proposed in (Hawkins 1974) and entails a scaling of each discarded score vector by its variance

$$T_d^2 = \mathbf{t}_d^T \mathbf{\Lambda}_d^{-1} \mathbf{t}_d = \mathbf{z}_0^T \mathbf{P}_d \mathbf{\Lambda}_d^{-1} \mathbf{P}_d^T \mathbf{z}_0 \tag{3.19}$$

and follows a χ^2 distribution with the control limit $T^2_{d_\alpha} = \chi^2_\alpha\left(n_z - n\right)$. If the data covariance matrix needs to be estimated, the control limit $T^2_{d_\alpha}$ is given by

$$T^2_{d_\alpha} = \frac{\left(n_z - n\right)\left(K^2 - 1\right)}{K\left(K - n_z + n\right)}\mathcal{F}_\alpha\left(n_z - n, K - n_z + n\right). \tag{3.20}$$

The diagonal matrix $\mathbf{\Lambda}_d$ contains the discarded eigenvalues of $\mathbf{S}_{z_0 z_0}$ and $\mathbf{t}_d = \mathbf{P}_d \mathbf{z}_0$. For process monitoring applications, however, a potential drawback of the residual T^2_d statistic is that some of the discarded eigenvalues, $\lambda_{n+1} \cdots \lambda_{n_z}$, may be very close or equal to zero. This issue, however, does not affect the construction of the Q statistic in (3.15).

Using the Q statistic for process monitoring, testing whether the process is in-statistical-control relies on the null hypothesis H_0, which is accepted if

$$H_0 \quad : \quad Q \leq Q_\alpha \tag{3.21}$$

and rejected if $Q > Q_\alpha$. On the other hand, the alternative hypothesis H_1, describing the out-of-statistical-control situation, is accepted if the null hypothesis H_0 is rejected

$$H_1 \quad : \quad Q > Q_\alpha. \tag{3.22}$$

Assuming that the fault condition is a bias of the mth sensor that has the form of a step, (3.15) becomes

$$\begin{aligned} Q_f = &\sum_{i=1}^{n_z} z^2_{0_i}\left(1 - \mathfrak{c}_{ii}\right) + \sum_{i=1}^{n_z} z_{0_i}\left(\sum_{j=1\neq i}^{n_z} z_{0_j}\mathfrak{c}_{ij}\right) + \\ &\Delta z^2_m\left(1 - \mathfrak{c}_{mm}\right) + \Delta z_m \sum_{i=1\neq m}^{n_z} z_{0_i}\mathfrak{c}_{mi}. \end{aligned} \tag{3.23}$$

Similar to the Hotelling's T^2 statistic, the step-type fault yields a Q statistic Q_f that includes the offset term $\Delta z^2_m\left(1 - \mathfrak{c}_{mm}\right)$, where Δz_i is the magnitude of the bias, and the Gaussian distributed term $\Delta z_m \sum_{k=1\neq i}^{n_z} z_k \mathfrak{c}_{ik}$. Figure 3.1 and (3.23) highlight that a larger bias leads to a more significant shift of the PDF for Q_f relative to the PDF for Q and therefore a smaller Type II error β. In contrast, a smaller and incipient sensor bias leads to a large Type II error and is therefore more difficult to detect.

3.1.2.2 PLS monitoring models

PLS and MRPLS models give rise to the generation of three univariate statistics. The ones for PLS models are presented first, followed by those for MRPLS models.

Monitoring statistics for PLS models. Similar to PCA, the retained t-score variables allow constructing a Hotelling's T^2 statistic, which according to (2.24) describes common cause variation

$$T^2 = \mathbf{t}^T \mathbf{S}_{\mathrm{tt}}^{-1} \mathbf{t} = \mathbf{x}_0^T \mathbf{R} \mathbf{S}_{\mathrm{tt}}^{-1} \mathbf{R}^T \mathbf{x}_0. \tag{3.24}$$

Here, $\mathbf{t} = \mathbf{R}^T \mathbf{x}_0$ and $\mathbf{S}_{\mathrm{tt}}$ is given in (3.2). Equations (3.5) or (3.6) show how to calculate the control limit for this statistic if $\mathbf{S}_{\mathrm{tt}}$ is not known *a priori*. If $\mathbf{S}_{x_0 x_0}$ and $\mathbf{S}_{x_0 y_0}$ are available the control limit is $T_\alpha^2 = \chi_\alpha^2(n)$. The Q statistic for the residual of the output variables is given by

$$Q_f = \mathbf{f}^T \mathbf{f}. \tag{3.25}$$

Here, $\mathbf{f} = \mathbf{y}_0 - \mathbf{Q}\mathbf{R}^T \mathbf{x}_0 = \mathbf{y}_s + \mathfrak{f} - \mathbf{Q}\mathbf{R}^T \mathbf{x}_0$.

The residuals of the input variables can either be used to construct a Hotelling's T^2 or a Q statistic, depending upon their variances. This follows from the discussion concerning the residual statistic proposed by Hawkins (1974). Very small residual variances can yield numerical problems in determining the inverse of the residual covariance matrix. If this is the case, it is advisable to construct a residual Q statistic

$$Q_e = \mathbf{e}^T \mathbf{e}, \tag{3.26}$$

where $\mathbf{e} = \left[\mathbf{I} - \mathbf{P}\mathbf{R}^T\right] \mathbf{x}_0$. In a similar fashion to PCA, the elements of the residual vector $\mathbf{e}$ are linear combinations of the input variables computed from the discarded r-weight vector stored in $\mathbf{R}_d$

$$\mathbf{t}_d = \mathbf{R}_d^T \mathbf{x}_0 \qquad \mathbf{e} = \mathbf{P}_d \mathbf{t}_d = \mathbf{P}_d \mathbf{R}_d^T \mathbf{x}_0. \tag{3.27}$$

Using the relationship in (3.27), equation (3.26) can be rewritten as follows

$$Q_e = \mathbf{t}_d^T \mathbf{P}_d^T \mathbf{P}_d \mathbf{t}_d. \tag{3.28}$$

For determining the control limits of the Q_e and Q_f statistics, it is possible to approximate the distribution functions of both non-negative quadratics by central χ^2 distributions[1]. Theorem 3.1 in Box (1954) describes this approximation, which allows the determination of control limits for a significance α

$$Q_{e_\alpha} = g_e \chi_\alpha^2\left(h_e\right) \qquad Q_{f_\alpha} = g_f \chi_\alpha^2\left(h_f\right) \tag{3.29}$$

where the g and h parameters are obtained such that the approximated distributions have the same first two moments as those of Q_e and Q_f in (3.25) and (3.28).

[1] For a central χ^2 distribution, the mean of each contributing element is assumed to be zero unlike a noncentral where this assumption is relaxed.

In other words, the mean and variance of $Q = g\chi^2_\alpha(h)$ $E\left\{g\chi^2_\alpha(h)\right\} = gh^2$ and $E\left\{\left(g\chi^2_\alpha(h)\right)^2\right\} = g^2 E\left\{\left(\chi^2_\alpha(h)\right)^2\right\} = 2g^2h$, so that g_e and g_f are

$$g_e = \frac{E\left\{\left(Q_e - E\left\{Q_e\right\}\right)^2\right\}}{2E\left\{Q_e\right\}} \qquad g_f = \frac{E\left\{\left(Q_f - E\left\{Q_f\right\}\right)^2\right\}}{2E\left\{Q_f\right\}} \tag{3.30}$$

and h_e and h_f are

$$h_e = 2\frac{\left(E\left\{Q_e\right\}\right)^2}{E\left\{\left(Q_e - E\left\{Q_e\right\}\right)^2\right\}} \qquad h_f = 2\frac{\left(E\left\{Q_f\right\}\right)^2}{E\left\{\left(Q_f - E\left\{Q_f\right\}\right)^2\right\}}. \tag{3.31}$$

For larger variances of the discarded t-score variables, although they do not contribute significantly to the prediction of the output variables, it is advisable to construct a second Hotelling's T^2 statistic instead of the Q_e statistic

$$T_d^2 = \sum_{i=1}^{n_x - n} \frac{\mathrm{t}_{d_i}^2}{\sigma_{\mathrm{t}_{d_i}}^2} = \mathbf{x}_0^T \mathbf{R}_d \mathbf{S}_{\mathrm{t}_d\mathrm{t}_d}^{-1} \mathbf{R}_d^T \mathbf{x}_0 \tag{3.32}$$

where $\mathbf{S}_{\mathrm{t}_d\mathrm{t}_d}^{-1} = \mathrm{diag}\left\{{}^1/\sigma^2_{\mathrm{t}_{d_1}} \quad {}^1/\sigma^2_{\mathrm{t}_{d_2}} \quad \cdots \quad {}^1/\sigma^2_{\mathrm{t}_{dn}}\right\}$. The Hotelling's T_d^2 statistic follows a central χ^2 statistic with $n_x - n$ degrees of freedom if the covariance and cross-covariance matrices are known *a priori* or a scaled F distribution if not. As before, the estimate, $\widehat{\mathbf{S}}_{\mathrm{t}_d\mathrm{t}_d}$, has to be obtained from an different reference set that has not been used to estimate the weight and loading matrices (Tracey *et al.* 1992).

The advantage of the statistics in (3.24) and (3.32) is that each score variable is scaled to unity variance and has the same contribution to the Hotelling's T^2 statistics. Hence, those score variables with smaller variances, usually the last few ones, are not overshadowed by those with significantly larger variances, typically the first few ones. This is of concern if a fault condition has a more profound effect upon score variables with a smaller variance. In this case, the residuals Q_e statistic may yield a larger Type II error compared to the Hotelling's T_d^2 statistic. Conversely, if the variances of the last few score variables are very close to zero numerical problems and an increase in the Type I error, particularly for small reference sets, may arise. In this case, it is not advisable to rely on the Hotelling's T_d^2 statistic.

Monitoring statistics for MRPLS models. With respect to (2.51), a Hotelling's T^2 statistic to monitor common cause variation relies on the t-score

---

[2] The mean and variance of $\chi^2(h)$, $E\left\{\chi^2(h)\right\}$ and $E\left\{\left(\chi^2(h) - h\right)^2\right\}$, are h and $2h$, respectively.

variables that are linear combinations of the source variables corrupted by error variables

$$T^2 = \mathbf{t}^T\mathbf{t} = \mathbf{x}_0^T\mathbf{R}\mathbf{R}^T\mathbf{x}_0. \quad (3.33)$$

The covariance matrix of the score variables is $E\left\{\mathbf{t}\mathbf{t}^T\right\} = \mathbf{R}^T\mathbf{S}_{x_0x_0}\mathbf{R} = \mathbf{I}$ and represents, in fact, the length constraint of the MRPLS objective function in (2.66). The Hotelling's T^2 statistic defined in (3.33) follows a χ^2 distribution with n degrees of freedom and its control limit is $\chi^2_\alpha(n)$.

The data model in (2.51) highlights that the residuals of the input variables that are not correlated with the output variables may still be significant and can also be seen as common cause variation but only for the input variable set. The vector of source variables $\mathbf{s}'$ describes this variation and allows the construction of a second Hotelling's T^2 statistic, denoted here as the Hotelling's T'^2 statistic

$$T'^2 = \mathbf{t}'^T\mathbf{S}_{t't'}^{-1}\mathbf{t}' = \mathbf{x}_0^T\mathbf{R}'\mathbf{S}_{t't'}^{-1}\mathbf{R}'^T\mathbf{x}_0. \quad (3.34)$$

Here, $\mathbf{R}'$ is the r-loading matrix containing the $n_x - n$ r-loading vectors for determining $\mathbf{t}'$. The t′-score variables are equal to the s′ source variables up to a similarity transformation and, similar to the t-score variables, $\mathbf{S}_{t't'} = \mathbf{R}'^T\mathbf{S}_{x_0x_0}\mathbf{R}'$ is a diagonal matrix. If the score covariance matrices need to be estimated Tracey *et al.* (1992) outlined that this has to be done from a different reference set that was not used to estimate the weight and loading matrices. If this is guaranteed, the Hotelling's T^2 statistic follows a scaled F-distribution with n and $K - n$ and the Hotelling's T'^2 statistics follows a scaled F distribution with $n_x - n$ and $K - n$ degrees of freedom.

If the variance of the last few t′-score variables is very close to zero, it is advisable to utilize a Q statistic rather than the Hotelling's T'^2 statistic. Assuming that each of the t′-score variables have a small variance, a Q statistic can be obtained that includes each score variable

$$Q_e = \mathbf{t}'^T\mathbf{t}' = \mathbf{x}_0\mathbf{R}'\mathbf{R}'^T\mathbf{x}_0. \quad (3.35)$$

If there are larger differences between the variances of the t′-score variables, it is advisable to utilize the Q_e statistic or to divide the t′-score variables into two sets, one that includes those with larger variances and the remaining ones with a small variance. This would enable the construction of two non-negative quadratic statistics. Finally, the residuals of the output variables form the Q_f statistic in (3.25) along with its control limit in (3.29) to (3.31).

3.2 Fault isolation and identification

After detecting abnormal process behavior, the next step is to determine what has caused this event and what is its root cause. Other issues are how significantly does this event affect product quality and what impact does it have on the general process operation? Another important question is can the process continue to run while the abnormal condition is removed or its impact minimized, or is it necessary to shut down the process immediately to remove the fault condition? The diagnosis of abnormal behavior, however, is difficult (Jackson 2003) and often requires substantial process knowledge and, particularly, about the interaction between individual operating units. It is therefore an issue that needs to be addressed by experienced process operators.

To assist plant personnel in identifying potential causes of abnormal behavior, MSPC offers charts that describe to what extent a particular process variable is affected by such an event. It can also offer time-based trends that estimate the effect of a fault condition upon a particular process variable. These trends are particularly useful if the impact of a fault condition becomes more significant over time. For a sensor or actuator bias or precision degradation, such charts provide useful information that can easily be interpreted by a plant operator. For more complex process faults, such as the performance deterioration of units, or the presence of unmeasured disturbances, these charts offer diagnostic information allowing experienced plant operators to narrow down potential root causes for a more detailed examination.

It is important to note, however, that such charts examine changes in the correlation between the recorded process variables but do not present direct causal information (MacGregor and Kourti 1995; MacGregor 1997; Yoon and MacGregor 2001). Section 3.3 analyzes associated problems of the charts discussed in this section. Before developing and discussing such diagnosis charts, we first need to introduce the terminology for diagnosing fault conditions in technical systems. Given that there are a number of competing definitions concerning fault diagnosis, this book uses the definitions introduced by Isermann and Ballé (1997), which are:

Fault isolation: Determination of the kind, location and time of detection of a fault. Follows fault detection.

Fault identification: Determination of the size and time-variant behavior of a fault. Follows fault isolation.

Fault diagnosis: Determination of the kind, size, location and time of detection of a fault. Follows fault detection. Includes fault isolation and identification.

The literature introduced different fault diagnosis charts and methods, including:

- *contribution charts*;
- charting the results of *residual-based tests*; and
- *variable reconstruction*.

Contribution charts, for example discussed by Koutri and MacGregor (1996) and Miller *et al.* (1988), indicate to what extent a certain variable is affected by a fault condition. Residual-based tests (Wise and Gallagher 1996; Wise *et al.* 1989a) examine changes in the residual variables of a sufficiently large data set describing an abnormal event, and variable reconstruction removes the fault condition from a set of variables (Dunia and Qin 1998; Lieftucht *et al.* 2009).

3.2.1 Contribution charts

Contribution charts reveal which of the recorded variable(s) has(have) changed the correlation structure among them. More precisely, these charts reveal how each of the recorded variables affects the computation of particular t-score variables. This, in turn, allows computing the effect of a particular process variable upon the Hotelling's T^2 and Q statistics if at least one of them detects an out-of-statistical-control situation. The introduction of contribution charts is designed here for PCA. The tutorial session at the end of this chapter offers a project for developing contribution charts for PLS and to contrast them with the PCA ones.

3.2.1.1 Variable contribution to the Hotelling's T^2 statistic

The contribution of the recorded variable set $\mathbf{z}_0$ upon the ith t-score variable that forms part of the Hotelling's T^2 statistic is given by Kourti and MacGregor (1996):

1. Determine which score variable is significantly affected by out-of-statistical-control situation by testing the alternative hypothesis of the normalized score variables

$$\frac{t_i}{\sqrt{\lambda_i}} \qquad i \in [1, 2, \ldots n], \tag{3.36}$$

which is as follows

$$H_1 \quad : \quad n\frac{t_i^2}{\lambda_i} > T_\alpha^2. \tag{3.37}$$

This yields $n^* \leq n$ score variables that are affected and earmarked for further inspection. Moreover, the index set $1, 2, \ldots, n$ can be divided into a subset of indices that contains the affected score variables, $\mathfrak{N}^*$ that contains n^* elements, and a subset $\mathfrak{N}$ that stores the remaining $n - n^*$ elements. The union of both subsets, $\mathfrak{N} \cup \mathfrak{N}^*$, is the index set $1, 2, \ldots, n$ and the intercept of both subsets contains no element, $\mathfrak{N} \cap \mathfrak{N}^* = \varnothing$.

2. Next, compute the contribution of each process variable z_{0_j}, $j \in \left[1, 2, \ldots, n_z\right]$, for each of the violating score variables, t_i, where $i \in \mathfrak{N}^*$

$$\Delta z_{0_k}^{(i)} = \frac{t_i}{\lambda_i} p_{ji} z_{0_j}. \tag{3.38}$$

3. Should the contribution of $\Delta z_{0_j}^{(i)}$ be negative, set it equal to zero.

4. The contribution of the jth process variable on the Hotelling's T^2 statistic is

$$\Delta z_{0_j} = \sum_{i \in \mathfrak{N}^*} \Delta z_{0_j}^{(i)} \tag{3.39}$$

5. The final step is to plot the values of Δz_{0_j} for each of the n_z variables in the form of a 2D bar chart for a specific sample or a 3D bar chart describing a time-based trend of Δz_{0_j} values.

For this procedure, p_{ji} is the entry of the loading matrix $\mathbf{P}$ stored in the jth row and the ith column.

The above procedure invites the following two questions. Why is the critical value for the hypothesis test in (3.37) T_α^2/n and why do we remove negative values for $\Delta z_{0_k}^{(i)}$? The answer to the first question lies in the construction of the Hotelling's T^2 statistic, which follows asymptotically a χ^2 distribution (Tracey *et al.* 1992). Assuming that $n^* < n$, the Hotelling's T^2 statistic can be divided into a part that is affected and a part that is unaffected by a fault condition

$$T^2 = \underbrace{\sum_{i \in \mathfrak{N}^*} \frac{t_i^2}{\lambda_i}}_{\substack{\text{affected by the} \\ \text{fault condition}}} + \underbrace{\sum_{j \in \mathfrak{N}} \frac{t_j^2}{\lambda_j}}_{\substack{\text{not affected by} \\ \text{the fault condition}}} . \tag{3.40}$$

The definition of the χ^2 PDF, however, describes the sum of statistically independent Gaussian distributed variables of zero mean and unity variance. In this regard, each element of this sum has the same contribution to the overall statistic. Consequently, the critical contribution of a particular element is the ratio of the control limit over the number of sum elements. On the other hand, testing the alternative hypothesis that a single t-score variable, which follows a Gaussian distribution of zero mean and unity variance and its squared values asymptotically follows a χ^2 distribution with 1 degree of freedom, against the control limit of just this one variable can yield a significant Type II error, which Figure 3.2 shows.

The answer to the second question lies in revisiting the term in (3.40)

$$\sum_{i \in \mathfrak{N}^*} \frac{t_i^2}{\lambda_i} = \sum_{i \in \mathfrak{N}^*} \frac{t_i}{\lambda_i} \left(\sum_{j=1}^{n_z} p_{ji} z_{0_j} \right) . \tag{3.41}$$

From the above equation, it follows that the contribution of the jth process variable upon the Hotelling's T^2 statistic detecting an abnormal condition is equal to

$$\Delta z_{0_j} = z_{0_j} \sum_{i \in \mathfrak{N}^*} \frac{t_i p_{ji}}{\lambda_i} . \tag{3.42}$$

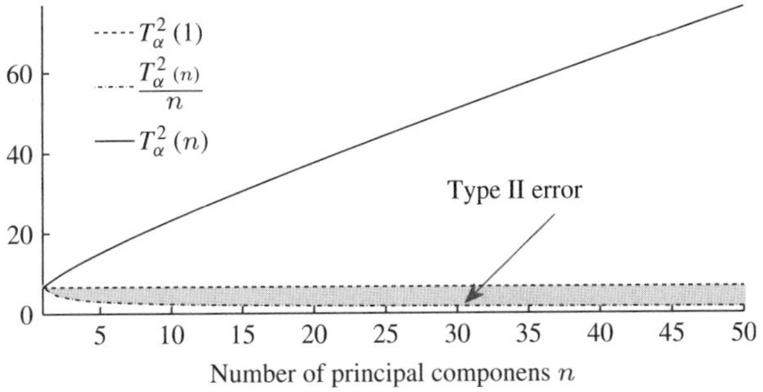

Figure 3.2 Type II error for incorrectly applying hypothesis test in Equation (3.37).

Now, including terms in the above sum that have a different sign to z_{0_k} reduces the overall value of Δz_{0_j}. Consequently, to identify the main contributors to the absolute value of Δz_{0_j} requires the removal of negative sum elements $t_i p_{ji} z_{0_j} / \lambda_i$.

3.2.1.2 Variable contribution to the Q statistic

Given that the Q statistic relies on the sum of the residuals for each variable, the variable contribution of the jth variable is simply (Yoon and MacGregor 2001):

$$\Delta z_{0_j} = g_j. \tag{3.43}$$

Alternative forms are discussed in Kourti (2005), $\Delta z_{0_j} = g_j^2$, or Chiang *et al.* (2001), $\Delta z_{0_j} = \frac{g_j^2}{E\left\{g_j^2\right\}}$. Since the variance for these residuals may vary, it is difficult to compare them without scaling. Furthermore, using squared values does not offer the possibility of evaluating whether whether a temperature reading is larger or smaller then expected, for example. This suggests that (3.43) should be

$$\Delta z_{0_j} = \frac{g_j}{\sqrt{E\left\{g_j^2\right\}}}. \tag{3.44}$$

3.2.1.3 Degree of reliability of contribution charts

Although successful diagnoses using contribution charts have been reported (Martin *et al.* 2002; Pranatyasto and Qin 2001; Vedam and Venkatasubramanien 1999; Yoon and MacGregor 2001), subsection 3.3.1 shows that the PCA residuals of the process variables are linearly dependent. The same analysis can also be applied to show that the recorded process variables have a linearly dependent contribution to the Hotelling's T^2 statistic. Moreover, Yoon and MacGregor (2001) discussed that contribution charts generally stem from an underlying correlation model of the recorded process variables which may not possess a causal relationship.

However, contribution charts can identify which group(s) of variables are mostly affected by a fault condition. Lieftucht *et al.* (2006a) highlighted that the number of degrees of freedom in the residual subspace is an important factor for assessing the reliability of Q contribution charts. The ratio $^{n}/_{n_z}$ is therefore an important index to determine the reliability of these charts. The smaller this ratio the larger is the dimension of the residual subspace and the less the degree of linear dependency among the variable contribution to the Q statistic.

Chapters 4 and 5 demonstrate how contribution charts can assist the diagnosis of fault conditions ranging from simple sensor or actuator faults to more complex process faults. It is important to note, however, that the magnitude of the fault condition can generally not be estimated through the use of contribution charts. Subsection 3.2.3 introduces the projection- and regression-based variable reconstruction that allows determining the kind and size of complex fault scenarios. The next subsection describes residual-based tests to diagnose abnormal process conditions.

3.2.2 Residual-based tests

Wise *et al.* (1989a) and Wise and Gallagher (1996) introduced residual-based tests that relate to the residuals of a PCA model. Residual-based tests for PLS models can be developed as a project in the tutorial session at the end of the chapter. Preliminaries for calculating the error variance are followed here by outlining the hypothesis tests for identifying which variable is affected by an abnormal operating condition.

3.2.2.1 Preliminaries

The residual variance for recovering the jth process variable, $g_j = z_{0_j} - \mathbf{c}_j^T \mathbf{z}_0$, is

$$E\left\{g_j^2\right\} = \sum_{i=n+1}^{n_z} p_{ij}^2 \sigma_{\mathrm{g}}^2, \tag{3.45}$$

where $\mathbf{c}_j^T$ is the jth row vector stored in $\mathfrak{C} = \mathbf{P}\mathbf{P}^T$. Equation (3.45) follows from the fact that $\mathbf{g} = \mathbf{P}_d \mathbf{t}_d$, that the t-score variables are statistically independent and that each error variable has the variance σ_{g}^2. If σ_{g}^2 is unknown, it can be estimated from $\widehat{\mathbf{S}}_{z_0 z_0}$,

$$\widehat{\sigma}_{\mathrm{g}_j}^2 = \frac{1}{n_z - n}\left(\sum_{i=1}^{n_z} \widehat{\lambda}_i - \sum_{i=1}^{n} \widehat{\lambda}_i\right)\left(1 - \sum_{i=1}^{n} \widehat{p}_{ij}^2\right). \tag{3.46}$$

That $\sum_{i=n+1}^{n_z} \widehat{p}_{ij}^2 = 1 - \sum_{i=1}^{n} \widehat{p}_{ij}^2$ follows from $\mathbf{I} - \mathfrak{C} = \mathbf{P}_d \mathbf{P}_d^T$ and the fact that the p-loading vectors are mutually orthonormal. Equation (2.122) points out that

$\sum_{i=1}^{n_z} \widehat{\lambda}_i$ is equal to the sum of the diagonal elements of $\widehat{\mathbf{S}}_{z_0 z_0}$, which yields

$$\widehat{\sigma}^2_{\mathfrak{g}_j} = \tfrac{1}{n_z - n}\left(\sum_{i=1}^{n_z} \widehat{\sigma}_i^2 - \sum_{i=1}^{n} \widehat{\lambda}_i\right)\left(1 - \sum_{i=1}^{n} \widehat{p}_{ij}^2\right). \tag{3.47}$$

Equation (3.47) only requires the first n eigenvalues and eigenvectors of $\widehat{\mathbf{S}}_{z_0 z_0}$.

Assuming that the availability of a data set of K_f samples describing an abnormal operating condition and a second data set of normal process behavior containing K samples, it is possible to calculate the following ratio which follows an F-distribution

$$\frac{\widehat{\sigma}^2_{\mathfrak{g}_j, f}}{\widehat{\sigma}^2_{\mathfrak{g}_j}} \sim F\left(K_f - n - 1, K - n - 1\right) \tag{3.48}$$

where both variances can be computed from their respective data sets. This allows testing the null hypothesis that the jth variable is not affected by a fault condition

$$H_0 \quad : \quad \frac{\widehat{\sigma}^2_{\mathfrak{g}_j, f}}{\widehat{\sigma}^2_{\mathfrak{g}_j}} \leq F_\alpha\left(K_f - n - 1, K - n - 1\right) \tag{3.49}$$

or the alternative hypothesis that this variable is affected

$$H_1 \quad : \quad \frac{\widehat{\sigma}^2_{\mathfrak{g}_j, f}}{\widehat{\sigma}^2_{\mathfrak{g}_j}} > F_\alpha\left(K_f - n - 1, K - n - 1\right). \tag{3.50}$$

Testing the null hypothesis can only be performed if the last $n_z - n$ eigenvalues are identical, that is, the variance of the error variables are identical. If this is not the case, however, the simplification leading to (3.47) cannot be applied and the variance of each residual must be estimated using computed residuals. A procedure for determining which variable is affected by a fault condition is given next.

3.2.2.2 Variable contribution to detected fault condition

The mean for each of the residuals is zero

$$E\{\mathbf{g}\} = E\{\mathbf{z} - \mathbf{P}\mathbf{t}\} = E\{\mathbf{z} - \mathfrak{C}\mathbf{z}\} = [\mathbf{I} - \mathfrak{C}]\, E\{\mathbf{z}\} = \mathbf{0}, \tag{3.51}$$

which can be taken advantage of in testing whether a statistically significant departure in mean occurred. This is a standard test that is based on the following statistic

$$\frac{-\widehat{\bar{g}}_{f_j}}{\widehat{\sigma}^2_{g_{f_j}}\sqrt{\frac{1}{K_f - n} + \frac{1}{K - n}}}, \tag{3.52}$$

where $\widehat{\bar{g}}_{f_j}$ is the mean estimated from the K_f samples describing a fault condition. This statistic follows a t-distribution with the following upper and lower control limit

$$\mathrm{UCL} = \mathcal{T}_{\alpha/2}\left(K_f + K - 2n - 2\right) \quad \mathrm{LCL} = -\mathrm{UCL} \tag{3.53}$$

for a significance α. The null hypothesis, H_0, is therefore

$$H_0 \quad : \quad \mathrm{LCL} \leq -\frac{\widehat{\bar{g}}_{f_j}}{\widehat{\sigma}^2_{g_{f_j}}\sqrt{\frac{1}{K_f - n} + \frac{1}{K - n}}} \leq \mathrm{UCL}. \tag{3.54}$$

The alternative hypothesis, H_1 is accepted if

$$H_1 \quad : \quad \left| -\frac{\widehat{\bar{g}}_{f_j}}{\widehat{\sigma}^2_{g_{f_j}}\sqrt{\frac{1}{K_f - n} + \frac{1}{K - n}}} \right| > \mathcal{T}_{\alpha/2}\left(K_f + K - 2n - 2\right). \tag{3.55}$$

By inspecting the above hypothesis tests in (3.49) and (3.55) for changes in the error variance and the mean of the error variables for recovering the recorded process variables, it is apparent that they require a set of recorded data that describes the fault condition. This, however, hampers the practical usefulness of these residual-based tests given that a root cause analysis should be conducted upon detection of a fault condition. Such immediate analysis can be offered by contribution charts and variable reconstruction charts that are discussed next.

3.2.3 Variable reconstruction

Variable reconstruction exploits the correlation among the recorded process variables and hence, variable interrelationships. This approach is closely related to the handling of missing data in multivariate data analysis (Arteaga and Ferrer 2002; Nelson *et al.* 1996; Nelson *et al.* 2006) and relies on recovering the variable set $\mathbf{z}_0$ and $\mathbf{x}_0$ based on the n source variables of the PCA and PLS data structures, respectively

$$\mathbf{z}_0 = \widehat{\mathbf{z}}_0 + \mathbf{g} = \mathfrak{C}\mathbf{z}_0 + \mathbf{g} \qquad \mathbf{x}_0 = \widehat{\mathbf{x}}_0 + \mathbf{e} = \mathbf{P}\mathbf{R}^T\mathbf{x}_0 + \mathbf{e}. \tag{3.56}$$

According to (2.2) and (2.24), $\widehat{\mathbf{z}}_0$ and $\widehat{\mathbf{x}}_0$ mainly describe the impact of the common-cause variation, $\boldsymbol{\Xi}\mathbf{s}$ and $\mathfrak{P}\mathbf{s}$, upon $\mathbf{z}_0$ and $\mathbf{x}_0$, respectively.

3.2.3.1 Principle of variable reconstruction

For PCA, the derivation below shows how to reconstruct a subset of variables in $\mathbf{z}_0$. A self-study project in the tutorial section aims to develop a reconstruction scheme for PLS. Under the assumption that $n < n_z$, it follows from (3.56) that

$$\widehat{\mathbf{z}}_0 = \left[\sum_{i=1}^{n} \mathbf{p}_i \mathbf{p}_i^T\right] \mathbf{z}_0 = \sum_{i=1}^{n} \mathbf{p}_i t_i = \begin{bmatrix} \mathbf{p}_1 & \mathbf{p}_2 & \cdots & \mathbf{p}_n \end{bmatrix} \begin{pmatrix} t_1 \\ t_2 \\ \vdots \\ t_n \end{pmatrix}, \tag{3.57}$$

which can be rewritten to become

$$\begin{pmatrix} \widehat{\mathbf{z}}_{0_1} \\ \widehat{\mathbf{z}}_{0_2} \end{pmatrix} = \begin{bmatrix} \mathbf{P}_1 \\ \mathbf{P}_2 \end{bmatrix} \mathbf{t} \tag{3.58}$$

where $\widehat{\mathbf{z}}_{0_1} \in \mathbb{R}^n$, $\widehat{\mathbf{z}}_{0_2} \in \mathbb{R}^{n_z - n}$, $\mathbf{P}_1 \in \mathbb{R}^{n \times n}$ and $\mathbf{P}_2 \in \mathbb{R}^{n \times n_z - n}$. The linear dependency between the elements of $\widehat{\mathbf{z}}_0$ follow from

$$\mathbf{t} = \mathbf{P}_1^{-1} \widehat{\mathbf{z}}_{0_1} \Rightarrow \widehat{\mathbf{z}}_{0_2} = \mathbf{P}_2 \mathbf{P}_1^{-1} \widehat{\mathbf{z}}_{0_1} \Rightarrow \widehat{\mathbf{z}}_0 = \begin{bmatrix} \mathbf{I} \\ \mathbf{P}_2 \mathbf{P}_1^{-1} \end{bmatrix} \widehat{\mathbf{z}}_{0_1}. \tag{3.59}$$

The above partition of the loading matrix into the first n rows and the remaining $n_z - n$ rows is arbitrary. By rearranging the row vectors of $\mathbf{P}$ along with the elements of $\widehat{\mathbf{z}}_0$ any $n_z - n$ elements in $\widehat{\mathbf{z}}_0$ are linearly dependent on the remaining n elements. Subsection 3.3.1 discusses this in more detail.

Nelson *et al.* (1996) described three different techniques to handle missing data. The analysis in Arteaga and Ferrer (2002) showed that two of them are projection-based while the third one uses regression. Variable reconstruction originates from the projection-based approach for missing data, which is outlined next. The regression-based reconstruction is then presented, which isolates deterministic fault signatures from the t-score variables and removes such signatures from the recorded data.

3.2.3.2 Projection-based variable reconstruction

The principle of projection-based variable reconstruction is best explained here by a simple example. This involves two highly correlated process variables, z_1 and z_2. For a high degree of correlation, Figure 2.1 outlines that a single component can recover the value of both variables without a significant loss of accuracy. The presentation of the simple introductory example is followed by a regression-based formulation of projecting samples optimally along predefined directions. This is then developed further to directly describe the effect of the sample projection upon the Q statistic. Next, the concept of reconstructing single variables is extended to include general directions that describe a fault subspace. Finally, the impact of reconstructing a particular variable upon the Hotelling's T^2 and Q statistics is analyzed and a regression-based reconstruction approach is introduced for isolating deterministic fault conditions.

An introductory example. Assuming that the sensor measuring variable z_1 produces a bias that takes the form of a step (sensor bias), the projection of the data vector $\mathbf{z}_{0_f}^T = \begin{pmatrix} z_{1_0} + \Delta z_1 & z_{2_0} \end{pmatrix}$ onto the semimajor of the control ellipse yields

$$t_f = \begin{pmatrix} p_{11} & p_{21} \end{pmatrix} \begin{pmatrix} z_{1_0} + \Delta z_1 \\ z_{2_0} \end{pmatrix}. \tag{3.60}$$

with Δz_1 being the magnitude of the sensor bias. Using the score variable t_f to recover the sensor readings gives rise to

$$\begin{pmatrix} \widehat{z}_{1_0} \\ \widehat{z}_{2_0} \end{pmatrix}_f = \begin{pmatrix} p_{11} \\ p_{21} \end{pmatrix} t_f = \begin{bmatrix} p_{11}^2 & p_{11}p_{21} \\ p_{21}p_{11} & p_{21}^2 \end{bmatrix} \begin{pmatrix} z_{1_0} + \Delta z_1 \\ z_{2_0} \end{pmatrix}. \tag{3.61}$$

The recovered values are therefore

$$\begin{aligned} \widehat{z}_{1_{0_f}} &= p_{11}^2 \left(z_{1_0} + \Delta z_1\right) + p_{11}p_{21}z_{2_0} \\ \widehat{z}_{2_{0_f}} &= p_{21}p_{11} \left(z_{1_0} + \Delta z_1\right) + p_{21}^2 z_{2_0}. \end{aligned} \tag{3.62}$$

This shows that the recovery of the value of both variables is affected by the sensor bias of z_1. The residuals of both variables are also affected by this fault

$$\begin{pmatrix} g_1 \\ g_2 \end{pmatrix}_f = \begin{bmatrix} 1 - p_{11}^2 & p_{11}p_{21} \\ p_{21}p_{11} & 1 - p_{21}^2 \end{bmatrix} \begin{pmatrix} z_{1_0} + \Delta z_1 \\ z_{2_0} \end{pmatrix} \tag{3.63}$$

since

$$\begin{pmatrix} g_1 \\ g_2 \end{pmatrix}_f = \begin{pmatrix} g_1 \\ g_2 \end{pmatrix} + \begin{pmatrix} \Delta g_1 \\ \Delta g_2 \end{pmatrix} = \begin{pmatrix} g_1 \\ g_2 \end{pmatrix} + \begin{pmatrix} 1 - p_{11}^2 \\ p_{21}p_{11} \end{pmatrix} \Delta z_1. \tag{3.64}$$

Subsection 3.3.1 examines dependencies in variable contributions to fault conditions, a general problem for constructing contribution charts (Lieftucht *et al.* 2006a; Lieftucht *et al.* 2006b).

The application of projection-based variable reconstruction, however, can overcome this problem if the fault direction is known *a priori* (Dunia *et al.* 1996; Dunia and Qin 1998) or can be estimated (Yue and Qin 2001) using a singular value decomposition. The principle is to remove the faulty sensor reading for z_1 by an estimate, producing the following iterative algorithm

$$\begin{pmatrix} \widehat{z}_{1_0} \\ \widehat{z}_{2_0} \end{pmatrix}_{j+1} = \begin{bmatrix} p_{11}^2 & p_{21}p_{11} \\ p_{21}p_{11} & p_{21}^2 \end{bmatrix} \begin{pmatrix} \widehat{z}_{1_{0_j}} \\ z_{2_0} \end{pmatrix}. \tag{3.65}$$

The iteration converges for $j \to \infty$ and yields the following estimate for $\widehat{z}_{1_0}$

$$\widehat{z}_{1_0} = p_{11}^2 \widehat{z}_{1_0} + p_{11}p_{21}z_{2_0} = \frac{p_{11}p_{21}}{1 - p_{11}^2} z_{2_0}. \tag{3.66}$$

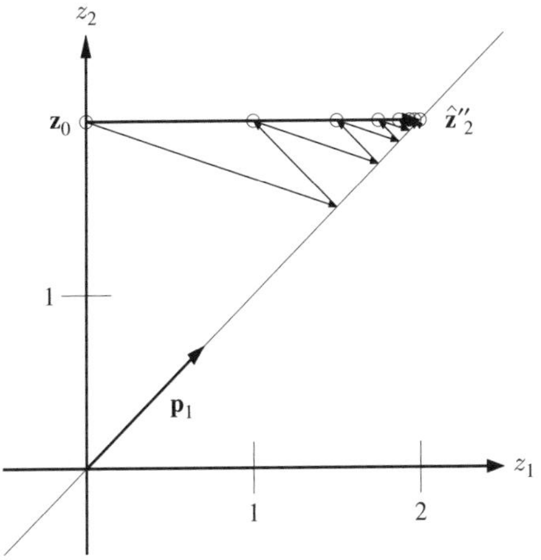

Figure 3.3 Illustrative example of projection-based variable reconstruction.

On the other hand, if a sensor bias is affecting z_2 the estimate $\widehat{z}_{2_0}$ is then

$$\widehat{z}_{2_0} = \frac{p_{21}p_{11}}{1 - p_{21}^2} z_{1_0}. \tag{3.67}$$

Figure 3.3 presents an example where the model subspace is spanned by $\mathbf{p}_1^T = \begin{pmatrix} 1/\sqrt{2} & 1/\sqrt{2} \end{pmatrix}$ and a data vector $\mathbf{z}_{0_f}^T = \begin{pmatrix} 0 & 2 \end{pmatrix}$ describes a sensor bias to the first variables, z_{1_0}. The series of arrows show the convergence of this iterative algorithm. Applying this scheme for $\mathbf{z}_{0_j} = \mathbf{z}_0$, Table 3.1 shows how $\widehat{z}_{1_{0_j}}$, $\widehat{z}_{2_{0_j}}$, $t_j = \mathbf{p}_1 \begin{pmatrix} \widehat{z}_{1_{0_j}} & z_{2_0} \end{pmatrix}^T$ and the convergence criterion $\varepsilon = z_{1_{0_{j+i}}} - z_{1_{0_j}}$ changes for the first few iteration steps. According to Figure 3.3 and Table 3.1, the 'corrected' or reconstructed data vector is $\widehat{\mathbf{z}}_0^T = \begin{pmatrix} 2 & 2 \end{pmatrix}$.

Table 3.1 Results of reconstructing z_{10} using the iterative projection-based method.

Iteration	$\widehat{z}_{1_{0_j}}$	$\widehat{z}_{2_0}$	t_j	ε
1	1.0000	2.0000	2.1213	1.0000
2	1.5000	2.0000	2.4749	0.5000
3	1.7500	2.0000	2.6517	0.2500
4	1.8750	2.0000	2.7400	0.1250
5	1.9375	2.0000	2.7842	0.0625
⋮	⋮	⋮	⋮	⋮
∞	2.0000	2.0000	2.8284	0.0000

Regression-based formulation of projection-based variable reconstruction. In a similar fashion to the residual-based tests in Subsection 3.2.2, it is possible to formulate the projection-based variable reconstruction scheme as a least squares problem under the assumptions that:

- the fault condition affects at most n sensors;
- the fault signature manifest itself as a step-type fault for the affected variables; and
- a recorded data set contains a sufficiently large set of samples representing the fault condition.

A sensor bias that affects the mth variable can be modeled as follows

$$\mathbf{z}_{0_f} = \mathbf{z}_0 + \boldsymbol{\upsilon}_m \Delta z_m, \tag{3.68}$$

where $\boldsymbol{\upsilon}_m \in \mathbb{R}^{n_z}$ is an Euclidian vector for which the mth element is 1 and the remaining ones are 0, and Δz_m is the magnitude of the sensor bias. The difference between the measured data vector $\mathbf{z}_{0_f}$ and $\boldsymbol{\upsilon}_m \Delta z_{0_m}$ is the Gaussian distributed vector $\mathbf{z}_0 \sim \mathcal{N}\left\{\mathbf{0}, \mathbf{S}_{z_0 z_0}\right\}$, whilst the $\mathbf{z}_{0_f}$ is Gaussian distributed with the same covariance matrix, that is, $\mathbf{z}_{0_f} \sim \mathcal{N}\left\{\boldsymbol{\upsilon}_m \Delta z_m, \mathbf{S}_{z_0 z_0}\right\}$, which follows from the data model in 2.2 and Table 2.1. This implies that the fault magnitude, Δz_m, can be estimated if the fault direction $\boldsymbol{\upsilon}_m$ is known by the following least squares formulation

$$\Delta \widehat{z}_m = \arg\min_{\zeta} \left\| \begin{pmatrix} z_{m_0}(1) \\ z_{m_0}(2) \\ \vdots \\ z_{m_0}(\widetilde{K}) \end{pmatrix}_f - \begin{pmatrix} 1 \\ 1 \\ \vdots \\ 1 \end{pmatrix} \zeta \right\|^2 . \tag{3.69}$$

The solution of (3.69) is the estimate of the mean value for z_{0_m}

$$\Delta \widehat{z}_m = \frac{\sum_{k=1}^{\widetilde{K}} z_{m_{0_f}}(k)}{\widetilde{K}} \tag{3.70}$$

which converges to the true fault magnitude as $\widetilde{K} \to \infty$.

Impact of fault condition upon Q statistic. Dunia and Qin (1998) showed that projecting samples along $\boldsymbol{\upsilon}_i$ has the following impact on the Q statistic

$$\Delta \widehat{z}_m = \arg\min_{\zeta} \sum_{k=1}^{\widetilde{K}} \left\| \mathbf{g}_f(k) - [\mathbf{I} - \mathfrak{C}]\, \boldsymbol{\upsilon}_m \zeta \right\|^2 , \tag{3.71}$$

Knowing that $\mathbf{g}_f = [\mathbf{I} - \mathfrak{C}]\left(\mathbf{z}_0 + \Delta\mathbf{z}\right) = \mathbf{g} + \Delta\mathbf{g}$, (3.71) becomes

$$\begin{aligned}\Delta\widehat{z}_m = \arg\min_{\zeta} = &\sum_{k=1}^{\widetilde{K}} Q_0\left(k\right) + 2\left(\Delta\mathbf{z}^T - \boldsymbol{\upsilon}_m^T\zeta\right)\mathbf{g} \\ &+ \left(\Delta\mathbf{z}^T - \boldsymbol{\upsilon}_m^T\zeta\right)[\mathbf{I} - \mathfrak{C}]\left(\Delta\mathbf{z} - \boldsymbol{\upsilon}_m\zeta\right),\end{aligned} \tag{3.72}$$

where $Q_0(k) = \mathbf{z}_0^T(k)\,[\mathbf{I} - \mathfrak{C}]\,\mathbf{z}_0(k)$. It follows from (3.72) that if

$$\Delta\mathbf{z} \propto \boldsymbol{\upsilon}_m. \tag{3.73}$$

Δz_m asymptotically converges to the true fault magnitude allowing a complete isolation between the step-type fault and normal stochastic process variation, hence

$$\frac{1}{\widetilde{K}}\sum_{k=1}^{\widetilde{K}} Q_0\left(k\right) \leq Q_\alpha. \tag{3.74}$$

The optimal solution of the objective function in (3.71) is given by

$$\Delta\widehat{z}_m = \left(\sum_{k=1}^{\widetilde{K}} \mathbf{g}_f^T\left(k\right)\right)\frac{[\mathbf{I} - \mathfrak{C}]\,\boldsymbol{\upsilon}_m}{\widetilde{K}\left(1 - \boldsymbol{\upsilon}_m^T\mathfrak{C}\boldsymbol{\upsilon}_m\right)}. \tag{3.75}$$

The estimates for Δz_m in (3.70) and (3.75) are identical. To see this, the sensor bias upon has the following impact on the residuals

$$\mathbf{g}_f = [\mathbf{I} - \mathfrak{C}]\left(\mathbf{z}_0 + \boldsymbol{\upsilon}_m\Delta z_m\right), \tag{3.76}$$

which we can substitute into the expectation of (3.75)

$$\begin{aligned}E\left\{\frac{\left(\mathbf{z}_0^T + \boldsymbol{\upsilon}_m^T\Delta z_m\right)[\mathbf{I} - \mathfrak{C}]\,\boldsymbol{\upsilon}_m}{\boldsymbol{\upsilon}_m^T\,[\mathbf{I} - \mathfrak{C}]\,\boldsymbol{\upsilon}_m}\right\} &= \underbrace{E\left\{\mathbf{z}_0^T\right\}}_{=\mathbf{0}}\frac{[\mathbf{I} - \mathfrak{C}]\,\boldsymbol{\upsilon}_m}{\boldsymbol{\upsilon}_m^T\left[\mathbf{I} - \mathfrak{C}\right]\boldsymbol{\upsilon}_m} + \\ &\qquad \Delta z_m\frac{\boldsymbol{\upsilon}_m^T\left[\mathbf{I} - \mathfrak{C}\right]\boldsymbol{\upsilon}_m}{\boldsymbol{\upsilon}_m^T\left[\mathbf{I} - \mathfrak{C}\right]\boldsymbol{\upsilon}_m} \\ E\left\{\frac{\left(\mathbf{z}_0^T + \boldsymbol{\upsilon}_m^T\Delta z_m\right)[\mathbf{I} - \mathfrak{C}]\,\boldsymbol{\upsilon}_m}{\boldsymbol{\upsilon}_m^T\,[\mathbf{I} - \mathfrak{C}]\,\boldsymbol{\upsilon}_m}\right\} &= \Delta z_m.\end{aligned} \tag{3.77}$$

Geometrically, the fault condition *moves* the samples along the direction $\boldsymbol{\upsilon}_m$ and this shift can be identified and removed from the recorded data.

Extension to multivariate fault directions. A straightforward extension is the assumption that $\mathfrak{m} \leq n$ sensor faults arise at the same time. The limitation

$\mathfrak{m} \leq n$ follows from the fact that the rank of $\mathbf{P}$ is n and is further elaborated in Subsection 3.3.2. The extension of reconstructing up to n variables would enable the description of more complex fault scenarios. This extension, however, is still restricted by the fact that the fault subspace, denoted here by $\boldsymbol{\Upsilon}$, is spanned by Euclidian vectors

$$\boldsymbol{\Upsilon} = \begin{bmatrix} \boldsymbol{\upsilon}_{\mathfrak{J}(1)} & \boldsymbol{\upsilon}_{\mathfrak{J}(2)} & \cdots & \boldsymbol{\upsilon}_{\mathfrak{J}(\mathfrak{m})} \end{bmatrix} \in \mathbb{R}^{n_z \times \mathfrak{m}} \tag{3.78}$$

where $\mathfrak{J}$ is an index set storing the variables to be reconstructed, for example $\mathfrak{J}_1 = 2$, $\mathfrak{J}_2 = 6$, $\cdots$, $\mathfrak{J}_\mathfrak{m} = 24$. The case of $\boldsymbol{\Upsilon}$ storing non-Euclidian vectors is elaborated below. Using the fault subspace $\boldsymbol{\Upsilon}$ projection-based variable reconstruction identifies the fault magnitude for each Euclidian base vector from the Q statistic

$$\Delta\widehat{\mathbf{z}}_{\mathfrak{J}} = \arg\min_{\boldsymbol{\zeta}} \sum_{k=1}^{\widetilde{K}} Q_f(k) - 2\mathbf{z}_{0_f}^T(k)\,\widetilde{\boldsymbol{\Upsilon}}\boldsymbol{\zeta} + \boldsymbol{\zeta}^T\widetilde{\boldsymbol{\Upsilon}}^T\widetilde{\boldsymbol{\Upsilon}}\boldsymbol{\zeta}. \tag{3.79}$$

Here, $\widetilde{\boldsymbol{\Upsilon}} = [\mathbf{I} - \mathfrak{C}]\,\boldsymbol{\Upsilon}$. The solution for the least squares objective function is

$$\Delta\widehat{\mathbf{z}}_{\mathfrak{J}} = \widetilde{\boldsymbol{\Upsilon}}^{\dagger} \frac{1}{\widetilde{K}} \sum_{k=1}^{\widetilde{K}} \mathbf{z}_{0_f}(k) \tag{3.80}$$

with $\widetilde{\boldsymbol{\Upsilon}}^{\dagger} = \left[\widetilde{\boldsymbol{\Upsilon}}^T\widetilde{\boldsymbol{\Upsilon}}\right]^{-1}\widetilde{\boldsymbol{\Upsilon}}^T$ being the generalized inverse of $\widetilde{\boldsymbol{\Upsilon}}$. If $\boldsymbol{\Upsilon}$ is the correct fault subspace, the above sum estimates the fault magnitude $\Delta\mathbf{z}_{0_f}$ consistently

$$\Delta\mathbf{z}_{\mathfrak{J}} = \lim_{\widetilde{K}\to\infty} \widetilde{\boldsymbol{\Upsilon}}^{\dagger} \frac{1}{\widetilde{K}} \sum_{k=1}^{\widetilde{K}} \left(\mathbf{z}_0(k) + \boldsymbol{\Upsilon}\Delta\mathbf{z}_{\mathfrak{J}}\right), \tag{3.81}$$

since

$$\widetilde{\boldsymbol{\Upsilon}}^{\dagger} \underbrace{E\left\{\mathbf{z}_0\right\}}_{\mathbf{0}} + \underbrace{\left[\boldsymbol{\Upsilon}^T[\mathbf{I} - \mathfrak{C}]\,\boldsymbol{\Upsilon}\right]^{-1}\boldsymbol{\Upsilon}^T[\mathbf{I} - \mathfrak{C}]\,\boldsymbol{\Upsilon}}_{\mathbf{I}}\,\Delta\mathbf{z}_{\mathfrak{J}} = \Delta\mathbf{z}_{\mathfrak{J}} \tag{3.82}$$

This, in turn, implies that the correct fault subspace allows identifying the correct fault magnitude and removing the fault information from the Q statistic

$$\lim_{\widetilde{K}\to\infty} \widehat{Q}_0(k) = \left(\mathbf{z}_{0_f}(k) - \boldsymbol{\Upsilon}\Delta\widehat{\mathbf{z}}_{\mathfrak{J}}\right)^T[\mathbf{I} - \mathfrak{C}]\left(\mathbf{z}_{0_f}(k) - \boldsymbol{\Upsilon}\Delta\widehat{\mathbf{z}}_{\mathfrak{J}}\right) \longrightarrow Q_0(k). \tag{3.83}$$

So far, we assumed that the fault subspace is spanned by Euclidian vectors that represent individual variables. This restriction can be removed by defining a set

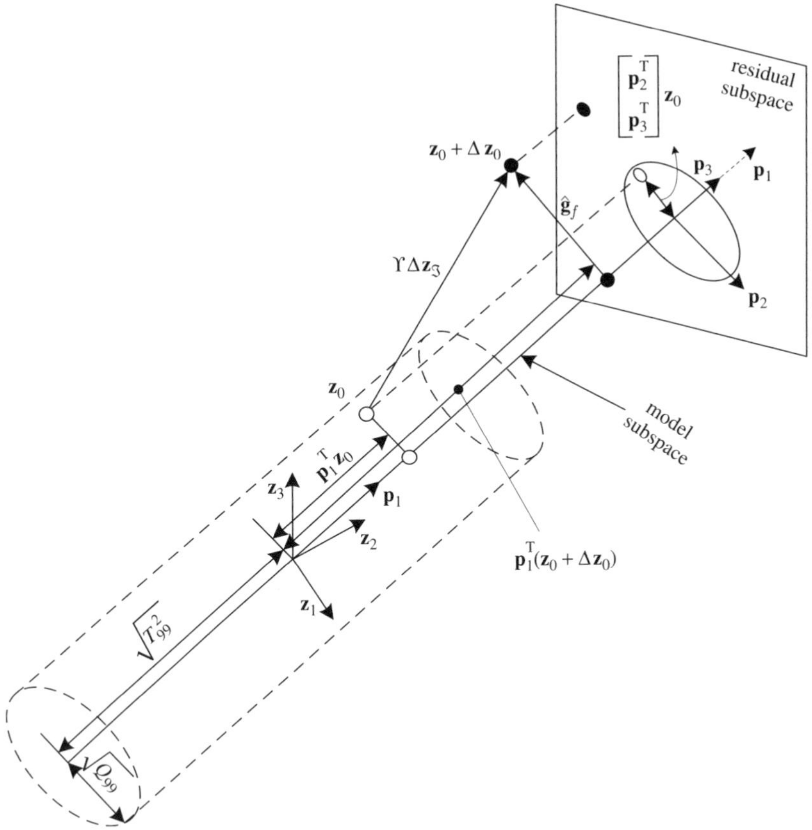

Figure 3.4 Graphical illustration of multivariate projection-based variable reconstruction along a predefined fault axis.

of up to $\mathfrak{m} \leq n$ linearly independent vectors $\boldsymbol{\upsilon}_1, \boldsymbol{\upsilon}_2, \ldots, \boldsymbol{\upsilon}_{\mathfrak{m}}$ of unit length, which yields the following model for describing fault conditions

$$\mathbf{z}_{0_f} = \mathbf{z}_0 + \boldsymbol{\Upsilon} \Delta \mathbf{z}_{\mathfrak{J}}. \tag{3.84}$$

The fault magnitude in the direction of vector $\boldsymbol{\upsilon}_i$ is $\Delta z_{\mathfrak{J}_i}$. The $\boldsymbol{\upsilon}_i$ vectors can be obtained by applying a singular value decomposition on the data set containing $\widetilde{K}$ samples describing the fault condition (Yue and Qin 2001). Figure 3.4 gives a graphical interpretation of this regression problem. Equation (3.79) presents the associated objective function for estimating $\Delta \mathbf{z}_{\mathfrak{J}}$. Different from the projection along Euclidian directions, if the fault subspace is constructed from non-Euclidian vectors, the projection-based variable reconstruction constitutes an oblique projection.

Projection-based variable reconstruction for a single sample. As for residual-based tests, the above projection-based approach requires a sufficient number of samples describing the fault condition, which hampers its practical usefulness. If an abnormal event is detected, the identification of potential root causes is vital in order to support process operators in carrying out appropriate responses. Despite the limitations of contribution charts, discussed in (Lieftucht *et al.* 2006a; Yoon and MacGregor 2001) and Subsection 3.3.1, they should be used as a first instrument.

Equations (3.60) to (3.66) show that a sensor bias can be removed if it is known which one is faulty. Dunia and Qin (1998), and Lieftucht *et al.* (2006b) pointed out that projection-based reconstruction and contribution charts can be applied together to estimate the impact of a fault condition upon individual process variables. A measure for assessing this impact is how much the reconstructed variable reduces the value of both non-negative quadratic statistics. For the Q statistic, this follows from (3.72) for $\widetilde{K} = 1$ and $\mathfrak{m} = 1$, and (3.79) for $1 < \mathfrak{m} \leq n$.

As highlighted in Lieftucht *et al.* (2006a), however, the projection-based reconstruction impacts the geometry of the PCA model and therefore the non-negative quadratic statistics. Subsection 3.3.2 describes the effect of reconstructing a variable set upon the geometry of the PCA model. In summary, the procedure to incorporate the effect of projecting a set of $\mathfrak{m}$ process variables is as follows.

1. Reconstruct the data covariance matrix by applying the following equation

$$\widetilde{\mathbf{S}}^{\star}_{z_0z_0} = \begin{bmatrix} \widetilde{\mathbf{S}}^{\star}_{z_0z_{0_1}} & \widetilde{\mathbf{S}}^{\star}_{z_0z_{0_2}} \\ \widetilde{\mathbf{S}}^{\star^T}_{z_0z_{0_2}} & \widetilde{\mathbf{S}}^{\star}_{z_0z_{0_3}} \end{bmatrix} = \begin{bmatrix} \mathbf{\Omega} \\ \mathbf{I} \end{bmatrix} \mathbf{S}^{\star}_{z_0z_{0_3}} \begin{bmatrix} \mathbf{\Omega}^T & \mathbf{I} \end{bmatrix} \tag{3.85}$$

where:

- $\mathbf{\Omega} = \left[\mathbf{I} - \mathfrak{C}^{\star}_1\right]^{-1} \mathfrak{C}^{\star}_2$;
- $\mathfrak{C}^{\star} = \begin{bmatrix} \mathfrak{C}^{\star}_1 & \mathfrak{C}^{\star}_2 \\ \mathfrak{C}^{\star^T}_2 & \mathfrak{C}^{\star}_3 \end{bmatrix}$;
- $\mathbf{z}^{\star^T}_0 = \left(\mathbf{z}^{\star^T}_{0_1} \; \mathbf{z}^{\star^T}_{0_2}\right)$ is a data vector that stores the variables to be reconstructed as the first $\mathfrak{m}$ elements, $\mathbf{z}^{\star}_{0_1} \in \mathbb{R}^{\mathfrak{m}}$, and the variables that are not reconstructed as the remaining $n_z - \mathfrak{m}$ elements, $\mathbf{z}^{\star}_{0_2} \in \mathbb{R}^{n_z - \mathfrak{m}}$;
- the matrices denoted by the superscript $\star$ refer to the variables stored in the rearranged vector $\mathbf{z}^{\star}_0$;
- the symbol $\sim$ refers to the reconstructed portions of the data covariance matrix describing the impact of reconstructing the set of $\mathfrak{m}$ process variables;
- $\mathbf{S}^{\star}_{z_0z_{0_3}}$ is the partition of the data covariance matrix $\mathbf{S}^{\star}_{z_0z_0}$ that relates to the variable set that is not reconstructed; and
- the indices 1, 2 and 3 are associated with the covariance matrix of the reconstructed variables, the cross covariance matrix that includes the

reconstructed and the remaining variables and the covariance matrix of the unreconstructed variables, respectively.

2. Calculate the eigendecomposition of $\widetilde{\mathbf{S}}^{\star}_{z_0 z_0}$, that is, calculate $\widetilde{\mathbf{\Lambda}}^{\star}$ and $\widetilde{\mathbf{P}}^{\star}$.

3. Compute the T^2 statistic from the retained score variables $\widetilde{\mathbf{t}}^{\star} = \widetilde{\mathbf{P}}^{\star^T} \widetilde{\mathbf{z}}^{\star}_0$ with $\widetilde{\mathbf{z}}^{\star}_0 = \begin{bmatrix} \mathbf{\Omega}^T & \mathbf{I} \end{bmatrix}^T \mathbf{z}^{\star}_{0_2}$, i.e. $\widetilde{T}^2 = \widetilde{\mathbf{t}}^{\star^T} \widetilde{\mathbf{\Lambda}}^{\star^{-1}} \widetilde{\mathbf{t}}^{\star}$.

4. Determine the Q statistic, $Q = \left(\mathbf{z}^{\star}_0 - \widetilde{\mathbf{P}}^{\star}\widetilde{\mathbf{t}}\right)^T \left(\mathbf{z}^{\star}_0 - \widetilde{\mathbf{P}}^{\star}\widetilde{\mathbf{t}}\right)$, and recompute the control limit by reapplying (3.16) using the discarded eigenvalues of $\widetilde{\mathbf{S}}^{\star}_{z_0 z_0}$.

Ignoring the effect the reconstruction process imposes upon the underlying PCA model is demonstrated in Chapter 4.

Limitations of projection-based variable reconstruction. Despite the simplicity and robustness of the projection-based variable reconstruction approach, it is prone to the following problems (Lieftucht *et al.* 2009).

Problem 3.2.1 *The maximum number of variables to be reconstructed equals the number of retained principal components Dunia and Qin* (1998). This limitation renders the technique difficult to apply for diagnosing complex process faults, which typically affect a larger number of variables.

Problem 3.2.2 *The reconstruction process reduces the dimensions of the residual subspace and if $n_z - n - \mathfrak{m} \leq 0$ the Q statistic does not exist (Lieftucht et al. 2006b).* This is a typical scenario for cases where the ratio n/n_z is close to 1.

Problem 3.2.3 *The fault condition is assumed to be described by the linear subspace* $\mathbf{\Upsilon}$. Therefore, a fault path that represents a curve or trajectory is not fully reconstructible.

These problems are demonstrated in Lieftucht *et al.* (2009) through the analysis of experimental data from a reactive distillation column. A regression-based variable reconstruction is introduced next to address these problems.

3.2.3.3 Regression-based variable reconstruction

This approach estimates and separates the fault signature from the recorded variables (Lieftucht *et al.* 2006a; Lieftucht *et al.* 2009). Different from projection-based reconstruction, the regression-based approach relies on the following assumptions:

- The fault signature in the score space, $f(k)$, is deterministic, that is, the signature for a particular process variable is a function of the sample index k; and

- The fault is superimposed onto the process variables, that is, it is added to the complete set of score variables

$$\mathbf{t}_f(k) = \begin{bmatrix} \mathbf{P}^T \\ \mathbf{P}_d^T \end{bmatrix} \left(\mathbf{z}_0(k) + \Delta\mathbf{z}(k)\right) \quad \mathbf{t}_f(k) = \mathbf{t}(k) + \mathrm{f}(k) \tag{3.86}$$

where the subscript $_f$ refers to the corrupted vectors,

- $\mathbf{t}(k) = \begin{bmatrix} \mathbf{P}^T \\ \mathbf{P}_d^T \end{bmatrix} \mathbf{z}_0(k)$; and
- $f(k) = \begin{bmatrix} \mathbf{P}^T \\ \mathbf{P}_d^T \end{bmatrix} \Delta\mathbf{z}(k)$.

In contrast to the regression-based technique for missing data, different assumptions are applied to the score variables leading to a method for variable reconstruction according to Figure 3.5. Based on the above assumptions, the fault signature can be described by a parametric curve that can be identified using various techniques, such as polynomials, principal curves and artificial neural networks (Walczak and Massart 1996). For simplicity, radial basis function networks (RBFNs) are utilized here.

According to Figure 3.5, $\widehat{f}(k)$ is subtracted from $\mathbf{t}_f(k)$. The benefit of using the score variables follows from Theorem 9.3.2. The score variables are statistically independent but the original process variables are highly correlated. Equations (3.61) and (3.63) describe the negative impact of variable correlation upon the PCA model, which not only identify a faulty sensor but may also suggest that other variables are affected by a sensor bias. Separating the fault signature from the corrupted samples on the basis of the score variables, however, circumvents the correlation issue.

The first block in Figure 3.5 produces the complete set of n_z score variables $\mathbf{f}(k)$ from the corrupted samples of the process variables $\mathbf{z}_0(k) + \Delta\mathbf{z}(k)$. The score variable set is then used to estimate the fault signature $\widehat{f}(k)$ as above which is subtracted from the score variables to estimate $\widehat{\mathbf{t}}(k)$. Since a fault signature may affect each of the score variables, all score variables need to be included to estimate the fault signature.

The output from a radial basis function network is defined by

$$\begin{aligned} \widehat{f}_1(k) &= \textstyle\sum_{i=1}^{R} l_{1i}\left(t_{1_f}(k), c_i, \varrho\right) a_{1i} \\ &\vdots \\ \widehat{f}_{n_z}(k) &= \textstyle\sum_{i=1}^{R} l_{n_z i}\left(t_{n_{z f}}(k), c_i, \varrho\right) a_{n_z i} \end{aligned} \quad . \tag{3.87}$$

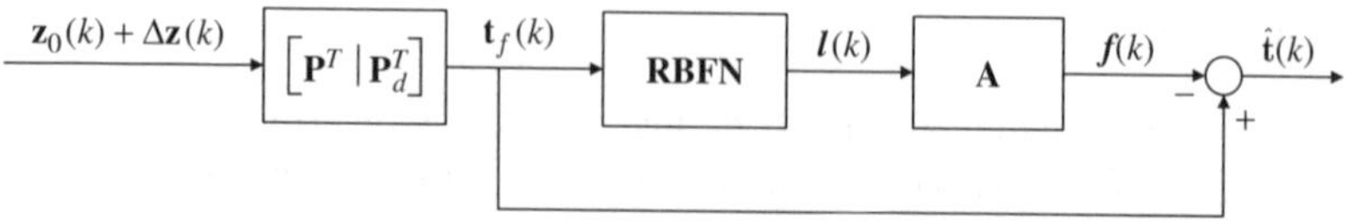

Figure 3.5 Structure of the regression based variable reconstruction technique.

Here, R is the number of network nodes, $l_{ji} = \exp(-(\frac{t_{jf}(k)-c_i}{\varrho})^2)$, $1 \leq j \leq n_z$, $1 \leq i \leq R$ is a Gaussian basis function for which c_i and ϱ are the center and the width, respectively, and $a_{11}, \ldots, a_{n_z R}$ are the network weights. By applying (3.87), the separation of the fault signature becomes

$$\mathbf{t}_f(k) = \mathbf{A}\boldsymbol{l}(k) + \widehat{\mathbf{t}}(k) = \widehat{f}(k) + \widehat{\mathbf{t}}(k), \tag{3.88}$$

where, $\boldsymbol{l}(k)$ is a vector storing the values for each network node for the kth sample, $\mathbf{A}$ is a parameter matrix storing the network weights and $\widehat{\mathbf{t}}(k)$ is the isolated stochastic part of the computed score variables $\mathbf{t}_f(k)$. For simplicity, the center of Gaussian basis functions is defined by a grid, that is, the distance between two neighboring basis functions is equivalent for each pair and their widths are assumed to be predefined. The training of the network therefore reduces to a least squares problem.

Chapter 4 shows that the regression-based reconstruction has the potential to provide a clear picture of the impact of a fault condition. Lieftucht *et al.* (2009) present another example that involves recorded data from an experimental reactive distillation unit at the University of Dortmund, Germany, where this technique could offer a good isolation of the fault signatures for a failure in the reflux preheater and multiple faults in cooling water and acid feed supplies.

3.3 Geometry of variable projections

For the projection-based reconstruction of a single sample, this section analyzes the geometry of variable projections, which involves orthogonal projections from the original variable space onto smaller dimensional reconstruction subspaces that capture significant variation in the original variable set. Subsection 3.3.1 examines the impact of such projections upon the contribution charts of the Q statistic for PCA and PLS and shows that the variable contributions are linearly dependent. This is particularly true if the number of retained components is close to the size of the original variable set. Subsection 3.3.2 then studies the impact of variable reconstruction upon the geometry of the PCA model.

3.3.1 Linear dependency of projection residuals

Given the PCA residual vector $\mathbf{g} \in \mathbb{R}^{n_z}$ and the centered data vector $\mathbf{z}_0 \in \mathbb{R}^{n_z}$, it is straightforward to show that the residual vector is orthogonal to the model subspace

$$\mathbf{g} = [\mathbf{I} - \mathfrak{C}]\,\mathbf{z}_0 \qquad \mathbf{P}^T\mathbf{g} = \mathbf{P}^T\,[\mathbf{I} - \mathfrak{C}]\,\mathbf{z}_0 = \left[\mathbf{P}^T - \mathbf{P}^T\right]\mathbf{z}_0 = \mathbf{0}. \tag{3.89}$$

Here, $\mathbf{P} \in \mathbb{R}^{n_z \times n}$, $n < n_z$, is the loading matrix, storing the eigenvectors of the data covariance matrix as column vectors, which span the model plane. The above relationship holds true, since the eigenvectors of a symmetric matrix are mutually orthonormal. Equation (3.89), however, can be further elaborated

$$\begin{bmatrix} p_{11} & p_{21} & \cdots & p_{n1} & p_{n+1,1} & p_{n+2,1} & \cdots & p_{n_z1} \\ p_{12} & p_{22} & \cdots & p_{n2} & p_{n+1,2} & p_{n+2,2} & \cdots & p_{n_z2} \\ \vdots & \vdots & & \vdots & \vdots & \vdots & & \vdots \\ p_{1n} & p_{2n} & \cdots & p_{nn} & p_{n+1,n} & p_{n+2,n} & \cdots & p_{n_zn} \end{bmatrix} \begin{pmatrix} g_1 \\ g_2 \\ \vdots \\ g_n \\ g_{n+1} \\ g_{n+2} \\ \vdots \\ g_{n_z} \end{pmatrix} = \begin{pmatrix} 0 \\ 0 \\ \vdots \\ 0 \end{pmatrix} \tag{3.90}$$

and partitioning it as follows

$$\mathbf{g}_1^T = \begin{pmatrix} g_1 \\ g_2 \\ \vdots \\ g_n \end{pmatrix} \in \mathbb{R}^n \qquad \mathbf{g}_2^T = \begin{pmatrix} g_{n+1} \\ g_{n+2} \\ \vdots \\ g_{n_z} \end{pmatrix} \in \mathbb{R}^{n_z-n}$$

$$\boldsymbol{\Psi}_1 = \begin{bmatrix} p_{11} & p_{21} & \cdots & p_{n1} \\ p_{12} & p_{22} & \cdots & p_{n2} \\ \vdots & \vdots & \ddots & \vdots \\ p_{1n} & p_{2n} & \cdots & p_{nn} \end{bmatrix} \in \mathbb{R}^{n\times n} \tag{3.91}$$

$$\boldsymbol{\Psi}_2 = \begin{bmatrix} p_{n+1,1} & p_{n+2,1} & \cdots & p_{n_z,1} \\ p_{n+1,2} & p_{n+2,2} & \cdots & p_{n_z,2} \\ \vdots & \vdots & \ddots & \vdots \\ p_{1n} & p_{2n} & \cdots & p_{n_zn} \end{bmatrix} \in \mathbb{R}^{n\times n_z-n}$$

which gives rise to

$$\begin{bmatrix} \boldsymbol{\Psi}_1 & \boldsymbol{\Psi}_2 \end{bmatrix} \begin{pmatrix} \mathbf{g}_1 \\ \mathbf{g}_2 \end{pmatrix} = \boldsymbol{\Psi}_1\mathbf{g}_1 + \boldsymbol{\Psi}_2\mathbf{g}_2 = \mathbf{0}. \tag{3.92}$$

$\boldsymbol{\Psi}_1$ is a square matrix and can be inverted, which yields

$$\mathbf{g}_1 = -\boldsymbol{\Psi}_1^{-1}\boldsymbol{\Psi}_2\mathbf{g}_2. \tag{3.93}$$

The above relationship is not dependent upon the number of source signals n. In any case there will be a linear combination of the variable contribution of the Q statistic.

3.3.2 Geometric analysis of variable reconstruction

The geometrical properties of the multidimensional reconstruction technique for a single sample are now analyzed. Some preliminary results for one-dimensional sensor faults were given in Lieftucht *et al.* (2004) and a more detailed analysis is given in Lieftucht *et al.* (2006a). For simplicity, the data vector $\mathbf{z}_0$ is rearranged such that the reconstructed variables are stored as the first $\mathfrak{m} \leq n$ elements. The rearranged vector and the corresponding covariance and loading matrix are denoted by $\mathbf{z}_0^\star$, $\mathbf{S}_{z_0z_0}^\star$ and $\mathbf{P}^\star$, respectively. Furthermore, the reconstructed vector and the reconstructed data covariance and loading matrices are referred to as $\widetilde{\mathbf{z}}_0^\star$, $\widetilde{\mathbf{S}}_{z_0z_0}^\star$ and $\widetilde{\mathbf{P}}^\star$, respectively.

3.3.2.1 Optimality of projection-based variable reconstruction

The following holds true for the reconstructed data vector and the model subspace.

Theorem 3.3.1 *After reconstructing* $\mathfrak{m} \leq n$ *variables, the reconstructed data vector lies in an* $n_z - \mathfrak{m}$ *dimensional subspace, such that the orthogonal distance between the reconstructed vector and the model subspace is minimized.*

Proof. The residual vector, $\widetilde{\mathbf{g}}^\star$ is equal to

$$\widetilde{\mathbf{g}}^\star = \left[\mathbf{I} - \mathfrak{C}^\star\right]\widetilde{\mathbf{z}}_0^\star, \tag{3.94}$$

where,

$$\mathfrak{C}^\star = \mathbf{P}^\star\mathbf{P}^{\star^T} \text{ and } \widetilde{\mathbf{z}}_0^\star = \begin{pmatrix} \boldsymbol{\Omega}\mathbf{z}_{0_2}^\star \\ \mathbf{z}_{0_2}^\star. \end{pmatrix}. \tag{3.95}$$

Here $\boldsymbol{\Omega}$ is a projection matrix, which is defined below

$$\boldsymbol{\Omega} = \begin{bmatrix} \boldsymbol{\omega}_1^T \\ \vdots \\ \boldsymbol{\omega}_i^T \\ \vdots \\ \boldsymbol{\omega}_{\mathfrak{m}}^T \end{bmatrix} \qquad \boldsymbol{\omega}_i \in \mathbb{R}^{n_z - \mathfrak{m}}. \tag{3.96}$$

The squared distance between the reconstructed point and the model subspace is

$$\widetilde{\mathbf{g}}^{\star^T}\widetilde{\mathbf{g}}^\star = \begin{pmatrix} \mathbf{z}_{0_2}^{\star^T}\boldsymbol{\Omega}^T & \mathbf{z}_{0_2}^{\star^T} \end{pmatrix} \begin{bmatrix} \mathbf{I} - \mathfrak{C}_1^\star & -\mathfrak{C}_2^\star \\ -\mathfrak{C}_2^{\star^T} & \mathbf{I} - \mathfrak{C}_3^\star \end{bmatrix} \begin{pmatrix} \boldsymbol{\Omega}\mathbf{z}_{0_2}^\star \\ \mathbf{z}_{0_2}^\star \end{pmatrix}. \tag{3.97}$$

Here $\mathfrak{C}_1^\star \in \mathbb{R}^{\mathfrak{m}\times\mathfrak{m}}$, $\mathfrak{C}_2^\star \in \mathbb{R}^{\mathfrak{m}\times n_z-\mathfrak{m}}$ and $\mathfrak{C}_3^\star \in \mathbb{R}^{n_z-\mathfrak{m}\times n_z-\mathfrak{m}}$. Defining the objective function

$$\begin{aligned} \boldsymbol{\Omega} &= \arg\min_{\boldsymbol{\Omega}} \mathbf{z}_{0_2}^{\star^T}\boldsymbol{\Omega}^T\left[\mathbf{I} - \mathfrak{C}_1^\star\right]\boldsymbol{\Omega}\mathbf{z}_{0_2}^\star - 2\mathbf{z}_{0_2}^{\star^T}\mathfrak{C}_2^{\star^T}\boldsymbol{\Omega}\mathbf{z}_{0_2}^\star + \mathbf{z}_{0_2}^{\star^T}\left[\mathbf{I} - \mathfrak{C}_3^\star\right]\mathbf{z}_{0_2}^\star \\ \boldsymbol{\Omega} &= \arg\min_{\boldsymbol{\Omega}} \mathbf{z}_{0_2}^{\star^T}\left[\boldsymbol{\Omega}^T\left[\mathbf{I} - \mathfrak{C}_1^\star\right]\boldsymbol{\Omega} - 2\boldsymbol{\Omega}^T\mathfrak{C}_2^\star + \left[\mathbf{I} - \mathfrak{C}_3^\star\right]\right]\mathbf{z}_{0_2}^\star \end{aligned} \tag{3.98}$$

yields the following minimum

$$\boxed{\mathbf{\Omega} = \left[\mathbf{I} - \mathfrak{C}_1^\star\right]^{-1} \mathfrak{C}_2^\star}. \tag{3.99}$$

The resulting expression for $\mathbf{\Omega}$ is equivalent to the projection matrix obtained from reconstructing the variable set $\mathbf{z}_{0_1}^\star$ using projection-based variable reconstruction. This follows from an extended version of (3.65) for $j \to \infty$

$$\begin{pmatrix} \widetilde{\mathbf{z}}_{0_1}^\star \\ \widetilde{\mathbf{z}}_{0_2}^\star \end{pmatrix} = \begin{bmatrix} \mathfrak{C}_1^\star & \mathfrak{C}_2^\star \\ \mathfrak{C}_2^{\star^T} & \mathfrak{C}_3^\star \end{bmatrix} \begin{pmatrix} \widetilde{\mathbf{z}}_{0_1}^\star \\ \mathbf{z}_{0_2}^\star \end{pmatrix}. \tag{3.100}$$

which can be simplified to

$$\boxed{\widetilde{\mathbf{z}}_{0_1}^\star = \mathfrak{C}_1^\star \widetilde{\mathbf{z}}_{0_1}^\star + \mathfrak{C}_2^\star \mathbf{z}_{0_2}^\star = \left[\mathbf{I} - \mathfrak{C}_1^\star\right]^{-1} \mathfrak{C}_2^\star \mathbf{z}_{0_2}^\star}. \tag{3.101}$$

3.3.2.2 Reconstruction subspace

The data vector $\mathbf{z}_0 \in \mathbb{R}^{n_z}$ but the reconstruction of $\mathbf{z}_{0_1}^\star$ to form $\widetilde{\mathbf{z}}_0^\star$ results in a projection of $\mathbf{z}_0^\star$ onto a $(n_z - \mathrm{m})$-dimensional subspace, which Theorem 3.3.2 formulates. Any sample $\mathbf{z}_0^\star$ is projected onto this *reconstruction subspace* along the m directions defined in $\mathbf{\Upsilon}$. As Theorem 3.3.1 points out, the distance of the reconstructed sample and the model subspace is minimal.

Theorem 3.3.2 *The reconstruction of $\mathbf{z}_0^\star$ is equivalent to the projection of $\mathbf{z}_0^\star$ onto a $(n_z - \mathrm{m})$-dimensional subspace $\mathbf{\Pi}$. This subspace is spanned by the column vectors of the matrix $\mathbf{\Pi}$*

$$\mathbf{\Pi} = \begin{bmatrix} \mathbf{\Omega} \\ \mathbf{I} \end{bmatrix}, \tag{3.102}$$

which includes the model and the residual subspace.

Proof. After reconstructing $\mathbf{z}_0^\star$, the subspace in which the projected samples lie is given by the following $n_z - \mathrm{m}$ base vectors. These can be extracted from $\widetilde{\mathbf{z}}_0^\star$

$$\widetilde{\mathbf{z}}_0^\star = \begin{pmatrix} \boldsymbol{\omega}_1^T \mathbf{z}_{0_2}^\star \\ \vdots \\ \boldsymbol{\omega}_{\mathrm{m}}^T \mathbf{z}_{0_2}^\star \\ z_{0_{2,1}}^\star \\ \vdots \\ z_{0_{2,n_z-\mathrm{m}}}^\star \end{pmatrix} \tag{3.103}$$

and are consequently given by:

$$\boldsymbol{\Pi} = \left[\begin{pmatrix} \omega_{11} \\ \vdots \\ \omega_{\mathrm{m}1} \\ 1 \\ 0 \\ 0 \\ \vdots \\ 0 \end{pmatrix} \begin{pmatrix} \omega_{12} \\ \vdots \\ \omega_{\mathrm{m}2} \\ 0 \\ 1 \\ 0 \\ \vdots \\ 0 \end{pmatrix} \cdots \begin{pmatrix} \omega_{1,n_z-\mathrm{m}} \\ \vdots \\ \omega_{\mathrm{m},n_z-\mathrm{m}} \\ 0 \\ 0 \\ 0 \\ \vdots \\ 1 \end{pmatrix} \right]. \tag{3.104}$$

Note that the base vectors defined in (3.104) are not of unit length.

3.3.2.3 Maximum dimension of fault subspace

Theorem 3.3.3 discusses the maximum dimension of the fault subspace.

Theorem 3.3.3 *The maximum dimension of the fault subspace $\boldsymbol{\Upsilon}$ is n irrespective of whether the columns made up of Euclidian or non-Euclidian vectors.*

Proof. **$\boldsymbol{\Upsilon}$ contains Euclidian vectors**. Following from (3.101), rewriting the squared matrix $\left[\mathbf{I} - \mathfrak{C}_1^\star\right]^{-1}$ using the matrix inversion lemma yields

$$\left[\mathbf{I} - \mathfrak{C}_1^\star\right]^{-1} = \mathbf{I} - \left[\left[\mathfrak{C}_1^\star\right]^{-1} + \mathbf{I}\right]^{-1} \tag{3.105}$$

and produces the following equivalent projection matrices

$$\left[\mathbf{I} - \mathfrak{C}_1^\star\right]^{-1} \mathfrak{C}_2^\star = \mathfrak{C}_2^\star - \left[\left[\mathfrak{C}_1^\star\right]^{-1} + \mathbf{I}\right]^{-1} \mathfrak{C}_2^\star. \tag{3.106}$$

Both sides, however, can only be equivalent if the inverse of $\mathfrak{C}_1^\star$ exists. Given that the rank of $\mathfrak{C}^\star$ is equal to n and assuming that any combination of n column vectors in $\mathfrak{C}^\star$ span a subspace of dimension n, the maximum partition in the upper left corner of $\mathfrak{C}^\star$ can be an n by n matrix. This result can also be obtained by reexamining the iteration procedure in (3.65)

$$\widetilde{\mathbf{z}}^\star_{0_{1_{j+1}}} = \mathfrak{C}_1^\star \widetilde{\mathbf{z}}^\star_{0_{1_j}} + \mathfrak{C}_2^\star \mathbf{z}^\star_{0_2}. \tag{3.107}$$

If the iteration is initiated by selecting $\widetilde{\mathbf{z}}^\star_{0_{1_0}} = \mathbf{0}$ and the iterative formulation in (3.107) is used to work backwards from $\widetilde{\mathbf{z}}^\star_{0_{1_j}}$ to $\widetilde{\mathbf{z}}^\star_{0_{1_0}}$, it becomes

$$\widetilde{\mathbf{z}}^\star_{0_{1_{j+1}}} = \left[\sum_{i=0}^{j} \mathfrak{C}_1^{\star^i}\right] \mathfrak{C}_2^\star \mathbf{z}^\star_{0_2}. \tag{3.108}$$

Asymptotically, however,

$$\boxed{\lim_{j\to\infty}\sum_{i=1}^{j}\mathfrak{C}_1^{\star^i} \to \left[\mathbf{I}-\mathfrak{C}_1^{\star}\right]^{-1} \quad \mathfrak{C}_1^{\star}\in\mathbb{R}^{\mathrm{m}\times\mathrm{m}} \quad \mathrm{m}\le n}. \tag{3.109}$$

The above relationships yield some interesting facts. First, the block matrix $\mathfrak{C}_1^{\star}$ is symmetric and positive definite if $\mathrm{m}\le n$. The symmetry follows from the symmetry of $\mathfrak{C}^{\star}$ and the positive definiteness results from the partitioning of

$$\mathbf{P}^{\star}=\begin{bmatrix}\mathbf{P}_1^{\star}\\ \mathbf{P}_2^{\star}\end{bmatrix} \text{ and hence, } \mathfrak{C}_1^{\star}=\mathbf{P}_1^{\star}\mathbf{P}_1^{\star^T}. \tag{3.110}$$

That $\mathbf{P}_1^{\star}\mathbf{P}_1^{\star^T}$ has positive eigenvalues follows from a singular value decomposition of $\mathbf{P}_1^{\star}$, which is shown in Section 9.1. Secondly, the eigenvalues of $\mathfrak{C}_1^{\star}$ must be between 0 and 1. More precisely, eigenvalues >0 follow from the positive definiteness of $\mathfrak{C}_1^{\star}$ and eigenvalues >1 would not yield convergence for $\sum_{i=1}^{j}\mathfrak{C}_1^{\star^i}$.

Proof. **$\boldsymbol{\Upsilon}$ contains non-Euclidian vectors**. If the model of the fault condition is $\mathbf{z}_{0_f}=\mathbf{z}_0+\boldsymbol{\Upsilon}\Delta\mathbf{z}_{\mathfrak{J}}$ an objective function, similar to (3.79), can be defined as follows

$$\Delta\widehat{\mathbf{z}}_{\mathfrak{J}}=\arg\min_{\Delta\mathbf{z}_{\mathfrak{J}}}\sum_{k=1}^{\widetilde{K}}\mathbf{z}_{0_f}^T(k)\mathfrak{C}\mathbf{z}_{0_f}(k)-\mathbf{z}_{0_f}^T(k)\mathfrak{C}\boldsymbol{\Upsilon}\Delta\mathbf{z}_{\mathfrak{J}}+\Delta\mathbf{z}_{\mathfrak{J}}^T\boldsymbol{\Upsilon}^T\mathfrak{C}\boldsymbol{\Upsilon}\Delta\mathbf{z}_{\mathfrak{J}} \tag{3.111}$$

which yields the following estimate of the fault magnitude

$$\Delta\widehat{\mathbf{z}}_{\mathfrak{J}}=\left[\boldsymbol{\Upsilon}^T\mathfrak{C}\boldsymbol{\Upsilon}\right]^{-1}\left[\boldsymbol{\Upsilon}^T\mathfrak{C}\right]\left(\frac{1}{\widetilde{K}}\sum_{k=1}^{\widetilde{K}}\mathbf{z}_{0_f}(k)\right)$$
$$\boxed{\boldsymbol{\Upsilon}^T\mathfrak{C}\boldsymbol{\Upsilon}\in\mathbb{R}^{\mathrm{m}\times\mathrm{m}} \quad \mathrm{m}\le n}. \tag{3.112}$$

Only $\mathrm{m}\le n$ guarantees that $\boldsymbol{\Upsilon}^T\mathfrak{C}\boldsymbol{\Upsilon}$ is invertible. If this is not the case, the projection-based reconstruction process does not yield a unique analytical solution.

3.3.2.4 Influence on the data covariance matrix

For simplicity, the analysis in the remaining part of this subsection assumes that $\boldsymbol{\Upsilon}$ stores Euclidian column vectors. It is, however, straightforward to describe the impact of the projection-based reconstruction upon the data covariance matrix if $\boldsymbol{\Upsilon}$ includes non-Euclidian column vectors, which is a project in the tutorial session. Variable reconstruction leads to changes of the data covariance matrix which therefore must be reconstructed. Partitioning the rearranged data covariance matrix $\mathbf{S}^{\star}_{z_0z_0}$

$$\mathbf{S}^{\star}_{z_0z_0} = \begin{bmatrix} \mathbf{S}^{\star}_{z_0z_{0_1}} & \mathbf{S}^{\star}_{z_0z_{0_2}} \\ \mathbf{S}^{\star^T}_{z_0z_{0_2}} & \mathbf{S}^{\star}_{z_0z_{0_3}} \end{bmatrix}, \tag{3.113}$$

where $\mathbf{S}^{\star}_{z_0z_{0_1}} \in \mathbb{R}^{\mathrm{m}\times\mathrm{m}}$ and $\mathbf{S}^{\star}_{z_0z_{0_2}} \in \mathbb{R}^{\mathrm{m}\times n_z-\mathrm{m}}$ and $\mathbf{S}^{\star}_{z_0z_{0_3}} \in \mathbb{R}^{n_z-\mathrm{m}\times n_z-\mathrm{m}}$, the reconstruction of $\mathbf{z}^{\star}_{0_1}$ affects the first two matrices, $\mathbf{S}^{\star}_{z_0z_{0_1}}$ and $\mathbf{S}^{\star}_{z_0z_{0_2}}$

$$\widetilde{\mathbf{S}}^{\star}_{z_0z_{0_1}} = E\left\{\widetilde{\mathbf{z}}^{\star}_{0_1}\widetilde{\mathbf{z}}^{\star^T}_{0_1}\right\} = \mathbf{\Omega} E\left\{\mathbf{z}^{\star}_{0_2}\mathbf{z}^{\star^T}_{0_2}\right\}\mathbf{\Omega}^T = \mathbf{\Omega}\mathbf{S}^{\star}_{z_0z_{0_3}}\mathbf{\Omega}^T \tag{3.114}$$

and

$$\widetilde{\mathbf{S}}^{\star}_{z_0z_{0_2}} = E\left\{\widetilde{\mathbf{z}}^{\star}_{0_1}\mathbf{z}^{\star^T}_{0_2}\right\} = \mathbf{\Omega} E\left\{\mathbf{z}^{\star}_{0_2}\mathbf{z}^{\star^T}_{0_2}\right\} = \mathbf{\Omega}\mathbf{S}^{\star}_{z_0z_{0_3}} \tag{3.115}$$

where $\mathbf{\Omega} = \left[\mathbf{I} - \mathfrak{C}^{\star}_1\right]^{-1}\mathfrak{C}^{\star}_2$, and $\widetilde{\mathbf{S}}^{\star}_{z_0z_{0_1}}$ and $\widetilde{\mathbf{S}}^{\star}_{z_0z_{0_2}}$ are the covariance and cross-covariance matrices involving $\widetilde{\mathbf{z}}^{\star}_{0_1}$. Replacing $\mathbf{S}^{\star}_{z_0z_{0_1}}$ by $\widetilde{\mathbf{S}}^{\star}_{z_0z_{0_1}}$ and $\mathbf{S}^{\star}_{z_0z_{0_2}}$ by $\widetilde{\mathbf{S}}^{\star}_{z_0z_{0_2}}$

$$\widetilde{\mathbf{S}}^{\star}_{z_0z_0} = \begin{bmatrix} \widetilde{\mathbf{S}}^{\star}_{z_0z_{0_1}} & \widetilde{\mathbf{S}}^{\star}_{z_0z_{0_2}} \\ \widetilde{\mathbf{S}}^{\star^T}_{z_0z_{0_2}} & \mathbf{S}^{\star}_{z_0z_{0_3}} \end{bmatrix} \tag{3.116}$$

yields the following corollary.

Corollary 3.3.4 *The rank of* $\widetilde{\mathbf{S}}^{\star}_{z_0z_0}$ *is* $(n_z - \mathrm{m})$, *as the block matrices* $\widetilde{\mathbf{S}}^{\star}_{z_0z_{0_1}}$ *and* $\widetilde{\mathbf{S}}^{\star}_{z_0z_{0_2}}$ *are linearly dependent on* $\mathbf{S}^{\star}_{z_0z_{0_3}}$.

The effect of variable reconstruction upon the model and residual subspaces, which are spanned by eigenvectors of $\widetilde{\mathbf{S}}^{\star}_{z_0z_0}$, is analyzed in the following subsections.

3.3.2.5 Influence on the model plane

Pearson (1901) showed that the squared length of the residual vector between K mean-centered and scaled data points of dimension n_z and a given model subspace of dimension n is minimized if the model subspace is spanned by the first-n dominant eigenvectors of the data covariance matrix. Theorems 3.3.1 and 3.3.2 highlight that projecting samples onto the subspace $\mathbf{\Pi}$ leads to a minimum distance between the projected points and the model subspace. This gives rise to the following theorem.

Theorem 3.3.5 *The reconstruction of the* m *variables does not influence the orientation of the model subspace.*

The above theorem follows from the work of Pearson (1901), given that the reconstructed samples have a minimum distance from the original model subspace.

Corollary 3.3.6 *That there is no affect upon the orientation of the model subspace does not imply that the n dominant eigenvectors of* $\widetilde{\mathbf{S}}^{\star}_{z_0 z_0}$ *and* $\mathbf{S}^{\star}_{z_0 z_0}$ *are identical.*

The above corollary is a result of the changes that the reconstruction procedure imposes on $\mathbf{S}^{\star}_{z_0 z_0}$, which may affect the eigenvectors and the eigenvalues.

Corollary 3.3.7 *The dominant eigenvectors and eigenvalues of* $\widetilde{\mathbf{S}}^{\star}_{z_0 z_0}$ *may differ from those of* $\mathbf{S}^{\star}_{z_0 z_0}$, *which implies that the directions for which the score variables have a maximum variance and the variance of each score variable may change.*

The influence of the projection onto the residual subspace is discussed next.

3.3.2.6 Influence on the residual subspace

Following from the preceding discussion, the reconstruction results in a shift of a sample in the direction of the fault subspace, such that the squared length of the residual vector is minimal (Theorem 3.3.1). Since the reconstruction procedure is, in fact, a projection of $\mathbf{z}_0$ onto $\mathbf{\Pi}$, which is of dimension $n_z - \mathfrak{m}$ (Theorem 3.3.2), it follows that the dimension of the residual subspace is $n_z - n - \mathfrak{m}$, because the dimension of the model subspace remains unchanged.

Since the model subspace is assumed to describe the linear relationships between the recorded and source variables, which follows from Equation (2.2), the $n_z - n - \mathfrak{m}$ discarded eigenvalues represent the cumulative variance of the residual vector. Moreover, given that $\widetilde{\mathbf{S}}^{\star}_{z_0 z_0}$ has a rank of $n_z - \mathfrak{m}$, $\mathfrak{m}$ eigenvalues are equal to zero.

Corollary 3.3.8 *If* $\mathbf{S}_{\mathfrak{gg}} = \sigma_{\mathfrak{g}}^2 \mathbf{I}$, *the cumulative variance of* $\mathbf{g}^{\star}$, $E\left\{\mathbf{g}^{\star^T}\mathbf{g}^{\star}\right\}$, *is equal to* $(n_z - n)\sigma_{\mathfrak{g}}^2$. *In contrast, the cumulative variance of* $\widetilde{\mathbf{g}}^{\star}$ *is* $(n_z - n - \mathfrak{m})\sigma_{\mathfrak{g}}^2$ *and hence,* $E\left\{\mathbf{g}^{\star^T}\mathbf{g}^{\star}\right\} > E\left\{\widetilde{\mathbf{g}}^{\star^T}\widetilde{\mathbf{g}}^{\star}\right\}$.

Corollary 3.3.9 *Variable reconstruction has a minor effect on the data covariance matrix if the ratio* $^n/_{n_z}$ *is small. Conversely, if* $^n/_{n_z}$ *is close to 1, the squared length of* $\widetilde{\mathbf{g}}^{\star}$ *can be significantly affected by the reconstruction process.*

An example to illustrate the effect of the above corollaries is given in Chapter 4. It is important to note that even if the projection-based variable reconstruction has only minor effect on the data covariance matrix, this influence will lead to changes of the monitoring statistics and their confidence limits, which is examined next.

3.3.2.7 Influence on the monitoring statistics

The impact of variable reconstruction manifests itself in the construction of $\widetilde{\mathbf{S}}^{\star}_{z_0 z_0}$, which yields a different eigendecomposition. Since the Hotelling's T^2 and the Q

statistic are based on this eigendecomposition, it is necessary to account for such changes when applying projection-based variable reconstruction. For reconstructing a total of $\mathfrak{m} \leq n$ variables, this requires the following steps to be carried out:

1. Reconstruct the covariance matrix by applying (3.114) to (3.116).
2. Compute the eigenvalues and eigenvectors of $\widetilde{\mathbf{S}}^{\star}_{z_0 z_0}$, that is, $\widetilde{\mathbf{\Lambda}}^{\star}$ and $\widetilde{\mathbf{P}}^{\star}$.
3. Calculate the Hotelling's T^2 statistic using the dominant n eigenvectors and eigenvalues of $\widetilde{\mathbf{S}}^{\star}_{z_0 z_0}$, that is, $\widetilde{\mathbf{t}} = \widetilde{\mathbf{P}}^{\star^T} \widetilde{\mathbf{z}}^{\star}_0$ and $\widetilde{T}^2 = \widetilde{\mathbf{t}}^T \widetilde{\mathbf{\Lambda}}^{\star^{-1}} \widetilde{\mathbf{t}}$, where $\widetilde{\mathbf{z}}^{\star}_0$ that is, defined in (3.95).
4. Compute the Q statistic, $\widetilde{Q} = \left(\widetilde{\mathbf{z}}^{\star}_0 - \widetilde{\mathbf{P}}^{\star}\widetilde{\mathbf{t}}\right)^T \left(\widetilde{\mathbf{z}}^{\star}_0 - \widetilde{\mathbf{P}}^{\star}\widetilde{\mathbf{t}}\right)$, and recalculate the confidence limits for the Q statistic by applying (3.16) using the discarded eigenvalues of $\widetilde{\mathbf{S}}^{\star}_{z_0 z_0}$.

This procedure allows establishing reliable monitoring statistics for using projection-based variable reconstruction. If the above procedure is not followed, the variance of the score variables, the loading vectors and the variance of the residuals are incorrect, which yields erroneous results (Lieftucht *et al.* 2006a).

3.4 Tutorial session

Question 1: What is the advantage of assuming a Gaussian distribution of the process variables for constructing monitoring charts?

Question 2: With respect to the data structure in (2.2), develop fault conditions that do not affect (i) the Hotelling's T^2 statistic and (ii) the residual Q statistic. Provide general fault conditions which neither statistic is sensitive to.

Question 3: Considering that the Hotelling's T^2 and Q statistics are established on the basis of the eigendecomposition of the data covariance matrix, is it possible to construct a fault condition that neither affects the Hotelling's T^2 nor the Q statistic?

Question 4: Provide a proof of (3.27).

Question 5: Provide proofs for (3.72) and (3.75).

Question 6: For projections along predefined Euclidean axes for single samples, why does variable reconstruction affect the underlying geometry of the model and residual subspaces?

Question 7: Following from Question 6, why is the projection-based approach for multiple samples not affected by variable reconstruction to the same extent as the single-sample approach? Analyze the asymptotic properties of variable reconstruction.

Question 8: Excluding the single-sample projection-based approach, what is the disadvantage of projection- and regression-based variable reconstruction over contribution charts?

Project 1: With respect to Subsection 3.2.2, develop a set of residual-based tests for PLS.

Project 2: For PCA, knowing that $\mathbf{t} = \mathbf{P}^T \mathbf{z}_0$, $\widehat{\mathbf{z}}_0 = \mathbf{P}\mathbf{t} = \mathfrak{C}\mathbf{z}_0$ and $\mathbf{g} = \left[\mathbf{I} - \mathfrak{C}\right] \mathbf{z}_0$, design an example that describes a sensor fault for the first variable but for which the contribution chart for the Q statistic identifies the second variable as the dominant contributor to this fault. Can the same problem also arise with the use of the residual-based tests in Subsection 3.2.2? Check your design by a simulation example.

Project 3: With respect to Subsection 3.2.3, develop a variable reconstruction scheme for the input variable set of a PLS model. How can a fault condition for the output variable set be diagnosed?

Project 4: Simulate an example involving 3 process variables that is superimposed by a fault condition described by $\mathbf{z}_{0_f} = \mathbf{z}_0 + \boldsymbol{\upsilon} \Delta z_{\upsilon}$, $\boldsymbol{\upsilon} \in \mathbb{R}^3$, $\|\boldsymbol{\upsilon}\| = 1$, $\Delta z_{\upsilon} \in \mathbb{R}$ and develop a reconstruction scheme to estimate the fault direction $\boldsymbol{\upsilon}$ and the fault magnitude Δz_{υ}.

Project 5: Using a simulation example that involves 3 process variables, analyze why projection-based variable reconstruction is not capable of estimating the fault signature is not of the form $\mathbf{z}_{0_f} = \mathbf{z}_0 + \boldsymbol{\upsilon} \Delta z_{\upsilon}$, i.e. $\mathbf{z}_{0_f} = \mathbf{z}_0 + \Delta \mathbf{z}(k)$, with $\Delta \mathbf{z}(k)$ being a deterministic sequence. Contrast the performance of the regression-based variable reconstruction scheme with that of the projection-based scheme? How can a fault condition be diagnosed if it is of the form $\Delta \mathbf{z}(k)$ but stochastic in nature?

PART II

APPLICATION STUDIES

4

Application to a chemical reaction process

This chapter summarizes an application of MSPC, described in Chapters 1 to 3, to recorded data from a chemical reaction process, producing solvent chemicals. Data from this process have been studied in the literature using PCA and PLS, for example (Chen and Kruger 2006; Kruger *et al.* 2001; Kruger and Dimitriadis 2008; Lieftucht *et al.* 2006a). Sections 4.1 and 4.2 provide a process description and show how to determine a PCA monitoring model, respectively. Finally, Section 4.3 demonstrates how to detect and diagnose a fluidization problem in one of the tubes.

4.1 Process description

This process produces two solvent chemicals, denoted as F and G, and consists of several unit operations. The core elements of this plant are five parallel fluidized bed reactors, each producing F and G by complex exothermic reactions. These reactors are fed with five different reactants. Figure 4.1 shows one of the parallel reactors.

Streams A, B and C represent fresh reactant feed supplying pure components A, B and C, while feedstream D is from an upstream unit. Stream E is plant recycle. Streams D and E are vaporized before entering the reactor. After leaving the reactors, the separation of components F and G is achieved by downstream distillation units.

Each reactor consists of a large shell and a number of vertically oriented tubes in which the chemical reaction is carried out, supported by fluidized catalyst. There is a thermocouple at the bottom of each tube to measure the temperature

Statistical Monitoring of Complex Multivariate Processes: With Applications in Industrial Process Control, First Edition. Uwe Kruger and Lei Xie.

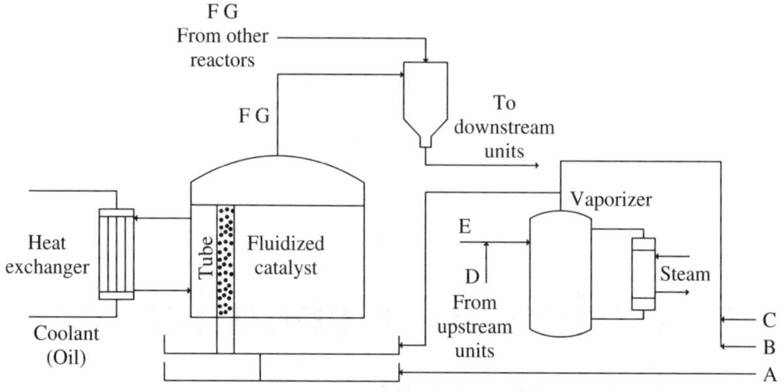

Figure 4.1 Schematic diagram of chemical reaction process.

of the fluidized bed. To remove the heat of the exothermic reaction, oil circulates around the tubes.

The ratio of components F and G is obtained from a lab analysis at eight hour intervals. Based on this analysis, operators adjust the F:G ratio by manipulating reactor feedrates. To keep the catalyst fluidized at all times, the fluidization velocity is maintained constant by adjusting reactor pressure relative to the total flow rate.

The chemical reaction is affected by unmeasured disturbances and changes in the fluidization of the catalyst. The most often observed disturbances relate to pressure upsets in the steam supply to the vaporizer and the coolant, which is provided by a separate unit. Fluidization problems appear if the catalyst density is considerably greater at the bottom of the tube, which additionally enhances chemical reaction in the tube resulting in a significant increase in the tube temperature.

During a period of several weeks, normal operating data as well as data describing abnormal process behavior were recorded for a single reactor. The reference data set had to be selected with care, to ensure that it did not capture disturbances as described above or fluidization problems of one or more tubes. Conversely, if the size of the reference data set was too small then common cause variation describing the reaction system may not be adequately represented.

4.2 Identification of a monitoring model

Since any disturbance leads to alterations in the reacting conditions and hence the tube temperatures, the analysis here is based on the recorded 35 tube temperatures. Thus, the data structure in (2.2) allows modeling of the recorded variables. Figure 4.2 shows time-based plots of the temperature readings for the reference data, recorded at 1 minute intervals.

A closer inspection of the 35 variables in Figure 4.2 suggests significant correlation between the tube temperatures, since most of them follow a similar

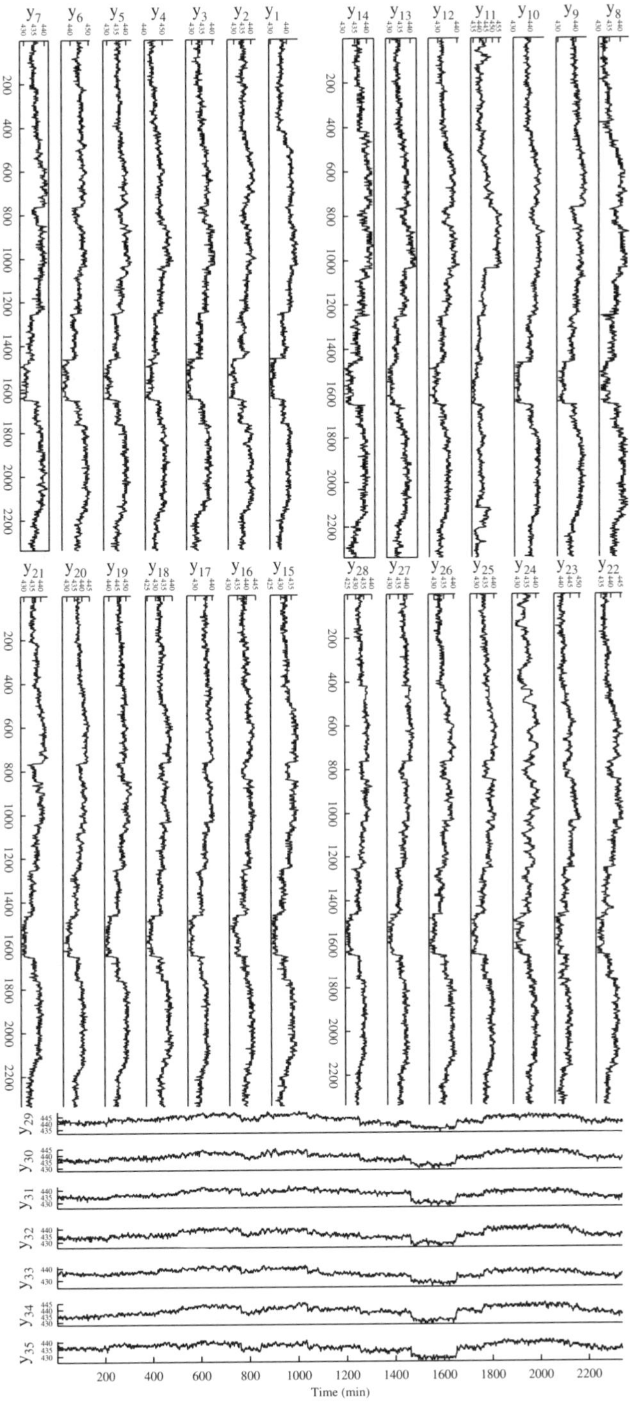

Figure 4.2 Recorded sequence of reference data for chemical reaction process.

pattern. The identification of a PCA model requires the estimation of the mean vector and the data covariance matrix using (2.13) and (2.14). The mean vector contains values between 330 and 340°C.

Dividing each of the mean-centered signals by its associated standard deviation allows the construction of the correlation matrix. The observation that the tube temperatures are highly correlated can be verified by analyzing the non-diagonal elements of the correlation matrix, $\mathcal{C}_{z_0z_0}$. Displaying the upper left block of this matrix of the first five tube temperatures

$$\mathcal{C}_{z_0z_0} = \begin{bmatrix} 1.0000 & 0.8258 & 0.8910 & 0.8184 & 0.8515 & \cdots \\ 0.8258 & 1.0000 & 0.8412 & 0.7370 & 0.8700 & \cdots \\ 0.8910 & 0.8412 & 1.0000 & 0.8417 & 0.8791 & \cdots \\ 0.8184 & 0.7370 & 0.8417 & 1.0000 & 0.8052 & \cdots \\ 0.8515 & 0.8700 & 0.8791 & 0.8052 & 1.0000 & \cdots \\ \vdots & \vdots & \vdots & \vdots & \vdots & \ddots \end{bmatrix} \tag{4.1}$$

confirms this, as most of the correlation coefficients are larger than 0.8 indicating significant correlation among the variables. Jackson (2003) outlined that it often makes negligible difference which matrix to use for constructing a PCA model in practice. The analysis of the reference data in Figure 4.3 confirm this later on by inspecting the distribution of the eigenvalues for both matrices.

Equation (4.2) shows the upper left block of the covariance matrix

$$\mathbf{S}_{z_0z_0} = \begin{bmatrix} 9.2469 & 7.5636 & 6.5811 & 7.1108 & 7.6650 & \cdots \\ 7.5636 & 8.7431 & 7.3084 & 6.9863 & 7.0903 & \cdots \\ 6.5811 & 7.3084 & 8.6240 & 6.3555 & 6.1245 & \cdots \\ 7.1108 & 6.9863 & 6.3555 & 7.2242 & 7.0179 & \cdots \\ 7.6650 & 7.0903 & 6.1245 & 7.0179 & 9.4869 & \cdots \\ \vdots & \vdots & \vdots & \vdots & \vdots & \ddots \end{bmatrix}. \tag{4.2}$$

Figure 4.3 plots the distribution of the eigenvalue for the covariance and the correlation matrix and shows that they are almost identical up to a scaling factor.

The construction of the covariance matrix is followed by determining the number of source signals. Section 2.4 outlined that the VRE method provides a consistent estimation of n under the assumption that $\mathbf{S}_{gg} = \sigma_g^2\mathbf{I}$ and is a computationally efficient method. Figure 4.4 shows the results when the covariance and correlation matrices are used. For both matrices, the minimum of the VRE criteria is for four source signals.

Figure 4.5 shows the time-based plots of these four score variables. With regards to (2.8), these score variables represent linear combinations of the score variables that are corrupted by the error vector. On the other hand, the score variables are the coordinates for the orthogonal projection of the samples onto the model plane according to Figure 2.2 and (2.5).

A comparison of the signals in Figure 4.5 with those of Figure 4.2 suggests that the 4 source signals can capture the main variation within the reference

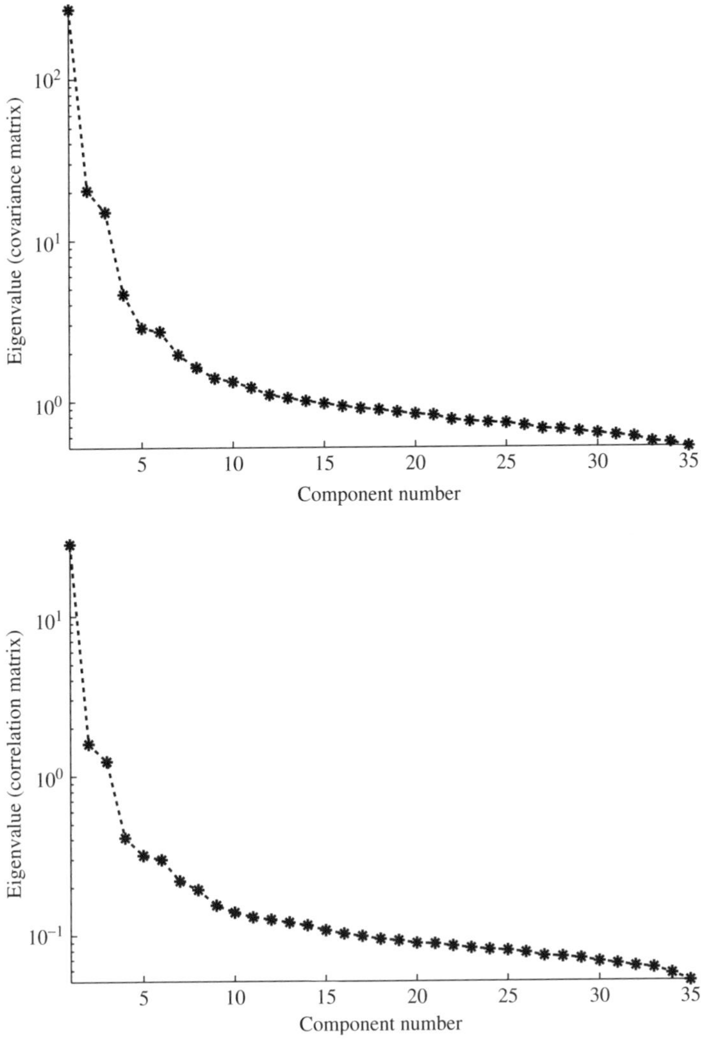

Figure 4.3 Eigenvalues distribution of data covariance (upper plot) and data correlation matrix (lower plot).

data. This can be verified more clearly by comparing the signals of the original tube temperatures with their projections onto the PCA model subspace, which Figure 4.6 illustrates for the first five temperature readings. The thick lines represent the recovered signals and the thin lines correspond to the five recorded signals.

The analysis conducted thus far suggests that the identified data structure models the reference data accurately. By inspecting Figures 4.2 and 4.5, however, the original process variables, and therefore the score variables, do not follow

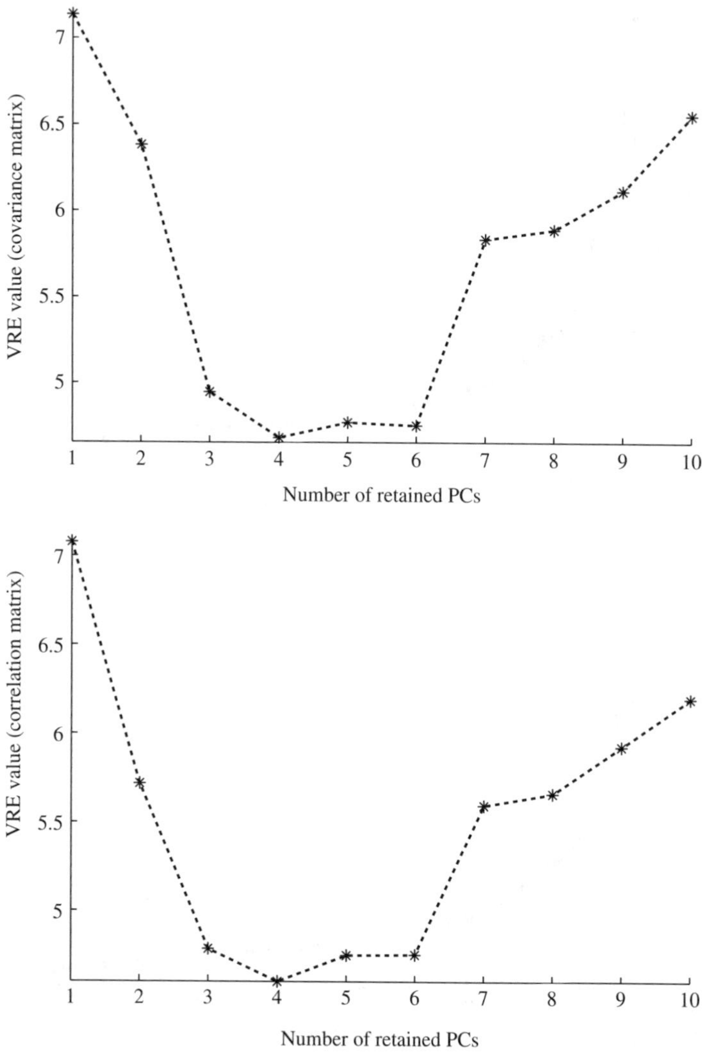

Figure 4.4 Selection of the number of retained PCs using the VRE technique.

a Gaussian distribution. Figure 4.7 confirms this by comparing the estimated distribution function with that of a Gaussian distribution. The comparison shows very significant departures, particularly for the first two score variables.

Theorem 9.3.4 outlines that the score variables are asymptotically Gaussian distributed, which follows from the central limit theorem (CLT). This, however, requires a significant number of variables. Chapter 8 discusses how to address non-Gaussian source signals and in Section 6.1.8 shows that the assumption $\mathbf{S}_{gg} = \sigma_g^2 \mathbf{I}$ is not met. More precisely, it must be assumed that each error variable has a

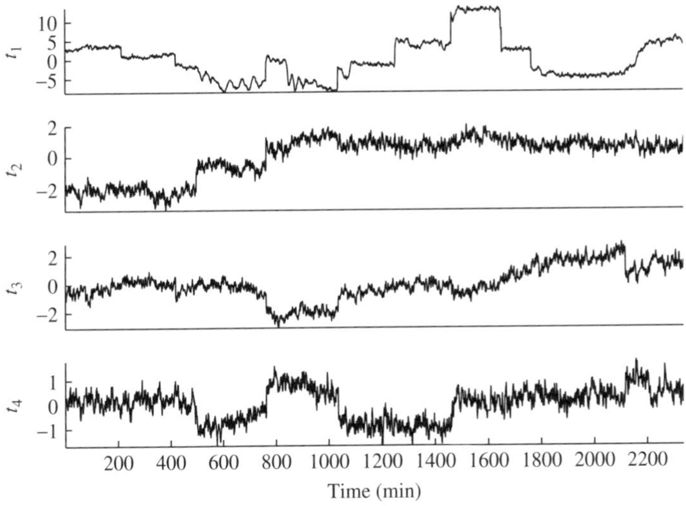

Figure 4.5 Time-based plot of the score variables for reference data.

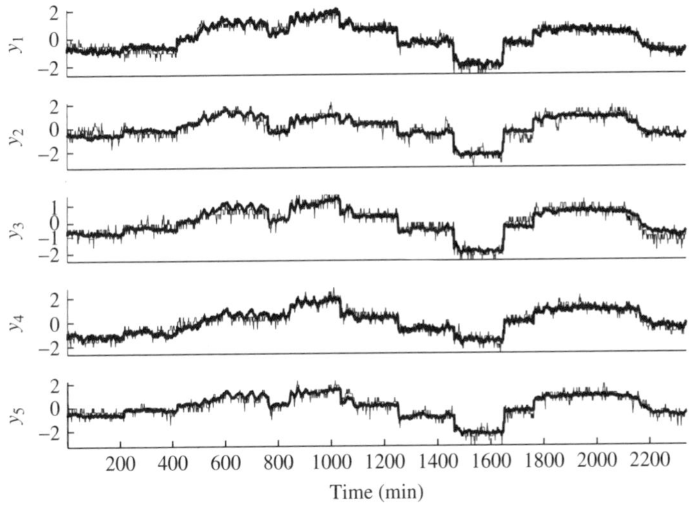

Figure 4.6 Time-based plot of the first five tube temperature readings and their projections onto the PCA model subspace.

slightly different variance and that the number of source signals is significantly larger than four.

For the reminder of this section, the statistical inference is based on the assumption that the score variables follow a multivariate Gaussian distribution to demonstrate the working of MSPC for detecting and diagnosing process faults. Chapter 3 highlights that process monitoring relates to the use of scatter diagrams as well as the Hotelling's T^2 and Q monitoring statistics.

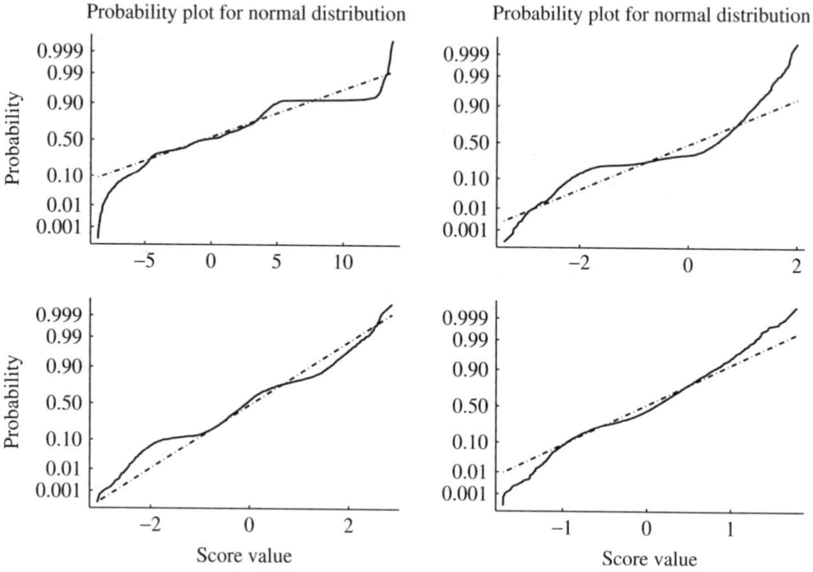

Figure 4.7 Comparison between normal distribution (dashed line) and estimated distribution for principal components (solid line).

In relation to Figure 4.7, however, it is imperative to investigate the effect of the non-Gaussian distributed score variables upon both nonnegative quadratic statistics. Figure 4.8 highlights, as expected, significant departures for the T^2 statistic, which is assumed to follow and F-distribution with 4 and 2334 degrees of freedom. However, the F-distribution with 31 and 2334 degrees of freedom is a good approximation of the Q statistic when constructed using Equation (3.19).

Figure 4.9 plots the resulting Hotelling's T^2 and Q statistics for the reference data. The effect of the 4 non-Gaussian source signals upon the Hotelling's T^2 statistic can clearly be noticed, as there are no violations of the control limit, which was determined for a significance of 1%. In contrast, the Q statistic violates the control limit a total of 18 times out of 2338, which implies that the number of Type I errors is roughly 1%. This result does not surprise given that the approximation of this statistic is very close to an F-distribution, which the lower plot in Figure 4.8 shows.

With regards to the Hotelling's T^2 statistic, the upper plot in Figure 4.8 gives a clear indication as to why there are no violations. The critical values of the empirical distribution function for $\alpha = 0.05$ and $\alpha = 0.99$ are 7.5666 and 8.9825, respectively. However, computing the control limit using (3.5) for a significance of $\alpha = 0.01$ yields 13.3412. This outlines that the Hotelling's T^2 statistic is prone to significant levels of Type II errors and may not be sensitive in detecting incipient fault conditions. The next subsection shows how to use the monitoring model to detect abnormal tube behavior and to identify which of the tube(s) behave anomalously.

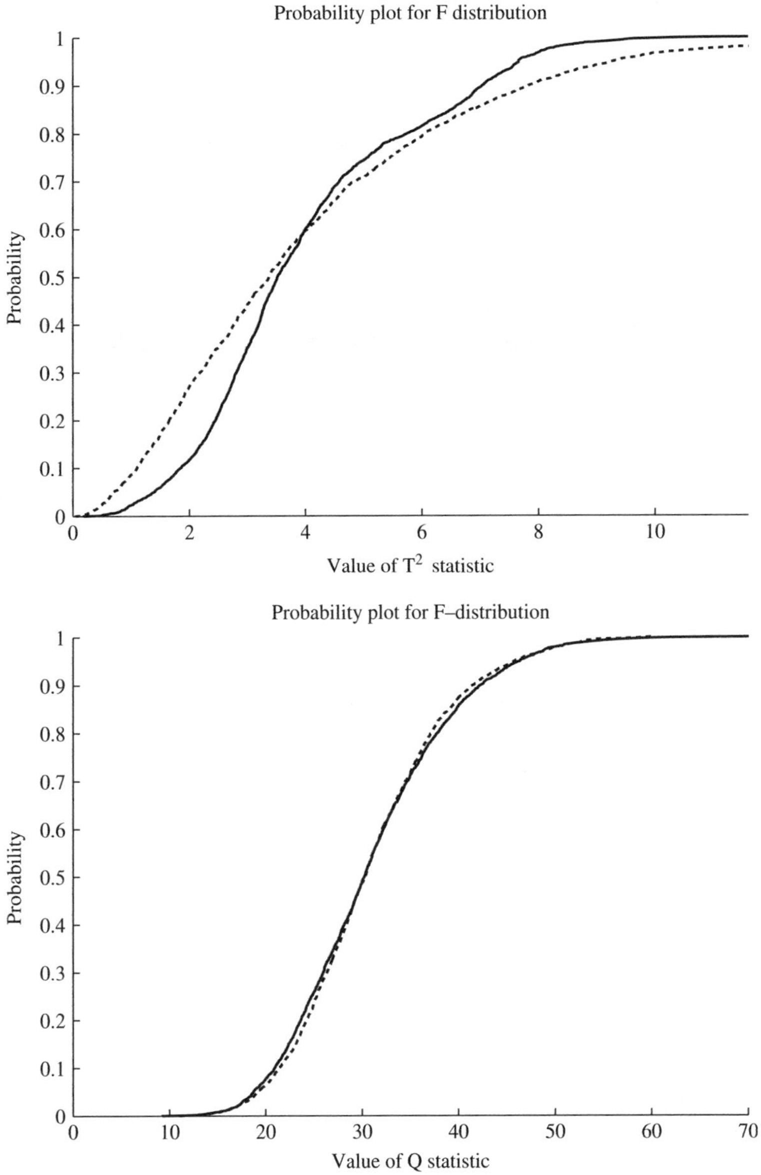

Figure 4.8 Comparison between F-distribution (dashed line) and estimated distribution for the Hotelling's T^2 (upper plot) and Q (lower plot) statistics.

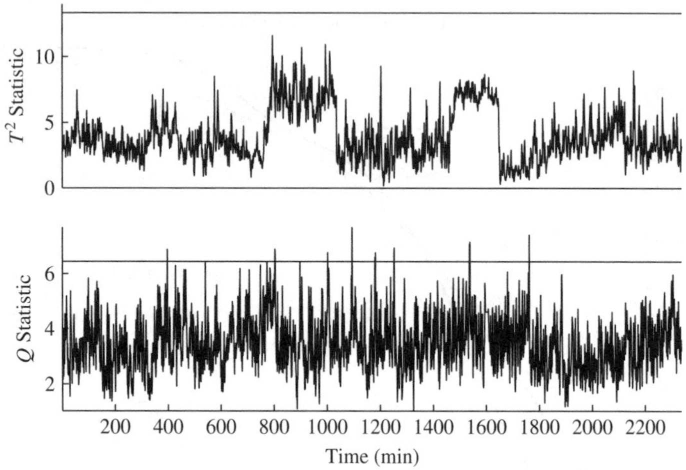

Figure 4.9 Time based plot of Hotelling's T^2 and Q statistics for reference data.

It should be noted that the usefulness of applying a multivariate analysis of the recorded data is not restricted to the generation of the Hotelling's T^2 and Q statistics. Section 2.1 outlined that the elements of different loading vectors can be plotted against each other. As discussed in Kaspar and Ray (1992), such *loading plots* can identify groups of variables that have a similar covariance structure and hence show similar time-based patterns of common cause variation that are driven by the source signals. Figure 4.10 shows the loading plot of the first three eigenvectors of $\mathcal{C}_{z_0 z_0}$.

The figure shows that most of the temperature variables fall within a small section, implying that they have very similar patterns on the basis of the first three components. A second and considerably smaller cluster of variables emerges that include the temperature sensors y_4, y_8, y_{23} and y_{24}. The analysis also shows that

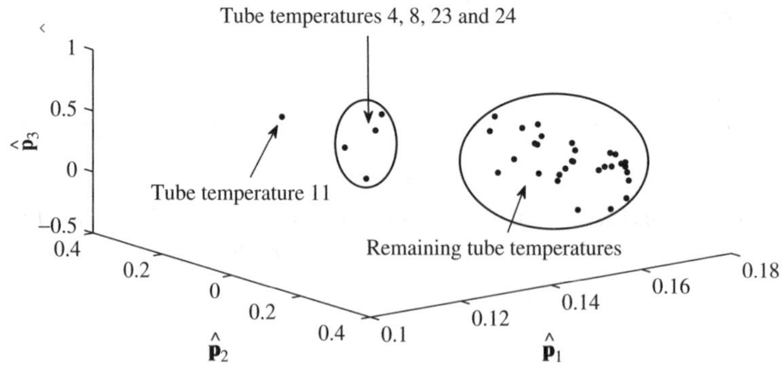

Figure 4.10 Loading plots of first three eigenvectors of estimated correlation matrix.

thermocouple #11 (y_{11}) is isolated from the other two clusters. A comparison between the variables that make up the second cluster and thermocouple #11 yields that the signal for y_{11} shows more distinct operating conditions. In contrast y_4, y_8, y_{23} and y_{24} show substantially more variation and only a few distinct steady state operating regions.

As the thermocouples measure the temperature at the bottom of each tube, the two distinct clusters in Figure 4.10 may be a result of different conditions affecting the chemical reaction within the tubes. This could, for example, be caused by differences in the fluidization of the catalyst or the concentration distribution of the five feeds among the tubes. The distinct behavior of tube #11 would suggest the need to conduct to conduct a further and more detailed analysis. It is interesting to note that this tube showed an abnormal behavior at a later stage and developed a severe fluidization problem. The analysis in Kruger *et al.* (2001) showed that the tube had to be shut down eventually.

Equation (2.2) is, in fact, a correlation model that assumes that the source signals describe common cause variation as the main contributor to the data covariance matrix. In contrast, the error variables have a minor contribution to this matrix. As the correlation-based model is non-causal, it cannot be concluded that the distinct characteristic of tube #11 relative to the other temperature readings is a precursor or an indication of a fluidization problem. However, the picture presented in Figure 4.10 suggests inspecting the performance of this tube in more detail.

4.3 Diagnosis of a fault condition

After generating a PCA-based monitoring model this section shows how to utilize this model for detecting and diagnosing an abnormal event resulting from a fluidization problem in one of the tubes. There are some manipulations a plant operator can carry out to improve the fluidization and hence bring the tube temperature back to a normal operating level. However, incipient fluidization problems often go undetected by plant operators, as illustrated by this example.

The recorded data showed a total of three cases of abnormal behavior in one of the tubes. Whilst the first two of these events went unnoticed by plant operators, the third one was more severe in nature and required the tube to be shut down (Kruger *et al.* 2001). The first occurrence is shown in Figure 4.11 by plotting temperature readings #9 to #13 confirming that tube temperature 11 performed abnormally about 100 samples into the data set, which lasts for about 1800 samples (30 hours).

Constructing non-negative quadratic statistics from this data produces the plots displayed in Figure 4.12. The Hotelling's T^2 and Q statistics detected this abnormal event about 1 hour and 30 minutes (90 samples) into the data set by a significant number of violations of the Q statistic. The violations covered a period of around 3 hours and 20 minutes (380 samples). The event then decayed in magnitude although significant violations of the Q statistic remained for about

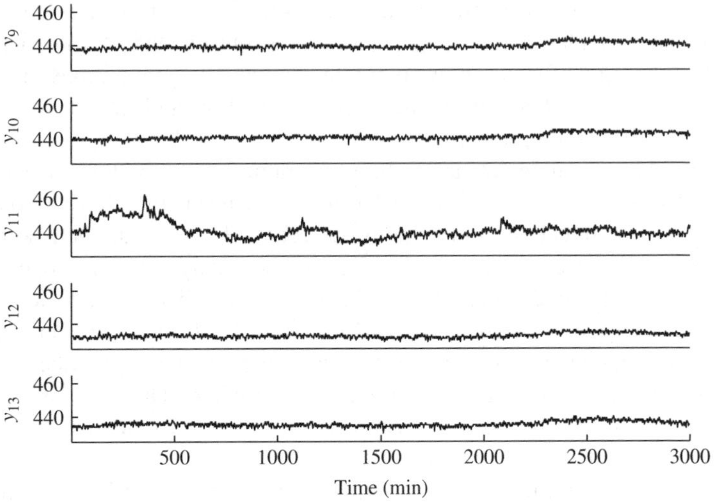

Figure 4.11 Time-based plots of tube temperatures 9 to 13.

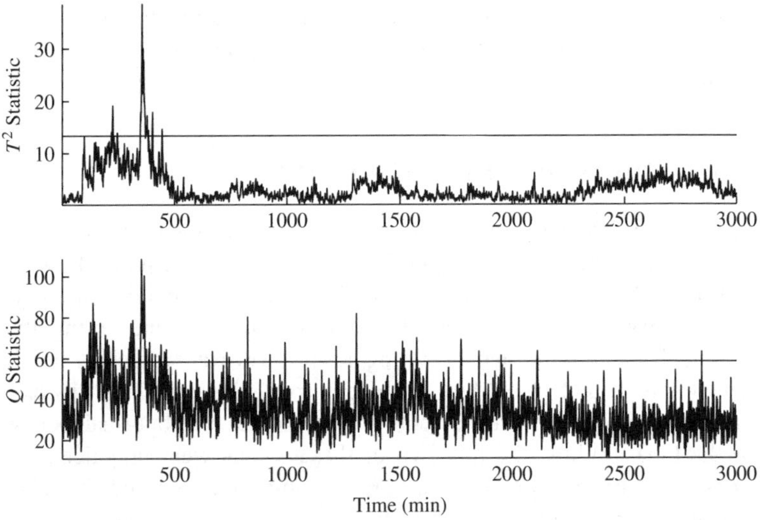

Figure 4.12 Univariate monitoring statistics representing abnormal tube behavior.

30 hours after which the number of violations were around 1% indicating in-statistical-control behavior.

For diagnosing this event, contribution charts are considered first. These include the score variable contribution to the T^2 statistic as well as the process variable contribution to the T^2 and Q statistics. Figure 4.13 shows these score and variable contributions between the 85th and 92nd samples, which yield that

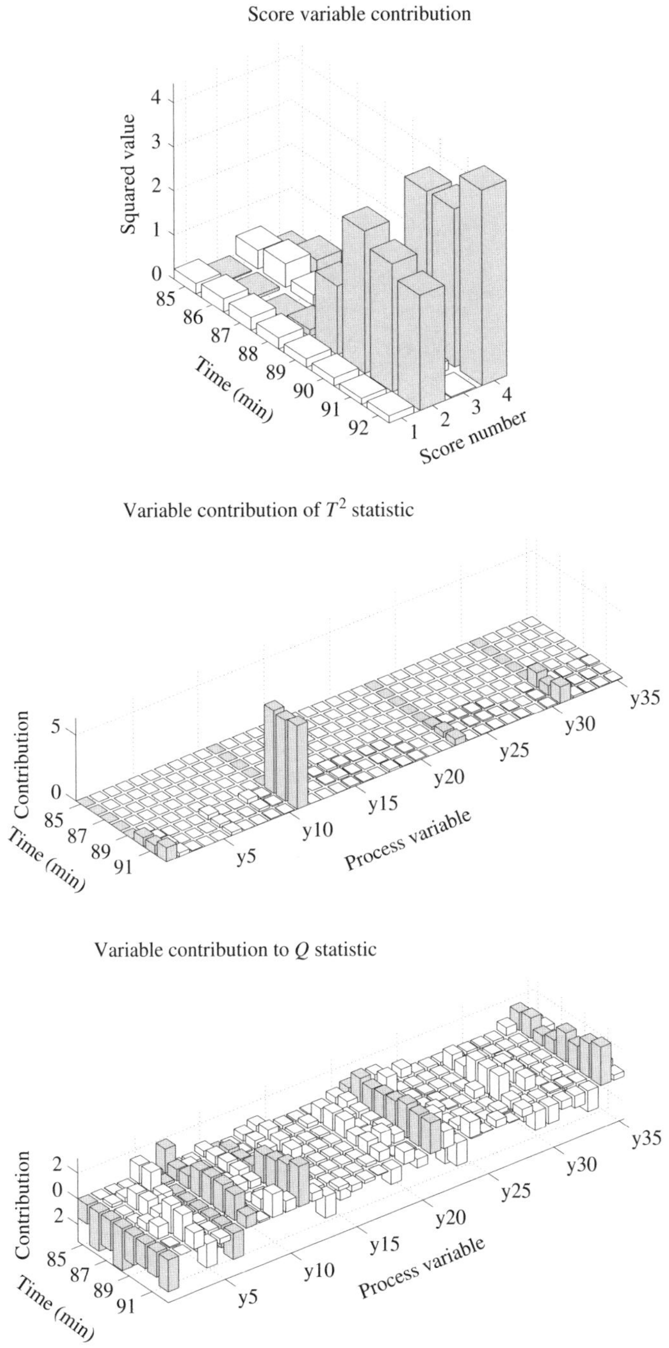

Figure 4.13 Contribution charts for samples 85 to 92.

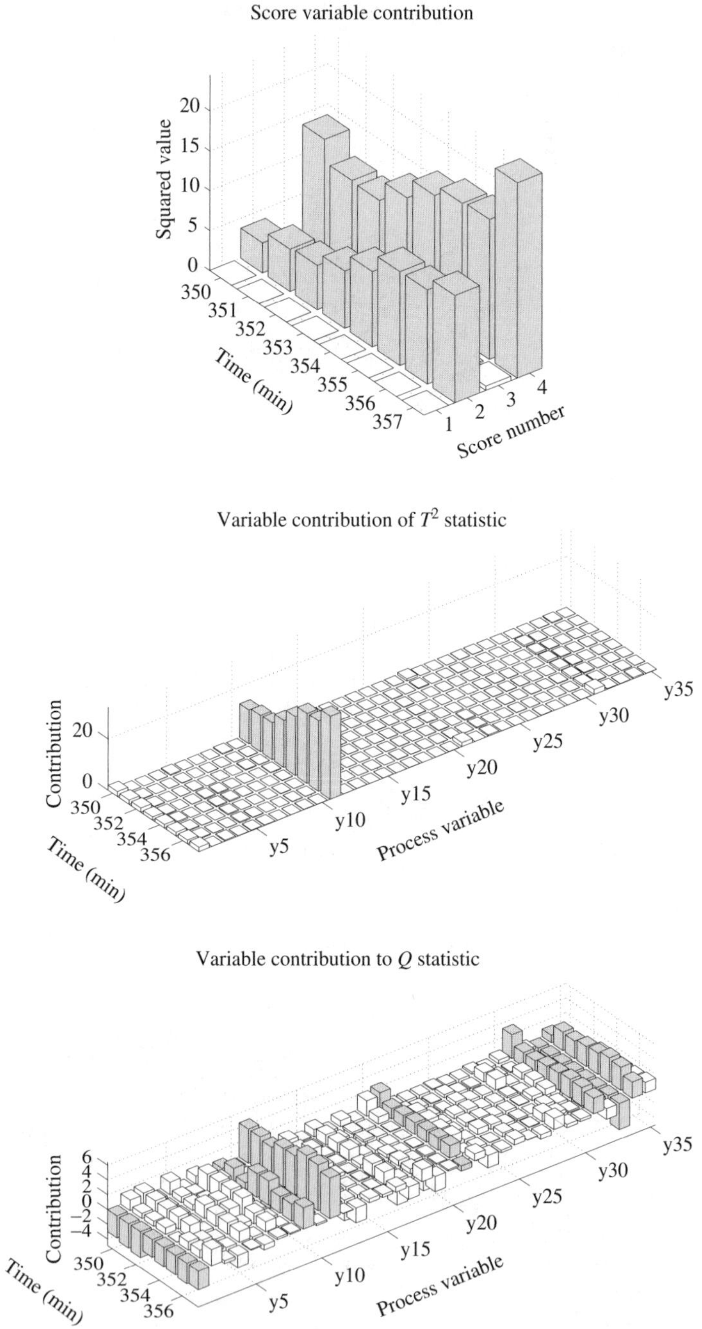

Figure 4.14 Contribution charts for samples 350 to 357.

score variables #2 and #4 were predominantly contributing to the T^2 statistic. The middle plot shows that the 11th tube temperature was predominantly contributing to the T^2 statistic from the 90th sample, which is the first instance where this abnormal event was detected. Furthermore, the variable contribution to the Q statistic does not provide a clear picture although tube temperature 11 was one of the contributing variables among the temperature readings #1, #6, #7, #21 and #34.

Figure 4.14 presents a second set of contribution charts between the 350th and 357th samples, which covers a period of larger T^2 and Q values. As before, score variables #2 and #4 showed a dominant contribution to this event. The middle and lower plots in Figure 4.14 outlined a dominant variable contribution of tube temperature #11 to the T^2 and Q statistic, respectively, whilst the lower plot also identified tube temperatures #1, #9, #21, #31, #33 and #34 as affected by this event.

The above pre-analysis based on contribution charts correctly suggested that tube temperature #11 is the most dominantly affected process variable. Reconstructing this tube temperature using the remaining 34 tube temperatures allows studying the impact of temperature #11 upon the T^2 and Q statistics. Subsection 3.2.3 describes how to reconstruct a set of variables using the remaining ones and how to recompute the monitoring statistics to account for the effects of this reconstruction. The reconstruction of tube temperature #11 required the following projection

$$\widehat{\widehat{z}}_{0_{11}} = \frac{1}{1 - \mathfrak{c}_{11,11}} \sum_{i=1 \neq 11}^{35} c_{11,i} z_{0_i} \tag{4.3}$$

where $c_{11,i}$ are the elements stored in the 11th row of $\mathfrak{C} = \mathbf{P}\mathbf{P}^T$. To assess the impact of the reconstruction process, Figure 4.14 shows the difference of the eigenvalues for the reconstructed and the original covariance matrix, which are computed as follows

$$\Delta\widehat{\lambda}_i = \frac{\widehat{\lambda}_i - \widehat{\widehat{\lambda}}_1}{\widehat{\lambda}_i} 100\%, \tag{4.4}$$

where $\widehat{\lambda}_i$ and $\widehat{\widehat{\lambda}}_i$ represent the eigenvalues of $\widehat{\mathbf{S}}_{z_0 z_0}$ and $\widehat{\widehat{\mathbf{S}}}_{z_0 z_0}$, respectively, and $\Delta\widehat{\lambda}_i$ is the percentage deviation. It follows from Figure 4.15 that the reconstruction of one tube temperature using the remaining ones produced percentage departures of around 5% or less for most eigenvalues. However, eigenvalues #2, #3, #4 and #6 showed departures of up to 45%, which outlines that it is essential to take these changes into account when recalculating the Hotelling's T^2 and Q statistics. A detailed discussion of this is given in Subsections 3.2.3 and 3.3.2. Applying the eigenvalues of $\widehat{\widehat{\mathbf{S}}}_{z_0 z_0}$ to determine the 'adapted' confidence limits yields that:

- the variances of the score variables changed to 266.4958, 15.1414, 8.0340 and 3.1524 from 269.2496, 20.4535, 14.9976 and 4.6112, respectively; and
- the control limit for the Q statistics changes to 55.0356 from 58.3550.

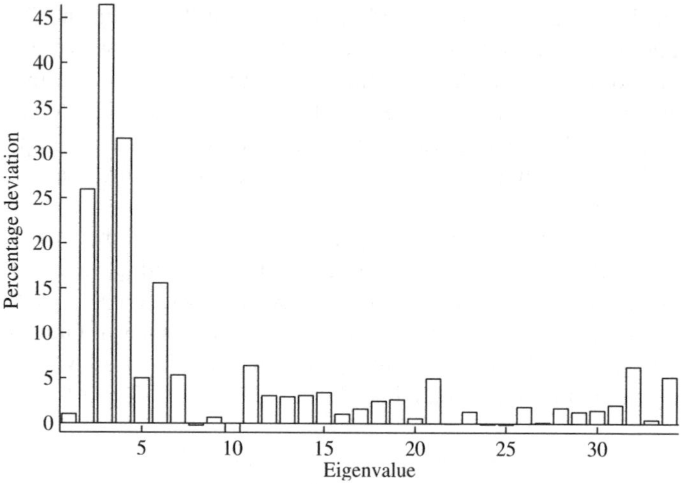

Figure 4.15 Percentage changes in the eigenvalues of the data covariance matrix for reconstructing tube temperature #11.

The resulting statistics after reconstructing tube temperature #11 are shown in Figure 4.16. Comparing them with those displayed in Figures 4.12 yields that reconstructing temperature reading #11 reduces the number of violations for the Q statistic significantly and removed the violations of the T^2 statistic. The remaining violations of the Q statistic are still indicative of an out-of-statistical-control situation. It can be concluded, however, that tube temperature #11 is significantly affected by this event.

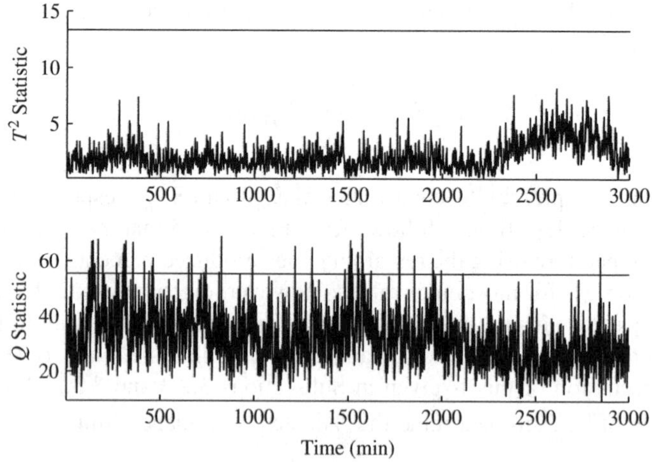

Figure 4.16 Univariate monitoring statistics after reconstruction of tube temperature #11.

To control the fluidization of the catalyst, an empirically determined fluidization velocity is monitored and regulated by adjusting the pressure within the reactor. However, the increase in one tube temperature has not been significant enough to show a noticeable effect upon most of the other tube temperatures by this feedback control mechanism, which Figure 4.11 shows. Nevertheless, the controller interaction and hence the changes in pressure within the reactor influenced the reaction conditions, which may contribute to the remaining violations of the Q statistic.

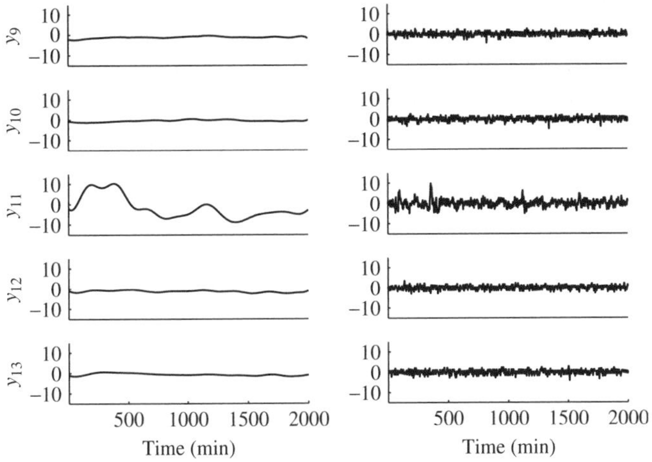

Figure 4.17 Estimated fault signature and remaining stochastic signal contribution for tube temperatures #9 to #13.

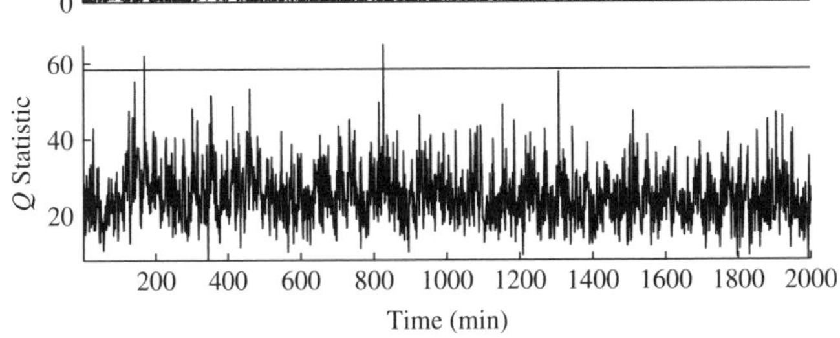

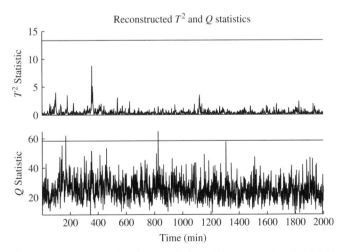

Figure 4.18 Univariate monitoring statistics after removing the fault signature from the recorded tube temperatures.

The analysis has so far only indicated that tube temperature #11 is the most dominant contributor to this out-of-statistical-control situation. An estimate of the fault signature and magnitude for each of the tube temperatures, however, could not be offered. Subsection 3.2.3 highlighted that regression-based variable reconstruction can provide such an estimate. Using a total of $R = 20$ network nodes and a radius of $\varrho = 0.1$, Figure 4.17 presents the separation of the recorded temperature readings into the fault signature and the stochastic signal components for the first 2000 samples. Given that the abnormal tube behavior did not affect the last 1000 samples, only the first 2000 samples were included in this reconstruction process.

The plots in Figure 4.17 show hardly any contribution from temperature variables #9, #10, #12 and #13 but a substantial fault signature associated with variable #11 that amounts to about 20°C in magnitude. Moreover, apart from very rapid alterations, noticeable by the spikes occurring in the middle left plot in Figure 4.17, the estimated fault signature accurately describes the abnormal tube temperature signal when compared with the original signal in Figure 4.11. Constructing the Hotelling's T^2 and Q statistics after the fault signatures have been removed from the recorded temperature readings produced the plots in Figure 4.18. In comparison with Figure 4.16, no statistically significant violations remained.

5

Application to a distillation process

This chapter presents a second application study, which involves recorded data from an industrial distillation process, which is a debutanizer unit. Recorded data from this process were analyzed for process monitoring purposes using PCA and PLS (Kruger and Dimitriadis 2008; Kruger *et al.* 2008a; Meronk 2001; Wang *et al.* 2003).

Section 5.1 provides a description of this process. Section 5.2 then describes the data pre-processing and the identification of an MRPLS-based monitoring model. Finally, Section 5.3 analyzes a recorded data set that describes a severe drop in the fresh feed flow to the unit and shows how to detect and diagnose this event.

5.1 Process description

This process, schematically shown in Figure 5.1, is designed to purify Butane from a fresh feed comprising of a mixture of hydrocarbons, mainly Butane (C4), Pentane (C5) and impurities of Propane (C3). The separation is achieved by 33 trays in the distillation column. The feed to the unit is provided by an upstream depropanizer unit, which enters the column between trays 13 and 14. At trays 2 and 31 temperature sensors are installed to measure the top and bottom temperatures of the column.

The lighter vaporized components C3 and C4 are liquefied as overhead product with large fan condensers. This overhead stream is captured in a reflux drum vessel and the remaining amount of C3 and C5 in the C4 product stream is

Statistical Monitoring of Complex Multivariate Processes: With Applications in Industrial Process Control, First Edition. Uwe Kruger and Lei Xie.

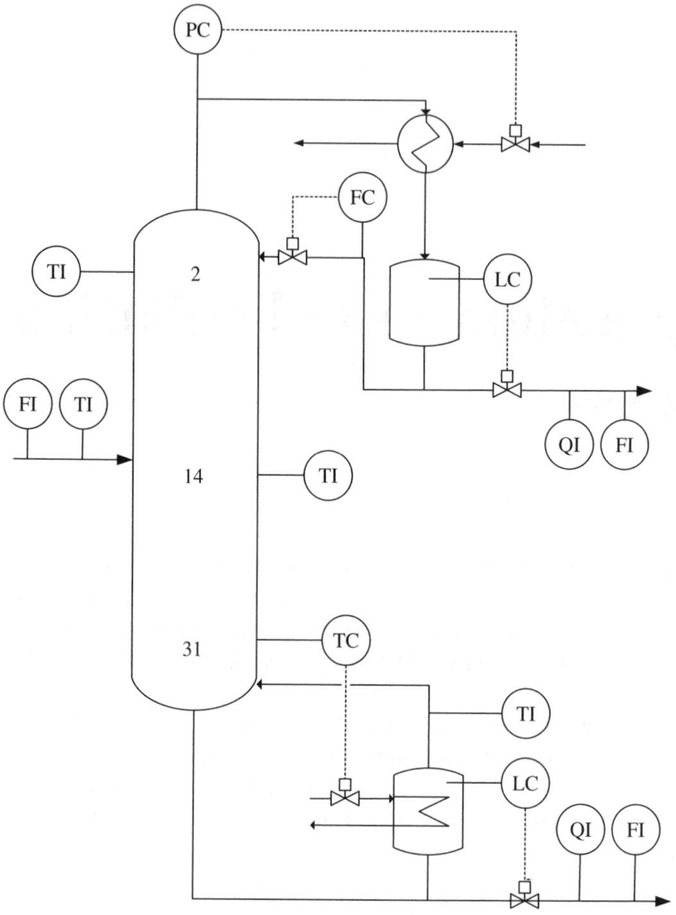

Figure 5.1 Schematic diagram of debutanizer unit.

measured through on-line analyzers. The product stream is divided into a product stream stored in butane tanks, and a reflux stream that returns to the column just above tray 1.

A second stream, taken from the bottom of the column, is divided between a feed entering the re-boiler and the bottom product. The temperature of the vaporized stream leaving the re-boiler and the C4 concentration in the bottom draw are measured. The bottom product is pumped to a mixing point, where it is blended into crude oil. Table 5.1 lists the recorded variables included in this study.

The reflux flow rate and the liquid level in the butane and reboiler tanks are omitted here, as they are either constant over the recording periods or controlled to a set-point and hence not affected by common cause variation. The remaining variable set includes the concentrations of C3 in C4, C5 in C4 (top draw) and C5

Table 5.1 Recorded process variables included in the MRPLS model.

Type	Number	Tag	Description	Unit
Input variable	1	x_1	Tray 14 temperature	$^\circ C$
	2	x_2	Column overhead pressure	bar
	3	x_3	Tray 2 temperature	$^\circ C$
	4	x_4	Fresh feed temperature	$^\circ C$
	5	x_5	Reboiler steam flow	$^t/_h$ [1]
	6	x_6	Tray 31 temperature	$^\circ C$
	7	x_7	Fresh feed flow	$^\circ C$
	8	x_8	Reboiler temperature	$^\circ C$
Output variable	9	y_1	Bottom draw	$^t/_h$ [1]
	10	y_2	Percentage C3 in C4	%
	11	y_3	Percentage C5 in C4	%
	12	y_4	Top draw	$^t/_h$ [1]
	13	y_5	Percentage C4 in C5	%

[1] tonnes per hour.

in C4 (bottom draw), flow rates of fresh feed, top and bottom draw, temperatures of the fresh feed, trays 2, 14 and 31, and the pressure at the top of the column.

This distillation unit operates in a closed-loop configuration. The composition control structure corresponds to the configuration discussed in Skogestad (2007):

- the top- and bottom draw flow rates control the holdup in the reflux drum and the reboiler vessel;
- the coolant flow into the condenser (not measured) controls the column pressure;
- the reflux flow rate, adjusted by a plant operator, controls the concentration of the distillate; and
- the temperature on tray 31 controls the concentration of the bottom product.

This configuration is known to achieve good performance for composition control but is also sensitive to upsets in the feed flow. Section 5.3 presents the detection and diagnosis of such an upset using the MSPC framework outlined in Chapters 1 to 3.

The variables related to product quality and the output of this process are the concentration measurements and the flow rates for top- and bottom draw. The remaining eight variables affect these output variable set and are therefore considered as the input variables. The next two sections describe the identification of a MRPLS model and the detection and diagnosis of a severe drop in the flow rate of the fresh feed. Figures 5.2 and 5.3 plot the reference data for the eight input and five output variables respectively, recorded at a sampling interval of 30 seconds over a period of around 165 hours.

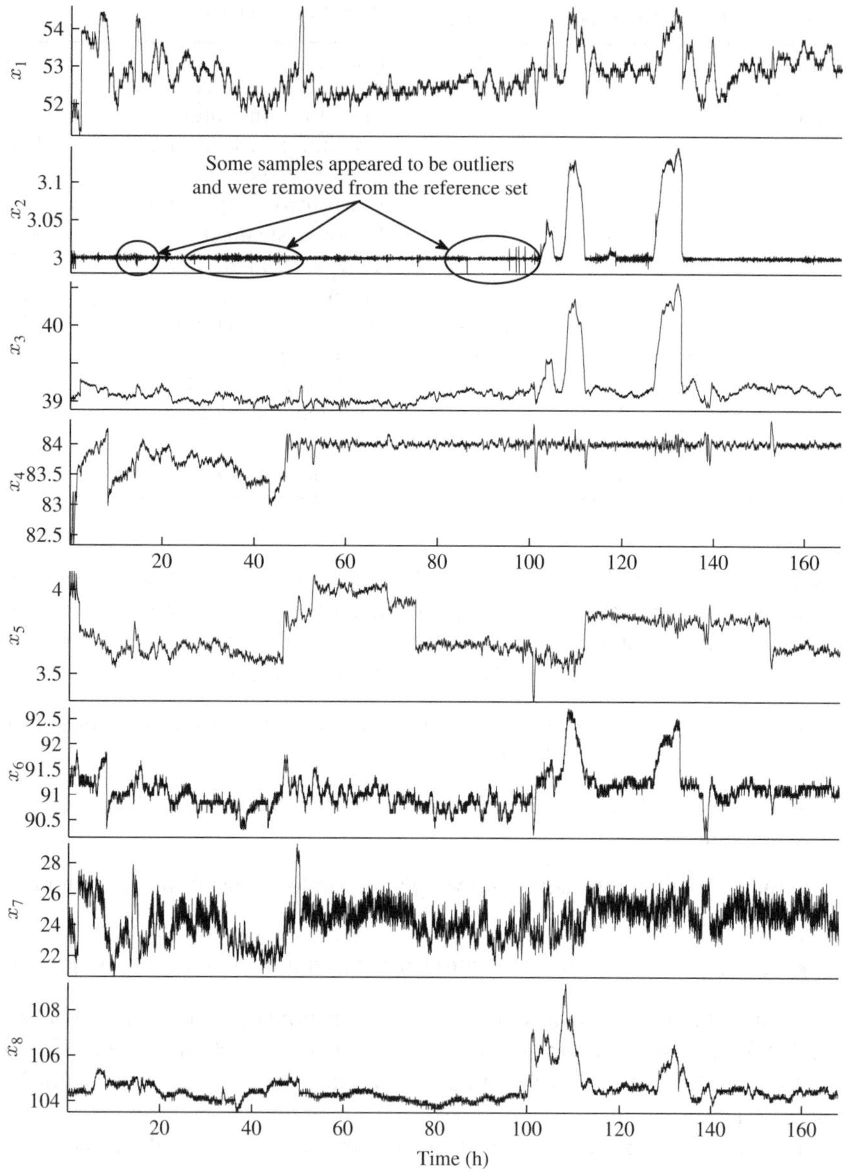

Figure 5.2 Plots of reference data for input variables set.

5.2 Identification of a monitoring model

The first step to establish a MRPLS model from the data shown in Figures 5.2 and 5.3 is to estimate the covariance and cross-covariance matrices $\widehat{\mathbf{S}}_{x_0x_0}$, $\widehat{\mathbf{S}}_{y_0y_0}$ and $\widehat{\mathbf{S}}_{y_0x_0}$. Including a total of 20,157 samples (around 165 hours), (5.2) to (5.4)

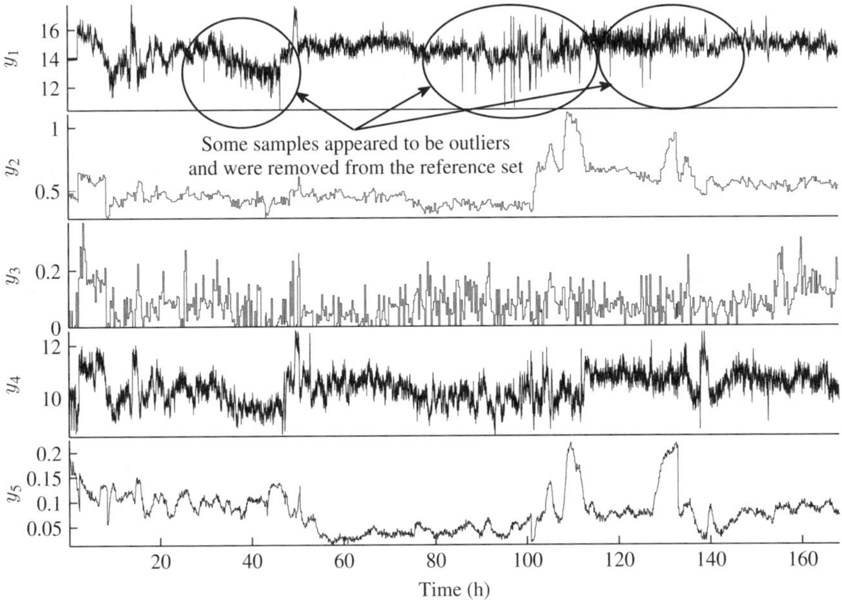

Figure 5.3 Plots of reference data for output variables set.

show these estimates. Prior to the estimation of these matrices, the mean value of each process variable was estimated and subtracted. In addition to that, the variance of each process variable was estimated and the process variables were scaled to unity variance. Equation (5.1) summarizes the pretreatment of the data

$$x_{0_i}(k) = \frac{x_i(k) - \widehat{\bar{x}}_i}{\widehat{\sigma}_{x_i}} \qquad y_{0_i}(k) = \frac{y_i(k) - \widehat{\bar{y}}_i}{\widehat{\sigma}_{y_i}}. \tag{5.1}$$

Equations (1.2), (1.3), (2.14) and (2.39) show how to estimate the variable mean and variance, as well as $\widehat{\mathbf{S}}_{x_0x_0}$, $\widehat{\mathbf{S}}_{y_0y_0}$ and $\widehat{\mathbf{S}}_{y_0x_0}$.

$$\widehat{\mathbf{S}}_{x_0x_0} = \begin{bmatrix} 1.000 & 0.505 & 0.636 & 0.264 & -0.205 & \cdots & 0.522 \\ 0.505 & 1.000 & 0.949 & 0.122 & -0.053 & \cdots & 0.619 \\ 0.638 & 0.949 & 1.000 & 0.210 & -0.126 & \cdots & 0.627 \\ 0.264 & 0.122 & 0.216 & 1.000 & 0.217 & \cdots & 0.075 \\ -0.205 & -0.053 & -0.126 & 0.217 & 1.000 & \cdots & -0.215 \\ 0.584 & 0.746 & 0.793 & 0.229 & 0.116 & \cdots & 0.671 \\ 0.535 & 0.091 & 0.162 & 0.389 & 0.471 & \cdots & 0.077 \\ 0.522 & 0.619 & 0.627 & 0.075 & -0.215 & \cdots & 1.000 \end{bmatrix}. \tag{5.2}$$

Inspection of the entries in these matrices yields that:

- the only significant correlation within the output variable set is between the bottom and the top draw flow rates, and between the C4 in C5 and the C3 in C4 concentrations, producing a correlation coefficient of around 0.75 and 0.55, respectively;

- there is a very significant correlation of 0.95 between the column overhead pressure and the tray 2 temperature variables;
- there are significant correlation coefficients of 0.5 to 0.7 between all temperature variables; and
- further significant correlation within the input variable set are between the overhead pressure and the reboiler temperature and between the temperature reading of tray 14 and the fresh feed flow, again with correlation coefficients of 0.5 to 0.7.

The input variable set, therefore, shows considerable correlation among the pressure, the temperature variables and the flow rate of the fresh feed. This is expected and follows from a steady state approximation of the first law of thermodynamics.

$$\widehat{\mathbf{S}}_{y_0y_0} = \begin{bmatrix} 1.000 & 0.270 & 0.126 & 0.743 & -0.082 \\ 0.270 & 1.000 & 0.176 & 0.257 & 0.548 \\ 0.126 & 0.176 & 1.000 & 0.130 & 0.168 \\ 0.743 & 0.257 & 0.130 & 1.000 & 0.010 \\ -0.082 & 0.548 & 0.168 & 0.010 & 1.000 \end{bmatrix} \tag{5.3}$$

$$\mathbf{S}_{x_0y_0} = \begin{bmatrix} 0.463 & 0.574 & 0.330 & 0.552 & 0.568 \\ 0.110 & 0.593 & -0.012 & 0.036 & 0.590 \\ 0.169 & 0.681 & 0.110 & 0.108 & 0.631 \\ 0.405 & 0.267 & 0.179 & 0.336 & -0.388 \\ 0.450 & 0.033 & -0.228 & 0.360 & -0.314 \\ 0.194 & 0.773 & 0.106 & 0.155 & 0.656 \\ 0.846 & 0.302 & 0.133 & 0.835 & -0.034 \\ 0.017 & 0.632 & -0.001 & 0.116 & 0.475 \end{bmatrix}. \tag{5.4}$$

Using the estimated variance and cross-covariance matrices, the next step is the identification of a MRPLS model. Given that there are a total of eight input and five output variables, MRPLS can obtain a maximum of five sets of latent variables, whilst PLS/PCA can extract three further latent component sets for the input variable set. According to the data structure in (2.51), the identification of a MRPLS model entails (i) the estimation of the source signals describing common cause variation of the input and output variable sets and (ii) the estimation of a second set of latent variables that describes the remaining variation of the input variable set.

Figure 5.4 shows the results of applying the leave-one-out cross validation stopping rule, discussed in Subsection (2.4.2). A minimum of the PRESS statistic is for $n = 4$ sets of latent variables, suggesting that there are four source signals describing common cause variation of the input and output variable sets.

Table 5.2 shows the cumulative contribution of the LV sets to $\widehat{\mathbf{S}}_{x_0x_0}$, $\widehat{\mathbf{S}}_{y_0y_0}$ and $\widehat{\mathbf{S}}_{y_0x_0}$. These contributions were computed using (2.102) to (2.104), which implies

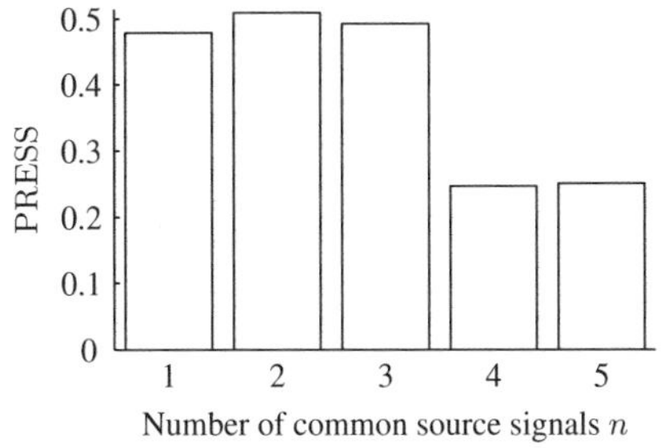

Figure 5.4 Cross-validation results for determining n.

Table 5.2 Performance of MRPLS model on reference data.

Number	t-score variable	cont$\{\mathbf{S}_{x_0y_0}\}$	cont$\{\mathbf{S}_{x_0x_0}\}$	cont$\{\mathbf{S}_{y_0y_0}\}$
1	t_1	0.4118	0.5052	0.4623
2	t_2	0.0516	0.1572	0.1820
3	t_3	0.0197	0.0889	0.1507
4	t_4	0.0023	0.0545	0.1176
5	t_1'	0.0001	0.0329	0.1176
6	t_2'	0.0000	0.0124	0.1176
7	t_3'	0.0000	0.0002	0.1176
8	t_4'	0.0000	0.0000	0.1176

that values of 1 are for the original covariance and cross-covariance matrices and values close to zero represent almost deflated matrices.

The first four rows in Table 5.2 describe LV contributions obtained from the MRPLS objective function in (2.68) and the deflation procedure in (2.74). The LV contribution representing rows five to eight relate to the PLS objective function in (2.76) and the deflation procedure in (10.18).

In terms of prediction accuracy, the rightmost column in Table 5.2 outlines that the first four LV sets rapidly reduce the squared sum of the elements in $\mathbf{S}_{ff}^{(i+1)}$, which is evident by the reduction to a value of around 0.1 for the covariance matrix $\widehat{\mathbf{S}}_{ff}^{(5)}$. A similar trend can be noticed for the cumulative sum of the elements in $\mathbf{S}_{ee}^{(i+1)}$ and $\mathbf{S}_{ef}^{(i+1)}$. Different from the simulation example in Section 2.3.4, the common cause variation within input variable set is largely shared with the output variables. More precisely, the squared sum of the elements in $\widehat{\mathbf{S}}_{ff}^{(5)}$ over those in $\widehat{\mathbf{S}}_{x_0x_0}$ is 0.055. For the cross-covariance matrix $\widehat{\mathbf{S}}_{ef}^{(5)}$, a negligible ratio of 0.002 remains, as expected.

The application of PLS to the deflated covariance and cross-covariance matrices then allows deflating the covariance matrix $\widehat{\mathbf{S}}_{x_0x_0}$ by determining four further LV sets for the input variables. The variances of first four t-score variables are equal to 1, which follows from (2.65) and (2.66). The variances

Table 5.3 Prediction accuracy of MRPLS model on referenced data.

n	$MSE\{f_{1_n}\}$	$MSE\{f_{2_n}\}$	$MSE\{f_{3_n}\}$	$MSE\{f_{4_n}\}$	$MSE\{f_{5_n}\}$
1	0.4008	0.6778	0.9537	0.3844	0.9036
2	0.2642	0.4600	0.9442	0.2692	0.1666
3	0.2642	0.3475	0.8711	0.2597	0.1311
4	0.2580	0.3222	0.7519	0.2561	0.0975

of the last four t-score variables $\widehat{t}_1$, $\widehat{t}_2$, $\widehat{t}_3$ and $\widehat{t}_4$ are 0.6261, 0.2891, 0.4821 and 0.0598, respectively.

Next, Table 5.3 lists the *mean squared error* (MSE) for predicting each of the output variable by including between 1 to 4 LV sets

$$\mathrm{MSE}\left\{f_{i_n}\right\} = \frac{1}{K}\sum_{k=1}^{K}\left(y_i(k) - \sum_{j=1}^{n}\widehat{q}_{ji}\widehat{t}_j(k)\right)^2 . \tag{5.5}$$

Apart from y_3, that is, C5 in C4 concentration, the MRPLS model is capable of providing a sufficiently accurate prediction model. The prediction accuracy is compared by plotting the measured and predicted signals for each output variables below.

Obtaining a fifth set of latent variables using the MRPLS objective function, would have resulted in a further but insignificant reduction of the MSE values for each output variables: $\mathrm{MSE}\{f_{1_5}\} = 0.2439$, $\mathrm{MSE}\{f_{2_5}\} = 0.3220$, $\mathrm{MSE}\{f_{3_5}\} = 0.7515$, $\mathrm{MSE}\{f_{4_5}\} = 0.2426$ and $\mathrm{MSE}\{f_{5_5}\} = 0.0974$. This confirms the selection of four source signals that describe common cause variation shared by the input and output variable set.

The issue of high correlation among the input variable set is further elaborated in Subsection 6.2.1, which outlines that the slight increase in prediction accuracy by increasing n from four to five may be at the expense of a considerable variance of the parameter estimation and, therefore, a poor predictive performance of the output variables for samples that are not included in the reference set.

Equations (5.6) and (5.7) show estimates of the r-weight and q-loading matrices, respectively.

$$\widehat{\mathbf{R}} = \begin{bmatrix} 0.41 & 0.02 & -0.93 & 0.30 & -0.24 & -0.25 & -0.47 & -0.21 \\ -0.27 & 0.17 & -0.16 & 1.37 & 0.30 & 0.28 & -0.53 & 0.22 \\ 0.28 & -0.30 & 0.54 & -1.06 & 0.31 & 0.38 & -0.41 & -0.43 \\ -0.04 & 0.42 & 0.37 & -0.75 & 0.22 & -0.32 & -0.41 & 0.46 \\ 0.16 & 0.38 & 0.39 & 0.62 & 0.42 & -0.55 & 0.17 & -0.33 \\ 0.15 & -0.84 & -0.04 & -0.41 & 0.15 & -0.41 & -0.07 & 0.41 \\ 0.56 & 0.23 & 0.01 & 0.05 & 0.06 & 0.38 & 0.25 & 0.49 \\ 0.03 & 0.13 & 0.81 & 0.09 & -0.71 & -0.10 & -0.26 & 0.04 \end{bmatrix} \tag{5.6}$$

$$\widehat{\mathbf{Q}} = \begin{bmatrix} 0.7741 & 0.3696 & 0.0038 & 0.0787 \\ 0.5676 & -0.4667 & 0.3354 & -0.1590 \\ 0.2150 & -0.0976 & -0.2705 & -0.3452 \\ 0.7846 & 0.3394 & -0.0976 & 0.0595 \\ 0.3105 & -0.8585 & -0.1886 & 0.1831 \end{bmatrix}. \quad (5.7)$$

To graphically interpret the relationship:

- between the computed t-score variables and the original input variables; and
- between the prediction of the output variables using the t-score variables

the elements of the r-weight and q-loading vectors can be plotted in individual loading plots, which Figure 5.5 shows. For the elements of the first three r-weight vectors (upper plot), most of the input variables contribute to a similar extent to

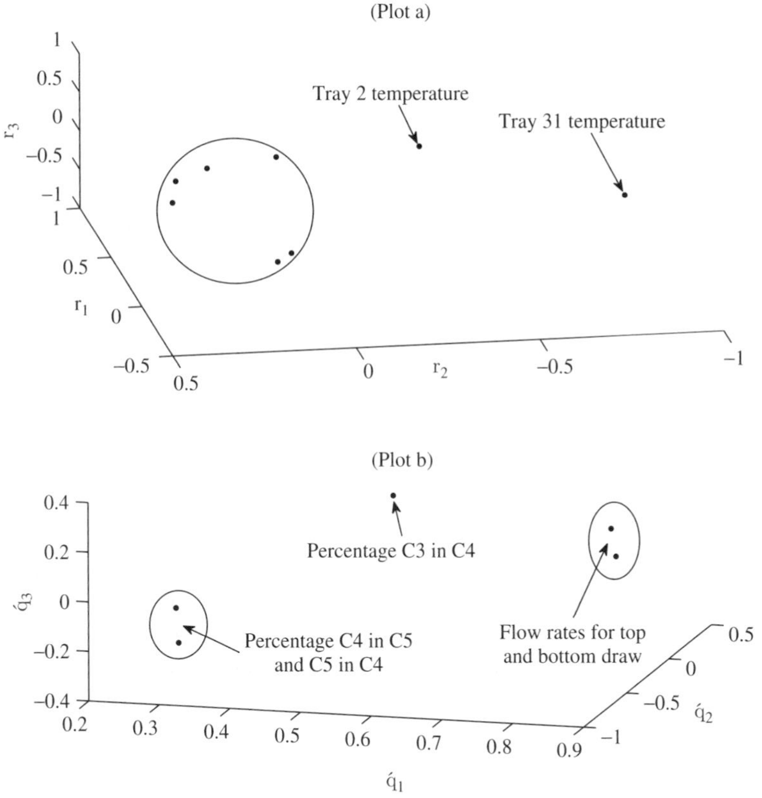

Figure 5.5 Loading plots of first three r-weight vectors and q-loading vectors.

the first three t-score variables, with the exception of the temperatures on trays 31 and 2.

The main cluster suggests that the tray 14 temperature, the overhead pressure, the reboiler steam flow and the reboiler temperature have a very similar contribution to the computation of the t-score variables and hence, the prediction of the output variables. The first r-weight vector shows a distinctive contribution from the fresh feed flow and the upper plot in Figure 5.5 indicates a distinctive contribution from tray 2 and 31 temperatures to the computation of the t-score variables.

The t-score variables are obtained to predict the output variables as accurately as possible, which follows from (2.62) to (2.66). This implies a distinctive contribution from input variables tray 2, tray 31 and fresh feed temperature. This analysis makes sense, physically, as an alteration in the tray 31 temperature affects the concentration of the impurities in the top draw, changes in the tray 2 temperature impact the level of impurities (C4 in C5) of the bottom draw and alterations in the feed temperature can affect the impurity levels in the top and/or the bottom draw.

Now, inspecting the bottom plot in Figure 5.5 yields that the first three q-loading vectors produce distinctive patterns for:

- the flow rates of the top and the bottom draw;
- the C5 in C4 and C4 in C5 concentrations of the top and bottom draw; and
- an isolated pattern for the C3 in C4 concentration of the top draw.

Although prediction of the C5 in C4 concentration using the linear MRPLS regression model is less accurate compared to the remaining four output variables, it is interesting to note that the elements in the q-loading vectors suggest a similar contribution for each of the t-score variables in terms of predicting the C5 in C4 and the C4 in C5 concentration. The distinct pattern for the top and bottom draw flow rates is also not surprising, given that slight variations in the feed rate (x_7) are mostly responsible for alterations in both output variables. This is analyzed in more detail below.

On the other hand, any variation in the C3 in C4 concentration is mostly related to the composition of the fresh feed. If the unmeasured concentration of C3 in this feed is highly correlated to the feed rate or temperature then those variables predict the C3 in C4 concentration. This outlines that the MSPC models are correlation-based representations of the data that do not directly relate to first-principle models representing causality, for example governed by conservation laws such as simple heat and material balance equations or more complex thermodynamic relationships. A detailed discussion of this may be found in Yoon and MacGregor (2000, 2001).

Assessing which input variable has a significant affect upon a specific output variable relies on analyzing the elements of the estimated regression matrix, which Table 5.4 presents. An inspection of these parameters yields that the fresh feed temperature has a major contribution to the C4 in C5 concentration but has hardly any effect on the remaining output variables. The negative sign of this

Table 5.4 Coefficients of regression model for $n = 4$.

b_{ij}	$i = 1$	$i = 2$	$i = 3$	$i = 4$	$i = 5$
$j = 1$	0.3432	−0.1400	0.2327	0.4352	0.3363
$j = 2$	−0.0353	−0.5032	−0.5067	−0.0540	0.0487
$j = 3$	0.0246	0.6499	0.3090	0.0025	0.0542
$j = 4$	0.0689	0.0247	0.1094	0.0326	−0.5815
$j = 5$	0.3140	−0.0546	−0.3199	0.2531	−0.2373
$j = 6$	−0.2272	0.5290	0.2679	−0.1882	0.7000
$j = 7$	0.5219	0.2088	0.0769	0.5185	−0.0144
$j = 8$	0.0825	0.2166	−0.2549	−0.0046	−0.2344

A row represents the coefficients for the prediction of the output variables using the jth input variable and the coefficients in a column are associated with the prediction of the ith output variable using the input variables.

parameter implies that the temperature increase results in a reduction of the C5 in C4 concentration.

The flow rates of the top and bottom draw are mainly affected, in order of significance, by the fresh feed flow, the tray 14 temperature, the reboiler steam flow and the tray 31 temperature. Following from the preceding discussion, the tray 31 temperature has a distinctive effect on calculating the t-score variables, whilst the remaining three variables have a similar contribution. Knowing that the fresh feed flow and temperature are input variables that physically affect the temperatures, pressures and flow rates within the column, the analysis here yields that the most significant influence upon both output streams are the feed level and the tray 31 temperature.

For predicting the C3 in C4 concentration the column overhead pressure and the tray 2 and 31 temperatures are the most dominant contributors. However, C3 is more volatile than C4 and C5 and significant changes in the C3 in C4 concentration must originate from variations of the C3 concentration in the fresh feed. Moreover, Table 5.3 yields correlation between the concentrations of C3 in C4 and C4 in C5, and between the C3 in C4 concentration and the flow rates of both output streams.

Given this correlation structure and the fact that each of these three variables can be accurately predicted by the MRPLS regression model, it is not surprising that the C3 in C4 concentration can be accurately predicted too. This, again, highlights the fact that the MSPC model exploits the correlation between the input and output variables as well as the correlation within the input and output variable sets.

The remaining two variables to be discussed are the C5 in C4 and the C4 in C5 concentrations. With the exception of the fresh feed temperature and flow rate, each of the remaining variables contributed to the prediction of the C5 in C4 concentration. For the C4 in C5 concentration, the overhead pressure, the tray 2 temperature and the flow rate of the fresh feed did not show a significant contribution. The most dominant contribution for predicting the C5 in C4 and the

C4 in C5 concentration is column overhead pressure and the tray 31 temperature, respectively.

A relationship between the overhead pressure and the C5 in C4 concentration is based on the estimated correlation structure. On the other hand, the positive parameter between the tray 31 temperature and the C4 in C5 concentration implies that an increase in this temperature coincides with an increase in the C4 in C5 temperature. The reason behind this observation, again, relates to the correlation structure that is encapsulated within the recorded variable set.

The analysis of the identified MRPLS model concludes with an inspection of the extracted source signals. Figure 5.6 shows the estimated sequences of the t-score variables that describe common cause variation of the input and output

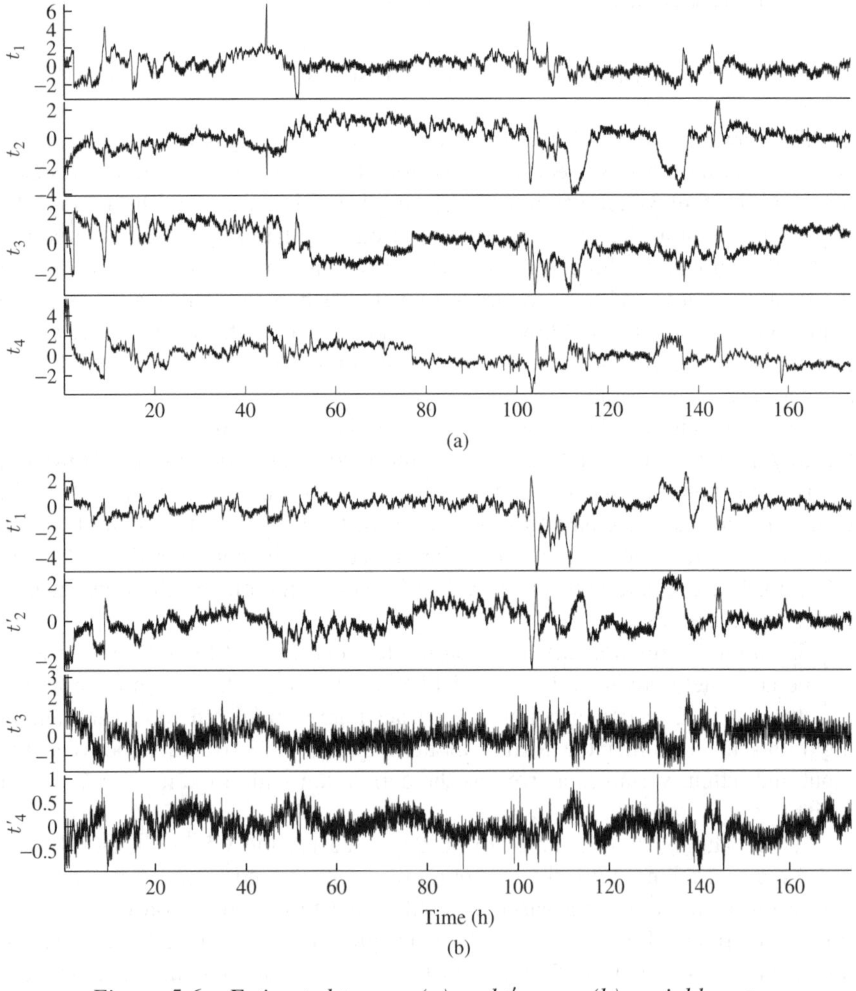

Figure 5.6 Estimated t-score (a) and t'-score (b) variable sets.

variable sets (Plot a) and sequences that describe the remaining variation within the input variable set only (Plot b). Following the MRPLS objective function, the t′-scores are not informative for predicting output variables.

Staying with the t-score variables Figure 5.7 shows the contribution of the estimated t-score variables (left plots) and the t′-score variables (right plots) to the input variables. As expected, with the exception of variables x_2, x_5 and x_8, the variance contribution of the t-score variables is significantly larger than that of the t′-score variables to the input variable set.

Evaluating the remaining variance of the input variables after subtracting the contribution of the four t-score variables, which Figure 5.8 shows, confirms this. An approximate estimation of the residual variances is given by[1]

$$\widehat{\boldsymbol{\sigma}}_{f_5} \approx \text{diag}\left\{\frac{1}{20{,}156}\sum_{k=1}^{20{,}157}\left(\mathbf{x}_0(k) - \widehat{\mathbf{P}}\widehat{\mathbf{t}}(k)\right)\left(\mathbf{x}_0(k) - \widehat{\mathbf{P}}\widehat{\mathbf{t}}(k)\right)^T\right\} \tag{5.8}$$

and stored in $\widehat{\boldsymbol{\sigma}}_{f_5}$. As before, $\widehat{\mathbf{P}} \in \mathbb{R}^{8\times 4}$ is the loading matrix and $\widehat{\mathbf{t}}(k) \in \mathbb{R}^4$ is the t-score vector for the kth sample. As the variance of each of the original input variables is 1, the smaller the displayed value of a particular variable in Figure 5.8 the more variance of this input variable is used for predicting the output variables.

The smallest value is for the flow rate of the fresh feed, followed by those of the tray 31 temperature, the tray 14 temperature and the temperature of the fresh feed. The other variables are less significantly contributing to the prediction of output variables. After assessing how the t-score and the t′-score variables contribute to the individual input variables, Figure 5.9 shows the prediction of the output variables (left column) and the residual variables for the reference data set.

Apart from y_3, describing the C5 in C4 concentration, the main common cause variation within the unit can be accurately described by the MRPLS model. Particularly the residuals of the output variables y_1, y_4 and y_5 show little correlation to the recorded output variables.

Figure 5.10 presents a direct comparison of the measured signals for variables y_3 to y_5 with their prediction using the MRPLS model. This comparison confirms the accurate prediction of y_4 and y_5 and highlights the prediction of the C5 in C4 concentration can describe long-term trends but not the short-term variation, which is the main factor for the high MSE value, listed in Table 5.3.

5.3 Diagnosis of a fault condition

The process is known for severe and abrupt drops in feed flow which can significantly affect product quality, for example the C5 in C4 concentration in the top draw. The propagation of this disturbance results from an interaction of process control structure, controller tuning and operator support. The characteristics of

[1] To ensure an unbiased estimation, the denominator may not be assumed to be K − 1 given that the loading matrix and the scores are estimates too. With the large number of samples available, however, the estimate given in this equation is sufficiently accurate.

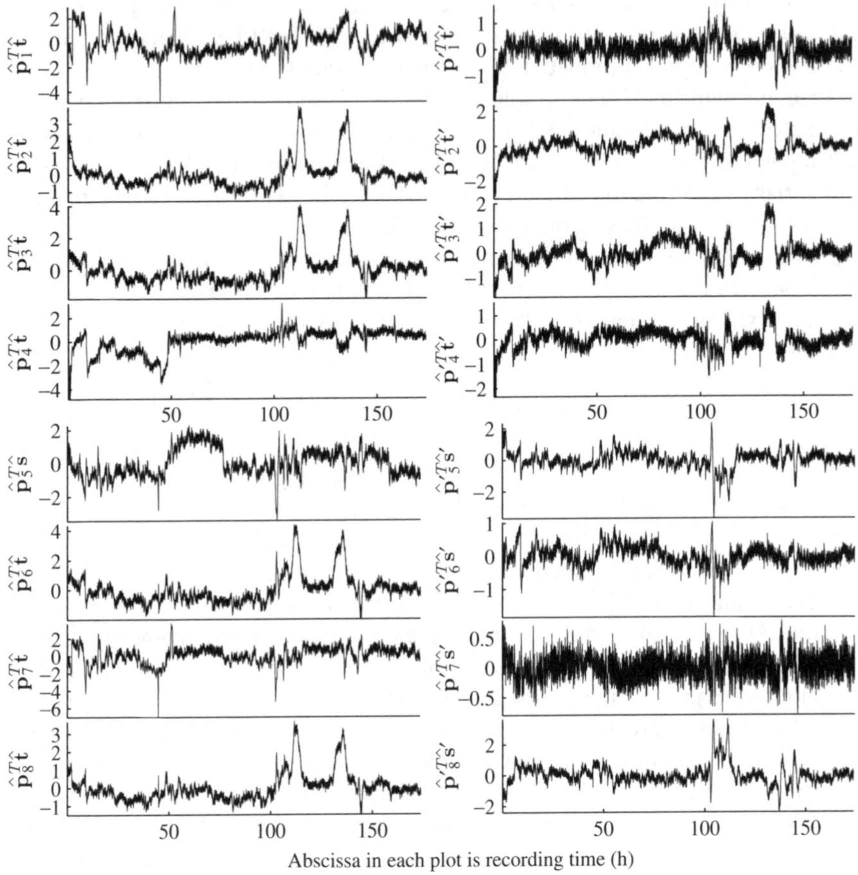

Figure 5.7 *Signal contributions of* **s** *(left plots) and* **s**′ *(right plots) to input variables.*

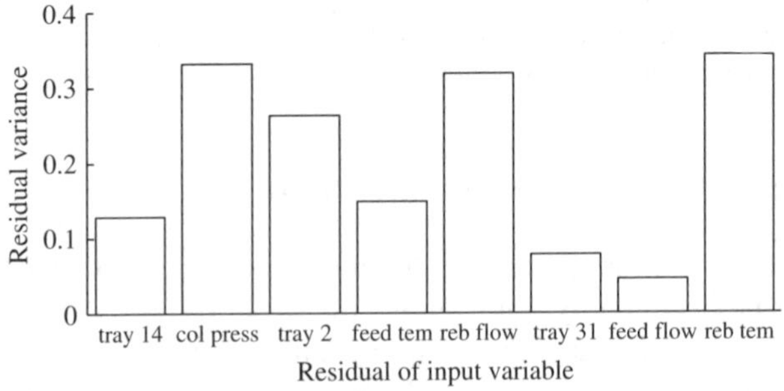

Figure 5.8 *Residual variance of input variables after deflation.*

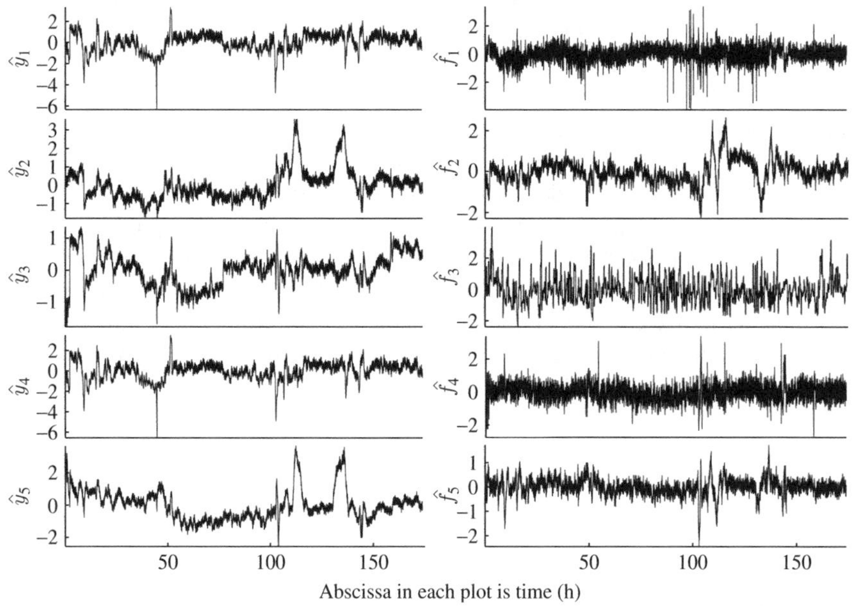

Figure 5.9 Model prediction (left column) and residuals (right column) of outputs.

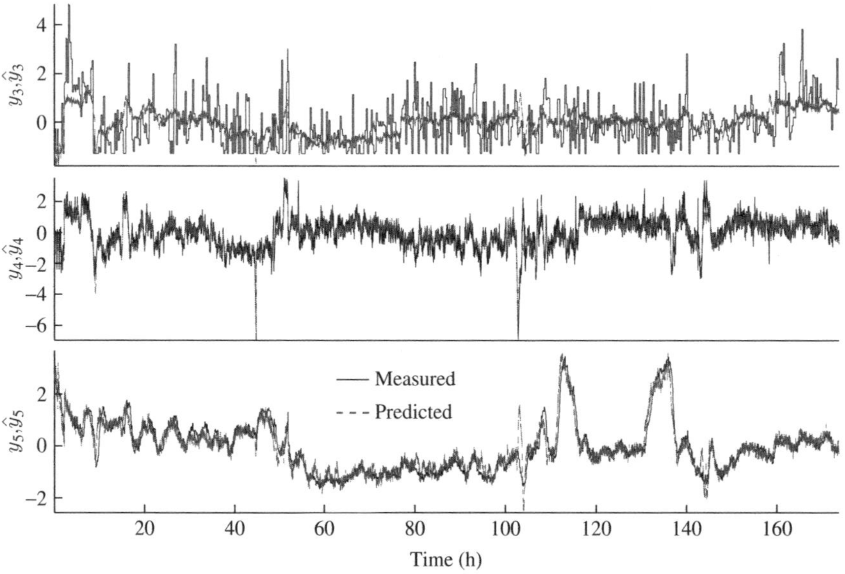

Figure 5.10 Measured and predicted signals of output variables y_3 to y_5.

this fault are described first, prior to an analysis of its detection and diagnosis. This fault is also analyzed in Kruger *et al.* (2007), Lieftucht *et al.* (2006a), Wang *et al.* (2003).

The feed flow, provided by a depropanizer unit, usually shows some variations that, according to Figure 5.2, range between 22 to $27^t/_h$. This flow, however, can abruptly decrease by up to 30%. An example of such a severe drop is studied here, highlighting that the identified monitoring model can discriminate between the severity of feed variations and whether they potentially affect product quality.

Figure 5.11 plots the recorded input variables for such an event, where a minor but prolonged drop in feed level is followed by a severe reduction that has significantly impacted product quality. The first minor drop in feed flow occurred around 23 hours into the data set with a reduction from 25 to $24^t/_h$ and lasted for about 3 hours. The second and severe reduction in feed level occurred around

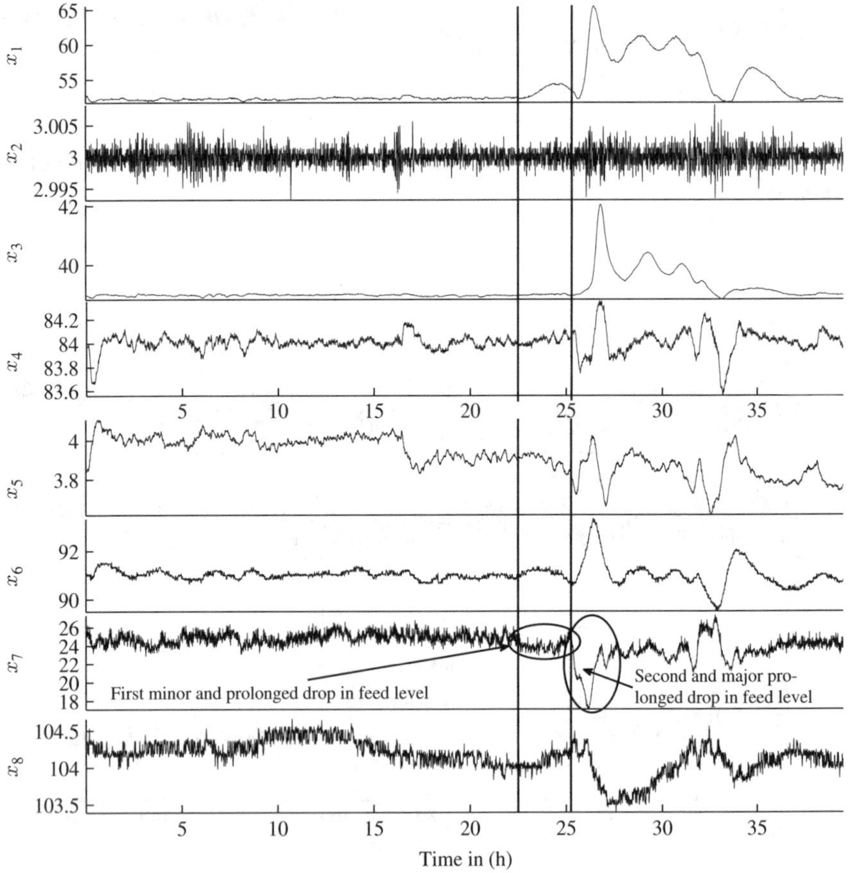

Figure 5.11 Data set describing impact of drop in feed level on input variable set.

26 hours into the data set and reduced the flow rate to around $17^t/_h$ and lasted for about 2 hours.

It should be noted that the level controllers in the bottom vessel, distillate drum and the column pressure can compensate the material balance within the column to fluctuations in the fresh feed level. The temperature controller, however, may be unable to adequately respond to these events by regulating the energy balance within the column if severe and prolonged reductions in feed arise. Figures 5.11 and 5.12 demonstrate this. The response of the column to the first prolonged drop is a slight increase the tray 14 temperature and a minor reduction in the reboiler steam flow.

The second drop that followed had a significantly greater impact on the operation of the column. More precisely, each of the column temperatures, that is, v_1, v_3 and v_6, increases sharply and almost instantly. Consequently, the C5 in C4 concentration (y_3) rose above its desired level about $1.5h$ after the feed level dropped substantially around 26h into the data set. The effect of this drop in feed level was also felt in the operation of the reboiler, as the reboiler temperature v_8 and the reboiler steam flow decreased. Another result of this event is a reduction of the flow rate of both output streams, y_1 and y_4. Figure 5.12 shows that the maximum drop of $7^t/_h$ is split into a reduction of 3 and $4^t/_h$ in the flow rate of top and bottom draw, respectively.

Figure 5.2 shows that the feed level can vary slowly and moderately, which will introduce common cause variation within the unit. The issue here is to determine when a severe drop in feed arises that has a profound impact upon product quality. This is of crucial importance, as any increase in the C5 in C4 concentration arises with a significant delay of around 90 minutes. In other words, monitoring the output variable y_3 or the input variable v_7 alone is insufficient for the monitoring task at hand. This is because variations in the feed level naturally

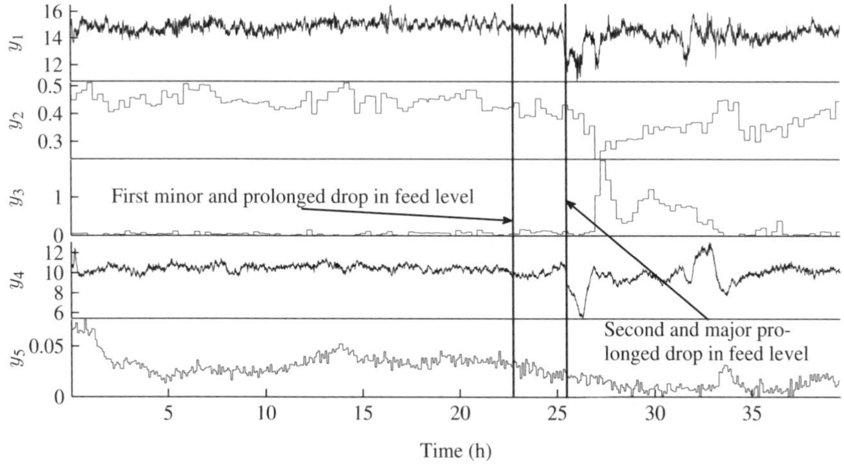

Figure 5.12 Data set describing impact of drop in feed level on output variable set.

occur and may not necessarily lead to an undesired effect on product quality. The monitoring of this process thus requires a multivariate analysis in order to alert process personnel when a severe feed drop occurs. Meronk (2001) discussed a number of potential remedies to prevent the undesired increase of the C5 in C4 concentration.

As outlined in Subsection 3.1.2, the multivariate analysis relies on scatter diagrams and non-negative quadratic statistics. Figure 5.13 shows the Hotelling's T^2 and T'^2, and the Q_f statistics. As expected, the first 22 hours into the data set did not yield an out-of-statistical-control situation. The drop in fresh feed was first noticed by the Hotelling's T'^2 statistic after about 23.5h, with the remaining two statistics showing no response up until 26h into the data set.

Figure 5.14 shows a zoom of each monitoring statistic in Figure 5.13 between 20 to 26h. The linear combinations of the input variables that are *predictive* for the output variables are not affected by the initial and minor drop in fresh feed. The Q_f statistic also confirms this, since it represents the mismatch between the measured and predicted output variables and did not suggest an out-of-statistical-control situation during the first minor drop in feed. However, the correlation between the input variables that is not informative for predicting the output variables, that is, the remaining variation that contributes to the input variables only, is affected by the prolonged drop in the feed level. Particularly the increase in the tray 14 temperature and the decrease in the reboiler steam flow, according to Figure 5.11, is uncharacteristic for normal process behavior and resulted in the out-of-statistical-control situation.

The diagnosis of this event is twofold. The first examination concentrates on the impact of the first minor reduction in the flow rate of the fresh feed. Given that only the Hotelling's T'^2 statistic was sensitive to this event, Figure 5.15 presents time-based plots of the four t′-scores variables between 20 to 25.5 hours into the data set. The plots in this figure are a graphical representation of (3.37).

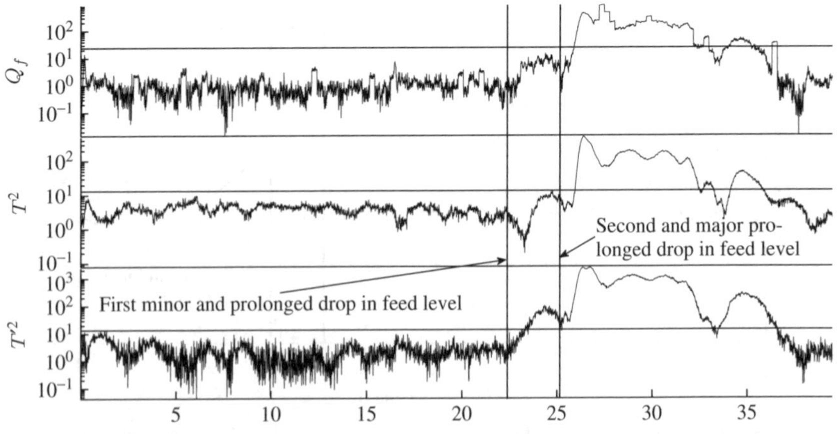

Figure 5.13 Monitoring statistics of data set covering drop in fresh feed level.

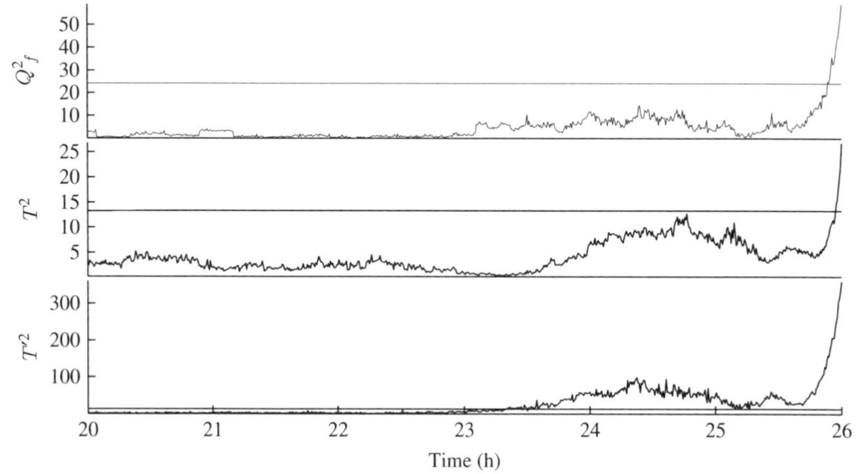

Figure 5.14 Monitoring statistics for feed drop between 20 to 26 h into the data set.

In plot (a) of Figure 5.15, a value of zero implies acceptance of the null hypothesis, that is, $t_i^2/\sigma_{t_i}^2 \leq T_\alpha^2/4$. Conversely, values larger than zero yield rejection of the null hypothesis. Plot (b) in Figure 5.15 shows the computed variable contribution by applying (3.38) and (3.39) for variables t_2' and t_3', and highlights that v_7 (fresh feed flow) is most dominantly contributing.

The second and severe drop developed over a period of 1 hour and resulted in an eventual reduction to just over $17^t/_h$. According to Figure 5.14, the Q_f statistic was sensitive around 25.5h into the data set, that is, before the fresh feed flow reached its bottom level. Around the same time, the Hotelling's T'^2 statistic also increased sharply in value and with a slight delay, the Hotelling's T^2 statistic violated its control limit.

The preceding discussion outlined that the first drop did not have an effect on the C5 in C4 concentration, that is, the product quality. However, the prolonged drop in feed level from 25 to $24^t/_h$ is (relative to the variations in feed of the large reference data set) unusual and affected the second and third t′-score variables. This suggests that monitoring these two variables along with the most dominant contributor for product quality, the first t-score variable, is an effective way of detecting minor but prolonged drops in feed.

Figure 5.16 shows the progression of the first drop using scatter diagrams in six stages. The first 23 hours and 20 minutes (2800 samples) are plotted in the upper left plot and indicate, as expected, that the scatter points cluster around the center of coordinate system. Each of the following plots, from the top right to the bottom right plot, adds 25 or 50 minutes of data and covers the additional range from 23 hours and 20 minutes to 25 hours and 50 minutes. Therefore, the bottom right plot describes the first stage of the second and sharp reduction in feed level.

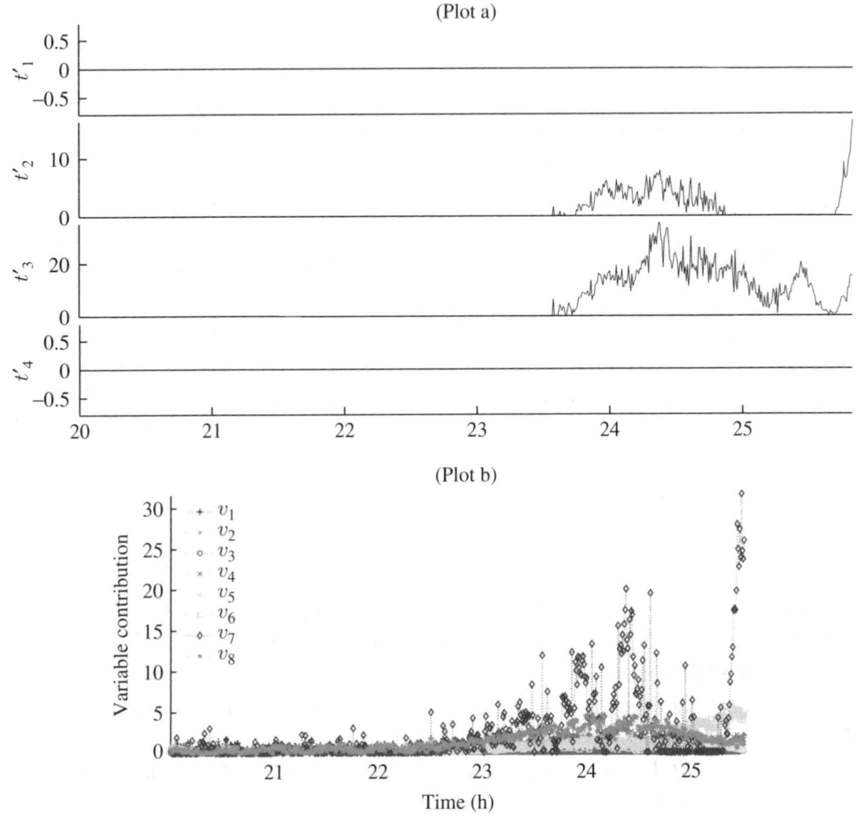

Figure 5.15 *t' scores and variable contributions to feed drop for 20 to 26 hours.*

First, the scatter points move away from the initial cluster, which represents an in-statistical-control behavior, in the direction of the t'_2- and t'_3-score variable and establishes a second cluster, which represents the first drop in feed. When the second and more severe drop in feed emerges, the scatter points move away from this second cluster and appear to establish a third cluster. It is interesting to note that this third cluster is shifted from the original one along the axis of the t_1-score variable.

Plot (a) in Figure 5.17 shows, graphically, the hypothesis test of (3.37) for each of the t-score variables, covering the second drop in feed. This time, as Figure 5.14 confirms, each monitoring statistic showed an out-of-statistical-control situation. Score variables t_2 and t_3 were initially most sensitive to the significant reduction in feed and after about 26 hours into the data set. Applying the procedure for determining the variable contributions in (3.38) and (3.39), on the basis of these score variables produced the plot (b) in Figure 5.17.

The main contributing variables to this event were the flow rate of the fresh feed v_7 and with a delay the tray 14 temperature v_1. Applying the same procedure

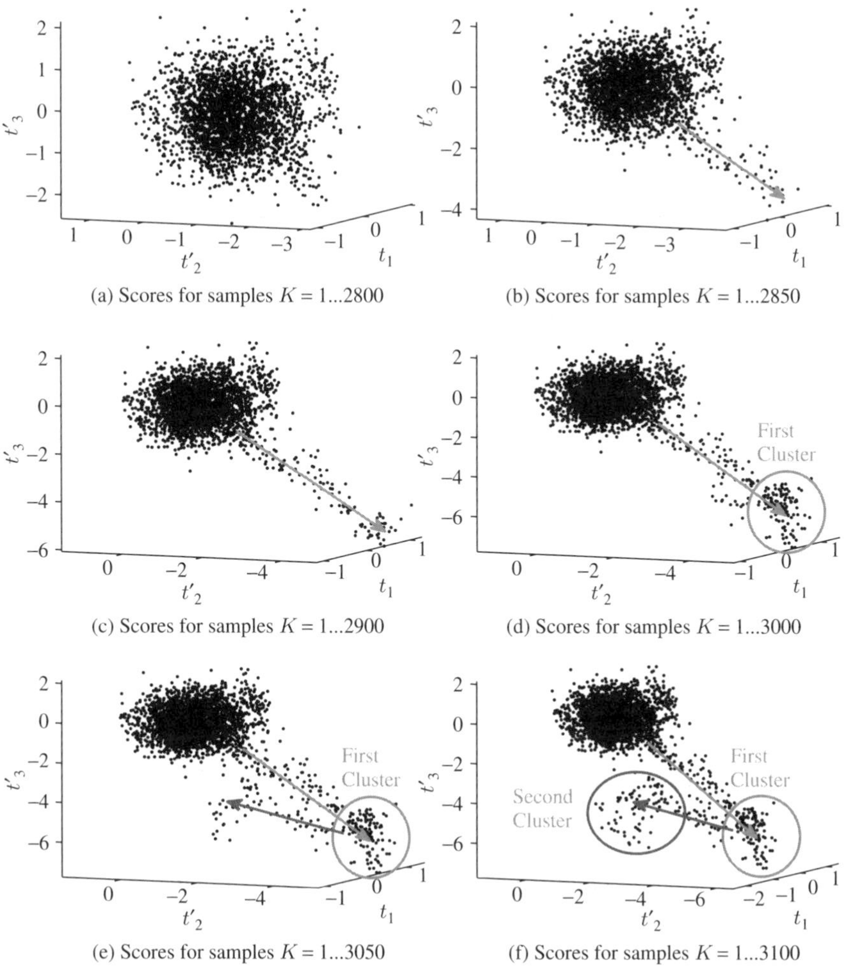

Figure 5.16 Scatter diagrams for t_1, t_2' and t_3' showing various stages in feed drop.

for the first t-score variable between 25h and 30min and 25h and 50min (not shown here) yields the tray 14 temperature as the main contributor. This results from the sharply decreasing value of this variable during this period, which Figure 5.11 confirms.

Recall that the MRPLS monitoring model is a steady state representation of this unit that is based on modeling the correlation structure between the input and output variables but also among the input and output variable sets. The observed decrease in the tray 14 temperature is a result of the energy balance within the column that may not be described by the static correlation-based model. On the basis of an identified dynamic model of this process, Schubert *et al.* (2011) argued that the increase in the impurity level of the top draw (C5 in

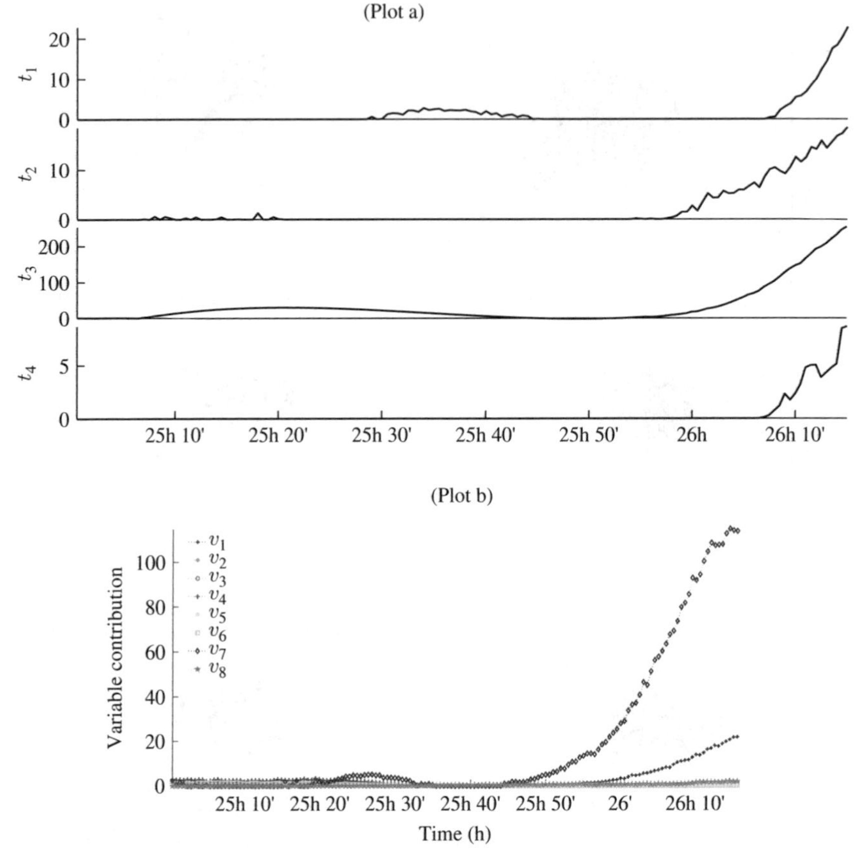

Figure 5.17 t scores and variable contributions to feed drop for 25 to 26 hours.

C4 concentration) originates from an upset in the energy balance. On the other hand, Buckley *et al.* (1985) pointed out that the flow rate of the fresh feed is the most likely disturbance upsetting the energy balance within the distillation column. Other but minor sources that may exhibit an undesired impact upon the energy balance is the fresh feed temperature and enthalpy.

In this regard, the application of the MRPLS monitoring model, however, correctly detected the reduction in feed level as the root cause triggering the undesired increase in the C5 in C4 concentration. Furthermore, the monitoring model was also capable of discriminating:

- between a minor reduction in feed level that was still abnormal relative to the reference data set but did not affect product quality; and
- a severe drop in this level that had a profound and undesired impact on product quality.

More precisely, the preceding analysis suggested that the t'-score variables and thus the Hotelling T'^2 statistic reject the null hypothesis, that is, the process is in-statistical-control, if abnormal changes in the feed level arise. On the other hand, if such changes are significant, the t-score variables and with them the Hotelling's T^2 and the Q_f statistics reject the null hypothesis. This confirms the benefits of utilizing the model structure in (2.51). From the point of the process operator, the Hotelling's T^2 and the Q_f statistics are therefore informative, as they can provide an early warning that a feed drop has arisen that has the potential to affect produce quality. The immediate implementation of the recommendations in Meronk (2001) can prevent the negative impact of this change in feed level upon produce quality. For a minor alteration in feed level, however, is important to assess its duration as the prolonged presence may also have the potential to upset the energy balance.

PART III

ADVANCES IN MULTIVARIATE STATISTICAL PROCESS CONTROL

6

Further modeling issues

Chapters 1 to 3 introduced the basic MSPC approach that is applied to the chemical reaction and the distillation processes in Chapters 4 and 5, respectively. This chapter extends the coverage of MSPC modeling methods by discussing the following and practically important aspects:

1. how to estimate PCA models if the error covariance matrix $\mathbf{S}_{gg} \neq \sigma_g^2\mathbf{I}$;
2. how to estimate PLS/MRPLS models if the input variable sets are also corrupted by an error vector;
3. how to estimate MSPC models if the reference data contain outliers; and
4. how to estimate MSPC models if only small reference sets are available.

Section 6.1 introduces a maximum likelihood formulation for simultaneously estimating an unknown diagonal error covariance matrix and the model subspace, and covers cases where $\mathbf{S}_{gg}$ is known but not of the form $\sigma_g^2\mathbf{I}$.

Section 6.2 discusses the accuracy of estimating PLS models and compares them with OLS models with respect to the relevant case that the input variables are highly correlated. The section then extends the data structure in (2.23), (2.24) and (2.51) by including an error term for the input variable set, which yields an *error-in-variable* (Söderström 2007) or *total least squares* (van Huffel and Vandewalle 1991) data structure. The section finally introduces a maximum likelihood formulation for PLS and MRPLS models to identify error-in-variable estimates of the LV sets.

Outliers, which are, at first glance, samples associated with a very large error or are simply different from the majority of samples, can profoundly affect the accuracy of statistical estimates (Rousseeuw and Hubert 2011). Section 6.3 summarizes methods for a robust estimation of PCA and PLS models by reducing

Statistical Monitoring of Complex Multivariate Processes: With Applications in Industrial Process Control, First Edition. Uwe Kruger and Lei Xie.

the impact of outliers upon the estimation procedure and trimming approaches that exclude outliers.

Section 6.4 describes how a small reference set, that is, a set that only contains few reference samples, can adversely affect the accuracy of the estimation of MSPC models. The section stresses the importance of statistical independence for determining the Hotelling's T^2 statistics and also discusses a cross-validatory approach for the residual-based Q statistics.

Finally, Section 6.5 provides a tutorial session including short questions and small projects to help familiarization with the material of this chapter. This enhances the learning outcomes, which describes important and practically relevant extensions of the conventional MSPC methodology, summarized in Chapters 1 to 3.

6.1 Accuracy of estimating PCA models

This section discusses how to consistently estimate PCA models if $\mathbf{S}_{\mathrm{gg}} \neq \sigma_{\mathrm{g}}^2 \mathbf{I}$, which includes the estimation of the model subspace and $\mathbf{S}_{\mathrm{gg}}$. The section first revises the underlying assumptions for consistently estimating a PCA model by applying the eigendecomposition of $\mathbf{S}_{z_0 z_0}$ in Subsection 6.1.1. Next, Subsection 6.1.2 presents two illustrative examples to demonstrate that a general structure of the error covariance matrix, that is, $\mathbf{S}_{\mathrm{gg}} \neq \sigma_{\mathrm{g}}^2 \mathbf{I}$ yields an inconsistent estimation of the model subspace.

Under the assumption that the error covariance matrix is known *a priori*, Subsection 6.1.3 develops a maximum likelihood formulation to consistently estimate the orientation of the model and residual subspaces. If $\mathbf{S}_{\mathrm{gg}}$ is unknown, Subsection 6.1.4 introduces an approach for a simultaneous estimation of the model subspace and $\mathbf{S}_{\mathrm{gg}}$ using a Cholesky decomposition. Subsection 6.1.5 then presents a simulation example to show a simultaneous estimation of the model subspace and $\mathbf{S}_{\mathrm{gg}}$ for a known number of source signals n. Assuming n is unknown, Subsection 6.1.6 then develops a stopping rule to estimate the number of source signals.

Subsection 6.1.7 revisits the maximum likelihood estimates of the model and residual subspaces and introduces a re-adjustment to ensure that the loading vectors, spanning both subspaces, point in the direction of maximum variance for the sample projections. Finally, Subsection 6.1.8 puts the material presented in this section together and revisits the application study of the chemical reaction process in Chapter 4. The revised analysis shows that the recorded variable set contains a larger number of source signals than the four signals previously suggested in Chapter 4.

6.1.1 Revisiting the eigendecomposition of $\mathbf{S}_{z_0 z_0}$

Equation (2.2) and Table 2.1 show that the data structure for recorded data is

$$\mathbf{z} = \Xi \mathbf{s} + \bar{\mathbf{z}} + \mathbf{g} = \mathbf{z}_s + \bar{\mathbf{z}} + \mathbf{g} = \mathbf{z}_0 + \bar{\mathbf{z}}. \tag{6.1}$$

Removing the mean from the recorded variable, the stochastic component is assumed to follow a zero mean multivariate Gaussian distribution with the covariance matrix

$$E\left\{\mathbf{z}_0\mathbf{z}_0^T\right\} = \mathbf{S}_{z_0z_0} = \boldsymbol{\Xi}\mathbf{S}_{ss}\boldsymbol{\Xi}^T + \sigma_{\mathrm{g}}^2\mathbf{I}. \tag{6.2}$$

Asymptotically, assuming that $E\left\{\mathbf{z}_s\mathbf{g}^T\right\} = \mathbf{0}$ the eigendecomposition of $\mathbf{S}_{z_0z_0}$

$$\lim_{K\to\infty}\widehat{\mathbf{S}}_{z_0z_0} = \lim_{K\to\infty}\tfrac{1}{K-1}\sum_{k=1}^{K}\left(\mathbf{z}(k)-\widehat{\bar{\mathbf{z}}}\right)\left(\mathbf{z}(k)-\widehat{\bar{\mathbf{z}}}\right)^T \to \mathbf{S}_{z_0z_0} \tag{6.3}$$

yields

$$\mathbf{S}_{z_0z_0} = \begin{bmatrix}\mathbf{P} & \mathbf{P}_d\end{bmatrix}\begin{bmatrix}\boldsymbol{\Lambda} & \mathbf{0}\\ \mathbf{0} & \boldsymbol{\Lambda}_d\end{bmatrix}\begin{bmatrix}\mathbf{P}^T\\ \mathbf{P}_d^T\end{bmatrix} = \sum_{i=1}^{n}\lambda_i\mathbf{p}_i\mathbf{p}_i^T + \sigma_{\mathrm{g}}^2\sum_{i=n+1}^{n_z}\mathbf{p}_i\mathbf{p}_i^T. \tag{6.4}$$

Given that $\mathbf{S}_{z_sz_s} = \mathbf{S}_{z_0z_0} - \sigma_{\mathrm{g}}^2\mathbf{I}$ and $\lim_{K\to\infty}\widehat{\mathbf{S}}_{z_0z_0} \to \mathbf{S}_{z_0z_0}$ the eigendecomposition of $\widehat{\mathbf{S}}_{z_0z_0}$ provides an asymptotic estimate of $\lim_{K\to\infty}\widehat{\sigma}_{\mathrm{g}}^2 \to \sigma_{\mathrm{g}}^2$ and allows extracting $\mathbf{S}_{z_sz_s}$

$$\mathbf{S}_{z_0z_0} = \underbrace{\sum_{i=1}^{n}\left(\lambda_i - \sigma_{\mathrm{g}}^2\right)\mathbf{p}_i\mathbf{p}_i^T}_{=\mathbf{S}_{z_sz_s}} + \underbrace{\sigma_{\mathrm{g}}^2\sum_{i=1}^{n_z}\mathbf{p}_i\mathbf{p}_i^T}_{=\sigma_{\mathrm{g}}^2\mathbf{I}}. \tag{6.5}$$

Since the matrix $\mathbf{P} = \begin{bmatrix}\mathbf{p}_1 & \mathbf{p}_2 & \cdots & \mathbf{p}_{n_z}\end{bmatrix}$ has orthonormal columns, which follows from Theorem 9.3.3, the term $\sigma_{\mathrm{g}}^2\sum_{i=1}^{n_z}\mathbf{p}_i\mathbf{p}_i^T$ reduces to $\sigma_{\mathrm{g}}^2\mathbf{I}$ and hence

$$\mathbf{S}_{z_0z_0} = \begin{bmatrix}\sigma_{z_{1_s}}^2 + \sigma_{\mathrm{g}}^2 & \sigma_{z_{1_s}z_{2_s}}^2 & \cdots & \sigma_{z_{1_s}z_{n_{z_s}}}^2\\ \sigma_{z_{2_s}z_{1_s}}^2 & \sigma_{z_{2_s}}^2 + \sigma_{\mathrm{g}}^2 & \cdots & \sigma_{z_{2_s}z_{n_{z_s}}}^2\\ \vdots & \vdots & \ddots & \vdots\\ \sigma_{z_{n_{z_s}}z_{1_s}}^2 & \sigma_{z_{n_{z_s}}z_{2_s}}^2 & \cdots & \sigma_{z_{n_{z_s}}}^2 + \sigma_{\mathrm{g}}^2\end{bmatrix}. \tag{6.6}$$

Under the above assumptions, the eigendecomposition of $\mathbf{S}_{z_0z_0}$ can be separated into $\mathbf{S}_{z_sz_s}$ and $\mathbf{S}_{\mathrm{gg}} = \sigma_{\mathrm{g}}^2\mathbf{I}$, where

$$\mathbf{S}_{z_sz_s} = \begin{bmatrix}\mathbf{p}_1 & \mathbf{p}_2 & \cdots & \mathbf{p}_n\end{bmatrix}\left[\boldsymbol{\Lambda} - \sigma_{\mathrm{g}}^2\mathbf{I}\right]\begin{bmatrix}\mathbf{p}_1^T\\ \mathbf{p}_2^T\\ \vdots\\ \mathbf{p}_n^T\end{bmatrix} \tag{6.7}$$

and

$$\mathbf{\Lambda} - \sigma_g^2 \mathbf{I} = \begin{bmatrix} \lambda_1 - \sigma_g^2 & 0 & \cdots & 0 \\ 0 & \lambda_2 - \sigma_g^2 & \cdots & 0 \\ \vdots & \vdots & \ddots & \vdots \\ 0 & 0 & \cdots & \lambda_n - \sigma_g^2 \end{bmatrix}. \tag{6.8}$$

Following the geometric analysis is Section 2.1, (2.2) to (2.5) and Figure 2.2, the model subspace, originally spanned by the column vectors of Ξ, can be spanned by the n retained loading vectors $\mathbf{p}_1, \mathbf{p}_2, \cdots, \mathbf{p}_n$, since

$$\mathbf{P}\left[\mathbf{\Lambda} - \sigma_g^2 \mathbf{I}\right] \mathbf{P}^T = \Xi \mathbf{S}_{ss} \Xi^T. \tag{6.9}$$

Determining the eigendecomposition of $\mathbf{S}_{ss}$ and substituting $\mathbf{S}_{ss} = \mathfrak{T}\mathfrak{L}\mathfrak{T}^T$ into (6.9) gives rise to

$$\mathbf{P}\left[\mathbf{\Lambda} - \sigma_g^2 \mathbf{I}\right] \mathbf{P}^T = \Xi \mathfrak{T}\mathfrak{L}\mathfrak{T}^T \Xi^T. \tag{6.10}$$

Next, re-scaling the eigenvalues of $\mathbf{S}_{ss}$ such that $\mathfrak{L} = \mathcal{L}^{-1/2}\left[\mathbf{\Lambda} - \sigma_g^2 \mathbf{I}\right] \mathcal{L}^{-1/2}$ yields

$$\mathbf{P}\left[\mathbf{\Lambda} - \sigma_g^2 \mathbf{I}\right] \mathbf{P}^T = \Xi \mathfrak{T} \mathcal{L}^{-1/2}\left[\mathbf{\Lambda} - \sigma_g^2 \mathbf{I}\right] \mathcal{L}^{-1/2} \mathfrak{T}^T \Xi^T. \tag{6.11}$$

Hence, $\mathbf{\Lambda} - \sigma_g^2 \mathbf{I} = \mathcal{L}^{1/2} \mathfrak{L} \mathcal{L}^{1/2}$, where $\mathcal{L}$ is a diagonal scaling matrix. The above relationship therefore shows that $\mathbf{P} = \Xi \mathfrak{T} \mathcal{L}^{-1/2}$ and hence, $\Xi = \mathbf{P} \mathcal{L}^{1/2} \mathfrak{T}^T$.

Now, multiplying this identity by $\mathbf{P}_d^T$ from the left gives rise to

$$\mathbf{P}_d^T \Xi = \mathbf{P}_d^T \mathbf{P} \mathcal{L}^{1/2} \mathfrak{T}^T = \mathbf{0}, \tag{6.12}$$

which follows from the fact that the PCA loading vectors are mutually orthonormal. That the discarded eigenvectors, spanning the residual subspace, are orthogonal to the column vectors of Ξ implies that the n eigenvectors stored as column vectors in $\mathbf{P}$ span the same model subspace. Consequently, the orientation of the model subspace can be estimated consistently by determining the dominant eigenvectors of $\mathbf{S}_{z_0 z_0}$

$$\lim_{K \to \infty} \mathbf{S}_{z_0 z_0} = \mathbf{P}\left[\mathbf{\Lambda} - \sigma_g^2 \mathbf{I}\right] \mathbf{P}^T + \sigma_g^2 \mathbf{I}. \tag{6.13}$$

In other words, the dominant n loading vectors present an orthonormal base that spans the model subspace under the PCA objective function of maximizing the variance of the score variables $t_i = \mathbf{p}_i^T \mathbf{z}_0$. It can therefore be concluded that the loading vectors present an asymptotic approximation of the model subspace, spanned by the column vectors in Ξ. However, this asymptotic property holds true only under the assumption that $\mathbf{S}_{gg}$ is a diagonal matrix with identical entries, which is shown next.

6.1.2 Two illustrative examples

The first example is based on the simulated process in Section 2.1, where three process variables are determined from two source signals that follow a multivariate Gaussian distribution, $\mathbf{s} \sim \mathcal{N}\left\{\mathbf{0}, \mathbf{S}_{ss}\right\}$. Equations (2.9) to (2.11) show the exact formulation of this simulation example. The error covariance matrix of (2.11) is therefore of the type $\sigma_{\mathrm{g}}^2\mathbf{I}$ so that the eigendecomposition of $\mathbf{S}_{z_0z_0}$ allows a consistent estimation of the model subspace, spanned by the two column vectors of Ξ, $\boldsymbol{\xi}_1^T = \begin{pmatrix} 0.2 & 0.8 & -0.3 \end{pmatrix}$ and $\boldsymbol{\xi}_2^T = \begin{pmatrix} -0.5 & 0.4 & -0.7 \end{pmatrix}$.

Constructing an error covariance matrix that is of a diagonal type but contains different diagonal elements, however, does not yield a consistent estimate of the model subspace according to the discussion in previous subsection. Let $\mathbf{S}_{\mathrm{gg}}$ be

$$\mathbf{S}_{\mathrm{gg}} = \begin{bmatrix} 0.075 & 0 & 0 \\ 0 & 0.01 & 0 \\ 0 & 0 & 0.15 \end{bmatrix}, \tag{6.14}$$

which produces the following covariance matrix of $\mathbf{z}_0$

$$\mathbf{S}_{z_0z_0} = \Xi\mathbf{S}_{ss}\Xi^T + \mathbf{S}_{\mathrm{gg}} = \begin{bmatrix} 0.4250 & 0.0560 & 0.2870 \\ 0.0560 & 0.6180 & -0.3160 \\ 0.2870 & -0.3160 & 0.6040 \end{bmatrix}. \tag{6.15}$$

The eigendecomposition of this covariance matrix is

$$\begin{aligned} \mathbf{P} &= \begin{bmatrix} -0.3213 & -0.7085 & -0.6283 \\ 0.5922 & -0.6681 & 0.4506 \\ -0.7390 & -0.2273 & 0.6342 \end{bmatrix} \\ \boldsymbol{\Lambda} &= \begin{bmatrix} 0.9820 & 0 & 0 \\ 0 & 0.5699 & 0 \\ 0 & 0 & 0.0951 \end{bmatrix}. \end{aligned} \tag{6.16}$$

To examine the accuracy of estimating the model subspace, the direction of the residual subspace, which is $\mathbf{n}^T = \begin{pmatrix} -0.6173 & 0.4068 & 0.6734 \end{pmatrix}$ according to (2.20), can be compared with the third column vector in $\mathbf{P}$

$$\cos\left(\varphi_{(\mathbf{n},\mathbf{p}_3)}\right) = \mathbf{n}^T\mathbf{p}_3 = 0.9982 \qquad \arccos(0.9982)\,\frac{180}{\pi} = 3.4249^\circ \tag{6.17}$$

As a result, the determined residual subspace departs by a minimum angle of 3.4249° from the correct one. Defining $\varphi_{(\mathbf{n},\mathbf{p}_3)}$ as a parameter the above analysis demonstrates that this parameter is not equal to 1. Hence, $\mathbf{n}$ can only be estimated with a bias (Ljung 1999). Asymptotically, $\varphi_{(\mathbf{n},\mathbf{p}_3)} = 1$ if $\mathbf{S}_{\mathrm{gg}} = \sigma_{\mathrm{g}}^2\mathbf{I}$ and else < 1.

A second example considers a Monte Carlo experiment where the variances for each of the three error variables are determined randomly within the range of $\begin{bmatrix} 0.04 & 0.06 \end{bmatrix}$. For a total of 100 experiments, Figure 6.1 shows the uniformly distributed values for each error variance. Applying the same calculation for

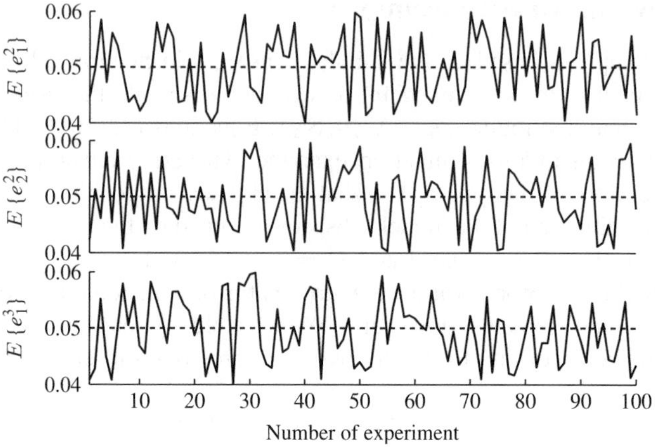

Figure 6.1 Variance for each of the three residual variables vs. number of experiment.

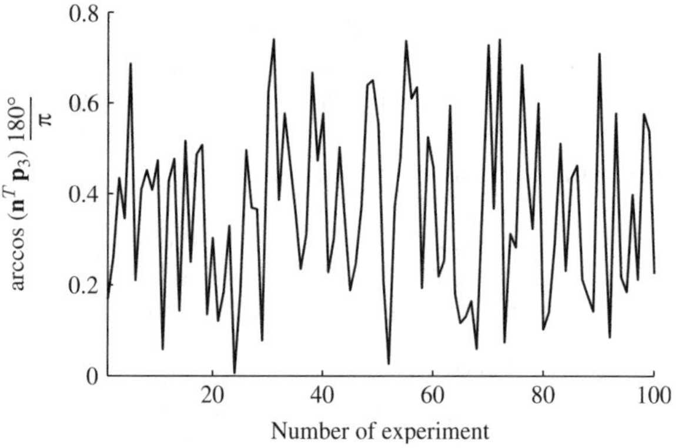

Figure 6.2 Angle between original and estimated eigenvector for residual subspace.

determining the minimum angle between $\mathbf{p}_3$ and $\mathbf{n}$ for each set of error variances yields the results shown in Figure 6.2. Angles close to zero, for example in experiments 23 and 51, relate to a set of error variances that are close to each other. On the other hand, larger angles, for example experiments 31, 53, 70, 72 and 90 are produced by significant differences between the error variances.

6.1.3 Maximum likelihood PCA for known $\mathbf{S}_{gg}$

Wentzell *et al.* (1997) introduced a maximum likelihood estimation (Aldrich 1997) for PCA under the assumption that $\mathbf{S}_{gg}$ is known. The maximum likelihood formulation, which is discussed in the next subsection, relies on the following formulation

$$J\left(\mathbf{g},\mathbf{S}_{gg}\right)=\frac{1}{(2\pi)^{n_z/2}\left|\mathbf{S}_{gg}\right|^{1/2}}\exp\left(-\tfrac{1}{2}\left(\mathbf{z}_0-\mathbf{z}_s\right)^T\mathbf{S}_{gg}^{-1}\left(\mathbf{z}_0-\mathbf{z}_s\right)\right)\tag{6.18}$$

where $J\left(\mathbf{g},\mathbf{S}_{gg}\right)>0$[1] is the likelihood of occurrence of the error vector $\mathbf{g}=\mathbf{z}_0-\widehat{\mathbf{z}}_0$, if the error vector follows $\mathbf{g}\sim\mathcal{N}\left\{\mathbf{0},\mathbf{S}_{gg}\right\}$. According to (2.2), $\mathbf{g}=\mathbf{z}_0-\boldsymbol{\Xi}\mathbf{s}$, $\boldsymbol{\Xi}\mathbf{s}=\mathbf{z}_s$. With k and l being sample indices, it is further assumed that $E\left\{\mathbf{g}(k)\,\mathbf{g}^T(l)\right\}=\delta_{kl}\mathbf{S}_{gg}$. If a total of K samples of $\mathbf{z}_0$ are available, $\mathbf{z}_0(1)$, $\ldots$, $\mathbf{z}_0(k)$, $\ldots$, $\mathbf{z}_0(K)$, the maximum likelihood objective function is given by

$$J=\prod_{k=1}^{K}J_k\left(\mathbf{z}_0(k)-\mathbf{z}_s(k),\mathbf{S}_{gg}\right),\tag{6.19}$$

where $J_k\left(\mathbf{z}_0(k)-\mathbf{z}_s(k),\mathbf{S}_{gg}\right)$ is defined by (6.18) when replacing $\mathbf{z}_0$ and $\mathbf{z}_s$ with $\mathbf{z}_0(k)$ and $\mathbf{z}_s(k)$, respectively. The above function is a product of likelihood values that is larger than zero. As the logarithm function is monotonously increasing, taking the natural logarithm of J allows redefining (6.19)

$$J^*=\sum_{k=1}^{K}\ln\left(J_k\left(\mathbf{z}_0(k)-\mathbf{z}_s(k),\mathbf{S}_{gg}\right)\right).\tag{6.20}$$

where $J^*=\ln(J)$. Substituting (6.18) into (6.20) yields

$$\begin{aligned}J^*=&-Kn_z\ln(2\pi)-\tfrac{K}{2}\ln\left(\left|\mathbf{S}_{gg}\right|\right)\\&-\tfrac{1}{2}\sum_{k=1}^{K}\left(\mathbf{z}_0(k)-\mathbf{z}_s(k)\right)^T\mathbf{S}_{gg}^{-1}\left(\mathbf{z}_0(k)-\mathbf{z}_s(k)\right).\end{aligned}\tag{6.21}$$

Multiplying both sides by -2 and omitting the constant terms $2Kn_z\ln(2\pi)$ and $K\ln\left(\left|\mathbf{S}_{gg}\right|\right)$ gives rise to

$$\widetilde{J}^*=\sum_{k=1}^{K}\left(\mathbf{z}_0(k)-\mathbf{z}_s(k)\right)^T\mathbf{S}_{gg}^{-1}\left(\mathbf{z}_0(k)-\mathbf{z}_s(k)\right)\tag{6.22}$$

where $\widetilde{J}^*=(-2)\,J^*-2Kn_z\ln(2\pi)-K\ln\left(\left|\mathbf{S}_{gg}\right|\right)$. A solution to the maximum likelihood objective function that is based on the reference set including K samples, $\widehat{\mathbf{Z}}_0^T=\left[\widehat{\mathbf{z}}_0(1)\quad\widehat{\mathbf{z}}_0(2)\quad\cdots\quad\widehat{\mathbf{z}}_0(K)\right]$, is the one that minimizes $\widetilde{J}^*$, which, in turn, maximizes J^* and hence J. Incorporating the data model $\mathbf{z}_0=\boldsymbol{\Xi}\mathbf{s}+\mathbf{g}$, Fuller (1987) introduced an optimum solution for estimating the parameter matrix $\widetilde{\boldsymbol{\Xi}}$

$$\mathbf{z}_0=\begin{bmatrix}\boldsymbol{\Xi}_1\\\boldsymbol{\Xi}_2\end{bmatrix}\mathbf{s}+\mathbf{g}=\begin{bmatrix}\mathbf{I}\\\boldsymbol{\Xi}_2\boldsymbol{\Xi}_1^{-1}\end{bmatrix}\boldsymbol{\Xi}_1\mathbf{s}+\mathbf{g}=\widetilde{\boldsymbol{\Xi}}\mathbf{z}_{1_s}+\mathbf{g}\tag{6.23}$$

[1] It is assumed that the absolute elements of $\mathbf{g}$ are bounded and hence $J(\cdot)>0$.

that minimizes $\widetilde{J}^*$. Here:[2]

- $\boldsymbol{\Xi}_1 \in \mathbb{R}^{n\times n}$, $\boldsymbol{\Xi}_2 \in \mathbb{R}^{(n_z-n)\times n}$ and $\widetilde{\boldsymbol{\Xi}}^T = \begin{bmatrix}\mathbf{I} & \boldsymbol{\Xi}_1^{-T}\boldsymbol{\Xi}_2^T\end{bmatrix}$;
- $\mathbf{z}_s^T = \begin{pmatrix}\mathbf{z}_{1_s}^T & \mathbf{z}_{2_s}^T\end{pmatrix}$;
- $\mathbf{z}_{1_s} = \begin{pmatrix}z_{1_s} & z_{2_s} & \cdots & z_{n_s}\end{pmatrix}^T = \boldsymbol{\Xi}_1\mathbf{s}$; and
- $\mathbf{z}_{2_s} = \begin{pmatrix}z_{n+1_s} & z_{n+2_s} & \cdots & z_{n_{zs}}\end{pmatrix}^T = \boldsymbol{\Xi}_2\mathbf{s}$.

An iterative and efficient maximum likelihood PCA formulation based on a singular value decomposition for determining $\widehat{\mathbf{Z}}_0$ to minimize (6.22) was proposed by Wentzell *et al.* (1997). Reexamining (6.23) for $\mathbf{g} \sim \mathcal{N}\left\{\mathbf{0}, \mathbf{S}_{\mathbf{gg}}\right\}$ suggests that the best linear unbiased estimate for $\mathbf{z}_{1_s}$, $\widehat{\mathbf{z}}_{1_s}$, is given by the generalized least squares solution of $\mathbf{z}_0 = \widetilde{\boldsymbol{\Xi}}\mathbf{z}_{1_s} + \mathbf{g}$ (Björck 1996)

$$\widehat{\mathbf{z}}_{1_s} = \left[\widetilde{\boldsymbol{\Xi}}^T\mathbf{S}_{\mathbf{gg}}^{-1}\widetilde{\boldsymbol{\Xi}}\right]^{-1}\widetilde{\boldsymbol{\Xi}}^T\mathbf{S}_{\mathbf{gg}}^{-1}\mathbf{z}_0 \quad \widehat{\mathbf{z}}_s = \widetilde{\boldsymbol{\Xi}}\left[\widetilde{\boldsymbol{\Xi}}^T\mathbf{S}_{\mathbf{gg}}^{-1}\widetilde{\boldsymbol{\Xi}}\right]^{-1}\widetilde{\boldsymbol{\Xi}}^T\mathbf{S}_{\mathbf{gg}}^{-1}\mathbf{z}_0. \tag{6.24}$$

In a PCA context, a singular value decomposition (SVD) of

$$\mathbf{Z}_0 = \begin{bmatrix}\widehat{\mathcal{U}} & \widehat{\mathfrak{U}}\end{bmatrix}\begin{bmatrix}\widehat{\mathcal{S}} & \mathbf{0}\\ \mathbf{0} & \widehat{\mathfrak{S}}\end{bmatrix}\begin{bmatrix}\widehat{\mathcal{V}}^T\\ \widehat{\mathfrak{V}}^T\end{bmatrix} = \widehat{\mathcal{U}}\widehat{\mathcal{S}}\widehat{\mathcal{V}}^T + \widehat{\mathfrak{U}}\widehat{\mathfrak{S}}\widehat{\mathfrak{V}}^T, \tag{6.25}$$

where:

- $\widehat{\mathcal{U}} \in \mathbb{R}^{K\times n}$, $\widehat{\mathfrak{U}} \in \mathbb{R}^{K\times(n_z-n)}$ and $\widehat{\mathcal{S}} \in \mathbb{R}^{n\times n}$; and
- $\widehat{\mathfrak{S}} \in \mathbb{R}^{(n_z-n)\times(n_z-n)}$, $\widehat{\mathcal{V}} \in \mathbb{R}^{n_z\times n}$ and $\widehat{\mathfrak{V}} \in \mathbb{R}^{n_z\times(n_z-n)}$,

yields in its transposed form

$$\widehat{\mathbf{Z}}_s^T = \begin{bmatrix}\widehat{\mathbf{Z}}_{1_s}^T\\ \widehat{\mathbf{Z}}_{2_s}^T\end{bmatrix} = \begin{bmatrix}\widehat{\mathcal{V}}_1\\ \widehat{\mathcal{V}}_2\end{bmatrix}\widehat{\mathcal{S}}\widehat{\mathcal{U}}^T = \underbrace{\begin{bmatrix}\mathbf{I}\\ \widehat{\mathcal{V}}_2\widehat{\mathcal{V}}_1^{-1}\end{bmatrix}}_{=\widehat{\mathcal{V}}\widehat{\mathcal{V}}_1^{-1}}\underbrace{\widehat{\mathcal{V}}_1\widehat{\mathcal{S}}\widehat{\mathcal{U}}^T}_{=\widehat{\mathbf{Z}}_{1_s}^T} = \widehat{\widetilde{\boldsymbol{\Xi}}}\widehat{\mathbf{Z}}_{1_s}^T, \tag{6.26}$$

where $\widehat{\widetilde{\boldsymbol{\Xi}}} = \widehat{\mathcal{V}}\widehat{\mathcal{V}}_1^{-1}$. Applying (6.24) to the above SVD produces

$$\widehat{\mathbf{Z}}_{1_s}^T = \widehat{\mathcal{V}}\widehat{\mathcal{V}}_1^{-1}\left[\left[\widehat{\mathcal{V}}\widehat{\mathcal{V}}_1^{-1}\right]^T\mathbf{S}_{\mathbf{gg}}^{-1}\widehat{\mathcal{V}}\widehat{\mathcal{V}}_1^{-1}\right]^{-1}\left[\widehat{\mathcal{V}}\widehat{\mathcal{V}}_1^{-1}\right]^T\mathbf{S}_{gg}^{-1}\mathbf{Z}_0^T \tag{6.27}$$

which can be simplified to

$$\widehat{\mathbf{Z}}_{1_s}^T = \widehat{\mathcal{V}}\left[\widehat{\mathcal{V}}^T\mathbf{S}_{\mathbf{gg}}^{-1}\widehat{\mathcal{V}}\right]^{-1}\widehat{\mathcal{V}}\mathbf{S}_{\mathbf{gg}}^{-1}\mathbf{Z}_0^T. \tag{6.28}$$

---

[2] It is assumed here that $\boldsymbol{\Xi}_1$ has full rank n.

Equations (6.26) to (6.28) exploit the row space of $\mathbf{Z}_0$. Under the assumption that the error covariance matrix is of diagonal type, that is, no correlation among the error terms, the row space of $\mathbf{Z}_0$ can be rewritten with respect to (6.22)

$$\widetilde{J}^* = \sum_{k=1}^{K} \left(\mathbf{z}_0(k) - \mathbf{z}_s(k)\right)^T \mathbf{S}_{gg}^{-1} \left(\mathbf{z}_0(k) - \mathbf{z}_s(k)\right) = \sum_{k=1}^{K}\sum_{j=1}^{n_z} \frac{\left(z_{0_j}(k) - z_{s_j}(k)\right)^2}{\sigma_{g_j}^2}. \tag{6.29}$$

Analyzing the column space of $\mathbf{Z}_0 = \left[\begin{array}{cccc} \zeta_{0_1} & \zeta_{0_2} & \cdots & \zeta_{0_{n_z}} \end{array}\right]$, Equation (6.22) can alternatively be rewritten as

$$\widetilde{J}^* = \sum_{j=1}^{n_z} \left(\zeta_0(j) - \widehat{\zeta}_s(j)\right)^T \boldsymbol{\Sigma}_{gg_j}^{-1} \left(\zeta_0(j) - \widehat{\zeta}_s(j)\right). \tag{6.30}$$

The definition of the error covariance matrices in the above equations is

- $\mathbf{S}_{gg} = \begin{bmatrix} \sigma_{g_1}^2 & 0 & \cdots & 0 \\ 0 & \sigma_{g_2}^2 & \cdots & 0 \\ \vdots & \vdots & \ddots & \vdots \\ 0 & 0 & \cdots & \sigma_{g_{n_z}}^2 \end{bmatrix} \in \mathbb{R}^{n_z \times n_z}$
- $\boldsymbol{\Sigma}_{gg_j} = \begin{bmatrix} \sigma_{g_j}^2 & 0 & \cdots & 0 \\ 0 & \sigma_{g_j}^2 & \cdots & 0 \\ \vdots & \vdots & \ddots & \vdots \\ 0 & 0 & \cdots & \sigma_{g_j}^2 \end{bmatrix} \in \mathbb{R}^{K \times K}$

Equation (6.22) and the singular value decomposition of $\mathbf{Z}_0$ allow constructing a generalized least squares model for the column vectors of $\mathbf{Z}_0$

$$\widehat{\mathbf{Z}}_s = \begin{bmatrix} \widehat{\mathcal{Z}}_{s_1} \\ \widehat{\mathcal{Z}}_{s_2} \end{bmatrix} = \underbrace{\begin{bmatrix} \mathbf{I} \\ \widehat{\mathcal{U}}_2 \widehat{\mathcal{U}}_1^{-1} \end{bmatrix}}_{=\widehat{\mathcal{U}}\widehat{\mathcal{U}}_1^{-1}} \underbrace{\widehat{\mathcal{U}}_1 \widehat{\mathcal{S}} \widehat{\mathcal{V}}^T}_{=\widehat{\mathcal{Z}}_{s_1}}. \tag{6.31}$$

Applying the same steps as those taken in (6.27) and (6.28) gives rise to

$$\widehat{\zeta}_{s_i} = \widehat{\mathcal{U}} \left[\widehat{\mathcal{U}}^T \boldsymbol{\Sigma}_{gg_i}^{-1} \widehat{\mathcal{U}}\right]^{-1} \widehat{\mathcal{U}}^T \boldsymbol{\Sigma}_{gg_i}^{-1} \zeta_{0_i}. \tag{6.32}$$

It should be noted that the error covariance matrix for the row space of $\mathbf{Z}_0$, $\mathbf{S}_{gg}$, is the same for each row, which follows from the assumption made earlier that $E\left\{\mathbf{g}(i)\mathbf{g}^T(j)\right\} = \delta_{ij}\mathbf{S}_{gg}$. However, the error covariance matrix for the column space or $\mathbf{Z}_0$ has different diagonal elements for each column. More precisely, $\mathbf{\Sigma}_{\mathfrak{gg}_j} = \sigma^2_{\mathfrak{g}_j}\mathbf{I}$ which implies that (6.32) is equal to

$$\widehat{\zeta}_{s_j} = \widehat{\mathcal{U}}\left[\widehat{\mathcal{U}}^T\widehat{\mathcal{U}}\right]^{-1}\widehat{\mathcal{U}}^T\zeta_{0_j} \tag{6.33}$$

and hence

$$\widehat{\mathbf{Z}}_s = \widehat{\mathcal{U}}\left[\widehat{\mathcal{U}}^T\widehat{\mathcal{U}}\right]^{-1}\widehat{\mathcal{U}}^T\mathbf{Z}_0. \tag{6.34}$$

Using (6.28) and (6.34), the following iterative procedure computes a maximum likelihood PCA, or MLPCA, model:

1. Carry out the SVD of $\mathbf{Z}_0$ to compute $\widehat{\mathcal{U}}$, $\widehat{\mathcal{S}}$ and $\widehat{\mathcal{V}}$ (Equation (6.25)).
2. Utilize (6.34) to calculate $\widehat{\mathbf{Z}}_s = \left[\widehat{\zeta}_{s_1} \quad \widehat{\zeta}_{s_2} \quad \cdots \quad \widehat{\zeta}_{s_{n_z}}\right]$.
3. Apply (6.30) to determine $\widetilde{J}_1^*$ using the estimate of $\widehat{\mathbf{Z}}_s$ from Step 2.
4. Take $\widehat{\mathbf{Z}}_s$ from Step 2 and carry out a SVD for recomputing $\widehat{\mathcal{U}}$, $\widehat{\mathcal{S}}$ and $\widehat{\mathcal{V}}$.
5. Employ (6.28) to determine $\widehat{\mathbf{Z}}_s^T = \left[\widehat{\mathbf{z}}_s(1) \quad \widehat{\mathbf{z}}_s(2) \quad \cdots \quad \widehat{\mathbf{z}}_s(K)\right]$ using $\mathbf{S}_{\mathfrak{gg}}$ and $\widehat{\mathcal{V}}$ from Step 4.
6. Apply (6.22) to determine $\widetilde{J}_2^*$ using $\widehat{\mathbf{Z}}_s$ from Step 5.
7. Take $\widehat{\mathbf{Z}}_s$ from Step 5 and carry out a SVD for recomputing $\widehat{\mathcal{U}}$, $\widehat{\mathcal{S}}$ and $\widehat{\mathcal{V}}$.
8. Check for convergence,[3] if $\epsilon = \left(\widetilde{J}_1^* - \widetilde{J}_2^*\right)/\widetilde{J}_2^* < 10^{-12}$ terminate else go to Step 2.

The performance of the iterative MLPCA approach is now tested for the three-variable example described in (2.9) and (2.11) and the error covariance matrix is defined in (6.14). Recall that the use of this error covariance matrix led to a biased estimation of the residual subspace, which departed from the true one by a minimum angle of almost 3.5°. The above MLPCA approach applied to a reference set of $K = 1000$ samples converged after nine times for a very tight threshold of 10^{-14}. Figure 6.3 shows that after the first three iteration steps, the minimum angle between the true and estimated model subspaces is close to zero.

In contrast to the discussion above, it should be noted that the work in Wentzell *et al.* (1997) also discusses cases where the error covariance matrix is symmetric and changes over time. In this regard, the algorithms in Tables 1 and 2 on page 348 and 350, respectively, in Wentzell *et al.* (1997) are of interest. The discussion in this book, however, assumes that the error covariance matrix remains constant over time.

---

[3] The value of 10^{-12} is a possible suggestion; practically, smaller thresholds can be selected without a significant loss of accuracy.

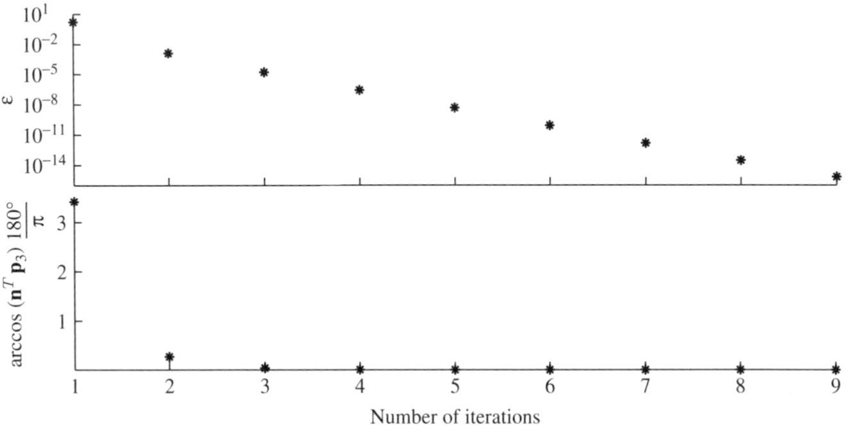

Figure 6.3 Convergence of the MLPCA algorithm for simulation example.

6.1.4 Maximum likelihood PCA for unknown $\mathbf{S}_{gg}$

Different from the method proposed by Wentzell *et al.* (1997), Narasimhan and Shah (2008) introduced a more efficient method for determining an estimate of the model subspace. If the error covariance matrix is known *a priori* and of full rank, a Cholesky decomposition of $\mathbf{S}_{gg} = \mathbf{L}\mathbf{L}^T$ can be obtained, which gives rise to

$$\mathbf{S}_{z_0 z_0} = \mathbf{S}_{z_s z_s} + \mathbf{L}\mathbf{L}^T \tag{6.35}$$

with $\mathbf{L}$ being a lower triangular matrix. Rewriting (6.35) as follows

$$\mathbf{L}^{-1}\mathbf{S}_{z_0 z_0}\mathbf{L}^{-T} = \mathbf{L}^{-1}\mathbf{S}_{z_s z_s}\mathbf{L}^{-T} + \mathbf{I} \tag{6.36}$$

yields a transformed error covariance matrix $\widetilde{\mathbf{S}}_{gg} = \mathbf{L}^{-1}\mathbf{S}_{gg}\mathbf{L}^{-T} = \mathbf{I}$ that is of the type $\sigma_g^2\mathbf{I}$ with $\sigma_g^2 = 1$. Hence, an eigendecomposition of $\mathbf{L}^{-1}\mathbf{S}_{z_0 z_0}\mathbf{L}^{-T}$ will provide a consistent estimation of the model subspace, which follows from (6.4) to (6.8). The dominant eigenvalues of $\widetilde{\mathbf{S}}_{z_s z_a}$ are equal to the dominant eigenvalues of $\widetilde{\mathbf{S}}_{z_0 z_0} = \mathbf{L}^{-1}\mathbf{S}_{z_0 z_0}\mathbf{L}^{-T}$ minus one, which the following relationship shows

$$\widetilde{\mathbf{S}}_{z_0 z_0} = \begin{bmatrix} \widetilde{\mathbf{P}} & \widetilde{\mathbf{P}}_d \end{bmatrix} \begin{bmatrix} \widetilde{\mathbf{\Lambda}} & \mathbf{0} \\ \mathbf{0} & \widetilde{\mathbf{\Lambda}}_d \end{bmatrix} \begin{bmatrix} \widetilde{\mathbf{P}}^T \\ \widetilde{\mathbf{P}}_d^T \end{bmatrix}. \tag{6.37}$$

By default, the diagonal elements of the matrices $\widetilde{\mathbf{\Lambda}}$ and $\widetilde{\mathbf{\Lambda}}_d$ are as follows

$$\underbrace{\widetilde{\lambda}_1 \quad \widetilde{\lambda}_2 \quad \cdots \quad \widetilde{\lambda}_n}_{>1} \quad \underbrace{\widetilde{\lambda}_{n+1} \quad \widetilde{\lambda}_{n+2} \quad \cdots \quad \widetilde{\lambda}_{n_z}}_{=1}. \tag{6.38}$$

Assuming that $E\left\{\mathbf{s}\mathbf{g}^T\right\} = \mathbf{0}$, it follows that

$$\widetilde{\mathbf{S}}_{z_0z_0} = \widetilde{\mathbf{P}}\left[\widetilde{\mathbf{\Lambda}} - \mathbf{I}\right]\widetilde{\mathbf{P}}^T + \left[\widetilde{\mathbf{P}} \quad \widetilde{\mathbf{P}}_d\right]\begin{bmatrix}\mathbf{I} & \mathbf{0}\\ \mathbf{0} & \mathbf{I}\end{bmatrix}\begin{bmatrix}\widetilde{\mathbf{P}}^T\\ \widetilde{\mathbf{P}}_d^T\end{bmatrix} \tag{6.39}$$

and hence

$$\widetilde{\mathbf{S}}_{z_0z_0} = \left[\widetilde{\mathbf{p}}_1 \quad \widetilde{\mathbf{p}}_2 \quad \cdots \quad \widetilde{\mathbf{p}}_n\right]\begin{bmatrix}\widetilde{\lambda}_1 - 1 & 0 & \cdots & 0\\ 0 & \widetilde{\lambda}_2 - 1 & \cdots & 0\\ \vdots & \vdots & \ddots & \vdots\\ 0 & 0 & \cdots & \widetilde{\lambda}_n - 1\end{bmatrix}\begin{bmatrix}\widetilde{\mathbf{p}}_1\\ \widetilde{\mathbf{p}}_2\\ \vdots\\ \widetilde{\mathbf{p}}_n\end{bmatrix} + \begin{bmatrix}1 & 0 & \cdots & 0\\ 0 & 1 & \cdots & 0\\ \vdots & \vdots & \ddots & \vdots\\ 0 & 0 & \cdots & 1\end{bmatrix}. \tag{6.40}$$

The determined eigenvectors of $\widetilde{\mathbf{S}}_{z_0z_0}$ are consequently a consistent estimation of base vectors spanning the model subspace. Despite the strong theoretical foundation, conceptual simplicity and computational efficiency of applying an eigendecomposition to (6.36), it does not produce an estimate of the model subspace in a PCA sense, which Subsection 6.1.7 highlights.

This approach, however, has been proposed by Narasimhan and Shah (2008) for developing an iterative approach that allows estimating $\widetilde{\mathbf{S}}_{\mathfrak{gg}}$ under the constraint in (6.46), which is discussed below. Revising (6.1) and evaluating the stochastic components

$$\mathbf{z}_0 = \mathbf{z}_s + \mathfrak{g} = \Xi\mathbf{s} + \mathfrak{g}, \tag{6.41}$$

where $\mathfrak{g} \sim \mathcal{N}\left\{\mathbf{0}, \mathbf{S}_{\mathfrak{gg}}\right\}$, gives rise to

$$\Xi^{\perp}\mathbf{z}_0 = \Xi^{\perp}\Xi\mathbf{s} + \Xi^{\perp}\mathfrak{g}. \tag{6.42}$$

Here $\Xi^{\perp}$ is a matrix that has orthogonal rows to the columns in Ξ and hence $\Xi^{\perp}\Xi = \mathbf{0}$. Consequently, (6.42) reduces to

$$\Xi^{\perp}\mathbf{z}_0 = \Xi^{\perp}\mathfrak{g} = \widetilde{\mathfrak{g}}. \tag{6.43}$$

The transformed error vector $\widetilde{\mathfrak{g}}$ has therefore the distribution function

$$\widetilde{\mathfrak{g}} \sim \mathcal{N}\left\{\mathbf{0}, \Xi^{\perp}\mathbf{S}_{\mathfrak{gg}}\Xi^{\perp^T}\right\} \tag{6.44}$$

since $E\left\{\widetilde{\mathfrak{g}}\widetilde{\mathfrak{g}}^T\right\} = E\left\{\Xi^{\perp}\mathfrak{g}\mathfrak{g}^T\Xi^{\perp^T}\right\} = \Xi^{\perp}E\left\{\mathfrak{g}\mathfrak{g}^T\right\}\Xi^{\perp^T} = \Xi^{\perp}\mathbf{S}_{\mathfrak{gg}}\Xi^{\perp^T}$. Using the maximum likelihood function in (6.21) to determine $\mathbf{S}_{\widetilde{\mathfrak{g}}\widetilde{\mathfrak{g}}} = \Xi^{\perp}\mathbf{S}_{\mathfrak{gg}}\Xi^{\perp^T}$ leads to the following objective function to be minimized

$$\widehat{\mathbf{S}}_{\mathfrak{gg}} = \arg\min_{\mathbf{S}_{\mathfrak{gg}}} K \ln \left| \Xi^{\perp} \mathbf{S}_{\mathfrak{gg}} \Xi^{\perp^T} \right| + \sum_{k=1}^{K} \widetilde{\mathfrak{g}}^T(k) \left[\Xi^{\perp} \mathbf{S}_{\mathfrak{gg}} \Xi^{\perp^T} \right]^{-1} \widetilde{\mathfrak{g}}(k). \quad (6.45)$$

It should be noted that the first term in (6.21), $Kn_z \ln(2\pi)$ is a constant and can therefore be omitted. In contrast to the method in Wentzell *et al.* (1997), where the second term $K \ln \left| \Xi^{\perp} \mathbf{S}_{\mathfrak{gg}} \Xi^{\perp^T} \right|$ could be ignored, the log likelihood function for the approach by Narasimhan and Shah (2008) requires the inclusion of this term as $\mathbf{S}_{\mathfrak{gg}}$ is an unknown symmetric and positive definite matrix.

Examining the maximum likelihood function of (6.45) or, more precisely, the error covariance matrix $\mathbf{S}_{\widetilde{\mathfrak{g}}\widetilde{\mathfrak{g}}}$ more closely, the rank of this matrix is $n_z - n$ and not n_z. This follows from the fact that $\Xi \in \mathbb{R}^{n_z \times n}$. Consequently, the size of the model subspace is n and the number of linearly independent row vectors in $\Xi^{\perp}$ that are orthogonal to the column vectors in Ξ is $n_z - n$. With this in mind, $\Xi^{\perp} \in \mathbb{R}^{n_z \times (n_z - n)}$ and $\Xi^{\perp} \mathbf{S}_{\mathfrak{gg}} \Xi^{\perp^T} \in \mathbb{R}^{(n_z - n) \times (n_z - n)}$. This translates into a constraint for determining the number of elements in the covariance matrix as the maximum number of independent parameters is ${}^{(n_z-n)(n_z-n+1)}/_2$.

Moreover, the symmetry of $\mathbf{S}_{\widetilde{\mathfrak{g}}\widetilde{\mathfrak{g}}}$ implies that only the upper or lower triangular elements must be estimated together with the diagonal ones. It is therefore imperative to constrain the number of estimated elements in $\mathbf{S}_{\mathfrak{gg}}$. A practically reasonable assumption is that the errors are not correlated so that $\mathbf{S}_{\mathfrak{gg}}$ reduces to a diagonal matrix. Thus, a complete set of diagonal elements can be obtained if $(n_z - n)(n_z - n + 1) \geq 2n_z$. The number of source signals must therefore not exceed

$$n \leq n_z + \frac{1}{2} - \sqrt{2n_z + \frac{1}{4}}. \quad (6.46)$$

Figure 6.4 illustrates that values for n must be below the graph $n = n_z + {}^1/_2 - \sqrt{2n_z + {}^1/_4}$ for a determination of a complete set of diagonal elements for $\mathbf{S}_{\mathfrak{gg}}$.

Narasimhan and Shah (2008) introduced an iterative algorithm for simultaneously estimating the model subspace and $\mathbf{S}_{\mathfrak{gg}}$ from an estimate of $\mathbf{S}_{z_0 z_0}$. This algorithm takes advantage of the fact that the model subspace and the residual space is spanned by the eigenvectors of $\widetilde{\mathbf{S}}_{z_0 z_0}$. The relationship below proposes a slightly different version of this algorithm, which commences by defining the initial error covariance matrix that stores 0.0001 times the diagonal elements of $\widehat{\mathbf{S}}_{z_0 z_0}$, then applies a Cholesky decomposition of $\widehat{\mathbf{S}}_{\mathfrak{gg}_i} = \mathbf{L}_i \mathbf{L}_i^T$ and subsequently (6.36).

Following an eigendecomposition of $\widehat{\widetilde{\mathbf{S}}}_{z_0 z_{0_i}} = \mathbf{L}_i^{-1} \widehat{\mathbf{S}}_{z_0 z_0} \mathbf{L}_i^{-T}$

$$\mathbf{L}_i^{-1} \widehat{\mathbf{S}}_{z_0 z_0} \mathbf{L}_i^{-T} = \begin{bmatrix} \widetilde{\mathbf{P}}_i & \widetilde{\mathbf{P}}_{d_i} \end{bmatrix} \begin{bmatrix} \widetilde{\mathbf{\Lambda}} & \mathbf{0} \\ \mathbf{0} & \widetilde{\mathbf{\Lambda}}_d \end{bmatrix} \begin{bmatrix} \widetilde{\mathbf{P}}_i^T \\ \widetilde{\mathbf{P}}_{d_i}^T \end{bmatrix} \quad (6.47)$$

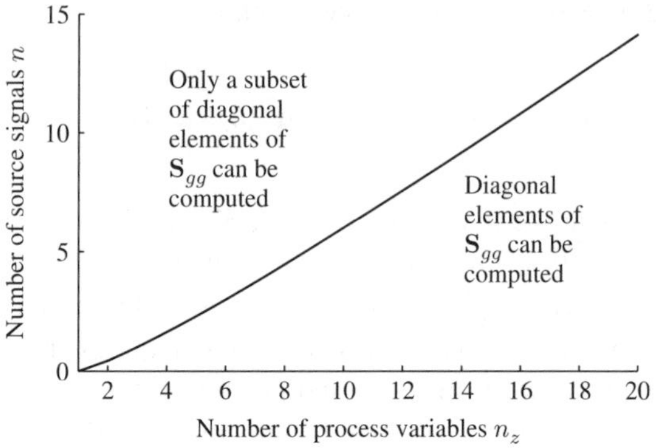

Figure 6.4 Graphical illustration of constraint in Equation (6.46).

an estimate of $\boldsymbol{\Xi}_i^{\perp}$ is given by $\widehat{\boldsymbol{\Xi}}_i^{\perp} = \widetilde{\mathbf{P}}_{d_i}^T \mathbf{L}_i^{-1}$, which follows from the fact that column vectors of $\boldsymbol{\Xi}$ span the same column space as the eigenvectors in $\widetilde{\mathbf{P}}_i$ after convergence. Given that $\widetilde{\mathbf{S}}_{z_0z_0} = \mathbf{L}^{-1}\mathbf{S}_{z_0z_0}\mathbf{L}^{-1}$ after convergence, it follows that

$$\widetilde{\mathbf{S}}_{z_0z_0} = \mathbf{L}^{-1}\mathbf{S}_{z_0z_0}\mathbf{L}^{-T} = \mathbf{L}^{-1} \underbrace{\mathbf{S}_{z_sz_s}}_{\mathbf{L}\widetilde{\mathbf{P}}\boldsymbol{\Lambda}\widetilde{\mathbf{P}}^T\mathbf{L}^T} \mathbf{L}^{-T} + \mathbf{I}. \tag{6.48}$$

Hence, $\widehat{\boldsymbol{\Xi}} = \mathbf{L}\widetilde{\mathbf{P}}$ and $\widehat{\boldsymbol{\Xi}}^{\perp} = \widetilde{\mathbf{P}}_d^T\mathbf{L}^{-1}$, since $\widehat{\boldsymbol{\Xi}}^{\perp}\widehat{\boldsymbol{\Xi}} = \widetilde{\mathbf{P}}_d^T\mathbf{L}^{-1}\mathbf{L}\widetilde{\mathbf{P}} = \widetilde{\mathbf{P}}_d^T\widetilde{\mathbf{P}} = \mathbf{0}$. The next step is the evaluation of the objective function in (6.45) for $\widetilde{\mathfrak{g}}(j) = \widehat{\boldsymbol{\Xi}}_i^{\perp}\mathbf{z}_0(j)$ prior to an update of $\widehat{\mathbf{S}}_{\mathfrak{g}\mathfrak{g}_i}$, $\widehat{\mathbf{S}}_{\mathfrak{g}\mathfrak{g}_{i+1}} = \widehat{\mathbf{S}}_{\mathfrak{g}\mathfrak{g}_i} + \Delta\widehat{\mathbf{S}}_{\mathfrak{g}\mathfrak{g}_i}$, using a gradient projection method (Byrd *et al.* 1995), a genetic algorithm (Sharma and Irwin 2003) or a particle swarm optimization (Coello *et al.* 2004).

Recomputing the Cholesky decomposition of $\widetilde{\mathbf{S}}_{\mathfrak{g}\mathfrak{g}_{i+1}}$ then starts the $(i+1)$th iteration step. The iteration converges if the difference of two consecutive values of $\widetilde{J}^*$ is smaller than a predefined threshold. Different to the algorithm in Narasimhan and Shah (2008), the proposed objective function here is of the following form

$$\widehat{\mathbf{S}}_{\mathfrak{g}\mathfrak{g}_i} = \arg\min_{\mathbf{S}_{\mathfrak{g}\mathfrak{g}_i}} a_1 \left(K \ln \left| \widehat{\boldsymbol{\Xi}}_i^{\perp}\mathbf{S}_{\mathfrak{g}\mathfrak{g}_i}\widehat{\boldsymbol{\Xi}}_i^{\perp^T} \right| + \sum_{j=1}^{K} \widetilde{\mathfrak{g}}^T(j) \left[\widehat{\boldsymbol{\Xi}}_i^{\perp}\mathbf{S}_{\mathfrak{g}\mathfrak{g}_i}\widehat{\boldsymbol{\Xi}}_i^{\perp^T} \right]^{-1} \widetilde{\mathfrak{g}}(j) \right)$$

$$+ a_2 \left\| \widehat{\boldsymbol{\Xi}}_i^{\perp}\mathbf{S}_{\mathfrak{g}\mathfrak{g}_i}\widehat{\boldsymbol{\Xi}}_i^{\perp^T} - \mathbf{I} \right\|^2 + a_3 \left\| \widehat{\boldsymbol{\Xi}}_i^{\perp} \left[\widehat{\mathbf{S}}_{z_0z_0} - \mathbf{S}_{\mathfrak{g}\mathfrak{g}_i} \right] \widehat{\boldsymbol{\Xi}}_i^{\perp^T} \right\|^2 \tag{6.49}$$

where $\|\cdot\|^2$ is the squared Frobenius norm of a matrix. The rationale behind this objective function is to ensure that the solution found satisfies the following constraints

$$\widehat{\boldsymbol{\Xi}}_i^{\perp}\widehat{\mathbf{S}}_{\mathfrak{g}\mathfrak{g}_i}\widehat{\boldsymbol{\Xi}}_i^{\perp^T} - \mathbf{I} = \mathbf{0} \qquad \widehat{\boldsymbol{\Xi}}_i^{\perp} \left[\widehat{\mathbf{S}}_{z_0z_0} - \widehat{\mathbf{S}}_{\mathfrak{g}\mathfrak{g}_i} \right] \widehat{\boldsymbol{\Xi}}_i^{\perp^T} = \mathbf{0}. \tag{6.50}$$

Note that Subsection 6.1.7 elaborates upon the geometric relationships, such as $\widehat{\boldsymbol{\Xi}}_i^{\perp} = \widehat{\widetilde{\mathbf{P}}}_{d_i}^T \mathbf{L}_i^{-1}$ in more detail. Since $\widehat{\boldsymbol{\Xi}}_i^{\perp}$ is orthogonal to the estimate of the model subspace, the following must hold true after the above iteration converged

$$\begin{aligned}
\widehat{\widetilde{\mathbf{S}}}_{z_0 z_0} &= \widehat{\widetilde{\mathbf{P}}}\widehat{\widetilde{\boldsymbol{\Lambda}}}\widehat{\widetilde{\mathbf{P}}}^T + \widehat{\widetilde{\mathbf{P}}}_d \widehat{\widetilde{\mathbf{P}}}_d^T \\
\underbrace{\widehat{\widetilde{\mathbf{P}}}_d^T \mathbf{L}^{-1}}_{=\widehat{\boldsymbol{\Xi}}^{\perp}} \widehat{\mathbf{S}}_{z_0 z_0} \underbrace{\mathbf{L}^{-T} \widehat{\widetilde{\mathbf{P}}}_d}_{=\widehat{\boldsymbol{\Xi}}^{\perp T}} &= \underbrace{\widehat{\widetilde{\mathbf{P}}}_d^T \widehat{\widetilde{\mathbf{P}}}}_{=\mathbf{0}} \widehat{\widetilde{\boldsymbol{\Lambda}}} \underbrace{\widehat{\widetilde{\mathbf{P}}}^T \widehat{\widetilde{\mathbf{P}}}_d}_{=\mathbf{0}} + \mathbf{I} \\
\widehat{\boldsymbol{\Xi}}^{\perp} \widehat{\mathbf{S}}_{z_0 z_0} \widehat{\boldsymbol{\Xi}}^{\perp} - \mathbf{I} &= \mathbf{0}
\end{aligned} \tag{6.51}$$

and

$$\begin{aligned}
\mathbf{L}^{-1}\widehat{\mathbf{S}}_{z_0 z_0}\mathbf{L}^{-T} &= \mathbf{L}^{-1}\widehat{\mathbf{S}}_{z_s z_s}\mathbf{L}^{-T} + \mathbf{L}^{-1}\widehat{\mathbf{S}}_{\mathbf{gg}}\mathbf{L}^{-T} \\
\mathbf{L}^{-1}\widehat{\mathbf{S}}_{z_s z_s}\mathbf{L}^{-T} &= \mathbf{L}^{-1}\left[\widehat{\mathbf{S}}_{z_0 z_0} - \widehat{\mathbf{S}}_{\mathbf{gg}}\right]\mathbf{L}^{-T} \\
\widehat{\widetilde{\mathbf{P}}}\widehat{\widetilde{\boldsymbol{\Lambda}}}\widehat{\widetilde{\mathbf{P}}}^T &= \mathbf{L}^{-1}\left[\widehat{\mathbf{S}}_{z_0 z_0} - \widehat{\mathbf{S}}_{\mathbf{gg}}\right]\mathbf{L}^{-T} \\
\underbrace{\widehat{\widetilde{\mathbf{P}}}_d^T \widehat{\widetilde{\mathbf{P}}}}_{=\mathbf{0}} \widehat{\widetilde{\boldsymbol{\Lambda}}} \underbrace{\widehat{\widetilde{\mathbf{P}}}^T \widehat{\widetilde{\mathbf{P}}}_d}_{=\mathbf{0}} &= \underbrace{\widehat{\widetilde{\mathbf{P}}}_d^T \mathbf{L}^{-1}}_{=\widehat{\boldsymbol{\Xi}}^{\perp}} \left[\widehat{\mathbf{S}}_{z_0 z_0} - \widehat{\mathbf{S}}_{\mathbf{gg}}\right] \underbrace{\mathbf{L}^{-T}\widehat{\widetilde{\mathbf{P}}}_d}_{=\widehat{\boldsymbol{\Xi}}^{\perp T}} \\
\widehat{\boldsymbol{\Xi}}^{\perp}\left[\widehat{\mathbf{S}}_{z_0 z_0} - \widehat{\mathbf{S}}_{\mathbf{gg}}\right]\widehat{\boldsymbol{\Xi}}^{\perp^T} &= \mathbf{0}
\end{aligned} \tag{6.52}$$

which is the second and third term in the objective function of (6.49). The coefficients a_1, a_2 and a_3 influence the solution and may need to be adjusted if the solution violates at least one of the above constraints or the value of the first term appears to be too high. Enforcing that the solution meets the constraints requires larger values for a_2 and a_3, which the simulation example below highlights. The steps of the above algorithm are now summarized below.

1. Set $\widehat{\mathbf{S}}_{gg_0}$ for $i = 0$ to store 0.0001 times the diagonal elements in $\widehat{\mathbf{S}}_{z_0 z_0}$.
2. Carry out a Cholesky decomposition of $\widehat{\mathbf{S}}_{\mathbf{gg}_0} = \mathbf{L}_0\mathbf{L}_0^T$.
3. Compute an eigendecomposition of

$$\widehat{\widetilde{\mathbf{S}}}_{z_0 z_0} = \mathbf{L}_0^{-1}\widehat{\mathbf{S}}_{z_0 z_0}\mathbf{L}_0^{-T} = \begin{bmatrix}\widetilde{\mathbf{P}}_0 & \widetilde{\mathbf{P}}_{d_0}\end{bmatrix}\begin{bmatrix}\widetilde{\boldsymbol{\Lambda}}_0 & \mathbf{0} \\ \mathbf{0} & \widetilde{\boldsymbol{\Lambda}}_{d_0}\end{bmatrix}\begin{bmatrix}\widetilde{\mathbf{P}}_0^T \\ \widetilde{\mathbf{P}}_{d_0}^T\end{bmatrix}.$$

4. Calculate initial estimate of residual subspace $\widehat{\boldsymbol{\Xi}}_0^{\perp} = \widetilde{\mathbf{P}}_{d_0}^T \mathbf{L}_0^{-1}$.
5. Evaluate initial value of objective function $\widetilde{J}_0^*$.
6. Update error covariance matrix $\widehat{\mathbf{S}}_{\mathbf{gg}_{i+1}} = \widehat{\mathbf{S}}_{\mathbf{gg}_i} + \Delta\widehat{\mathbf{S}}_{\mathbf{gg}_i}$.
7. Carry out Cholesky decomposition of $\widehat{\mathbf{S}}_{\mathbf{gg}_{i+1}} = \mathbf{L}_{i+1}\mathbf{L}_{i+1}^T$.

8. Determine the eigendecomposition of $\widehat{\widetilde{\mathbf{S}}}_{z_0z_0} = \widetilde{\mathbf{P}}_{i+1}\widetilde{\mathbf{\Lambda}}_{i+1}\widetilde{\mathbf{P}}_{i+1}^T + \widetilde{\mathbf{P}}_{d_{i+1}}\widetilde{\mathbf{\Lambda}}_{d_{i+1}}\widetilde{\mathbf{P}}_{d_{i+1}}^T$.
9. Get $(i+1)$th estimate for residual subspace $\widetilde{\mathbf{P}}_{d_{i+1}}^T\mathbf{L}_{i+1}^{-1}$.
10. Evaluate $(i+1)$th objective function using (6.49), $\widetilde{J}_{i+1}^*$.
11. Check for convergence[4], if $\left|\widetilde{J}_{i+1}^* - \widetilde{J}_i^*\right| < 10^{-12}$ terminate else go to Step 6.

To demonstrate the performance of the above algorithm, the next subsection presents an example. Section 6.2 describes a similar maximum likelihood algorithm for PLS models that relies on the inclusion of an additional error term for the input variables.

6.1.5 A simulation example

The three-variable example used previously in this chapter cannot be used here since three variables and two source signals leave only one parameter of $\mathbf{S}_{gg}$ to be estimated. The process studied here contains 14 variables that are described by the data model

$$\mathbf{z}_0 = \Xi\mathbf{s} + \mathbf{g} \qquad \mathbf{s} \sim \mathcal{N}\{\mathbf{0}, \mathbf{I}\} \qquad \mathbf{g} \sim \mathcal{N}\left\{\mathbf{0}, \mathbf{S}_{gg}\right\} \tag{6.53}$$

where $\mathbf{z}_0 \in \mathbb{R}^{14}$, $\mathbf{s} \in \mathbb{R}^4$, $\Xi \in \mathbb{R}^{14\times 4}$

$$\Xi = \begin{bmatrix} 1.00 & 0.00 & 0.00 & 0.00 \\ 0.00 & 1.00 & 0.00 & 0.00 \\ 0.00 & 0.00 & 1.00 & 0.00 \\ 0.00 & 0.00 & 0.00 & 1.00 \\ 0.50 & 0.00 & 0.25 & 0.25 \\ 0.00 & 0.25 & 0.50 & 0.25 \\ 0.25 & 0.25 & 0.25 & 0.25 \\ -0.25 & -0.50 & 0.00 & 0.25 \\ 0.00 & 0.50 & -0.05 & 0.00 \\ 0.25 & -0.25 & 0.25 & -0.25 \\ 0.75 & 0.00 & 0.25 & 0.00 \\ 0.00 & 0.25 & 0.00 & 0.75 \\ -0.50 & -0.50 & 0.00 & 0.00 \\ 0.00 & 0.25 & 0.00 & 0.75 \end{bmatrix}. \tag{6.54}$$

[4] The value of 10^{-12} is a possible suggestion; practically, smaller thresholds can be selected without a substantial loss of accuracy.

and, $\mathbf{g} \in \mathbb{R}^{14}$

$$\operatorname{diag}\left\{\mathbf{S}_{\mathfrak{gg}}\right\} = \begin{pmatrix} 0.15 \\ 0.10 \\ 0.25 \\ 0.05 \\ 0.20 \\ 0.50 \\ 0.35 \\ 0.40 \\ 0.30 \\ 0.45 \\ 0.10 \\ 0.25 \\ 0.15 \\ 0.05 \end{pmatrix}. \tag{6.55}$$

Recording 1000 samples from this process, setting the parameters for $\widetilde{J}^*$ to be

1. (Case 1) : $a_1 = 1$, $a_2 = 50$, $a_3 = 10$;
2. (Case 2) : $a_1 = 1$, $a_2 = 50$, $a_3 = 0$;
3. (Case 3) : $a_1 = 1$, $a_2 = 0$, $a_3 = 0$; and
4. (Case 4) : $a_1 = 0$, $a_2 = 50$, $a_3 = 0$,

and the boundaries for the 14 diagonal elements to be $\left[0.01 \quad 1\right]$, produced the results summarized in Tables 6.1 to 6.4 for Cases 1 to 4, respectively. Each table contains the resultant minimum of the objective function in (6.49), and the values for each of the three terms, $\widetilde{J}_1$, $\widetilde{J}_2$ and $\widetilde{J}_3$ for the inclusion of one to nine source signals. Note that $n = 9$ is the largest number that satisfies (6.46).

Table 6.1 Results for $a_1 = 1$, $a_2 = 50$, $a_3 = 10$.

n	$\widetilde{J}_1^*$	$\widetilde{J}_2^*$	$\widetilde{J}_3^*$	$\widetilde{J}^*$
1	9.728837	9.903143	9.194113	596.827139
2	10.623477	6.669295	4.730763	391.395848
3	11.730156	2.641624	1.282012	156.631463
4	**9.905117**	**0.679832**	**0.163441**	**45.531136**
5	8.776626	0.473485	0.116165	33.612547
6	7.802560	0.354886	0.073548	26.282350
7	6.812775	0.284251	0.051213	21.537448
8	5.857080	0.193634	0.032697	15.865753
9	4.952640	0.061742	0.017830	8.218051

Table 6.2 Results for $a_1 = 1$, $a_2 = 50$, $a_3 = 0$.

n	$\tilde{J}_1^*$	$\tilde{J}_2^*$	$\tilde{J}_3^*$	$\tilde{J}^*$
1	11.813837	10.088751	8.802784	516.251410
2	11.252778	6.167079	5.142943	319.606724
3	11.658961	2.642207	1.291209	143.769308
4	**9.901660**	**0.679877**	**0.163663**	**43.895523**
5	8.777040	0.473469	0.116278	32.450474
6	7.804200	0.354804	0.073785	25.544384
7	6.800655	0.275357	0.058046	20.568490
8	5.954737	0.144070	0.030562	13.158227
9	4.921906	0.080916	0.012328	8.967687

Table 6.3 Results for $a_1 = 1$, $a_2 = 0$, $a_3 = 0$.

n	$\tilde{J}_1^*$	$\tilde{J}_2^*$	$\tilde{J}_3^*$	$\tilde{J}^*$
1	8.454279	10.423839	10.423839	8.454279
2	6.036969	9.004625	9.004625	6.036969
3	3.941432	7.908937	7.908937	3.941432
4	**2.515947**	**7.484690**	**6.959272**	**2.515947**
5	2.032702	6.966966	6.071480	2.032702
6	1.626393	6.373496	5.211879	1.626393
7	1.244929	5.754994	4.385551	1.244929
8	0.891558	5.108477	3.568636	0.891558
9	0.659721	4.340301	2.800460	0.659721

Table 6.4 Results for $a_1 = 0$, $a_2 = 50$, $a_3 = 0$.

n	$\tilde{J}_1^*$	$\tilde{J}_2^*$	$\tilde{J}_3^*$	$\tilde{J}^*$
1	9.190492	9.859769	9.763201	492.988434
2	11.293882	6.166689	5.129166	308.334444
3	11.736554	2.641426	1.283513	132.071309
4	**9.927134**	**0.679582**	**0.163400**	**33.979092**
5	8.783186	0.473409	0.116265	23.670455
6	7.813840	0.354734	0.073990	17.736676
7	7.029985	0.167131	0.043608	8.356558
8	5.931184	0.138934	0.026496	6.946680
9	4.988845	0.102498	0.016111	5.124890

The results were obtained using the constraint nonlinear minimization function ‘fmincon’ of the Matlab^{TM} optimization toolbox, version 7.11.0.584 (R2010b). The results for Cases 1 and 2 do not differ substantially. This follows from the supplementary character of the constraints, which (6.51) and (6.52)

Table 6.5 Resultant estimates for $\mathbf{S}_{\mathfrak{gg}}$.

Variable	True	Case 1	Case 2	Case 3	Case 4
g_1	0.15	0.1801	0.1797	1.0000	0.1796
g_2	0.10	0.0832	0.0831	1.0000	0.0834
g_3	0.25	0.2290	0.2302	1.0000	0.2292
g_4	0.05	0.0473	0.0473	1.0000	0.0471
g_5	0.20	0.1915	0.1918	1.0000	0.1906
g_6	0.50	0.5390	0.5388	1.0000	0.5363
g_7	0.35	0.3828	0.3829	1.0000	0.3816
g_8	0.40	0.4265	0.4267	1.0000	0.4246
g_9	0.30	0.3041	0.3041	1.0000	0.3034
g_{10}	0.45	0.4372	0.4377	1.0000	0.4371
g_{11}	0.10	0.0899	0.0900	1.0000	0.0900
g_{12}	0.25	0.2283	0.2285	1.0000	0.2277
g_{13}	0.15	0.1562	0.1564	1.0000	0.1553
g_{14}	0.05	0.0527	0.0527	1.0000	0.0530

Case 1: $a_1 = 1, a_2 = 50, a_3 = 10$;
Case 2: $a_1 = 1, a_2 = 50, a_3 = 0$;
Case 3: $a_1 = 1, a_2 = 0, a_3 = 0$; and
Case 4: $a_1 = 0, a_2 = 50, a_3 = 0$.

show

$$
\begin{aligned}
\widehat{\widehat{\mathbf{S}}}_{z_0 z_0} &= \widehat{\widehat{\mathbf{S}}}_{z_s z_s} + \mathbf{I} \\
\mathbf{L}^{-1}\widehat{\widehat{\mathbf{S}}}_{z_0 z_0}\mathbf{L}^{-T} &= \mathbf{L}^{-1}\widehat{\mathbf{S}}_{z_0 z_0}\mathbf{L}^{-T} - \mathbf{L}^{-1}\widehat{\mathbf{S}}_{\mathfrak{gg}}\mathbf{L}^{-T} + \mathbf{I} \\
\mathbf{L}^{-1}\widehat{\mathbf{S}}_{z_0 z_0}\mathbf{L}^{-T} - \mathbf{I} &= \mathbf{L}^{-1}\left[\widehat{\mathbf{S}}_{z_0 z_0} - \widehat{\mathbf{S}}_{\mathfrak{gg}}\right]\mathbf{L}^{-T}.
\end{aligned}
\tag{6.56}
$$

Selecting a large a_2 value for the second term in (6.49) addresses the case of small discarded eigenvalues for $\widehat{\mathbf{S}}_{z_0 z_0} - \mathbf{S}_{\mathfrak{gg}}$ and suggests that the third term may be removed. Its presence, however, balances between the second and third terms and circumvents a suboptimal solution for larger process variable sets that yields discarded eigenvalues which are close to 1 but may not satisfy the 3rd constraint.

That Case 3 showed a poor performance is not surprising given that the only contributor to the first term is $\sum_{k=1}^{K} \widetilde{\mathfrak{g}}^T(k)[\widehat{\Xi}^{\perp}\widehat{\mathbf{S}}_{\mathfrak{gg}}\widehat{\Xi}^{\perp^T}]^{-1}\widetilde{\mathfrak{g}}(k)$. To produce small values in this case, the diagonal elements of $\widehat{\mathbf{S}}_{\mathfrak{gg}}^{-1}$ need to be small, which, in turn, suggests that larger error variance values are required. A comparison of the estimated error variances in Table 6.5 confirms this and stresses that the case of minimizing the log likelihood function only is insufficient for estimating the error covariance matrix.

Another interesting observation is that Case 4 (Table 6.4) produced a small value for the objective function after four components were retained. In fact, Table 6.5 highlights that the selection of the parameters for Case 4 produced a comparable accuracy in estimating the diagonal elements $\mathbf{S}_{\mathfrak{gg}}$. This would suggest omitting the contribution of the log likelihood function to the objective function and concentrating on terms two and three only. Inspecting Table 6.5 supports this conclusion, as most of the error variances are as accurately estimated as in Cases 1 and 2. However, the application for larger variable sets may yield suboptimal solutions, which the inclusion of the first term, the objective function in Equation (6.49), may circumvent.

It is not only important to estimate $\mathbf{S}_{gg}$ accurately but also to estimate the model subspace consistently, which has not been looked at thus far. The simplified analysis in (6.17) for $n_z = 3$ and $n = 2$ cannot, of course, be utilized in a general context. Moreover, the column space of Ξ can only be estimated up to a similarity transformation, which does not allow a comparison of the column vectors either.

The residual subspace is orthogonal to Ξ, which allows testing whether the estimated residuals subspace, spanned by the column vectors of $\widehat{\Xi}^{\perp} = \widehat{\widehat{\mathbf{P}}}_d\mathbf{L}^{-1}$, is perpendicular to the column space in Ξ. If so, $\widehat{\Xi}^{\perp}\Xi$ asymptotically converges to $\mathbf{0}$. Using $\widehat{\Xi}^{\perp}$, obtained for $a_1 = 1$, $a_2 = 50$ and $a_3 = 10$ this product is

$$\widehat{\Xi}^{\perp}\Xi = \begin{bmatrix} -0.043 & -0.003 & 0.026 & -0.014 \\ -0.010 & -0.020 & 0.001 & -0.004 \\ -0.005 & 0.048 & 0.018 & 0.013 \\ 0.029 & -0.032 & -0.079 & -0.023 \\ 0.028 & 0.007 & 0.013 & -0.016 \\ 0.098 & -0.023 & -0.004 & 0.034 \\ 0.024 & 0.018 & -0.024 & 0.020 \\ 0.041 & -0.016 & -0.003 & 0.008 \\ 0.057 & 0.018 & -0.037 & 0.041 \\ 0.031 & -0.011 & 0.048 & -0.068 \end{bmatrix}. \tag{6.57}$$

The small values in the above matrix indicate an accurate estimation of the model and residual subspace by the MLPCA algorithm. A comparison of the accuracy of estimating the model subspace by the MLPCA model with that of the PCA model yields, surprisingly, very similar results. More precisely, the matrix product $\widehat{\mathbf{P}}_d^T\Xi$, where $\widehat{\mathbf{P}}_d$ stores the last 10 eigenvectors of $\mathbf{S}_{z_0z_0}$, is equal to

$$\widehat{\mathbf{P}}_d^T\Xi = \begin{bmatrix} -0.032 & 0.058 & 0.077 & 0.012 \\ -0.034 & 0.029 & -0.034 & 0.065 \\ 0.011 & 0.031 & 0.004 & -0.044 \\ 0.033 & -0.029 & -0.014 & -0.005 \\ -0.018 & 0.081 & -0.005 & 0.008 \\ 0.006 & -0.008 & -0.005 & 0.051 \\ -0.022 & 0.035 & 0.013 & 0.037 \\ -0.003 & -0.036 & -0.006 & -0.009 \\ 0.001 & -0.011 & -0.034 & 0.014 \\ 0.014 & -0.011 & -0.012 & 0.012 \end{bmatrix}. \tag{6.58}$$

Increasing the error variance and the differences between the individual elements as well as the number of reference samples, however, will increase the difference

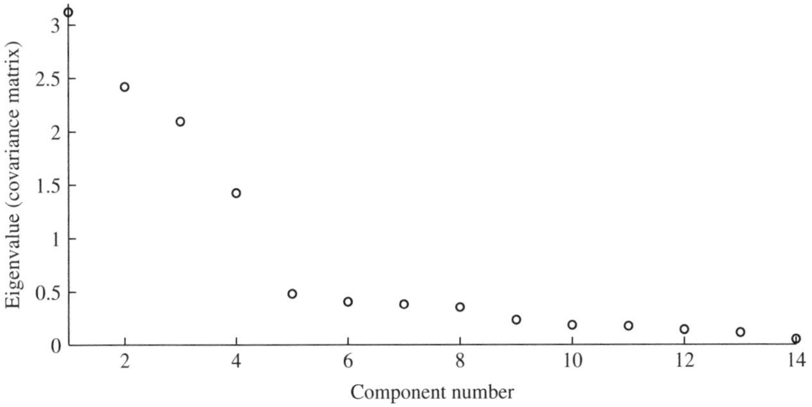

Figure 6.5 Plot of eigenvalues of $\widehat{\mathbf{S}}_{z_0 z_0}$.

between both estimates. A detailed study regarding this issue is proposed in the tutorial session of this chapter (Project 1). It is also important to note that PCA is unable to provide estimates of the error covariance matrix. To demonstrate this Figure 6.5 shows the distribution of eigenvalues of $\widehat{\mathbf{S}}_{z_0 z_0}$.

The next section introduces a stopping rule for MLPCA models. It is interesting to note that applying this rule for determining n yields a value of 1601.293 for (6.59), whilst the threshold is 85.965. This would clearly reject the hypothesis that the discarded 10 eigenvalues are equal. In fact, the application of this rule would not identify any acceptable value for $\mathfrak{n}$.

6.1.6 A stopping rule for maximum likelihood PCA models

Most stopping rules summarized in Subsection 2.4.1 estimate n based on the assumption that $\mathbf{S}_{\mathfrak{gg}} = \sigma_{\mathfrak{g}}^2 \mathbf{I}$ or analyze the variance of the recorded samples projected onto the residuals subspace. The discussion in this section, however, has outlined that the model subspace is only estimated consistently for $\mathbf{S}_{\mathfrak{gg}} = \sigma_{\mathfrak{g}}^2 \mathbf{I}$, which requires a different stopping rule for estimating n.

Feital *et al.* (2010) introduced a stopping rule if $\mathbf{S}_{\mathfrak{gg}} \neq \sigma_{\mathfrak{g}}^2 \mathbf{I}$. This rule relies on a hypothesis for testing the equality of the discarded eigenvalues. Equations (6.36) and (6.38) outline that these eigenvalues are 1 after applying the Cholesky decomposition to $\mathbf{S}_{z_0 z_0}$. To test whether the $n_z - n$ discarded eigenvalues are equal, Section 11.7.3 in Anderson (2003) presents the following statistic, which has a limiting χ^2 distribution with $\frac{1}{2}\left(n_z - \mathfrak{n} + 2\right)\left(n_z - \mathfrak{n} + 1\right)$ degrees of freedom

$$\kappa^2(\mathfrak{n}) = (K-1)\left(\left(n_z - \mathfrak{n}\right) \ln\left(\frac{1}{n_z - \mathfrak{n}} \sum_{k=\mathfrak{n}+1}^{n_z} \widehat{\widetilde{\lambda}}_k\right) - \ln\left(\prod_{k=\mathfrak{n}+1}^{n_z} \widehat{\widetilde{\lambda}}_k\right)\right). \tag{6.59}$$

It should be noted that the estimated eigenvalues $\widehat{\widetilde{\lambda}}_k$ are those of the scaled covariance matrix $\widetilde{\mathbf{S}}_{z_0 z_0} = \mathbf{L}^{-1} \mathbf{S}_{z_0 z_0} \mathbf{L}^{-T}$. According to the test statistic in (6.59), the null hypothesis is that the $n_z - \mathfrak{n}$ eigenvalues are equal. The alternative hypothesis is that the discarded $n_z - \mathfrak{n}$ eigenvalues are not identical and $n > \mathfrak{n}$.

The critical value of the χ^2 distribution for a significance α depends on the number of degrees of freedom for the χ^2 distribution. The statistic κ^2 must be compared against the critical value for $\chi^2_\alpha(dof)$, where dof represents the number of degrees of freedom. The null hypothesis H_0 is therefore accepted if

$$H_0 \quad : \quad \kappa^2(\mathfrak{n}) \leq \chi^2_\alpha\left(\tfrac{1}{2}\left(n_z - \mathfrak{n} + 2\right)\left(n_z - \mathfrak{n} + 1\right)\right) \tag{6.60}$$

and rejected if

$$H_1 \quad : \quad \kappa^2(\mathfrak{n}) > \chi^2_\alpha\left(\tfrac{1}{2}\left(n_z - \mathfrak{n} + 2\right)\left(n_z - \mathfrak{n} + 1\right)\right). \tag{6.61}$$

While H_0 describes the equality of the discarded $n_z - n$ eigenvalues, H_1 represents the case of a statistically significant difference between these eigenvalues.

The formulation of the stopping rule is therefore as follows. Start with $\mathfrak{n} = 1$ and obtain an MLPCA model. Then, compute the κ^2 value for (6.59) along with the critical value of a χ^2 distribution for $\frac{1}{2}\left(n_z + 1\right)n_z$ degrees of freedom and a significance of α. Accepting H_0 yields $n = 1$ and this model includes the estimate of the model subspace $\mathbf{L}\widetilde{\mathbf{P}}$ and its orthogonal complement $\widetilde{\mathbf{P}}_d\mathbf{L}^{-1}$. For rejecting H_0, iteratively increment $\mathfrak{n}$, $\mathfrak{n} = \mathfrak{n} + 1$, compute a MLPCA model and test H_0 until $\kappa^2(\mathfrak{n}) \leq \chi^2_\alpha\left(\frac{1}{2}\left(n_z - \mathfrak{n} + 2\right)\left(n_z - \mathfrak{n} + 1\right)\right)$.

To simplify the iterative sequence of hypothesis tests, κ^2 can be divided by χ^2_α

$$\widetilde{\kappa}^2(\mathfrak{n}) = \frac{\kappa^2(\mathfrak{n})}{\chi^2_\alpha\left(\frac{1}{2}\left(n_z - \mathfrak{n} + 2\right)\left(n_z - \mathfrak{n} + 1\right)\right)} \tag{6.62}$$

which gives rise to the following formulation of the stopping rule

$$H_0 \quad : \quad \widetilde{\kappa}^2(\mathfrak{n}) \leq 1 \tag{6.63}$$

and

$$H_1 \quad : \quad \widetilde{\kappa}^2(\mathfrak{n}) > 1. \tag{6.64}$$

The introduction of the stopping rule is now followed by an application study to the simulated process described in (6.53) to (6.55). This requires the application of (6.59), (6.62) and (6.63) to the MLPCA model for a varying number of estimated source signals, starting from 1. Table 6.6 shows the results of this series of hypothesis tests for $1 \leq \mathfrak{n} \leq 9$ for a significance of $\alpha = 0.05$.

The results in Table 6.6 confirm that $\widetilde{\kappa}^2(\mathfrak{n}) > \chi^2_\alpha\left(\frac{1}{2}\left(n_z - \mathfrak{n} + 2\right)\left(n_z - \mathfrak{n} + 1\right)\right)$ for $\mathfrak{n} \leq 3$. For $\mathfrak{n} = 4$, the null hypothesis is accepted and hence, the ten discarded eigenvalues are equivalent. Increasing $\mathfrak{n}$ further up to $\mathfrak{n} = 9$ also yields equivalent eigenvalues, which is not surprising either. For the sequence of nine hypothesis tests in Table 6.6, it is important to note that the first acceptance of H_0 is the estimate for n.

Table 6.6 Results for estimating n.

n	κ^2	dof	χ^2_α	$\widetilde{\kappa}^2$
1	7888.909466	105	129.917955	60.722242
2	4076.291059	91	114.267868	35.673117
3	783.559750	78	99.616927	7.865729
4	**50.016546**	**66**	**85.964907**	**0.581825**
5	28.513181	55	73.311493	0.388932
6	20.478358	45	61.656233	0.332138
7	4.223217	36	50.998460	0.082811
8	2.713276	28	41.337138	0.065638
9	1.471231	21	32.670573	0.045032

dof $=$ dof$(n_z, n) = \frac{1}{2}\left(n_z - n + 2\right)\left(n_z - n + 1\right)$ is the number of degrees of freedom of the κ^2 statistic.

6.1.7 Properties of model and residual subspace estimates

After introducing how to estimate the column space of Ξ and its complementary residual subspace $\Xi^\perp$, the next question is what are the geometric properties of these estimates. The preceding discussion has shown that the estimates for column space of Ξ, the generalized inverse[5] and its orthogonal complement are

$$\begin{aligned} \text{Estimate of column space of } \Xi \quad &: \quad \widehat{\mathbf{L}}\widehat{\widetilde{\mathbf{P}}} \\ \text{Estimate of generalized inverse } \Xi^\dagger \quad &: \quad \widehat{\widetilde{\mathbf{P}}}^T\widehat{\mathbf{L}}^{-1} \\ \text{Estimate of orthogonal complement } \Xi^\perp \quad &: \quad \widehat{\widetilde{\mathbf{P}}}_d^T\widehat{\mathbf{L}}^{-1}, \end{aligned} \tag{6.65}$$

where $\widehat{\mathbf{L}}\widehat{\mathbf{L}}^T = \widehat{\mathbf{S}}_{gg}$, $\widehat{\widetilde{\mathbf{P}}}$ and $\widehat{\widetilde{\mathbf{P}}}_d$ store the n and the remaining $n_z - n$ eigenvectors of $\widehat{\mathbf{L}}^{-1}\widehat{\mathbf{S}}_{z_0z_0}\widehat{\mathbf{L}}^{-1}$ associated with eigenvalues larger than 1 and equal to 1, respectively.

The missing proofs of the relationships in (6.65) are provided next, which commences by reformulating the relationship between the known covariance matrices of the recorded data vector, the uncorrupted data vector and the error vector

$$\begin{aligned} \mathbf{S}_{z_0z_0} &= \mathbf{S}_{z_sz_s} + \mathbf{S}_{gg} \\ \mathbf{S}_{z_0z_0} &= \mathbf{S}_{z_sz_s} + \mathbf{L}\mathbf{L}^T \\ \mathbf{L}^{-1}\mathbf{S}_{z_0z_0}\mathbf{L}^{-T} &= \mathbf{L}^{-1}\mathbf{S}_{z_sz_s}\mathbf{L}^{-T} + \mathbf{I}. \end{aligned} \tag{6.66}$$

For simplicity, it is assumed that each of the covariance matrices are available. Carrying out the eigendecomposition of $\widetilde{\mathbf{S}}_{z_0z_0} = \mathbf{L}^{-1}\mathbf{S}_{z_0z_0}\mathbf{L}^{-T}$ and comparing it

[5] The generalized inverse of a matrix is often referred to as the Moore-Penrose pseudo inverse.

to the right hand side of (6.66) gives rise to

$$\begin{aligned}\widetilde{\mathbf{P}}\widetilde{\mathbf{\Lambda}}\widetilde{\mathbf{P}}^T + \widetilde{\mathbf{P}}_d\widetilde{\mathbf{P}}_d^T &= \mathbf{L}^{-1}\mathbf{S}_{z_s z_s}\mathbf{L}^{-T} + \mathbf{I}\\ \widetilde{\mathbf{P}}\left[\widetilde{\mathbf{\Lambda}} - \mathbf{I}\right]\widetilde{\mathbf{P}}^T + \mathbf{I} &= \mathbf{L}^{-1}\mathbf{S}_{z_s z_s}\mathbf{L}^{-T} + \mathbf{I}\\ \widetilde{\mathbf{P}}\left[\widetilde{\mathbf{\Lambda}} - \mathbf{I}\right]\widetilde{\mathbf{P}}^T &= \mathbf{L}^{-1}\Xi\mathbf{S}_{ss}\Xi^T\mathbf{L}^{-T}.\end{aligned} \tag{6.67}$$

Pre- and post-multiplying (6.67) by $\mathbf{L}$ and $\mathbf{L}^T$ yields

$$\mathbf{L}\widetilde{\mathbf{P}}\left[\widetilde{\mathbf{\Lambda}} - \mathbf{I}\right]\widetilde{\mathbf{P}}^T\mathbf{L}^T = \Xi\mathbf{S}_{ss}\Xi^T. \tag{6.68}$$

It follows from (6.9) to (6.11) that the column space of Ξ is given by $\mathbf{L}\widetilde{\mathbf{P}}$. With regards to (6.65), $\widetilde{\mathbf{P}}_d^T\mathbf{L}^{-1}$ is the orthogonal complement of Ξ, since

$$\Xi^{\perp}\Xi = \widetilde{\mathbf{P}}_d\mathbf{L}^{-1}\Xi = \widetilde{\mathbf{P}}_d^T\mathbf{L}^{-1}\mathbf{L}\widetilde{\mathbf{P}} = \widetilde{\mathbf{P}}_d^T\widetilde{\mathbf{P}} = \mathbf{0}. \tag{6.69}$$

Finally, that $\widetilde{\mathbf{P}}^T\mathbf{L}^{-1}$ is the generalized inverse of $\mathbf{L}\widetilde{\mathbf{P}}$ follows from

$$\Xi^{\dagger}\Xi = \widetilde{\mathbf{P}}^T\mathbf{L}^{-1}\mathbf{L}\widetilde{\mathbf{P}} = \widetilde{\mathbf{P}}^T\widetilde{\mathbf{P}} = \mathbf{I}. \tag{6.70}$$

Geometrically, the estimate Ξ and its orthogonal complement $\Xi^{\perp}$ are estimates of the model and residual subspaces, respectively. The generalized inverse $\Xi^{\dagger}$ and the orthogonal complement $\Xi^{\perp}$ allow the estimation of linear combinations of the source signals and linear combinations of the error variables, respectively, since

$$\mathbf{z}_0 = \Xi\mathbf{s} + \mathbf{g} = \mathbf{L}\widetilde{\mathbf{P}\mathbf{t}} + \mathbf{g}. \tag{6.71}$$

With regards to (6.71), there is a direct relationship between the source signals and the components determined by the PCA model in the noise-free case

$$\widetilde{\mathbf{t}} = \widetilde{\mathbf{P}}^T\mathbf{L}^{-1}\Xi\mathbf{s} \Rightarrow \mathbf{s} = \left[\widetilde{\mathbf{P}}^T\mathbf{L}^{-1}\Xi\right]^{-1}\widetilde{\mathbf{t}}. \tag{6.72}$$

For the case $\mathbf{g} \neq \mathbf{0}$, it follows that

$$\begin{aligned}\widetilde{\mathbf{t}} &= \widetilde{\mathbf{P}}^T\mathbf{L}^{-1}\left(\Xi\mathbf{s} + \mathbf{g}\right)\\ \mathbf{s} &= \left[\widetilde{\mathbf{P}}^T\mathbf{L}^{-1}\Xi\right]^{-1}\widetilde{\mathbf{t}} - \left[\widetilde{\mathbf{P}}^T\mathbf{L}^{-1}\Xi\right]^{-1}\widetilde{\mathbf{P}}^T\mathbf{L}^{-1}\mathbf{g}\\ \mathbf{s} &\approx \left[\widetilde{\mathbf{P}}^T\mathbf{L}^{-1}\Xi\right]^{-1}\widetilde{\mathbf{t}}.\end{aligned} \tag{6.73}$$

Despite the fact that the source signals could be recovered for $\mathbf{g} = \mathbf{0}$ and approximated for $\mathbf{g} \neq \mathbf{0}$ and $E\left\{g_i^2\right\} \ll E\{(\boldsymbol{\xi}_i^T\mathbf{s})^2\}$, the following two problems remain.

Problem 6.1.1 *If* $\mathbf{S}_{gg} \neq \sigma_g^2\mathbf{I}$, *the application of the scaling mechanism, based on the Cholesky decomposition of* $\mathbf{S}_{gg} = \mathbf{L}\mathbf{L}^T$, *does not guarantee that*

- *the loading vectors point in directions that produce a maximum variance for each score variable; and*
- *the loading vectors may not have unit length.*

In addition to the above points, Feital *et al.* (2010) highlighted that the score variables may not be statistically independent either, that is, the score vectors may not be orthogonal as is the case for PCA. This is best demonstrated by comparing the score variables computed by applying the generalized inverse $\widetilde{\mathbf{P}}^T\mathbf{L}^{-1}$

$$\widetilde{\mathbf{t}} = \widetilde{\mathbf{P}}^T\mathbf{L}^{-1}\mathbf{z}_0 \Rightarrow E\left\{\widetilde{\mathbf{t}}\widetilde{\mathbf{t}}^T\right\} = \widetilde{\mathbf{P}}^T\widetilde{\mathbf{S}}_{z_0z_0}\widetilde{\mathbf{P}} = \widetilde{\mathbf{\Lambda}} \tag{6.74}$$

with those determined by an eigendecomposition of $\mathbf{S}_{z_0z_0} - \mathbf{S}_{gg}$

$$\mathbf{P}^T\left[\mathbf{S}_{z_0z_0} - \mathbf{S}_{gg}\right]\mathbf{P} = \mathbf{\Lambda}_s. \tag{6.75}$$

Removing the impact of the error covariance matrix from (6.74) allows a direct comparison with (6.75)

$$\widetilde{\mathbf{P}}^T\left[\widetilde{\mathbf{S}}_{z_0z_0} - \mathbf{I}\right]\widetilde{\mathbf{P}} = \widetilde{\mathbf{\Lambda}} - \mathbf{I} = \widetilde{\mathbf{\Lambda}}_s \tag{6.76}$$

which yields:

- that it generally cannot be assumed that the eigenvectors of $\widetilde{\mathbf{S}}_{z_0z_0} - \mathbf{I}$ are equal to those of $\mathbf{S}_{z_0z_0} - \mathbf{S}_{gg}$; and
- that it can also generally not be assumed that the eigenvalues of $\widetilde{\mathbf{S}}_{z_0z_0} - \mathbf{I}$ are equal to those of $\mathbf{S}_{z_0z_0} - \mathbf{S}_{gg}$

The subscript s in (6.75) and (6.76) refers to the source signals. Finally, the matrix product $\widetilde{\mathbf{P}}^T\mathbf{L}^{-1}\mathbf{L}^{-T}\widetilde{\mathbf{P}}$ is only a diagonal matrix if $\mathbf{S}_{gg}$ is diagonal and hence, $\mathbf{L}$ is of diagonal type. $\mathbf{S}_{gg}$, however, is assumed to be diagonal in (6.46). In any case, the row vectors in $\widetilde{\mathbf{P}}^T\mathbf{L}^{-1}$ do not have unit length, as the elements in $\mathbf{L}^{-2} = \mathbf{S}_{gg}^{-1}$ are not generally 1. Moreover, if $\mathbf{S}_{gg}$ is not a diagonal matrix, $\widetilde{\mathbf{P}}^T\mathbf{L}^{-1}$ does not, generally, have orthogonal column vectors.

Feital *et al.* (2010) and Ge *et al.* (2011) discussed two different methods for determining loading vectors of unit length that produce score variables that have a maximum variance, and are statistically independent irrespective of whether $\mathbf{S}_{gg}$ is a diagonal matrix or not. The first method has been proposed in Hyvarinen (1999); Yang and Guo (2008) and is to determine the eigendecomposition of $\mathbf{S}_{z_0z_0} - \mathbf{S}_{gg}$, which yields the loading vectors stored in $\mathbf{P}$. It is important to note, however, that the eigenvalues of $\mathbf{S}_{z_0z_0} - \mathbf{S}_{gg}$ are not those of the computed score variables.

This issue has been addressed in Feital *et al.* (2010) by introducing a constraint NIPALS algorithm. Table 6.7 summarizes an algorithm similar to that proposed in Feital *et al.* (2010). This algorithm utilizes the estimated model subspace, spanned by the column vectors of $\widehat{\widetilde{\mathbf{P}}}$ under the assumption that $\mathbf{S}_{gg}$ is of diagonal type.

Table 6.7 Constraint NIPALS algorithm.

Step	Description	Equation
1	Initiate iteration	$i = 1,\ \mathbf{Z}^{(1)} = \mathbf{Z}_0$
2	Set up projection matrix	$\mathfrak{C} = \widetilde{\widetilde{\mathbf{P}}}\widetilde{\widetilde{\mathbf{P}}}^T$
3	Define initial score vector	${}_0\mathbf{t}_i = \mathbf{Z}^{(i)}\,(:, 1)$
4	Determine loading vector	$\mathbf{p}_i = \mathfrak{C}\mathbf{Z}^{(i)^T}\left({}_0\mathbf{t}_i\right)$
5	Scale loading vector	$\widehat{\mathbf{p}}_i = {}^{\mathbf{p}_i}/_{\lVert\mathbf{p}_i\rVert}$
6	Calculate score vector	${}_1\mathbf{t}_i = \mathbf{Z}^{(i)}\widehat{\mathbf{p}}_i$
7	Compute eigenvalue	$\lambda_i = \left\lVert {}_1\mathbf{t}_i \right\rVert^2$
8	Check for convergence	If $\left\lVert {}_1\mathbf{t}_i -{}_0\mathbf{t}_i \right\rVert > \varepsilon$, set ${}_0\mathbf{t}_i ={}_1\mathbf{t}_i$ and go to Step 4 else set $\widehat{\mathbf{t}}_i ={}_1\mathbf{t}_i$ and go to Step 9
9	Scale eigenvalue	$\widehat{\lambda}_i = {}^{\lambda_i}/_{K-1}$
10	Deflate data matrix	$\mathbf{Z}^{(i+1)} = \mathbf{Z}^{(i)} - \widehat{\mathbf{t}}_i\widehat{\mathbf{p}}_i^T$
11	Check for dimension	If $i < n$ set $i = i + 1$ and go to Step 3 else terminate iteration procedure

In order to outline the working of this algorithm, setting $\mathfrak{C} = \mathbf{I}$ in Step 2 reduces the algorithm in Table 6.7 to the conventional NIPALS algorithm (Geladi and Kowalski 1986). The conventional algorithm, however, produces an eigendecomposition of $\widehat{\mathbf{S}}_{z_0z_0}$ and the associated score vectors for $\mathbf{Z}_0$.

Setting $\mathfrak{C} = \widetilde{\widetilde{\mathbf{P}}}\widetilde{\widetilde{\mathbf{P}}}^T$, however, forces the eigenvectors to lie within the estimated model subspace. To see this, the following matrix projects any vector of dimension n_z to lie within the column space of $\widetilde{\widetilde{\mathbf{P}}}$ (Golub and van Loan 1996)

$$\widetilde{\widetilde{\mathbf{P}}}\left[\widetilde{\widetilde{\mathbf{P}}}^T\widetilde{\widetilde{\mathbf{P}}}\right]^{-1}\widetilde{\widetilde{\mathbf{P}}}^T = \widetilde{\widetilde{\mathbf{P}}}\widetilde{\widetilde{\mathbf{P}}}^T = \mathfrak{C}. \tag{6.77}$$

Lemma 2.1.1 and particularly (2.5) in Section 2.1 confirm that (6.77) projects any vector orthogonally onto the model plane. Figure 2.2 gives a schematic illustration of this orthogonal projection. Step 4 in Table 6.7, therefore, guarantees that the eigenvectors of $\mathbf{S}_{z_0z_0}$ lie in the column space of $\widetilde{\widetilde{\mathbf{P}}}$.

Step 5 ensures that the loading vectors are of unit length, whilst Step 6 records the squared length of the t-score vector, which is $K - 1$ times its variance since the samples stored in the data matrix have been mean centered. Upon convergence, Step 9 determines the variance of the ith score vector and Step 10 deflates the data matrix. It is shown in Section 9.1 that the deflation procedure gives rise to orthonormal p-loading vectors and orthogonal t-score vectors, and that the power method converges to the most dominant eigenvector (Golub and van Loan 1996).

The working of this constraint NIPALS algorithm is now demonstrated using data from the simulation example in Subsection 6.1.5. Subsection 6.1.8 revisits the application study of the chemical reaction process in Chapter 4 by identifying an MLPCA model including an estimate of the number of source signals and a rearrangement of the loading vectors by applying the constraint NIPALS algorithm.

6.1.7.1 Application to data from the simulated process in subsection 6.1.5

By using a total of $K = 1000$ simulated samples from this process and including $n = 4$ source signals, the application of MLPCA yields the following loading matrix

$$\widehat{\widehat{\mathbf{P}}} = \begin{bmatrix} -0.0056 & 0.5038 & 0.4102 & -0.3142 \\ 0.0347 & 0.4384 & -0.7203 & 0.0187 \\ 0.0100 & 0.0733 & 0.0632 & 0.8188 \\ 0.7843 & -0.1140 & 0.1414 & -0.0463 \\ 0.0942 & 0.2199 & 0.2158 & 0.1288 \\ 0.0674 & 0.0675 & -0.0369 & 0.3649 \\ 0.0678 & 0.1434 & -0.0036 & 0.1697 \\ 0.0518 & -0.1995 & 0.1274 & 0.0148 \\ 0.0035 & 0.1135 & -0.2085 & -0.0219 \\ -0.0709 & 0.0559 & 0.1272 & 0.1304 \\ -0.0039 & 0.4772 & 0.3603 & 0.1248 \\ 0.2495 & 0.0408 & -0.0616 & 0.0034 \\ -0.0122 & -0.4136 & 0.0963 & 0.1262 \\ 0.5435 & 0.0771 & -0.1486 & -0.0235 \end{bmatrix}. \tag{6.78}$$

Applying the constraint NIPALS algorithm, however, yields a different loading matrix

$$\widehat{\mathbf{P}} = \begin{bmatrix} 0.2594 & 0.4274 & 0.2556 & 0.4158 \\ 0.2822 & 0.1105 & -0.6545 & -0.1012 \\ 0.2189 & 0.1087 & 0.2546 & -0.7437 \\ 0.3691 & -0.4708 & 0.1859 & 0.1741 \\ 0.2813 & 0.1212 & 0.2505 & 0.0652 \\ 0.2805 & -0.0391 & 0.0360 & -0.3932 \\ 0.2865 & 0.0554 & 0.0143 & -0.0969 \\ -0.1316 & -0.2815 & 0.3003 & 0.0097 \\ 0.1161 & 0.0507 & -0.3330 & -0.0323 \\ -0.0409 & 0.2439 & 0.2033 & -0.1067 \\ 0.2603 & 0.3597 & 0.2433 & 0.1013 \\ 0.3640 & -0.3247 & -0.0127 & 0.1001 \\ -0.2739 & -0.2675 & 0.2011 & -0.1460 \\ 0.3484 & -0.3235 & -0.0222 & 0.1057 \end{bmatrix}. \tag{6.79}$$

Finally, taking the loading matrix obtained from the constraint NIPALS algorithm and comparing the estimated covariance matrix of the score variables

$$\widehat{\mathbf{S}}_{tt} = \begin{bmatrix} 3.0002 & -0.0000 & -0.0000 & -0.0000 \\ -0.0000 & 2.5670 & 0.0000 & -0.0000 \\ -0.0000 & 0.0000 & 1.9740 & 0.0000 \\ -0.0000 & -0.0000 & 0.0000 & 1.5160 \end{bmatrix} \tag{6.80}$$

with those obtained from the loading matrix determined from the original data covariance matrix, i.e. $\widehat{\mathbf{S}}_{z_0z_0} = \widehat{\mathbf{P}}\widehat{\mathbf{\Lambda}}\widehat{\mathbf{P}}^T + \widehat{\mathbf{P}}_d\widehat{\mathbf{\Lambda}}_d\widehat{\mathbf{P}}_d^T$ and $\widehat{\mathbf{t}} = \widehat{\mathbf{P}}^T\mathbf{z}_0$

$$\widehat{\mathbf{S}}_{tt} = \begin{bmatrix} 3.0043 & -0.0000 & -0.0000 & -0.0000 \\ -0.0000 & 2.5738 & 0.0000 & -0.0000 \\ -0.0000 & 0.0000 & 1.9804 & -0.0000 \\ -0.0000 & -0.0000 & -0.0000 & 1.5213 \end{bmatrix} \tag{6.81}$$

yields that the diagonal elements that are very close to the theoretical maximum for conventional PCA. The incorporation of the constraint (Step 4 of the constraint NIPALS algorithm in Table 6.7) clearly impacts the maximum value but achieves:

- an estimated model subspace is that obtained from the MLPCA algorithm; and
- loading vectors that produce score variables which have a maximum variance.

6.1.8 Application to a chemical reaction process – revisited

To present a more challenging and practically relevant application study, this subsection revisits the application study of the chemical reaction process. Recall that the application of PCA relied on the following assumptions outlined in Section 2.1

- $\mathbf{z} = \Xi\mathbf{s} + \mathbf{g} + \bar{\mathbf{z}}$, where:
- $\mathbf{s} \sim \mathcal{N}\left\{\mathbf{0}, \mathbf{S}_{ss}\right\}$;
- $\mathbf{g} \sim \mathcal{N}\left\{\mathbf{0}, \mathbf{S}_{gg}\right\}$;
- with i and j being two sample indices

$$E\left\{\begin{pmatrix} \mathbf{s}(k) \\ \mathbf{g}(k) \end{pmatrix}\begin{pmatrix} \mathbf{s}^T(l) & \mathbf{g}^T(l) \end{pmatrix}\right\} = \delta_{kl}\begin{bmatrix} \mathbf{S}_{ss} & \mathbf{0} \\ \mathbf{0} & \mathbf{S}_{gg} \end{bmatrix} \qquad \mathbf{S}_{gg} = \sigma_g^2\mathbf{I}; \text{ and}$$

- the covariance matrices $\mathbf{S}_{ss}$ and $\mathbf{S}_{gg}$ have full rank n and n_z, respectively.

Determining the number of source signals. Under these assumptions, the application of the VRE technique suggested that the data model has four source signals (Figure 4.4). Inspecting the eigenvalue plot in Figure 4.3, however, does not support the assumption that the remaining 31 eigenvalues have the same value even without applying (6.59) and carrying out the hypothesis test for H_0 in (6.59) and (6.63).

According to (6.46), the maximum number of source signals for a complete estimation of the diagonal elements of $\mathbf{S}_{gg}$ is 27. Different to the suggested number of four source signals using the VRE criterion, the application of the hypothesis test in Subsection 6.1.6 yields a total of 20 source signals.

Table 6.8 lists the results for estimating the MLPCA model, including the optimal value of the objective function in (6.49), $\widetilde{J}^*$, the three contributing terms, $\widetilde{J}_1^*$, $\widetilde{J}_2^*$ and $\widetilde{J}_3$, the κ^2 values of (6.59), its number of degrees of freedom (dof) and its critical value χ^2_α for $n = 1, \ldots, 27$.

For (6.49), the diagonal elements of the error covariance matrix were constrained to be within $\begin{bmatrix} 0.01 & 0.5 \end{bmatrix}$, which related to the pretreatment of the data.

Table 6.8 Estimation results for MLPCA model (chemical reaction process).

n	$\widetilde{J}_1$	$\widetilde{J}_2$	$\widetilde{J}_3$	$\widetilde{J}$	κ^2	dof	χ^2_α	$\widetilde{\kappa}^2$
1	27.915	23.945	6.732	2489.7	33794	630	689.50	49.012
2	31.865	17.062	3.449	1772.5	19217	595	652.86	29.435
3	32.037	11.627	1.486	1209.6	8185.8	561	617.21	13.263
4	31.091	9.489	1.071	990.70	5652.1	528	582.56	9.702
5	29.534	7.540	0.815	791.73	3598.3	496	548.92	6.555
6	28.265	6.174	0.629	651.92	2387.3	465	516.27	4.624
7	27.098	5.249	0.522	557.22	1790.9	435	484.63	3.695
8	25.821	4.666	0.460	496.99	1429.0	406	453.98	3.148
9	25.172	4.224	0.404	451.64	1223.4	378	424.33	2.883
10	24.275	3.822	0.352	409.97	1037.8	351	395.69	2.623
11	23.258	3.456	0.312	371.97	900.52	325	368.04	2.447
12	22.079	3.078	0.277	332.69	741.21	300	341.40	2.171
13	21.153	2.735	0.243	297.06	637.50	276	315.75	2.019
14	20.531	2.402	0.212	262.85	555.73	253	291.10	1.909
15	19.284	2.116	0.181	232.71	445.25	231	267.45	1.665
16	18.545	1.827	0.156	202.76	368.32	210	244.81	1.505
17	17.507	1.571	0.131	175.94	299.79	190	223.16	1.343
18	16.721	1.393	0.116	157.23	252.85	171	202.51	1.249
19	15.502	1.171	0.092	133.57	192.61	153	182.86	1.053
20	**14.714**	**0.937**	**0.072**	**109.12**	**140.75**	**136**	**164.22**	**0.857**
21	13.652	0.862	0.066	100.54	125.30	120	146.57	0.855
22	12.707	0.661	0.052	79.301	85.430	105	129.92	0.658
23	11.735	0.530	0.040	65.176	57.806	91	114.27	0.506
24	10.886	0.378	0.029	49.007	32.955	78	99.617	0.331
25	9.823	0.327	0.026	42.773	33.511	66	85.965	0.390
26	8.930	0.217	0.017	30.824	17.062	55	73.311	0.233
27	7.980	0.080	0.006	16.017	2.965	45	61.656	0.048

Parameters for objective function in Equation (6.49): $a_0 = 1$, $a_1 = 100$, $a_3 = 10$.

Each temperature variable was mean centered and scaled to unity variance. Consequently, a measurement uncertainty of each thermocouples exceeding 50% of its variance was not expected and the selection of a too small lower boundary might have resulted in numerical problems in computing the inverse of the lower triangular matrix of the Cholesky decomposition, according to (6.36). The parameters for $\widetilde{J}_1$, $\widetilde{J}_2$ and $\widetilde{J}_3$ were $a_1 = 1$, $a_2 = 100$ and $a_3 = 10$, respectively.

Table 6.9 lists the elements of $\widehat{\mathbf{S}}_{gg}$ for $n = 20$. It should be noted that most error variances are between 0.05 and 0.13 with the exception of thermocouple 22 and 24. When comparing the results with PCA, the estimated model subspace for MLPCA is significantly larger. However, the application of MLPCA has shown here that estimating the model subspace simply by computing the eigendecomposition of $\mathbf{S}_{z_0z_0}$ has relied on an incorrect data structure. According to the results in Table 6.8, retaining just four PCs could not produce equal eigenvalues even under the assumption of unequal diagonal elements of $\mathbf{S}_{gg}$.

Chapter 4 discussed the distribution function of the source signals and showed that the first four score variables are, in fact, non-Gaussian. Whilst it was still possible to construct the Hotelling's T^2 and Q statistics that were able to detect an abnormal behavior, the issue of non-Gaussian source signals is again discussed in Chapter 8. Next, the adjustment of the base vectors spanning the model subspace is considered.

Readjustment of the base vector spanning the model subspace. Table 6.10 lists the eigenvectors obtained by the constraint NIPALS algorithm. Table 6.11 shows the differences in the eigenvalues of $\widehat{\mathbf{S}}_{z_0z_0}$ and those obtained by the constraint NIPALS algorithm. Figure 6.6 presents a clearer picture for describing the impact of the constraint NIPALS algorithm. The first four eigenvalues and eigenvectors show a negligible difference but the remaining ones depart significantly by up to 90° for the eigenvectors and up to 10% for the eigenvalues.

Summary of the application of MLPCA. Relying on the assumption that $\mathbf{S}_{gg} = \sigma_g^2\mathbf{I}$ suggested a relatively low number of source signals. Removing the assumption, however, presented a different picture and yielded a significantly larger number of source signals. A direct inspection of Figure 4.3 confirmed that the discarded components do not have an equal variance and the equivalence of the eigenvalues for the MLPCA has been tested in a statistically sound manner. The incorporation of the identified model subspace into the determination of the eigendecomposition of $\widehat{\mathbf{S}}_{z_0z_0}$ yielded a negligible difference for the first four eigenvalues and eigenvectors but significant differences for the remaining 16 eigenpairs. This application study, therefore, shows the need for revisiting and testing the validity of the assumptions imposed on the data models. Next, we examine the performance of the revised monitoring statistics in detecting the abnormal behavior of Tube 11 compared to the monitoring model utilized in Chapter 4.

Table 6.9 Estimated diagonal elements of $\mathbf{S}_{gg}$.

Variable (diagonal element of $\hat{\mathbf{S}}_{gg}$)	Error variance
g_1^2	0.0542
g_2^2	0.1073
g_3^2	0.0858
g_4^2	0.0774
g_5^2	0.0675
g_6^2	0.0690
g_7^2	0.0941
g_8^2	0.0685
g_9^2	0.0743
g_{10}^2	0.0467
g_{11}^2	0.1038
g_{12}^2	0.0798
g_{13}^2	0.0611
g_{14}^2	0.0748
g_{15}^2	0.0531
g_{16}^2	0.1163
g_{17}^2	0.0475
g_{18}^2	0.0688
g_{19}^2	0.0688
g_{20}^2	0.0792
g_{21}^2	0.0553
g_{22}^2	0.0311
g_{23}^2	0.1263
g_{24}^2	0.2179
g_{25}^2	0.0794
g_{26}^2	0.0764
g_{27}^2	0.0688
g_{28}^2	0.0648
g_{29}^2	0.0802
g_{30}^2	0.0816
g_{31}^2	0.0672
g_{32}^2	0.0777
g_{33}^2	0.0643
g_{34}^2	0.0714
g_{35}^2	0.0835

Table 6.10 Eigenvectors associated with first seven dominant eigenvalues.

p_{ij}	$j=1$	$j=2$	$j=3$	$j=4$	$j=5$	$j=6$	$j=7$
$i=1$	0.1702	0.1553	0.2058	0.1895	0.0222	0.2005	0.1886
$i=2$	0.1727	−0.0156	−0.0927	0.1797	−0.1123	0.1482	0.2831
$i=3$	0.1751	0.0843	0.0718	0.0605	−0.2154	−0.0718	0.0896
$i=4$	0.1572	0.3301	0.0252	−0.2317	−0.2953	−0.1760	−0.0208
$i=5$	0.1773	0.0158	−0.0001	−0.1754	−0.2077	0.1446	0.1575
$i=6$	0.1647	−0.1562	−0.2558	−0.2788	−0.1934	0.1082	−0.1509
$i=7$	0.1756	0.0030	−0.0652	−0.0388	−0.0837	−0.2560	−0.0589
$i=8$	0.1534	0.3681	−0.1251	−0.0523	0.1696	−0.2490	0.1557
$i=9$	0.1733	−0.1904	−0.0434	0.1436	−0.0531	−0.1325	0.0484
$i=10$	0.1774	0.1335	−0.0463	0.0377	−0.0231	0.2629	0.1216
$i=11$	0.1181	0.0896	0.6003	−0.4352	0.2118	0.1884	0.1074
$i=12$	0.1723	0.0576	0.2118	0.1905	0.1391	−0.1298	−0.0918
$i=13$	0.1732	0.1149	0.1266	−0.2004	−0.0590	−0.0771	−0.1354
$i=14$	0.1711	−0.0563	0.1625	0.0149	0.1574	−0.2397	−0.2727
$i=15$	0.1791	−0.0608	0.0327	0.0717	0.0860	−0.0952	0.0693
$i=16$	0.1702	0.0288	−0.1549	0.0488	−0.2022	0.1204	−0.2727
$i=17$	0.1651	−0.2928	0.1657	−0.0236	−0.0754	0.0623	0.0727
$i=18$	0.1722	0.0001	−0.2041	0.0381	0.1687	0.0859	0.3436
$i=19$	0.1750	−0.0980	−0.0218	−0.1741	−0.2880	0.1954	0.0097
$i=20$	0.1744	−0.1659	0.0419	0.0388	0.2008	−0.0721	−0.0430
$i=21$	0.1733	−0.1410	0.0287	0.1802	0.1174	−0.3004	−0.1751
$i=22$	0.1726	0.2010	0.0303	0.0155	−0.0175	−0.2110	−0.0486
$i=23$	0.1532	0.2597	0.2438	0.2137	−0.2477	0.1148	−0.3089
$i=24$	0.1472	0.2257	−0.2163	0.0111	0.4567	0.4498	−0.4446
$i=25$	0.1770	−0.0554	−0.0996	0.0112	0.0395	−0.0226	0.0237
$i=26$	0.1686	0.1257	−0.2211	−0.1773	0.2349	−0.1066	0.1424
$i=27$	0.1607	−0.3476	0.0303	0.0236	0.0620	−0.0680	−0.0761
$i=28$	0.1752	−0.1564	0.0643	0.0429	0.0733	0.1053	−0.0180
$i=29$	0.1770	−0.1036	0.0720	−0.0805	−0.0977	−0.1432	0.0124
$i=30$	0.1782	0.0326	−0.1117	−0.0917	−0.0257	−0.0103	−0.1105
$i=31$	0.1706	0.0383	0.0711	0.4773	−0.1093	0.1696	0.0911
$i=32$	0.1744	0.0233	−0.1877	−0.1674	0.1580	−0.0712	0.1956
$i=33$	0.1686	−0.2307	0.1605	−0.0271	0.1832	0.1150	0.1322
$i=34$	0.1734	0.1413	−0.1548	0.1652	0.0014	−0.0454	0.0842
$i=35$	0.1650	−0.2458	−0.1743	−0.1453	−0.0701	0.1162	−0.1727

The estimated number of source signals is $n=20$.

Detecting the abnormal behavior in tube 11. Figure 6.7 shows the Hotelling's T^2 and Q statistics for both data sets. Comparing Figure 4.10 with the upper plots in Figure 6.7, outlines that the inclusion of a larger number set of source signals does not yield the same 'distinct' regions, for example between 800 to 1100 minutes and between 1400 and 1600 minutes into the data set.

Table 6.11 Variances of score variables.

Component	Eigenvalue of $\widehat{\mathbf{S}}_{z_0z_0}$	Eigenvalue after adjustment
1	28.2959	28.2959
2	1.5940	1.5937
3	1.2371	1.2368
4	0.4101	0.4098
5	0.3169	0.3090
6	0.2981	0.2945
7	0.2187	0.2127
8	0.1929	0.1918
9	0.1539	0.1487
10	0.1388	0.1368
11	0.1297	0.1258
12	0.1251	0.1199
13	0.1199	0.1150
14	0.1148	0.1120
15	0.1067	0.1033
16	0.1015	0.0999
17	0.0980	0.0967
18	0.0939	0.0849
19	0.0919	0.0847
20	0.0884	0.0828

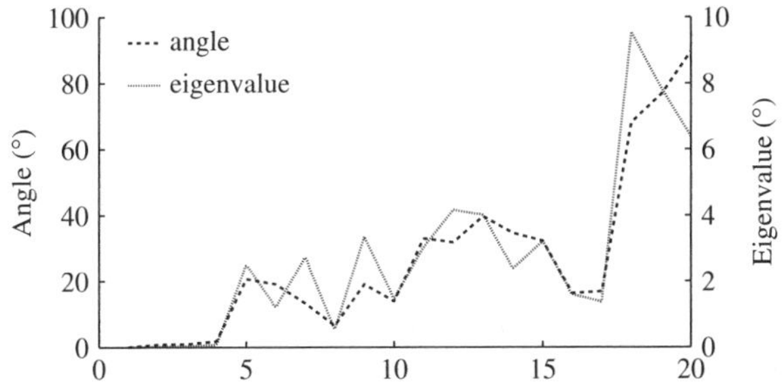

Figure 6.6 Percentage change in angle of eigenvectors and eigenvalues.

To qualify this observation, Figure 6.8 compares the F-distribution function with the empirical one, which shows a considerably closer agreement when contrasted with the PCA-based comparison in Figure 4.8. The upper plot in Figure 4.8 shows significant departures between the theoretical and the estimated distribution functions for the Hotelling's T^2 statistic. In contrast, the same plot in Figure 6.8 shows a close agreement for the MLPCA-based statistic. The residual-based Q statistics for the PCA and MLPCA models are accurately approximated by an F-distribution, when constructed with respect to (3.20), that is, $Q = T_d^2$.

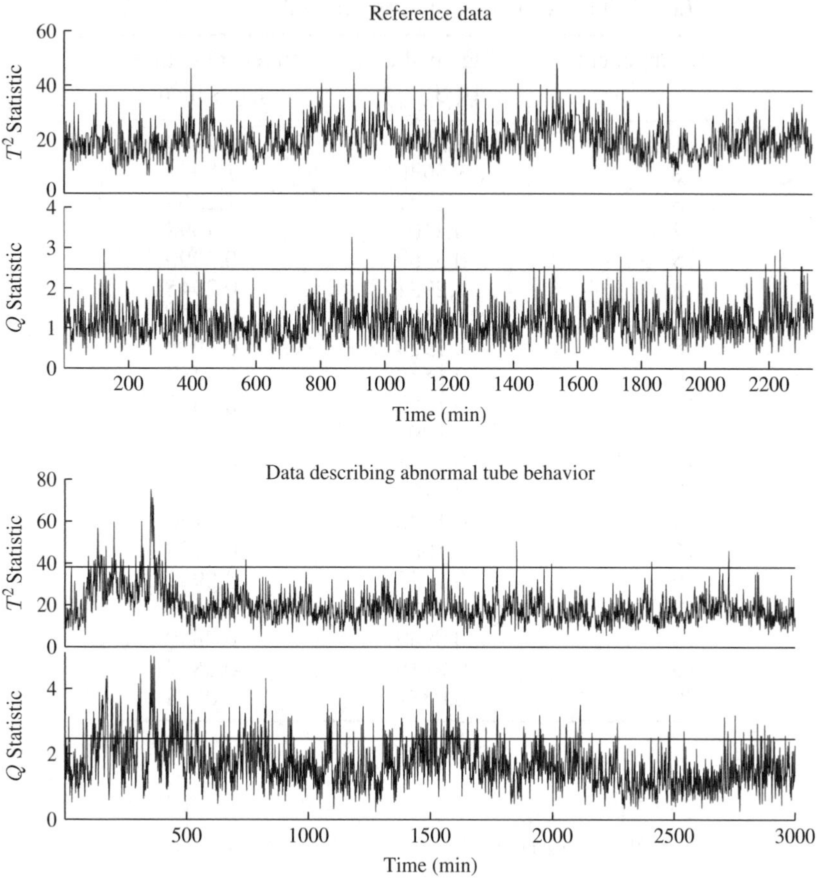

Figure 6.7 MLPCA-based monitoring statistics.

The reason that the MLPCA-based Hotelling's T^2 statistic is more accurately approximated by an F-distribution with 2338 and 20 degrees of freedom than the PCA-based one by an F-distribution with 2338 and 4 degrees of freedom is as follows. Whilst the first four components are strongly non-Gaussian, the remaining ones show significantly smaller departures from a Gaussian distribution. Figure 6.9 confirms this by comparing the estimated distribution function with the Gaussian one for score variables 5, 10, 15 and 20. Moreover, the construction of the Hotelling's T^2 statistic in (3.9) implies that each of the first four non-Gaussian score variables has the same contribution compared to the remaining 16 score variables. The strong impact of the first four highly non-Gaussian score variables to the distribution function of the Hotelling's T^2 statistic therefore becomes reduced for $n = 20$.

Analyzing the sensitivity of the MLPCA monitoring model in detecting the abnormal tube behavior requires comparing Figure 4.10 with the lower plots

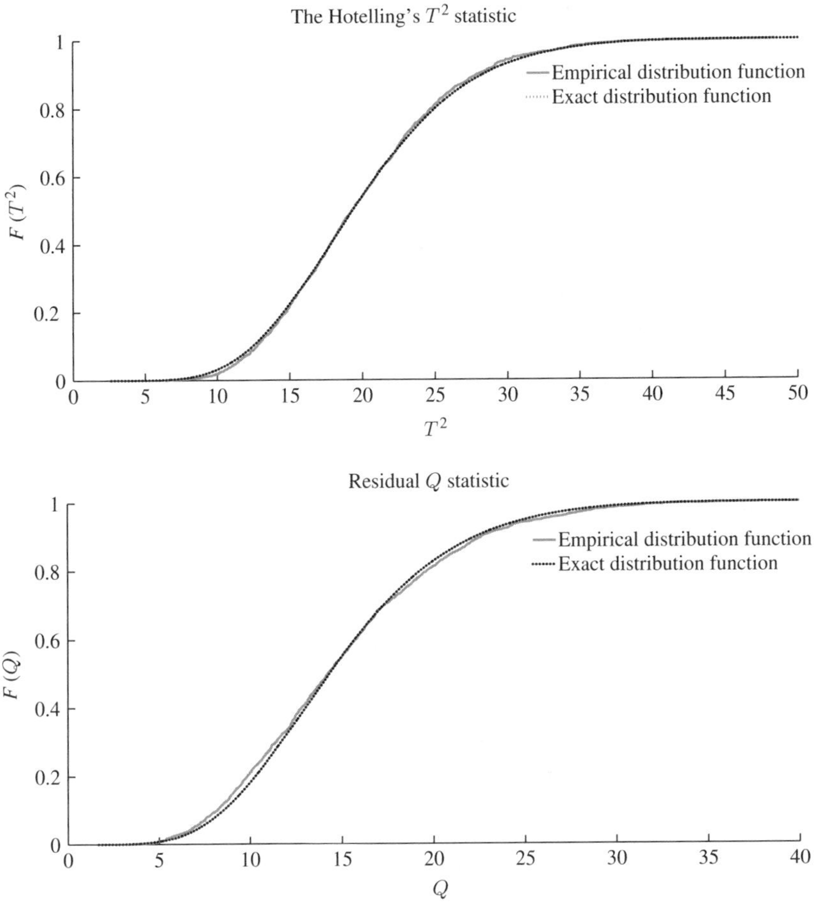

Figure 6.8 F-distribution (dotted line) and estimated distribution functions.

in Figure 6.7. This comparison yields a stronger response of both MLPCA-based non-negative squared monitoring statistics. In other words, the violation of the control limits, particularly by the MLPCA-Q statistic, is more pronounced. The inspection of Figure 4.17 highlights that the estimated fault signature for temperature variable #11 is not confined to the first third of the data set but instead spans over approximately two thirds of the recorded set. More precisely, the violation of the control limit by the MLPCA-based Q statistic corresponds more closely to the extracted fault signature.

In summary, the revised application study of the chemical reaction process outlined the advantage of MLPCA over PCA, namely a more accurate model estimation with respect to the data structure in (2.2). In contrast, the PCA model violated the assumption of $\mathbf{S}_{gg} = \sigma_g^2\mathbf{I}$. From the point of detecting the abnormal tube behavior, this translated into an increased sensitivity of both non-negative

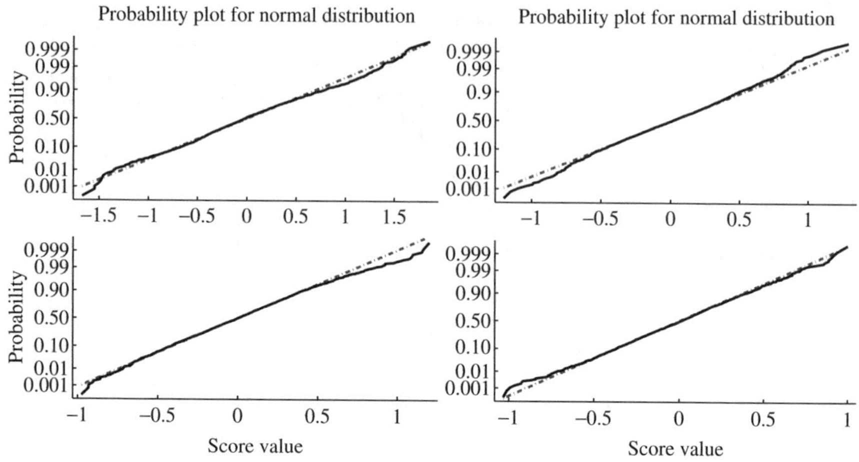

Figure 6.9 Comparison between Gaussian distribution (dashed line) and estimated distribution function for score variables 5 (upper left plot), 10 (upper right plot), 15 (lower left plot) and 20 (lower right plot).

quadratic monitoring statistics by comparing Figures 4.12 and 6.7. Despite the increased accuracy in estimating a data model for this process, the problem that the first four score variables do not follow a Gaussian distribution remains. Chapter 8 introduces a different construction of monitoring statistics that asymptotically follow a Gaussian distribution irrespective of the distribution function of the individual process variables and, therefore, addresses this remaining issue.

6.2 Accuracy of estimating PLS models

This section discusses the accuracy of estimating the weight and loading vectors as well as the regression matrix of PLS models. In this regard, the issue of high degrees of correlation among and between the input and output variable sets is revisited. Section 6.2.1 first summarizes the concept of bias and variance in estimating a set of unknown parameters. Using a simulation example, Subsection 6.2.2 then demonstrates that high correlation can yield a considerable variance of the parameter estimation when using OLS and outlines that PLS circumvents this large variance by including a reduced set of LVs in the regression model (Wold *et al.* 1984).

This, again, underlines the benefits of using MSPC methods in this context, which decompose the variation encapsulated in the highly correlated variable sets into source signals and error terms. For the identification of suitable models for model predictive control application, this is also an important issue. A number of research articles outline that PLS can outperform OLS and other multivariate regression techniques such as PCR and CCR (Dayal and MacGregor 1997b; Duchesne and MacGregor 2001) unless specific penalty terms are included in regularized least square (Dayal and MacGregor 1996) which, however, require prior knowledge of how to penalize changes in the lagged parameters of the input variables.

Finally, Subsection 6.2.3 shows how to obtain a consistent estimation of the LV sets and the parametric regression matrix if the data structure is assumed to be $\mathbf{y}_0 = \mathcal{B}^T \mathbf{x}_s + \mathfrak{f}$, whilst $\mathbf{x}_0 = \mathbf{x}_s + \mathfrak{e} = \mathfrak{Q}\mathbf{s} + \mathfrak{e}$ where $\mathfrak{e}$ is an error vector for the input variables.

6.2.1 Bias and variance of parameter estimation

According to (2.24) and (2.51), the number of source signals n must be smaller or equal to n_x. It is important to note, however, that if $n < n_x$ a unique ordinary least squares solution for estimating $\mathcal{B}$, $\widehat{\mathcal{B}} = \widehat{\mathbf{S}}_{x_0x_0}^{-1}\widehat{\mathbf{S}}_{x_0y_0}$, does not exist. More precisely, if $n < n_x$ the covariance matrix for the input variables is asymptotically ill conditioned and the linear equation $\mathbf{S}_{x_0x_0}\mathcal{B} = \mathbf{S}_{x_0y_0}$ yields an infinite number of solutions. On the other hand, if the condition number of the estimated covariance matrix $\widehat{\mathbf{S}}_{x_0x_0}$ is very large, the estimation variance of the elements in $\mathcal{B}$ can become very large too. This is now analyzed in more detail.

The OLS estimation is the best linear unbiased estimator if the error covariance matrix is of diagonal type $\mathbf{S}_{\mathfrak{f}\mathfrak{f}} = \mathrm{diag}\left\{\sigma_{\mathfrak{f}_1}^2 \quad \sigma_{\mathfrak{f}_2}^2 \quad \cdots \quad \sigma_{\mathfrak{f}_{n_y}}^2\right\}$

$$
\begin{aligned}
\widehat{\mathcal{B}} &= \widehat{\mathbf{S}}_{x_0x_0}^{-1}\widehat{\mathbf{S}}_{x_0y_0} = \left[\sum_{k=1}^{K} \mathbf{x}_0(k)\,\mathbf{x}_0(k)\right]^{-1}\left[\sum_{k=1}^{K} \mathbf{x}_0(k)\,\mathbf{y}_0(k)\right] \\
\widehat{\mathcal{B}} &= \left[\sum_{k=1}^{K} \mathbf{x}_0(k)\,\mathbf{x}_0^T(k)\right]^{-1}\left[\sum_{k=1}^{K} \mathbf{x}_0(k)\,\mathbf{x}_0^T(k)\,\mathcal{B} + \sum_{k=1}^{K} \mathbf{x}_0(k)\,\mathfrak{f}^T(k)\right] \\
E\left\{\widehat{\mathcal{B}}\right\} &= \mathcal{B} + \left[\sum_{k=1}^{K} \mathbf{x}_0(k)\,\mathbf{x}_0^T(k)\right]^{-1} E\left\{\sum_{k=1}^{K} \mathbf{x}_0(k)\,\mathfrak{f}^T(k)\right\} \qquad (6.82) \\
E\left\{\widehat{\mathcal{B}}\right\} &= \mathcal{B} + \left[\sum_{k=1}^{K} \mathbf{x}_0(k)\,\mathbf{x}_0^T(k)\right]^{-1}\left[\sum_{k=1}^{K} \mathbf{x}_0(k)\,E\left\{\mathfrak{f}^T(k)\right\}\right] \\
E\left\{\widehat{\mathcal{B}}\right\} &- \mathcal{B} = \mathbf{0}.
\end{aligned}
$$

It is important to note the data structures in (2.24) and (2.51) do not include any stochastic error terms for the input variables. Although the input and, therefore, the uncorrupted output variables are also assumed to follow multivariate Gaussian distributions, the K observations are assumed to be known. Hence, the only unknown stochastic element in the above relationship is $\mathfrak{f}(k)$, which has an expectation of zero. Hence the OLS solution is unbiased.

The next step is to examine the covariance matrix of the parameter estimation for each column vector of $\mathcal{B}$. For the ith column of $\mathcal{B}$, the corresponding covariance matrix can be constructed from $\widehat{\mathfrak{b}}_i - \mathfrak{b}_i = [\sum_{k=1}^{K} \mathbf{x}_0(k)\mathbf{x}_0^T k)]^{-1}[\sum_{k=1}^{K} \mathbf{x}_0(k)\mathfrak{f}_i(k)]$, which follows from (6.82)

$$E\left\{\left(\widehat{\mathfrak{b}}_i - \mathfrak{b}_i\right)\left(\widehat{\mathfrak{b}}_i - \mathfrak{b}_i\right)^T\right\} = \left[\sum_{k=1}^{K} \mathbf{x}_0(k)\,\mathbf{x}_0^T(k)\right]^{-1} E\left\{\left[\sum_{k=1}^{K} (k)\,\mathbf{x}_0(k)\,\mathfrak{f}_i(k)\right]\right.$$
$$\left.\left[\sum_{k=1}^{K} \mathfrak{f}_i(k)\,\mathbf{x}_0^T(k)\right]\right\}\left[\sum_{k=1}^{K} \mathbf{x}_0(k)\,\mathbf{x}_0^T(k)\right]^{-1} \quad (6.83)$$

which can be simplified to

$$E\left\{\left(\widehat{\mathfrak{b}}_i - \mathfrak{b}_i\right)\left(\widehat{\mathfrak{b}}_i - \mathfrak{b}_i\right)^T\right\} = \left[\sum_{k=1}^{K} \mathbf{x}_0(k)\,\mathbf{x}_0^T(k)\right]^{-1}$$
$$\left[\sum_{k=1}^{K} E\left\{\mathbf{x}_0(k)\,\mathfrak{f}_i(k)\sum_{k=1}^{K} \mathfrak{f}_i(k)\,\mathbf{x}_0^T(k)\right\}\right]$$
$$\left[\sum_{k=1}^{K} \mathbf{x}_0(k)\,\mathbf{x}_0^T(k)\right]^{-1}. \quad (6.84)$$

It follows from the Isserlis theorem (Isserlis 1918), that

$$E\left\{\mathbf{x}_0(k)\mathfrak{f}_i(k)\mathfrak{f}_i(l)\mathbf{x}_0^T(l)\right\} = \underbrace{E\left\{\mathbf{x}_0(k)\mathfrak{f}_i(k)\right\}E\left\{\mathfrak{f}_i(l)\mathbf{x}_0^T(l)\right\}}_{=\mathbf{0}}$$
$$+ \underbrace{E\left\{\mathbf{x}_0(k)\mathfrak{f}_k(l)\right\}E\left\{\mathfrak{f}_i(k)\mathbf{x}_0^T(l)\right\}}_{=\mathbf{0}} \quad (6.85)$$
$$+ \mathbf{x}_0(k)\mathbf{x}_0^T(l)\underbrace{E\left\{\mathfrak{f}_i(k)\mathfrak{f}_i(l)\right\}}_{=\mathbf{0}\text{ for all } k\neq l}.$$

Incorporating the fact that:

- $E\left\{\mathbf{x}_0(k)\mathfrak{f}_i(k)\mathfrak{f}_i(l)\mathbf{x}_0^T(l)\right\} = \mathbf{0}$ for all $k \neq l$; and
- $E\left\{\mathbf{x}_0(k)\mathfrak{f}_i(k)\mathfrak{f}_i(l)\mathbf{x}_0^T(l)\right\} = \mathbf{x}_0(k)\mathbf{x}_0^T(l)\sigma_{\mathfrak{f}_i}^2$ if $k = l$

allows simplifying (6.84) to become (Ljung 1999)

$$E\left\{\left(\widehat{\mathfrak{b}}_i - \mathfrak{b}_i\right)\left(\widehat{\mathfrak{b}}_i - \mathfrak{b}_i\right)^T\right\} = \sigma_{\mathfrak{f}_i}^2\left[\sum_{k=1}^{K} \mathbf{x}_0(k)\,\mathbf{x}_0^T(k)\right]^{-1}$$

$$\left[\sum_{k=1}^{K}\mathbf{x}_0(k)\,\mathbf{x}_0^T(k)\right]\left[\sum_{k=1}^{K}\mathbf{x}_0(k)\,\mathbf{x}_0^T(k)\right]^{-1}$$

$$E\left\{\left(\widehat{\mathfrak{b}}_i-\mathfrak{b}_i\right)\left(\widehat{\mathfrak{b}}_i-\mathfrak{b}_i\right)^T\right\}=\frac{\sigma_{\mathfrak{f}}^2}{K-1}\widehat{\mathbf{S}}_{x_0x_0}^{-1}. \tag{6.86}$$

That $E\left\{\mathfrak{f}_i(k)\mathfrak{f}_i(l)\right\}=\delta_{kl}\sigma_{\mathfrak{f}_i}^2$ follows from the assumption that the error variables are independently distributed and do not possess any serial- or autocorrelation. Furthermore, the error variables are statistically independent of the input variables. At first glance, it is important to note that a large sample size results in a small variance for the parameter estimation.

It is also important to note, however, that the condition number of the estimated covariance matrix $\widehat{\mathbf{S}}_{x_0x_0}$ has a significant impact upon the variance of the parameter estimation. To see this, using the eigendecomposition of $\widehat{\mathbf{S}}_{x_0x_0}=\mathcal{USU}^T$, its inverse becomes $\widehat{\mathbf{S}}_{x_0x_0}^{-1}=\mathcal{US}^{-1}\mathcal{U}^T$. If there is at least one eigenvalue that is close to zero, some of elements of the inverse matrix become very large, since $\mathfrak{s}_{n_x}^{-1}\mathfrak{u}_{n_x}\mathfrak{u}_{n_x}^T$ contains some large values which depend on the elements in n_xth eigenvector $\mathfrak{u}_{n_x}$.

With regards to the data structure in (2.24), PLS can provide an estimate of the parameter matrix that predicts the output variables $\mathbf{y}_0$ based on the t-score variables and hence circumvents the problem of a large estimation variance for determining the regression matrix $\mathcal{B}$ using OLS. This is now demonstrated using a simulation example.

6.2.2 Comparing accuracy of PLS and OLS regression models

The example includes one output variable and ten highly correlated input variables

$$\mathbf{x}_0=\mathbf{Ps}+\mathbf{P}'\mathbf{s}', \tag{6.87}$$

where $\mathbf{P}\in\mathbb{R}^{10\times4}$, $\mathbf{P}'\in\mathbb{R}^{10\times6}$, $\mathbf{s}\in\mathbb{R}^4$ and $\mathbf{s}'\in\mathbb{R}^6$. Furthermore, $\mathbf{s}$ and $\mathbf{s}'$ are statistically independent, i.i.d. and follow a multivariate Gaussian distribution with diagonal covariance matrices. The diagonal elements of $\mathbf{S}_{ss}$ and $\mathbf{S}_{s's'}$ are 1 and 0.075, respectively. The output variable is a linear combination of the ten input variables and corrupted by an error variable

$$y_0=\mathfrak{b}^T\mathbf{x}_0+\mathfrak{f}. \tag{6.88}$$

The elements of the parameter matrices $\mathbf{P}$ and $\mathbf{P}'$ as well as the parameter vector $\mathfrak{b}$, shown in (6.89a) to (6.89c), were randomly selected to be within $\begin{bmatrix}-1 & 1\end{bmatrix}$ from a uniform distribution. The variance of the error term was $\sigma_{\mathfrak{f}}=0.2$. It should be noted that the data structure in this example is different from that in

(2.51), as both types of source signals influence the output variables.

$$\mathbf{P} = \begin{bmatrix} -0.047 & -0.125 & -0.593 & -0.061 \\ 0.824 & -0.440 & 0.089 & -0.173 \\ -0.970 & 0.971 & 0.750 & 0.006 \\ -0.687 & 0.218 & -0.758 & -0.749 \\ -0.057 & -0.493 & 0.713 & -0.735 \\ 0.086 & -0.735 & 0.800 & 0.741 \\ -0.881 & 0.090 & -0.564 & 0.206 \\ 0.316 & 0.656 & -0.846 & -0.469 \\ 0.779 & 0.674 & -0.052 & 0.730 \\ -0.781 & 0.667 & 0.670 & -0.884 \end{bmatrix} \tag{6.89a}$$

$$\mathbf{P}' = \begin{bmatrix} -0.042 & 0.365 & 0.400 & -0.432 & 0.375 & 0.129 \\ 0.222 & -0.080 & -0.435 & -0.415 & -0.151 & 0.378 \\ -0.161 & -0.260 & -0.164 & -0.432 & -0.458 & 0.162 \\ -0.099 & 0.098 & -0.496 & -0.090 & -0.358 & 0.375 \\ 0.027 & -0.021 & 0.328 & -0.377 & -0.423 & -0.033 \\ 0.394 & 0.399 & 0.007 & -0.057 & 0.241 & -0.359 \\ 0.278 & 0.435 & -0.134 & 0.399 & -0.044 & -0.432 \\ -0.431 & 0.318 & -0.273 & -0.146 & 0.168 & 0.214 \\ -0.221 & 0.209 & 0.035 & -0.380 & 0.199 & -0.192 \\ -0.121 & 0.243 & -0.211 & 0.069 & 0.071 & 0.171 \end{bmatrix} \tag{6.89b}$$

$$\mathfrak{b} = \begin{pmatrix} -0.427 \\ 0.952 \\ -0.053 \\ 0.586 \\ 0.355 \\ 0.779 \\ -0.182 \\ 0.991 \\ 0.322 \\ 0.808 \end{pmatrix}. \tag{6.89c}$$

With respect to (6.87) to (6.89c), the covariance matrix of $\mathbf{x}_0$ is

$$\mathbf{S}_{x_0x_0} = \mathbf{P}\mathbf{P}^T + \sigma_{s'}^2\mathbf{P}'\mathbf{P}'^T \tag{6.90}$$

$$= \begin{bmatrix} 0.420 & -0.029 & -0.529 & 0.485 & -0.304 & \cdots & -0.387 \\ -0.029 & 0.953 & -1.133 & -0.568 & 0.365 & \cdots & -0.719 \\ -0.529 & -1.133 & 2.486 & 0.330 & 0.130 & \cdots & 1.899 \\ 0.485 & -0.568 & 0.330 & 1.695 & -0.057 & \cdots & 0.848 \\ -0.304 & 0.365 & 0.130 & -0.057 & 1.327 & \cdots & 0.833 \\ \vdots & \vdots & \vdots & \vdots & \vdots & \ddots & \vdots \\ -0.387 & -0.719 & 1.899 & 0.848 & 0.833 & \cdots & 2.296 \end{bmatrix}.$$

Equation (6.86) shows that the variance of the parameter estimation for the OLS solution is proportional to $\sigma_{\mathfrak{f}}^2/K-1$ but also depends on the estimated covariance matrix. With respect to the true covariance matrix in (6.90), it is possible to approximate the covariance matrix for the parameter estimation using OLS

$$E\left\{\left(\widehat{\mathfrak{b}}-\mathfrak{b}\right)\left(\widehat{\mathfrak{b}}-\mathfrak{b}\right)^T\right\} \approx \frac{\sigma_{\mathfrak{f}}^2}{K-1}\mathbf{S}_{x_0x_0}^{-1}. \tag{6.91}$$

As discussed in the previous subsection, the examination of the impact of $\mathbf{S}_{x_0x_0}^{-1}$ relies on its eigendecomposition

$$E\left\{\left(\widehat{\mathfrak{b}}-\mathfrak{b}\right)\left(\widehat{\mathfrak{b}}-\mathfrak{b}\right)^T\right\} \approx \frac{\sigma_{\mathfrak{f}}^2}{K-1}\left[\sum_{i=1}^{10}\frac{\mathfrak{u}_i\mathfrak{u}_i^T}{\mathfrak{s}_i}\right]. \tag{6.92}$$

Given that the eigenvalues of $\mathbf{S}_{x_0x_0}$ are

$$\mathbf{\Lambda} = \mathrm{diag}\left\{\begin{pmatrix} 6.058308 \\ 4.393462 \\ 2.644292 \\ 1.944088 \\ 0.082358 \\ 0.062873 \\ 0.054620 \\ 0.020089 \\ 0.003175 \\ 0.000021 \end{pmatrix}\right\} \tag{6.93}$$

the condition number of $\mathbf{S}_{x_0x_0}$ is 2.9066×10^5, which highlights that this matrix is indeed badly conditioned. On the basis of (6.92), Figure 6.10 shows the approximated variances for estimating the ten parameters, that is, the diagonal elements of $E\{(\widehat{\mathfrak{b}}-\mathfrak{b})(\widehat{\mathfrak{b}}-\mathfrak{b})^T\}$. The largest curves in Figure 6.10 are those for parameters $\mathfrak{b}_8$, $\mathfrak{b}_4$, $\mathfrak{b}_9$, $\mathfrak{b}_3$ (from largest to smallest). The remaining curves represent smaller but still significant variances for $\mathfrak{b}_1$, $\mathfrak{b}_2$, $\mathfrak{b}_5$, $\mathfrak{b}_6$, $\mathfrak{b}_7$ and $\mathfrak{b}_{10}$. Even for a sample size of $K = 1000$, variances of the parameter estimation in the region of five can arise. The impact of such a large variance for the parameter estimation is now demonstrated using a Monte Carlo experiment.

The experiment includes a sample size of $K = 200$ and a total number of 1000 repetitions. The comparison here is based on the parameter estimation using OLS and the estimation of latent variable sets using PLS. For each of these sets, the application of OLS and PLS produced estimates of the regression parameters and estimates of sets of LVs, respectively. Analyzing the 1000 estimated parameter sets for OLS and PLS then allow determining histograms of individual values for each parameter set, for example the OLS regression coefficients.

Figure 6.11 shows histograms for each of the ten regression parameters obtained using OLS. In each plot, the abscissa relates to the value of the estimated parameter and the ordinate shows the relative frequency of a particular

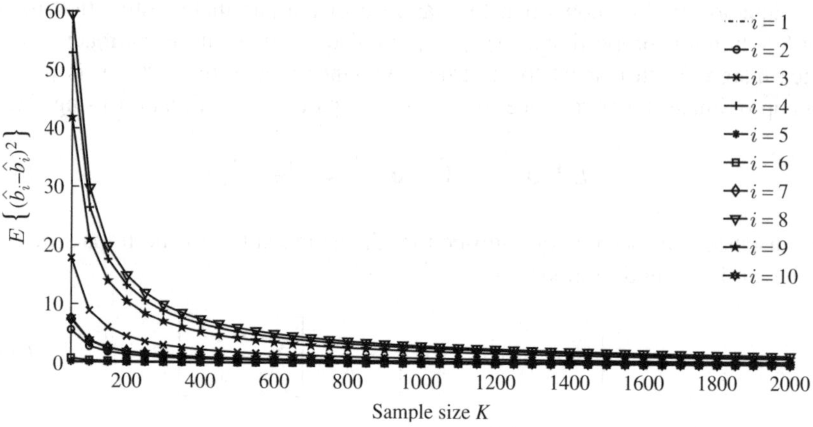

Figure 6.10 Variance of parameter estimation (OLS model) for various sample sizes.

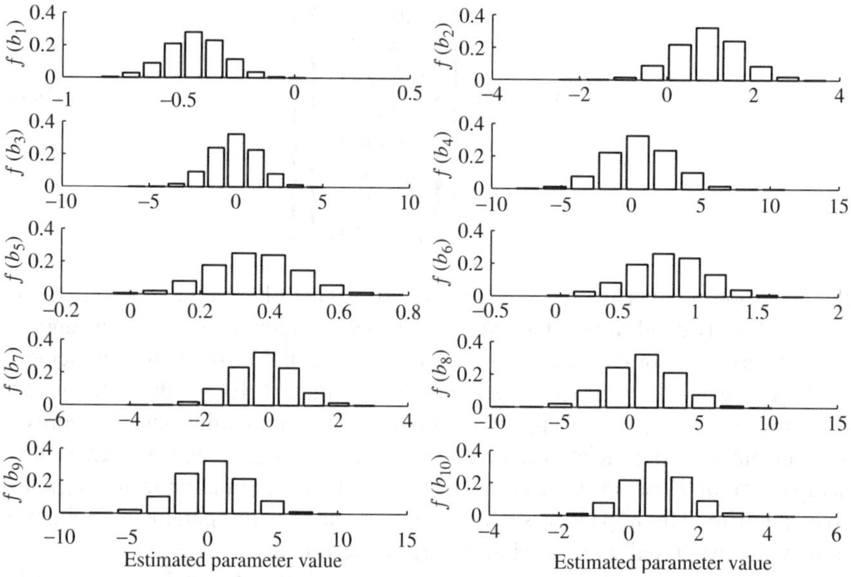

Figure 6.11 Histograms for parameter estimation of regression coefficients (OLS).

parameter value. According to Figure 6.10, for $K = 200$, the largest estimation variance is in the region of 16 for the eighth parameter.

It follows from the central limit theorem that the parameter estimation follows approximately a Gaussian distribution with the mean value being the true parameter vector (unbiased estimation) and the covariance matrix given in (6.86). With this in mind, the estimated variance of 16 for the eighth parameter implies

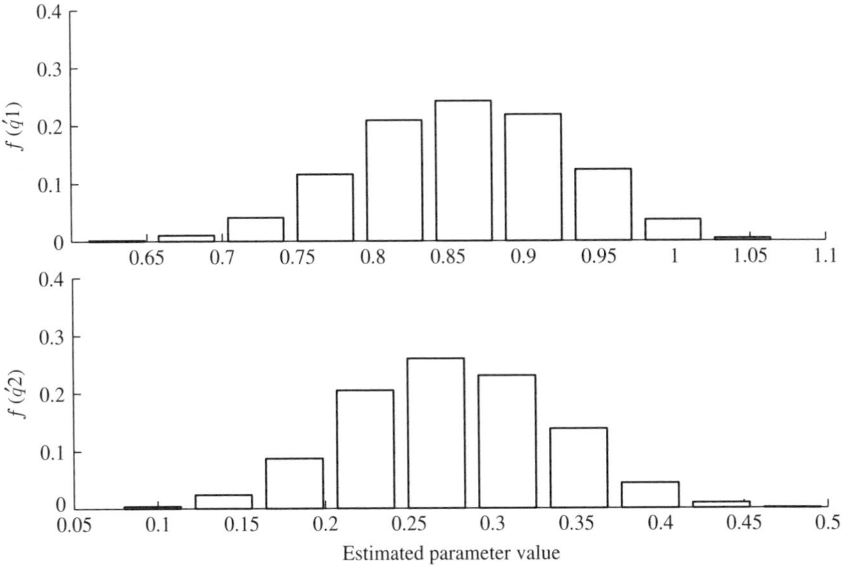

Figure 6.12 Histograms for parameter estimation of q́-loading coefficients (PLS).

that around 68% of estimated parameters for $\mathfrak{b}_8$ are within the range 0.991 ± 4 and around 95% of estimated parameters fall in the range of 0.991 ± 8, which Figure 6.10 confirms.

The Monte Carlo simulation also shows larger variances for the parameter estimation for $\mathfrak{b}_3$, $\mathfrak{b}_4$ and $\mathfrak{b}_9$. The ranges for estimating the remaining parameters, however, are still significant. For example, the smallest range is for estimating parameter $\mathfrak{b}_1$, which is bounded roughly by $\begin{bmatrix}-0.9 & 0\end{bmatrix}$. The above analysis therefore illustrates that the values of the parameter estimation can vary substantially and strongly depend on the recorded samples. Höskuldsson (1988) pointed out that PLS is to be preferred over OLS as it produces a *more stable estimation* of the regression parameters in the presence of highly correlated input variables. This is examined next.

In contrast to OLS, PLS regression relates to an estimated parametric model between the extracted t-score and the output variables, $y_0 = \widehat{\mathbf{q}}^T \widehat{\mathfrak{t}} + f$. Figure 6.12, plotting the histograms for estimating the parameters of the first two q-loading values, does not show large variances for the parameter estimation. More precisely, the computed variance for the 1000 estimates of $\acute{q}_1$ and $\acute{q}_2$ are 0.0049 and 0.0038, respectively. Based on the original covariance matrix, constructed from the covariance matrix in (6.90) and $\mathbf{s}_{x_0 y_0} = \mathbf{P}\mathbf{P}^T \mathfrak{b} + \sigma_{s'}^2 \mathbf{P}' \mathbf{P}'^T \mathfrak{b}$, the mean values for $\acute{q}_1$ and $\acute{q}_2$ are 0.8580 and 0.2761, respectively. The estimation variance for $\acute{q}_1$ and $\acute{q}_2$, therefore, compares favorably to the large estimation variances for $\widehat{\mathfrak{b}}$, produced by applying OLS.

The small estimation variance for the first and second ǵ-loading value, however, does not take into consideration the computation of the t-score variables. According to Lemma 10.4.7, the t-score variables can be obtained by the scalar product of the r-loading and the input variables, i.e. $t_i = \mathbf{r}_i^T \mathbf{x}_0$. For the first two r-loading vectors, Figure 6.13, again, suggests a small variance for each of the elements in $\mathbf{r}_1$ and $\mathbf{r}_2$, Table 6.12 lists the estimated mean and variance for each element of the two vectors. The largest variance is 0.0140 for element r_{52}.

Equation (10.60) shows that the PLS estimate for the parameter matrix $\mathcal{B}$ is $\widehat{\mathcal{B}} = \widehat{\mathbf{R}}\widehat{\mathbf{Q}}^T$. Retaining the first two latent variable pairs, Figure 6.14 shows the histograms of the elements of the PLS regression vectors. In contrast to the histograms of the loading and weight vectors in Figures 6.12 and 6.13, respectively, the histograms in Figure 6.14 can be directly compared to those in Figure 6.11. As expected, the variance of the parameter estimation for models obtained using PLS is significantly smaller compared to those computed by OLS. This is a result of the relatively small variance of the parameter estimation for the latent loading and weight vectors.

A more qualitative analysis is offered in Table 6.13. It is interesting to note that the estimation using OLS is more accurate for parameter $\mathfrak{b}_6$, which follows from the fact that the confidence region of this parameter, obtained from the estimates in Table 6.13 for a significance of 0.05, is $\bar{\mathfrak{b}}_{ols_6} \pm 1.96\widehat{\sigma}_{\mathfrak{b}_{ols_6}} = 0.7714 \pm 0.5324$[6] for OLS, which compares to -0.2774 ± 0.0297 for the PLS models. In the worst case, the estimated parameter for OLS becomes 0.2390 for OLS, whilst the closest estimate for PLS is -0.2477. For almost all other parameters, the confidence regions for the estimated parameters using OLS include those of the PLS estimates. This is known as the bias/variance tradeoff, where an estimation bias by the regression tool is accepted in favor of a significantly reduced estimation variance.

The application of PLS in this Monte Carlo study yielded, therefore, a better estimation of the parameters apart from $\mathfrak{b}_6$ with the retention of just two sets of latent variables compared to the application of OLS. Given that the PLS regression model is identical to the OLS regression model if all of the n_x LV sets are included, increasing $\mathfrak{n}$, consequently, reduces the estimation bias. On the other hand, the increase in $\mathfrak{n}$ significantly increases the variance of the parameter estimation.

Whilst the retention of only two latent variable sets yielded a biased parameter estimation that resulted in a substantial reduction in the variance of the estimation and hence, a more accurate parameter estimation, the final question is how accurate is the prediction of this model. Using the variance of the error term over the variance of the output variable as a measure of accuracy

$$\frac{E\left\{\mathfrak{f}^2\right\}}{E\left\{y_0^2\right\}} = \frac{\sigma_{\mathfrak{f}}^2}{\mathfrak{b}^T\left[\mathbf{P}\mathbf{P}^T + \mathbf{P}'\mathbf{S}_{s's'}\mathbf{P}'^T\right]\mathfrak{b} + \sigma_{\mathfrak{f}}^2} = 0.0744. \tag{6.94}$$

[6] Assumed here to be 0.7714, whilst the true $\mathfrak{b}_6$ is 0.779 according to (6.89c).

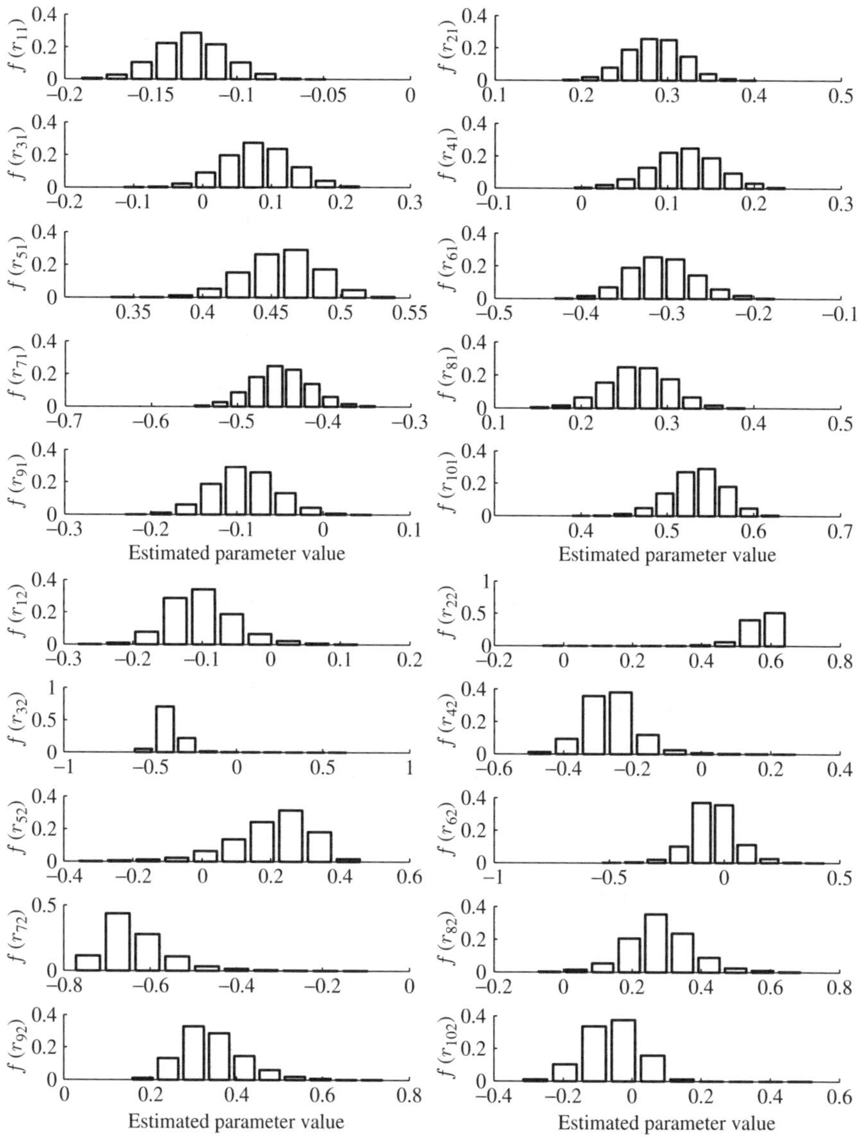

Figure 6.13 Histograms for parameter estimation of r-weight coefficients (PLS).

Figure 6.15 indicates that including just one set of latent variables, the estimated mean value of the statistic in (6.94) for the PLS regression models is 0.2528 and 0.1289 if $n = 2$. For the retention of further latent variable sets, the estimated mean for this statistic becomes 0.1205, 0.1127, 0.0742, 0.0736, 0.0733, 0.0732, 0.0731, 0.0730. As analyzed above, however, an increase in n will reduce the accuracy of the parameter estimation, whilst increasing the

Table 6.12 Mean and variance for estimating $\mathbf{r}_1$ and $\mathbf{r}_2$.

Element r_{ij}	$\bar{r}_{ij} = E\{r_{ij}\}$	$\sigma^2_{r_{ij}} = E\left\{\left(r_{ij} - \bar{r}_{ij}\right)^2\right\}$
r_{11}	−0.1261	0.0004
r_{21}	0.2859	0.0010
r_{31}	0.0782	0.0022
r_{41}	0.1196	0.0015
r_{51}	0.4576	0.0007
r_{61}	−0.3060	0.0014
r_{71}	−0.4498	0.0011
r_{81}	0.2664	0.0014
r_{91}	−0.0913	0.0015
r_{101}	0.5341	0.0010
r_{12}	−0.1001	0.0022
r_{22}	0.5645	0.0023
r_{32}	−0.3861	0.0043
r_{42}	−0.2651	0.0058
r_{52}	0.2000	0.0140
r_{62}	−0.0472	0.0094
r_{72}	−0.6329	0.0060
r_{82}	0.2816	0.0090
r_{92}	0.3470	0.0060
r_{102}	−0.0600	0.0059

Table 6.13 Comparing accuracy of OLS and PLS regression models.

Parameter $\mathfrak{b}_i$	$\widehat{\mathfrak{b}}_{pls_i}$	$\widehat{\mathfrak{b}}_{ols_i}$	$\widehat{\sigma}^2_{\mathfrak{b}_{pls_i}}$	$\widehat{\sigma}^2_{\mathfrak{b}_{ols_i}}$	$\widehat{\mathfrak{b}}_{ols_i} - \widehat{\mathfrak{b}}_{pls_i}$
$\mathfrak{b}_1$	−0.1367	−0.4255	0.000144	0.0145	−0.2888
$\mathfrak{b}_2$	0.4046	0.9467	0.000154	0.5678	0.5421
$\mathfrak{b}_3$	−0.0404	−0.0704	0.000398	1.8401	−0.0300
$\mathfrak{b}_4$	0.0298	0.6135	0.000200	5.4602	0.5837
$\mathfrak{b}_5$	0.4540	0.3540	0.000184	0.0166	−0.1000
$\mathfrak{b}_6$	−0.2774	0.7714	0.000229	0.0738	1.0488
$\mathfrak{b}_7$	−0.5659	−0.1898	0.000192	0.7608	0.3761
$\mathfrak{b}_8$	0.3070	0.9552	0.000299	6.1609	0.6482
$\mathfrak{b}_9$	0.0148	0.3501	0.000538	4.3249	0.3353
$\mathfrak{b}_{10}$	0.4453	0.8208	0.000224	0.7859	0.3755

predictive performance of the resulting regression model for the reference data. A further study of this example is encouraged in the tutorial session of this chapter (Project 3).

6.2.3 Impact of error-in-variables structure upon PLS models

After outlining the benefits of utilizing PLS as a technique for determining regression parameters in the presence of highly correlated input variables, we

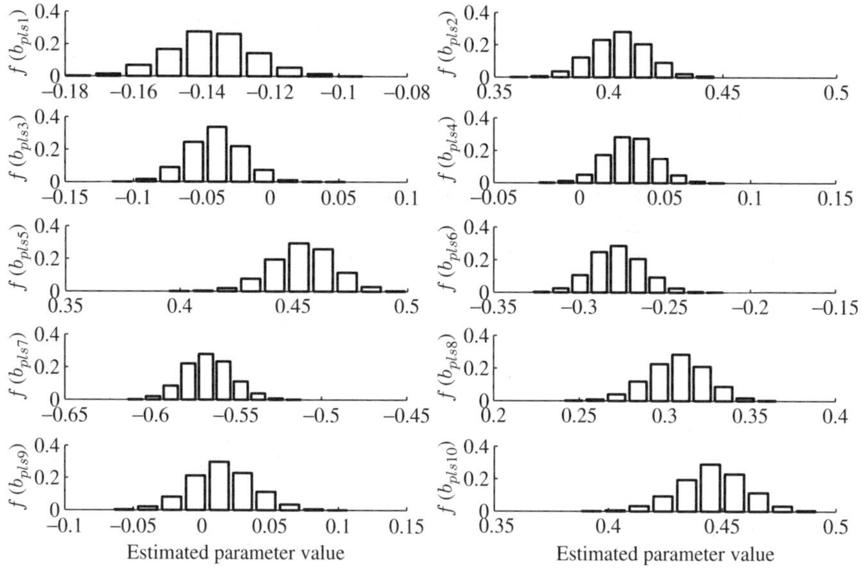

Figure 6.14 Histograms for parameter estimation of regression coefficients (PLS).

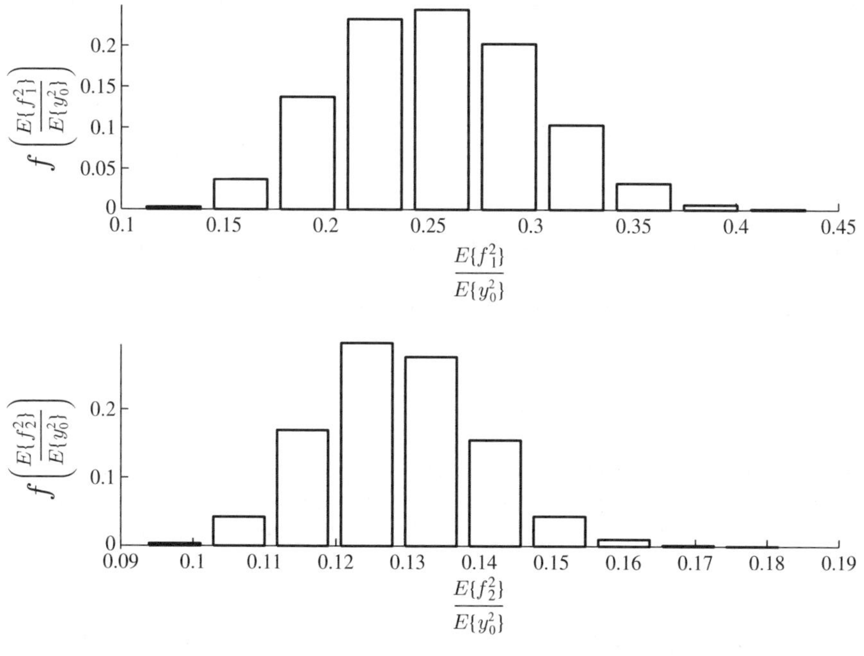

Figure 6.15 Histograms of accuracy of PLS models for retaining $\mathfrak{n} = 1, 2$.

now focus on the accuracy of estimating the latent variable sets if the input variables are also corrupted by an error vector. Recall that the data structures in (2.24) and (2.51) only include the error term $\mathfrak{f}$ for the output variables. Such extended data structures are often referred to as error-in-variable or EIV structures (Söderström 2007).

Error-in-variable data structure for PLS models. Equations (6.95) and (6.96) introduce the extended EIV data structures for PLS and MRPLS models, respectively, that include the error vector $\mathfrak{e}$ for the input variables

$$\begin{aligned} \mathbf{y} &= \mathbf{y}_0 + \bar{\mathbf{y}} & \mathbf{y}_0 &= \mathbf{y}_s + \mathfrak{f} & \mathbf{y}_s &= \mathfrak{Q}\mathbf{s} \\ \mathbf{x} &= \mathbf{x_0} + \bar{\mathbf{x}} & \mathbf{x}_0 &= \mathbf{x}_s + \mathfrak{e} & \mathbf{x}_s &= \mathfrak{P}\mathbf{s} \end{aligned} \tag{6.95}$$

$$\begin{aligned} \mathbf{y} &= \mathbf{y}_0 + \bar{\mathbf{y}} & \mathbf{y}_0 &= \mathbf{y}_s + \mathfrak{f} & \mathbf{y}_s &= \mathfrak{Q}\mathbf{s} \\ \mathbf{x} &= \mathbf{x_0} + \bar{\mathbf{x}} & \mathbf{x}_0 &= \mathbf{x}_s + \mathfrak{e} & \mathbf{x}_s &= \mathfrak{P}\mathbf{s} + \mathfrak{P}'\mathbf{s}'. \end{aligned} \tag{6.96}$$

The following assumptions are imposed on $\mathbf{s} \in \mathbb{R}^n$, $\mathbf{s}' \in \mathbb{R}^m$, $\mathfrak{f} \in \mathbb{R}^{n_y}$ and $\mathfrak{e} \in \mathbb{R}^{n_x}$

$$\mathbf{s} \sim \mathcal{N}\left\{\mathbf{0}, \mathbf{S}_{ss}\right\} \quad \mathbf{s} \sim \mathcal{N}\left\{\mathbf{0}, \mathbf{S}_{s's'}\right\} \quad \mathfrak{f} \sim \mathcal{N}\left\{\mathbf{0}, \mathbf{S}_{\mathfrak{ff}}\right\} \quad \mathfrak{e} \sim \mathcal{N}\left\{\mathbf{0}, \mathbf{S}_{\mathfrak{ee}}\right\}. \tag{6.97}$$

Defining k and l as sampling indices, the joint covariance matrix is assumed to be

$$E\left\{\begin{pmatrix} \mathbf{s}(k) \\ \mathbf{s}'(k) \\ \mathfrak{f}(k) \\ \mathfrak{e}(k) \end{pmatrix} \begin{pmatrix} \mathbf{s}^T(l) & \mathbf{s}'^T(l) & \mathfrak{f}^T(l) & \mathfrak{e}^T(l) \end{pmatrix}\right\} = \delta_{kl} \begin{bmatrix} \mathbf{S}_{ss} & \mathbf{0} & \mathbf{0} & \mathbf{0} \\ \mathbf{0} & \mathbf{S}_{s's'} & \mathbf{0} & \mathbf{0} \\ \mathbf{0} & \mathbf{0} & \mathbf{S}_{\mathfrak{ff}} & \mathbf{0} \\ \mathbf{0} & \mathbf{0} & \mathbf{0} & \mathbf{S}_{\mathfrak{ee}} \end{bmatrix}. \tag{6.98}$$

The data structure in (6.95) does not include $\mathbf{s}'$. In this case, the covariance matrix of the joint variable sets only includes the stochastic vectors $\mathbf{s}$, $\mathfrak{f}$ and $\mathfrak{e}$. Moreover, the following linear parametric relationship exists between the $\mathbf{y}_s$ and $\mathbf{x}_s$

$$\mathbf{y}_s = \mathcal{B}^T\mathbf{x}_s \qquad \mathcal{B} = \mathbf{S}_{x_s x_s}^{-1}\mathbf{S}_{x_s y_s}. \tag{6.99}$$

The next few pages examine the impact of $\mathfrak{e}$ upon the computation of the LVs, commencing with the influence upon the covariance and cross-covariance matrices. Subsections 6.2.4 and 6.2.5 then discuss how to remove this undesired impact.

Impact upon $\mathbf{S}_{x_0x_0}$ and $\mathbf{S}_{x_0y_0}$. The examination of the impact of $\mathfrak{e}$ upon the accuracy of the weight and loading vectors requires studying the impact of $\mathfrak{e}$ upon the covariance matrix $\mathbf{S}_{x_0x_0}$ and the cross-covariance matrix $\mathbf{S}_{x_0y_0}$ first. According to the data structures in (6.95) and (6.96), the covariance and cross-covariance matrices are given by

$$\mathbf{S}_{x_0x_0} = \mathbf{P}\mathbf{S}_{ss}\mathbf{P}^T + \mathbf{P}'^T\mathbf{S}_{s's'}\mathbf{P}'^T + \mathbf{S}_{\mathfrak{ee}} \qquad \mathbf{S}_{x_0y_0} = \mathbf{P}\mathbf{S}_{ss}\mathbf{Q}^T. \tag{6.100}$$

With PLS being an iterative algorithm, the analysis commences with the first set of weight and loading vectors. It is important to note that a linear parametric relationship between $\mathbf{x}_s$ and $\mathbf{y}_s$ can be established irrespective of whether $\mathbf{x}_s = \mathbf{P}\mathbf{s}$ or $\mathbf{x}_s = \mathbf{P}\mathbf{s} + \mathbf{P}'\mathbf{s}'$, provided that $\mathbf{S}_{x_sx_s}$ has full rank, which (6.99) shows.

Impact on first set of weight vectors. This set of weight vectors is the solution to the following objective function

$$\begin{pmatrix}\mathbf{w}_1\\ \mathbf{q}_1\end{pmatrix} = \arg\max_{\mathbf{w},\mathbf{q}} \mathbf{w}^T\mathbf{S}_{x_0y_0}\mathbf{q} - \tfrac{1}{2}\lambda\left(\mathbf{w}^T\mathbf{w} - 1\right) - \tfrac{1}{2}\lambda\left(\mathbf{q}^T\mathbf{q} - 1\right), \tag{6.101}$$

which confirms that the inclusion of $\mathfrak{e}$ does not affect the first set of weight vectors.

Impact on first set of loading vectors. Equation (6.102) shows the calculation of the first pair of loading vectors

$$\begin{pmatrix}\mathbf{p}_1\\ \acute{\mathbf{q}}_1\end{pmatrix} = \arg\min_{\mathbf{p},\acute{\mathbf{q}}} \begin{pmatrix}\mathbf{p}^T & \acute{\mathbf{q}}^T\end{pmatrix}\begin{bmatrix}\mathbf{S}_{x_0x_0}\\ \mathbf{S}_{y_0x_0}\end{bmatrix}\mathbf{w}_1 - \tfrac{1}{2}\left(\mathbf{p}^T\mathbf{p} + \acute{\mathbf{q}}^T\acute{\mathbf{q}}\right)\mathbf{w}_1^T\mathbf{S}_{x_0x_0}\mathbf{w}_1, \tag{6.102}$$

which directly follows from (10.12) in Subsection 10.2. Compared to the analysis for the weight vector, however, a different picture emerges when analyzing the objective function in (6.102), since

$$\mathbf{p}_1 = \frac{\left[\mathbf{S}_{x_sx_s} + \mathbf{S}_{\mathfrak{ee}}\right]\mathbf{w}_1}{\mathbf{w}_1^T\left[\mathbf{S}_{x_sx_s} + \mathbf{S}_{\mathfrak{ee}}\right]\mathbf{w}_1} \qquad \acute{\mathbf{q}}_1 = \frac{\mathbf{S}_{y_0x_0}\mathbf{w}_1}{\mathbf{w}_1^T\left[\mathbf{S}_{x_sx_s} + \mathbf{S}_{\mathfrak{ee}}\right]\mathbf{w}_1}, \tag{6.103}$$

where $\mathbf{S}_{x_sx_s}$ is the covariance matrix of the source signals, that is, without the inclusion of $\mathfrak{e}$. Without the presence of this term, the first pair of loading vectors are equal to

$$\mathbf{p}_1^* = \frac{\mathbf{S}_{x_sx_s}\mathbf{w}_1}{\mathbf{w}_1^T\mathbf{S}_{x_sx_s}\mathbf{w}_1} \qquad \acute{\mathbf{q}}_1^* = \frac{\mathbf{S}_{y_0x_0}\mathbf{w}_1}{\mathbf{w}_1^T\mathbf{S}_{x_sx_s}\mathbf{w}_1}. \tag{6.104}$$

Here, the superscript * refers to the loading vectors determined from $\mathbf{S}_{x_sx_s}$. The difference between the two pairs of loading vectors is therefore

$$\Delta\mathbf{p}_1 = \frac{\gamma_{\mathfrak{e}}}{1 + \gamma_{\mathfrak{e}}}\left(\mathbf{p}_1^* - \frac{\mathbf{S}_{\mathfrak{ee}}\mathbf{w}_1}{\mathbf{w}_1^T\mathbf{S}_{\mathfrak{ee}}\mathbf{w}_1}\right) \tag{6.105}$$

and

$$\Delta \acute{\mathbf{q}}_1 = \acute{\mathbf{q}}_1^* - \acute{\mathbf{q}}_1 = \frac{\gamma_\mathfrak{e}}{\left(1+\gamma_\mathfrak{e}\right)\mathbf{w}_1^T\mathbf{S}_{\mathfrak{e}\mathfrak{e}}\mathbf{w}_1}\mathbf{q}_1^* \tag{6.106}$$

where $\gamma_\mathfrak{e} = \frac{\mathbf{w}_1^T\mathbf{S}_{\mathfrak{e}\mathfrak{e}}\mathbf{w}_1}{\mathbf{w}_1^T\mathbf{S}_{x_sx_s}\mathbf{w}_1} > 0$, since both covariance matrices are symmetric and positive definite. Equations (6.103) to (6.106) highlight that:

1. the direction and length of the p-loading vectors is affected;
2. the scalar product of $\mathbf{w}_1^T\Delta\mathbf{p}_1 = 0$;
3. the direction of the q́-loading vector remains unchanged; and
4. the length of the q́-loading vector reduces

by the presence of $\mathfrak{e}$. The reduction in length of the q́-loading vector follows from the fact that $\mathbf{S}_{\mathfrak{e}\mathfrak{e}}$ is a symmetric and positive definite matrix of rank n_x. Moreover, the scalar product $\mathbf{w}_i^T\mathbf{p}_i = 1$, which follows from Lemma 10.4.10 irrespective of whether the input variables are corrupted by the noise term or not. Consequently, the scalar product $\mathbf{w}_1^T\Delta\mathbf{p}_1 = 0$. In other words, $\mathbf{w}_1^T\left(\mathbf{p}_1^* - \mathbf{p}_1\right) = 0$.

Impact upon deflation of $\mathbf{S}_{x_0y_0}$. Using (6.103) and Theorem 10.4.6 shows that the deflation of the cross-covariance matrix can be expressed as follows

$$\mathbf{S}_{ef}^{(2)} = \left[\mathbf{I} - \mathbf{S}_{x_0x_0}\frac{\mathbf{w}_1\mathbf{w}_1^T}{\mathbf{w}_1^T\mathbf{S}_{x_0x_0}\mathbf{w}_1}\right]\mathbf{S}_{x_0y_0}. \tag{6.107}$$

Given that the deflation of $\mathbf{S}_{x_0y_0}$ using the uncorrupted input variables is equal to

$$\mathbf{S}_{ef}^{(2)^*} = \left[\mathbf{I} - \mathbf{S}_{x_sx_s}\frac{\mathbf{w}_1\mathbf{w}_1^T}{\mathbf{w}_1^T\mathbf{S}_{x_sx_s}\mathbf{w}_1}\right]\mathbf{S}_{x_0y_0}, \tag{6.108}$$

the difference between $\mathbf{S}_{ef}^{(2)^*}$ and $\mathbf{S}_{ef}^{(2)}$ becomes

$$\Delta\mathbf{S}_{ef}^{(2)} = \left[\frac{\mathbf{S}_{\mathfrak{e}\mathfrak{e}} - \gamma_\mathfrak{e}\mathbf{S}_{x_sx_s}}{1+\gamma_\mathfrak{e}}\right]\mathbf{w}_1\mathbf{q}_1^{*^T}. \tag{6.109}$$

Impact on subsequent pairs of weight and loading vectors. After deflating the cross-covariance matrix, $\mathbf{w}_2$, $\mathbf{q}_2$, $\mathbf{p}_2$ and $\acute{\mathbf{q}}_2$ can be computed. Different from $\mathbf{w}_1$ and $\mathbf{q}_1$, the computation of $\mathbf{w}_2$ and $\mathbf{q}_2$ is affected by $\mathfrak{e}$, as they are the dominant left and right singular vectors of $\mathbf{S}_{ef}^{(2)}$ (Kaspar and Ray 1993), which follows from (6.109). In summary, each of the subsequent sets of LVs differs in the presence of the additional error term.

Impact upon regression model. Theorem 10.4.15 highlights that the identified parameter matrix is equal to the OLS estimate if all n_x LV sets are included. The asymptotic OLS estimate is given by

$$\mathcal{B} = \mathbf{S}_{x_0x_0}^{-1}\mathbf{S}_{x_0y_0} \tag{6.110}$$

and for uncorrupted input variables

$$\mathcal{B}^* = \mathbf{S}_{x_sx_s}^{-1}\mathbf{S}_{x_0y_0}. \tag{6.111}$$

The estimation bias is therefore

$$\begin{aligned}\Delta\mathcal{B} &= \mathcal{B}^* - \mathcal{B}\\ \Delta\mathcal{B} &= \left[\mathbf{S}_{x_sx_s}^{-1} - \left[\mathbf{S}_{x_sx_s} + \mathbf{S}_{ee}\right]^{-1}\right]\mathbf{S}_{x_0y_0}\\ \Delta\mathcal{B} &= \mathbf{S}_{x_sx_s}^{-1}\left[\mathbf{S}_{x_sx_s}^{-1} + \mathbf{S}_{ee}^{-1}\right]^{-1}\mathcal{B}^*.\end{aligned} \tag{6.112}$$

The above relationship relies on the application of the matrix-inversion lemma, i.e. $\left[\mathbf{S}_{x_sx_s} + \mathbf{S}_{ee}\right]^{-1} = \mathbf{S}_{x_sx_s}^{-1} - \mathbf{S}_{x_sx_s}^{-1}\left[\mathbf{S}_{x_sx_s}^{-1} - \mathbf{S}_{ee}^{-1}\right]^{-1}\mathbf{S}_{x_sx_s}^{-1}$.

The analysis in (6.102) to (6.109) also applies for the MRPLS. However, the MRPLS cost function for determining the weight vectors is equal to

$$\begin{pmatrix}\mathbf{w}_1\\ \mathbf{q}_1\end{pmatrix} = \arg\max_{\mathbf{w},\mathbf{q}} \mathbf{w}^T\mathbf{S}_{x_0y_0}\mathbf{q} - \tfrac{1}{2}\lambda\left(\mathbf{w}^T\left[\mathbf{S}_{x_sx_s} + \mathbf{S}_{ee}\right]\mathbf{w} - 1\right) - \tfrac{1}{2}\lambda\left(\mathbf{q}^T\mathbf{q} - 1\right). \tag{6.113}$$

Consequently, the additional term $\mathbf{w}_1^T\mathbf{S}_{ee}\mathbf{w}_1$ will affect the resultant first set of weight vectors. Equations (6.114) and (6.115) show this in more detail

$$\left[\mathbf{S}_{x_sx_s} + \mathbf{S}_{ee}\right]^{-1}\mathbf{S}_{x_0y_0}\mathbf{S}_{y_0x_0}\mathbf{w}_1 = \lambda^2\mathbf{w}_1 \tag{6.114}$$

and

$$\mathbf{S}_{y_0x_0}\left[\mathbf{S}_{x_sx_s} + \mathbf{S}_{ee}\right]^{-1}\mathbf{S}_{x_0y_0}\mathbf{q}_1 = \lambda^2\mathbf{q}_1. \tag{6.115}$$

It is possible to substitute the computation of the weight vectors into (6.102) to (6.109) to examine the impact of $\mathbf{e}$ upon the loading vectors and the deflation procedure, which is the subject of a project in the tutorial session of this chapter (Question 4). Different from PLS, however, the loading vectors are computed as follows

$$\mathbf{p}_i = \left[\mathbf{S}_{x_sx_s} + \mathbf{S}_{ee}\right]\mathbf{w}_i \qquad \acute{\mathbf{q}}_i = \mathbf{S}_{fe}^{(i)}\mathbf{w}_i \tag{6.116}$$

and the deflation procedure reduces to

$$\mathbf{S}_{ef}^{(i+1)} = \left[\mathbf{I} - \left[\mathbf{S}_{x_sx_s} + \mathbf{S}_{ee}\right]\mathbf{w}_i\mathbf{w}_i^T\right]\mathbf{S}_{ef}^{(i)}. \tag{6.117}$$

6.2.4 Error-in-variable estimate for known $\mathbf{S}_{\mathfrak{ee}}$

Assuming that the error covariance matrices are known, this would allow determining the covariance matrix of the uncorrupted input variables

$$\mathbf{S}_{x_0 x_0} - \mathbf{S}_{\mathfrak{ee}} = \mathbf{S}_{x_s x_s}. \tag{6.118}$$

Applying the PLS and MRPLS algorithms with respect to the error correction of $\mathbf{S}_{x_0 x_0}$ using (6.118) and $\mathbf{S}_{x_0 y_0}$ produces now an unbiased and, therefore, a consistent estimation of the weight and loading vectors. For process monitoring, it is important to note that the t-score variables can be obtained in the same way as discussed in Subsections 3.1.1 and 3.1.2. The construction of scatter diagrams, the Hotelling's T^2 and the two Q statistics for fault detection also follow from the discussion in Subsection 3.1.1. Moreover, the presence of an error term does not affect the application of the fault diagnosis methods discussed in Subsection 3.2. The geometric effect of the inclusion of the error vector $\mathfrak{e}$ follows from the analogy of the data structures for MLPCA and EIV PLS, which is briefly discussed next.

Analogy between PLS error-in-variable and MLPCA data structures. Compared to the PCA model subspace, the EIV PLS algorithm also allows the definition of a model subspace when combining the input and output variables as one data set, i.e. $\mathbf{z}_0^T = \begin{pmatrix} \mathbf{x}_0^T & \mathbf{y}_0^T \end{pmatrix}$. This model subspace is consistently estimated after carrying out the error correction of the covariance matrix $\mathbf{S}_{x_0 x_0}$

$$\begin{pmatrix} \mathbf{x}_0 \\ \mathbf{y}_0 \end{pmatrix} = \begin{bmatrix} \mathcal{P} \\ \acute{\mathbf{Q}} \end{bmatrix} \mathbf{t} + \begin{pmatrix} \mathbf{e} \\ \mathbf{f} \end{pmatrix} + \begin{pmatrix} \mathfrak{e} \\ \mathfrak{f} \end{pmatrix}. \tag{6.119}$$

To distinguish between the p-loading matrices produced by PCA and PLS, the loading matrix determined by PLS is denoted by $\mathcal{P}$ for the remainder of this section.

It should be noted that the residuals vectors $\mathbf{e}$ and $\mathbf{f}$ become zero for $n = n_x$. Hence, the orientation of this model subspace is asymptotically identical to that obtained by the loading matrix obtained by the MLPCA, discussed in Subsections 6.1.3 and 6.1.4. The generalized inverse of $\Xi^T = \begin{bmatrix} \mathcal{P}^T & \acute{\mathbf{Q}}^T \end{bmatrix}$ is given by

$$\Xi^{\dagger} = \begin{bmatrix} \mathbf{R}^T & \mathbf{0} \end{bmatrix} = \begin{bmatrix} \mathcal{P}^T \mathcal{P} + \acute{\mathbf{Q}}^T \acute{\mathbf{Q}} \end{bmatrix}^{-1} \begin{bmatrix} \mathcal{P}^T & \acute{\mathbf{Q}}^T \end{bmatrix}. \tag{6.120}$$

An orthogonal complement for Ξ can be constructed as follows

$$\Xi^{\perp} = \begin{bmatrix} \mathbf{W}^T & \mathbf{Q}^T \end{bmatrix} \left[\mathbf{I} - \begin{bmatrix} \mathcal{P} \\ \acute{\mathbf{Q}} \end{bmatrix} \begin{bmatrix} \mathbf{R}^T & \mathbf{0} \end{bmatrix} \right]. \tag{6.121}$$

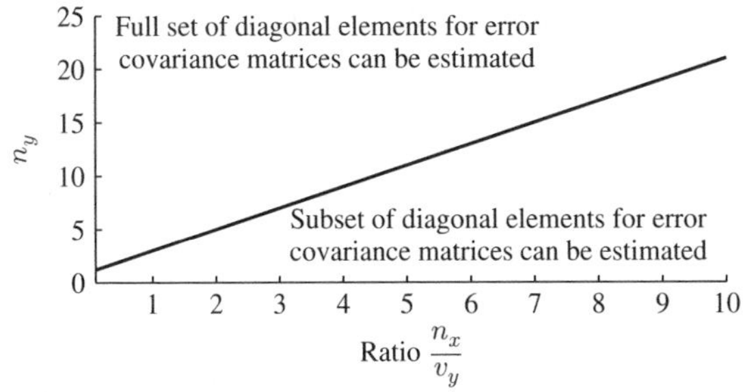

Figure 6.16 Relationship between the minimum number for n_y and the ratio n_x/n_y.

6.2.5 Error-in-variable estimate for unknown $\mathbf{S}_{ee}$

The previous subsection highlighted the analogy between the MLPCA and the EIV PLS data structures. For unknown error covariance matrices, it is consequently possible to develop a maximum likelihood PLS (MLPLS) algorithm on the basis of the MLPCA algorithm, discussed in Subsection 6.1.4, provided that the constraint of (6.46) is not violated. This constraint implies that $n_y^2 + n_y \geq 2\left(n_y + n_x\right)$. This gives rise to the following relationship between the minimum number of required output variables and the ratio n_x/n_y

$$n_y \geq 1 + 2\frac{n_x}{n_y} \tag{6.122}$$

which Figure 6.16 graphically analyzes. Different from MLPCA, PLS is a regression technique, which allows simplifying the objective function in (6.49) by decomposing the covariance matrix of the output variables, $\mathbf{S}_{y_0 y_0}$

$$\mathbf{S}_{y_0 y_0} = \mathcal{B}\left[\mathbf{S}_{x_0 x_0} - \mathbf{S}_{ee}\right]\mathcal{B}^T + \mathbf{S}_{ff} = \mathbf{S}_{y_0 x_0}\mathcal{B} + \mathbf{S}_{ff}. \tag{6.123}$$

This simplification follows from (6.95) and (6.99). This, in turn, implies that the following constraint can be formulated

$$\mathbf{S}_{y_0 y_0} - \mathbf{S}_{ff} - \mathbf{S}_{y_0 x_0}\mathcal{B} = \mathbf{0}. \tag{6.124}$$

On the other hand, the relationship of the extended covariance matrix of the variable sets $\mathbf{y}_0$ and $\mathbf{x}_0$ may be described as follows

$$\begin{bmatrix} \mathbf{S}_{x_0 x_0} & \mathbf{S}_{x_0 y_0} \\ \mathbf{S}_{y_0 x_0} & \mathbf{S}_{y_0 y_0} \end{bmatrix} - \begin{bmatrix} \mathbf{S}_{ee} & \mathbf{0} \\ \mathbf{0} & \mathbf{S}_{ff} \end{bmatrix} = \begin{bmatrix} \mathbf{S}_{x_s x_s} & \mathbf{S}_{x_s x_s}\mathcal{B} \\ \mathcal{B}^T\mathbf{S}_{x_s x_s} & \mathcal{B}^T\mathbf{S}_{x_s x_s}\mathcal{B} \end{bmatrix}. \tag{6.125}$$

The rank of the above matrix is equal to n_x, which results from the fact that

$$\begin{bmatrix} \mathbf{S}_{x_s x_s} & \mathbf{S}_{x_s x_s}\mathcal{B} \\ \mathcal{B}^T\mathbf{S}_{x_s x_s} & \mathcal{B}^T\mathbf{S}_{x_s x_s}\mathcal{B} \end{bmatrix} = \begin{bmatrix} \mathbf{I} \\ \mathcal{B}^T \end{bmatrix}\mathbf{S}_{x_s x_s}\begin{bmatrix} \mathbf{I} & \mathcal{B} \end{bmatrix} = \mathbf{S}_{z_s z_s}, \tag{6.126}$$

where $\mathbf{z}_s^T = \begin{pmatrix} \mathbf{x}_s^T & \mathbf{y}_s^T \end{pmatrix}$. Consequently, the eigendecomposition of $\mathbf{S}_{z_s z_s}$ yields a total of n_x nonzero eigenvalues and the associated n_x eigenvectors that span the same column space as $\Xi^T = \begin{bmatrix} \mathcal{P}^T & \acute{\mathbf{Q}}^T \end{bmatrix}$. Equation (6.121) defines the orthogonal complement of the estimated model subspace. A correct estimate of $\mathbf{S}_{\mathfrak{ee}}$ and $\mathbf{S}_{\mathfrak{ff}}$ satisfies the constraint in (6.124) and yields n_y zero eigenvalues for $\mathbf{S}_{z_s z_s}$.

$$\mathbf{S}_{z_s z_s} = \begin{bmatrix} \mathbf{S}_{x_s x_s} & \mathbf{S}_{x_s x_s}\mathcal{B} \\ \mathcal{B}^T\mathbf{S}_{x_s x_s} & \mathcal{B}^T\mathbf{S}_{x_s x_s}\mathcal{B} \end{bmatrix} = \begin{bmatrix} \mathbf{P} & \mathbf{P}_d \end{bmatrix} \begin{bmatrix} \mathbf{\Lambda} & \mathbf{0} \\ \mathbf{0} & \mathbf{0} \end{bmatrix} \begin{bmatrix} \mathbf{P}^T \\ \mathbf{P}_d^T \end{bmatrix}. \tag{6.127}$$

The column space of $\mathbf{P}$ defines the model subspace, whilst the column space of $\mathbf{P}_d$ defines the complementary residual subspace. The orthogonal complement to the model subspace Ξ is consequently given by

$$\Xi^{\perp} = \mathbf{P}_d^T. \tag{6.128}$$

In analogy to MLPCA and for conveniently presenting the determination of the residuals, the application of this orthogonal complement yields the following linear combinations of the error vector $\mathfrak{g}^T = \begin{pmatrix} \mathfrak{e}^T & \mathfrak{f}^T \end{pmatrix}$

$$\Xi^{\perp}\mathbf{z}_0 = \Xi^{\perp}\mathfrak{g} = \mathbf{g}, \tag{6.129}$$

which has the following error covariance matrix

$$E\left\{\mathbf{g}\mathbf{g}^T\right\} = \Xi^{\perp} \begin{bmatrix} \mathbf{S}_{\mathfrak{ee}} & \mathbf{0} \\ \mathbf{0} & \mathbf{S}_{\mathfrak{ff}} \end{bmatrix} \Xi^{\perp^T}. \tag{6.130}$$

Given that the two error vectors are statistically independent, that is, $E\left\{\mathfrak{e}\mathfrak{f}^T\right\} = \mathbf{0}$, (6.129) and (6.130) can be reformulated as follows

$$\Xi^{\perp} = \mathbf{P}_d^T = \begin{bmatrix} \Xi_{x_0}^{\perp} & \Xi_{y_0}^{\perp} \end{bmatrix}. \tag{6.131}$$

where $\Xi_{x_0}^{\perp} \in \mathbb{R}^{n_y \times n_x}$ and $\Xi_{y_0}^{\perp} \in \mathbb{R}^{n_y \times n_y}$. Applying the block matrices $\Xi_{x_0}^{\perp}$ and $\Xi_{y_0}^{\perp}$, the scaled residuals for the input and output variables are then

$$\Xi_{x_0}^{\perp}\mathbf{x}_0 = \Xi_{x_0}^{\perp}\mathfrak{e} = \mathbf{e} \qquad \Xi_{y_0}^{\perp}\mathbf{y}_0 = \Xi_{y_0}^{\perp}\mathfrak{f} = \mathbf{f}. \tag{6.132}$$

It follows from the assumption of statistical independence between the error vectors that the covariance matrices of $\mathbf{e}$ and $\mathbf{f}$ are

$$\mathbf{S}_{ee} = \Xi_{x_0}^{\perp}\mathbf{S}_{\mathfrak{ee}}\Xi_{x_0}^{\perp^T} \qquad \mathbf{S}_{ff} = \Xi_{y_0}^{\perp}\mathbf{S}_{\mathfrak{ff}}\Xi_{y_0}^{\perp^T}. \tag{6.133}$$

The utilization of (6.131) and (6.133) now allows determining the value of the maximum likelihood objective function of (6.45)

$$\begin{bmatrix}\mathbf{S}_{ee}\\ \mathbf{S}_{ff}\end{bmatrix} = \arg\min_{\mathbf{S}_{ee},\mathbf{S}_{ff}} K\ln\left|\Xi_{x_0}^{\perp}\mathbf{S}_{ee}\Xi_{x_o}^{\perp^T} + \Xi_{y_0}^{\perp}\mathbf{S}_{ff}\Xi_{y_o}^{\perp^T}\right| + \sum_{k=1}^{K}\mathbf{e}^T(k)\Xi_{x_0}^{\perp}\mathbf{S}_{ee}\Xi_{x_0}^{\perp^T}\mathbf{e}(k) + \sum_{k=1}^{K}\mathbf{f}^T(k)\Xi_{y_0}^{\perp}\mathbf{S}_{ff}\Xi_{y_0}^{\perp^T}\mathbf{f}(k). \tag{6.134}$$

Including the constraint in (6.124) and a second constraint based on the last n_y eigenvalues of $\mathbf{S}_{z_s z_s}$, the objective function for estimating $\mathbf{S}_{ee}$ and $\mathbf{S}_{ff}$ becomes

$$\begin{bmatrix}\widehat{\mathbf{S}}_{ee}\\ \widehat{\mathbf{S}}_{ff}\end{bmatrix} = \arg\min_{\mathbf{S}_{ee},\mathbf{S}_{ff}} a_1 K\ln\left|\widehat{\Xi}_{x_0}^{\perp}\mathbf{S}_{ee}\widehat{\Xi}_{v_o}^{\perp^T} + \widehat{\Xi}_{y_0}^{\perp}\mathbf{S}_{ff}\widehat{\Xi}_{y_o}^{\perp^T}\right| + \sum_{k=1}^{K}\mathbf{e}^T(k)\widehat{\Xi}_{x_0}^{\perp}\mathbf{S}_{ee}\widehat{\Xi}_{x_0}^{\perp^T}\mathbf{e}(k) + \sum_{k=1}^{K}\mathbf{f}^T(k)\widehat{\Xi}_{y_0}^{\perp}\mathbf{S}_{ff}\widehat{\Xi}_{y_0}^{\perp^T}\mathbf{f}(k) + a_2\left\|\widehat{\mathbf{S}}_{y_0y_0} - \mathbf{S}_{ff} - \widehat{\mathbf{S}}_{y_0x_0}\widehat{\mathcal{B}}\right\| + a_3\sum_{i=n_x+1}^{n_x+n_y}\widehat{\lambda}_i. \tag{6.135}$$

Note that the above MLPLS objective function relies on estimates of $\mathbf{S}_{x_0x_0}$, $\mathbf{S}_{y_0y_0}$ and $\mathbf{S}_{x_0y_0}$ and is similar to that of (6.49). The steps of the iterative MLPLS algorithms that rely on the developed equations above are now listed below.

1. Set diagonal elements of initial error covariance matrices, $\widehat{\mathbf{S}}_{ee_0}$ and $\widehat{\mathbf{S}}_{ff_0}$ to be 0.0001 times the diagonal elements of $\widehat{\mathbf{S}}_{x_0x_0}$ and $\widehat{\mathbf{S}}_{y_0y_0}$, respectively.

2. Compute the initial EIV estimate of $\mathcal{B}$, $\widehat{\mathcal{B}}_0 = \left[\widehat{\mathbf{S}}_{x_0x_0} - \widehat{\mathbf{S}}_{ee_0}\right]^{-1}\widehat{\mathbf{S}}_{x_0y_0}$.

3. Carry out eigendecomposition of extended covariance matrix in (6.125)

$$\begin{bmatrix}\widehat{\mathbf{S}}_{x_0x_0} & \widehat{\mathbf{S}}_{x_0y_0}\\ \widehat{\mathbf{S}}_{y_0x_0} & \widehat{\mathbf{S}}_{y_0y_0}\end{bmatrix} - \begin{bmatrix}\widehat{\mathbf{S}}_{ee_0} & \mathbf{0}\\ \mathbf{0} & \widehat{\mathbf{S}}_{ff_0}\end{bmatrix} = \begin{bmatrix}\widehat{\mathbf{P}}_0 & \widehat{\mathbf{P}}_{d_0}\end{bmatrix}\begin{bmatrix}\widehat{\mathbf{\Lambda}}_0 & \mathbf{0}\\ \mathbf{0} & \widehat{\mathbf{\Lambda}}_{d_0}\end{bmatrix}\begin{bmatrix}\widehat{\mathbf{P}}_0^T\\ \widehat{\mathbf{P}}_{d_0}^T\end{bmatrix}.$$

4. Extract orthogonal complements of (6.131), $\left[\widehat{\Xi}_{x_0}^{\perp}\right]_0$ and $\left[\widehat{\Xi}_{y_0}^{\perp}\right]_0$.

5. Use estimates for $\widehat{\mathbf{S}}_{\mathfrak{ee}_0}$, $\widehat{\mathbf{S}}_{\mathfrak{ff}_0}$, $\left[\widehat{\Xi}_{x_0}^{\perp}\right]_0$, $\left[\widehat{\Xi}_{y_0}^{\perp}\right]_0$ and $\widehat{\mathbf{\Lambda}}_{d_0}$ to work out initial value of the objective function in Equation (6.135), J_0.
6. Update the error covariance matrices, $\widehat{\mathbf{S}}_{\mathfrak{ee}_{i+1}} = \widehat{\mathbf{S}}_{\mathfrak{ee}_i} + \Delta\mathbf{S}_{\mathfrak{ee}_i}$ and $\widehat{\mathbf{S}}_{\mathfrak{ff}_{i+1}} = \widehat{\mathbf{S}}_{\mathfrak{ff}_i} + \Delta\mathbf{S}_{\mathfrak{ff}_i}$.
7. Compute EIV estimate of $\mathcal{B}$, $\widehat{\mathcal{B}}_{i+1} = \left[\widehat{\mathbf{S}}_{x_0x_0} - \widehat{\mathbf{S}}_{\mathfrak{ee}_{i+1}}\right]^{-1}\widehat{\mathbf{S}}_{x_0y_0}$.
8. carry out eigendecomposition of extended covariance matrix in (6.125)

$$\begin{bmatrix}\widehat{\mathbf{S}}_{x_0x_0} & \widehat{\mathbf{S}}_{x_0y_0}\\ \widehat{\mathbf{S}}_{y_0x_0} & \widehat{\mathbf{S}}_{y_0y_0}\end{bmatrix} - \begin{bmatrix}\widehat{\mathbf{S}}_{\mathfrak{ee}_{i+1}} & \mathbf{0}\\ \mathbf{0} & \widehat{\mathbf{S}}_{\mathfrak{ff}_{i+1}}\end{bmatrix} = \begin{bmatrix}\widehat{\mathbf{P}}_{i+1} & \widehat{\mathbf{P}}_{d_{i+1}}\end{bmatrix} \begin{bmatrix}\widehat{\mathbf{\Lambda}}_{i+1} & \mathbf{0}\\ \mathbf{0} & \widehat{\mathbf{\Lambda}}_{d_{i+1}}\end{bmatrix}\begin{bmatrix}\widehat{\mathbf{P}}_{i+1}^T\\ \widehat{\mathbf{P}}_{d_{i+1}}^T\end{bmatrix}.$$

9. extract orthogonal complements of (6.131), $\left[\widehat{\Xi}_{v_0}^{\perp}\right]_{i+1}$ and $\left[\widehat{\Xi}_{y_0}^{\perp}\right]_{i+1}$.
10. use estimates for $\widehat{\mathbf{S}}_{\mathfrak{ee}_{i+1}}$, $\widehat{\mathbf{S}}_{\mathfrak{ff}_{i+1}}$, $\left[\widehat{\Xi}_{x_0}^{\perp}\right]_{i+1}$, $\left[\widehat{\Xi}_{y_0}^{\perp}\right]_{i+1}$ and $\widehat{\mathbf{\Lambda}}_{d_{i+1}}$ to work out $(i+1)$th value of the objective function in (6.135), J_{i+1}.
11. check for convergence[7], if $\left|J_{i+1} - J_i\right| < 10^{-12}$ terminate, else go to Step 6.

It is interesting to compare the MLPLS with the MLPCA algorithm, discussed in Subsection 6.1.4. The main differences between both algorithms are:

- the MLPLS algorithm does not require the computation of a Cholesky decomposition of the diagonal matrix $\widehat{\mathbf{S}}_{gg_i}$, which is of dimension $n_x + n_y$;
- the MLPLS algorithm relies on the inverse of the symmetric positive definite matrix $\widehat{\mathbf{S}}_{x_0x_0} - \widehat{\mathbf{S}}_{\mathfrak{ee}_i}$ of dimension n_x;
- the MLPCA algorithm requires the inverse of a diagonal matrix $\mathbf{L}_i$ of dimension $n_x + n_y$;
- the MLPCA and MLPLS algorithms require a subsequent application of the constrained NIPALS or PLS algorithms, respectively, in order to compute the sets of latent variables;
- the MLPLS algorithm produces an EIV estimate of the regression matrix $\widehat{\mathcal{B}}$ together with estimates of the error covariance matrices $\widehat{\mathbf{S}}_{\mathfrak{ee}}$ and $\widehat{\mathbf{S}}_{\mathfrak{ff}}$; and
- the MLPCA algorithm produces an estimate of the PCA model subspace and an estimate of the error covariance matrix $\widehat{\mathbf{S}}_{gg}$.

[7] The value of 10^{-12} is a possible suggestion; practically, smaller thresholds can be selected without a substantial loss of accuracy.

6.2.6 Application to a distillation process – revisited

This subsection applies the MLPLS algorithm to determine an EIV model for the reference data of the distillation process. The MRPLS model, estimated in Section 5.2, relied on the data structure in (2.51) that did not include an error term for the input variables. Selecting the parameters for the MLPLS objective function in (6.135), a_1, a_2 and a_3, to be 0.05, 50 and 100, respectively, (6.136a) shows the estimate error variances of the input variables and (6.136b) gives estimates of the error variance of the five output variables.

$$\mathrm{diag}\left\{\widehat{\mathbf{S}}_{\mathfrak{ee}}\right\} = \begin{pmatrix} 0.0514 \\ 0.0375 \\ 0.0328 \\ 0.0716 \\ 0.0547 \\ 0.0643 \\ 0.0833 \\ 0.0487 \end{pmatrix} \tag{6.136a}$$

$$\mathrm{diag}\left\{\widehat{\mathbf{S}}_{\mathfrak{ff}}\right\} = \begin{pmatrix} 0.1221 \\ 0.2027 \\ 0.7147 \\ 0.1149 \\ 0.0496 \end{pmatrix}. \tag{6.136b}$$

Comparing the error variance for $\mathbf{y}_0$ obtained by the PLS/MRPLS model in Table 5.3 with the EIV estimate in (6.136b), the inclusion of $\mathfrak{e}$ for $\mathbf{x}_0$ gives rise to a more accurate prediction of the output variables. Moreover, the estimated error variances of the input variables as well as output variables y_1, y_4 and y_5 are around 0.05 to 0.1 with the exception of x_2 and x_3 (column overhead pressure and tray 2 temperature), which have slightly smaller error variances. In contrast the error variance of the y_2 and y_3 concentration is significant, particularly the C5 in C4 one.

Removing the impact of the error terms from the covariance matrices now allows estimating the LV sets. Equations (6.137) and (6.138) show the estimated r-weight and q-loading matrices. In a similar fashion to the MRPLS estimates for the r-weight matrix in (5.6), the EIV estimate outlines that the temperature of the fresh feed and the reboiler temperature do not significantly contribute to the computation of the four t-score variables. Moreover, the dominant contributions for computing each of the individual score variables are:

- the fresh feed level for the first t-score variable;
- temperature readings of tray 31 and 2 for the second t-score variable;
- tray 14 temperature and the fresh feed level for the third t-score variable; and

- the reboiler steam flow, the tray 31 and the fresh feed level for the fourth t-score variable.

In (6.137), these and other more minor contributing variables to each of the t-score variables are marked in bold.

$$\widehat{\mathbf{R}} = \begin{bmatrix} -0.3314 & \mathbf{-0.7117} & \mathbf{2.5391} & \mathbf{-1.3881} & 1.5328 & 0.4058 & 2.3383 & -1.0347 \\ -0.2922 & \mathbf{-0.8224} & \mathbf{-1.1530} & 0.2366 & 0.4846 & 1.2736 & -1.3067 & 1.9369 \\ 0.3071 & \mathbf{0.9939} & \mathbf{1.2286} & \mathbf{-1.0681} & -0.5008 & -0.1896 & 1.3403 & -2.3023 \\ -0.1198 & -0.5201 & -0.6048 & 0.7035 & 0.3488 & 0.3577 & 0.1126 & 0.1422 \\ -0.4176 & \mathbf{-0.7674} & \mathbf{1.3090} & \mathbf{-2.1575} & 0.8293 & 0.1229 & 2.1785 & -1.3282 \\ 0.5900 & \mathbf{1.2467} & \mathbf{-1.4786} & \mathbf{1.8131} & -1.5507 & -0.7475 & -1.3582 & 1.1693 \\ \mathbf{1.3535} & 0.4461 & \mathbf{-1.9038} & \mathbf{1.6360} & -1.4056 & -0.4250 & -2.5927 & 1.3318 \\ -0.0858 & -0.2066 & -0.2777 & -0.7939 & 1.1138 & -0.4765 & 0.3223 & -0.4025 \end{bmatrix}. \quad (6.137)$$

From the parameters of the q-loading matrix, the individual t-score variables contribute to the prediction of the output variables as follows:

- t-score variable 1 has the most significant contribution to the flow rates of both output streams and to a lesser extend the C3 in C4 concentration in the top draw;
- the most dominant contribution of the second and third t-score variable is to the prediction of the C4 in C5 and the C3 in C4 concentrations; and
- t-score variable 4 is a dominant contributor of the C5 in C4 concentration.

$$\widehat{\mathbf{Q}} = \begin{bmatrix} \mathbf{0.8358} & -0.3595 & -0.0296 & -0.0486 \\ \mathbf{0.5521} & \mathbf{0.5714} & \mathbf{-0.3807} & 0.0795 \\ 0.2456 & 0.1172 & 0.1942 & \mathbf{0.4168} \\ \mathbf{0.8121} & -0.3820 & 0.1279 & -0.0692 \\ 0.3088 & \mathbf{0.8626} & \mathbf{0.2701} & -0.1602 \end{bmatrix}. \quad (6.138)$$

The next questions relate to the selection of the number of source signals that the input and output variables commonly share and what the contribution of each set of latent variables explains to the covariance and cross-covariance matrices. An answer to both of these questions lies in evaluating (2.102) to (2.104). Figure 6.17 plots the ratios produced by these equations for $k = 1 \ldots 8$ for the maximum likelihood and the standard MRPLS algorithms. It is important

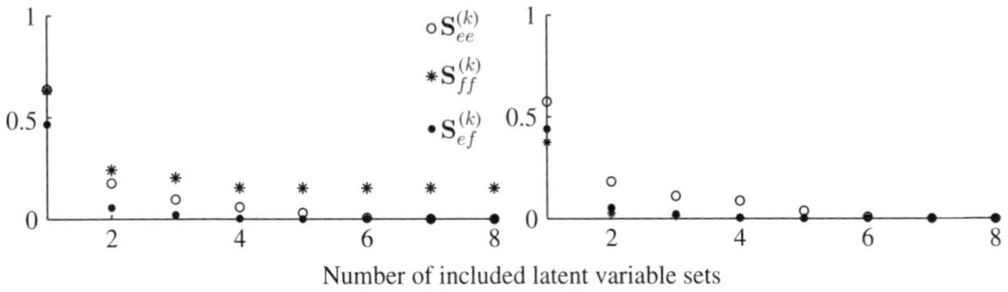

Figure 6.17 Deflation of $\mathbf{S}_{x_0x_0}$, $\mathbf{S}_{y_0y_0}$ and $\mathbf{S}_{x_0y_0}$ using extracted latent variable sets (left plot → MRPLS model; right plot → maximum likelihood MRPLS model).

to note that the maximum likelihood MRPLS algorithm relies on covariance matrices for which the variance of each error term is subtracted from the estimated covariance matrices.

This is different for the original MRPLS algorithm, which relies on the estimated covariance and cross-covariance matrices. It is also important to note that the deflation of the covariance matrices is not required for the computationally efficient MRPLS algorithm in Table 10.3. However, in order to compute the contribution of each set of latent variables from these matrices, a deflation procedure after the model building phase was carried out.

Addressing the first question, both plots in Figure 6.17 confirm that subtracting the contribution of the first four t-score variables maximally exhausted the squared sum of the elements of the cross-covariance matrix. For the maximum likelihood MRPLS model, the squared sum of the elements of this matrix are very close to zero, implying that there is no common cause variation left between both variable sets that requires the inclusion of a fifth source variable. The selection of $n = 4$ is therefore confirmed.

Different to its maximum likelihood counterpart, there is a remaining value of about 0.15 for the covariance matrix $\widehat{\mathbf{S}}_{ff}^{(5)}$.[8] This is not surprising, given that the error variables are assumed to be uncorrelated to the input variables. The decrease in the squared sum of the covariance matrix $\widehat{\mathbf{S}}_{ee}^{(k)}$ is similar for both models. That these values are slightly smaller for the maximum likelihood MRPLS algorithm is, again, a result of excluding the estimated variance of the error terms.

Finally, the regression model obtained by the maximum likelihood MRPLS algorithm for $n = 4$ can be compared to that computed by the MRPLS one in Table 5.4. Significant differences between both regression matrices are that the maximum likelihood MRPLS regression model confirms:

- that column pressure x_2, tray 2 temperature x_3 and fresh feed temperature x_4 have the most significant impact on the C4 in C5 concentration y_5;
- that the reboiler stream flow x_5 is mostly affecting the impurities y_2 and y_3;

[8] After deflating the four sets of latent variables computed by the MRPLS algorithm.

Table 6.14 Coefficients of regression model for $n = 4$.

b_{ij}	$i = 1$	$i = 2$	$i = 3$	$i = 4$	$i = 5$
$j = 1$	−0.1877	**−1.6272**	−0.2610	**0.5726**	0.1672
$j = 2$	−0.1849	−0.1091	−0.3110	0.1565	**−1.1893**
$j = 3$	0.3211	0.0839	0.0128	**−0.2812**	**1.5184**
$j = 4$	0.0107	−0.0623	0.0813	0.0317	**−0.7710**
$j = 5$	−0.0180	**−1.3362**	**−0.8382**	**0.2808**	−0.0935
$j = 6$	0.0386	**1.7358**	**0.7621**	**−0.3473**	**0.5738**
$j = 7$	**1.0484**	**1.8321**	**0.7037**	**0.4777**	0.0422
$j = 8$	−0.0858	−0.0892	**−0.4392**	0.1557	−0.1737

A row represents the coefficients for the prediction of the output variables using the jth input variable and the coefficients in a column are associated with the prediction of the ith output variable using the input variables.

- that the flow rate of the fresh feed x_7 impacts not only the flow rate of the output stream but also C3 in C4 and the C5 in C4 concentrations; and
- that the reboiler temperature mainly affects the C5 in C4 concentration y_3.

Both regression matrices, however, suggest that the tray 31 temperature x_6 has an affect on the concentrations of the top and bottom draw.

The information that can be extracted from the EIV estimate of the regression matrix describes the underlying causal relationships between the input and output variables correctly. It is important to recall that the static MRPLS model does not represent a causal dynamic mechanistic model that describes the physical and chemical relationships between the process variables. However, the steady state relationships that can be extracted from the regression matrix in Table 6.14 describe a correct relationship between the input and output variables.

6.3 Robust model estimation

Besides process monitoring, all branches of data chemometrics and analytical chemistry, for example, in industrial and academic research deal with large amounts of data, which can be subjected to errors, including bias, for example resulting from the poor calibration of measurement devices, and sporadic outliers, that can arise for any number of reasons. The first type is usually related to small persistent residual parts (offset) during the measurements being taken, whilst the second one is associated with large residuals and most of the time affect only single observations.

Detecting a bias is straightforward by carefully inspecting the recorded data and applying *a priori* knowledge and experience. The same cannot be said for outliers, as they infrequently arise, may easily be overlooked in large data sets and can have a profound and undesired impact upon the accuracy of the estimated parameters, for example the estimation of the data covariance matrix or the

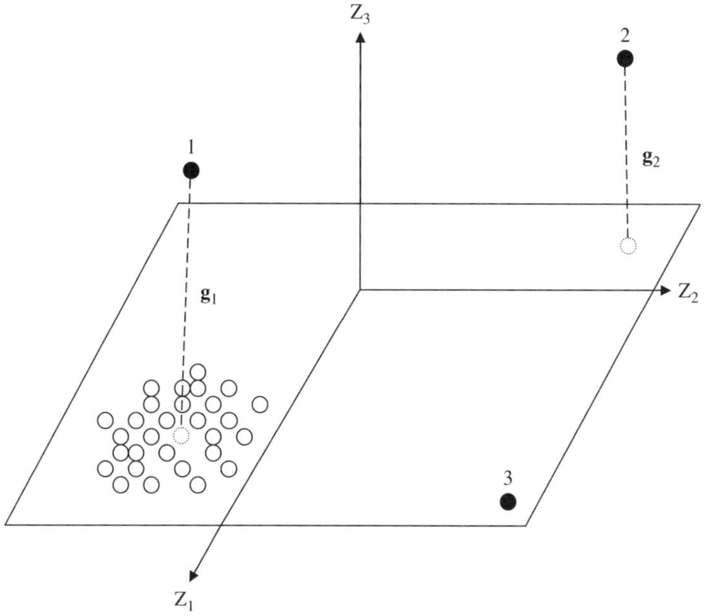

Figure 6.18 Illustration of the three different types of outliers in a data space.

control limits for univariate monitoring statistics. In general, outliers can be divided into three categories (Møller *et al.* 2005):

1. orthogonal outliers that have a large residual part but a small variance part;
2. 'bad' leverage points (large variance and residual parts); and
3. 'good' leverage points (large variance but small residual part).

Figure 6.18 gives a graphical account of each type of outlier for three process variables that can be described, according to (2.2), by two source signals and a superimposed error vector. Whilst the normal or 'sound' samples locate in the vicinity of the model subspace, the outliers have large departures either from the model subspace (large residuals $\mathbf{g}_1$ and $\mathbf{g}_2$ for the first and second outliers) and/or the cluster of normal samples. A robust estimation of parameters entails the removal or reduction of the impact of outliers upon the estimation and the aim of this section is to summarize research work, including recent trends, reported in the literature.

It should be noted that outliers in recorded reference data are identifiable using the covariance matrix $\mathbf{S}_{z_0 z_0}$ and the error covariance matrix $\mathbf{S}_{gg}$ if known *a priori*. As Figure 6.18 illustrates that outliers 1 and 2 possess a large residual part by definition, whilst outliers 2 and 3 are associated with a large variance part. In case the covariance matrices are available it is sufficient to evaluate a statistical test over the reference samples to determine whether a particular sample is an outlier or not. However, these matrices, particularly the error covariance matrix, are usually unknown and need to be estimated from the reference set. Over the past few decades, a plethora of methods have been proposed to produce robust estimates for parameters, such as variable mean and covariance matrix.

The discussion of robust methods can be roughly classified into two categories:

- accept all the data points and try to find a robust estimator which reduces the impact of outliers; and
- maintain the traditional estimators and try to eliminate the outliers (trimming) before the estimation by using some cluster property.

The literature regards the first approach as *robust regression* or *robust parameter estimation*, whilst the second one can be categorized as the *trimming approach*. The next two subsections summarize associated methods for both of these approaches. The aim of this subsection is to give a brief overview of existing methods. A more detailed and specific treatment of this topic is available in references Daszykowski (2007); Liang and Kvalheim (1996) and Møller *et al.* (2005) for example.

6.3.1 Robust parameter estimation

Robust regression methods can be further divided into (i) robust estimates of the moments, (ii) projection pursuit, (iii) M estimators and (iv) least median of squares.

6.3.1.1 Robust estimation of the moments

The definition of processes the produce variables that follow a Gaussian distribution require the estimation of the data location (mean vector) and spread (covariance matrix). The mean vector can be viewed as a least squares estimator

$$\min_{\bar{\mathbf{z}}} \sum_{k=1}^{K} \left\| \mathbf{z}(k) - \bar{\mathbf{z}} \right\|^2 \qquad (6.139)$$

which includes squared summation elements and is, accordingly, sensitive to the presence of outliers. A robust alternative is the use of the median of the samples

$$\min_{\bar{\mathbf{z}}} \sum_{k=1}^{K} \left\| \mathbf{z}(k) - \bar{\mathbf{z}} \right\| \qquad (6.140)$$

or the Stahel-Donoho location estimator (Donoho 1982; Stahel 1981)

$$\widehat{\mathbf{z}} = \frac{\sum_{i=1}^{K} d_k \mathbf{z}(k)}{\sum_{k=1}^{K} d_k} \tag{6.141}$$

where $d_k = d\left(r\left(\mathbf{z}(k), \mathbf{Z}\right)\right)$ is a weighting function, such as the iteratively reweighted least squares function (Phillips and Eyring 1983), and $r\left(\mathbf{z}(k), \mathbf{Z}\right)$ is defined as follows

$$r\left(\mathbf{z}(k), \mathbf{Z}\right) = \sup_{\|\boldsymbol{\delta}\|=1} \left\{ \frac{\left|\boldsymbol{\delta}^T \mathbf{z}(k) - \widetilde{\mu}\left(\boldsymbol{\delta}^T \mathbf{Z}\right)\right|}{\widetilde{\sigma}\left(\boldsymbol{\delta}^T \mathbf{Z}\right)} \right\}. \tag{6.142}$$

Here, $\widetilde{\mu}(\boldsymbol{\delta}^T \mathbf{Z})$ and $\widetilde{\sigma}(\boldsymbol{\delta}^T \mathbf{Z})$ are the median and the median absolute deviation of the projections of the samples stored in $\mathbf{Z}$ onto $\boldsymbol{\delta}$, respectively, and $|\cdot|$ is the absolute value. For PLS, Kruger *et al.* (2008a,b,c) proposed an iterative algorithm on the basis of (6.142) to discriminate outliers from sound observation.

The variance, or the second order moments of a variable, is also calculated as the sum of squares, and therefore affected by outliers. The literature proposed a number of methods for providing a robust estimation of variance, where the median absolute deviation (MAD) and the more efficient Sn and Qn estimators are among the most popular ones (Hampel 1974; Rousseeuw and Croux 1993)

$$\begin{aligned} \text{MAD} \quad \widehat{\sigma} &= b \cdot \text{med}_j\left(\left|z(j) - \text{med}_k\left(z(k)\right)\right|\right) \\ \text{Sn} \quad \widehat{\sigma} &= c \cdot \text{med}_j\left(\text{med}_k\left(\left|z(j) - z(k)\right|\right)\right) \\ \text{Qn} \quad \widehat{\sigma} &= d \cdot \text{med}\left(z(j) - z(k);\ j < k|\right) \end{aligned} \tag{6.143}$$

with $b = 1.4826$, $c = 1.1926$, $d = 2.219$ and med abbreviates median. A detailed discussion of these estimators is given in Rousseeuw and Croux (1993). With the availability of robust estimates for mean and variance, a robust pretreatment such as of the recorded data, such as mean centering and scaling, can be carried out. A direct estimation of the covariance matrix can be determined as the Stahel-Donoho scale estimator

$$\widehat{\mathbf{S}}_{z_0 z_0} = \frac{\sum_{k=1}^{K} d_k \left(\mathbf{z}(k) - \widehat{\mathbf{z}}\right)\left(\mathbf{z}(k) - \widehat{\mathbf{z}}\right)^T}{\sum_{k=1}^{K} d_k}. \tag{6.144}$$

The mean vector can be obtained using (6.141).

6.3.1.2 Projection pursuit

The projection pursuit approach substitutes a traditional objective function by a robust one (Daszykowski 2007). In the case of PCA, Section 2.1 pointed out that the associated objective function involves the maximization of the variance of the ith score variable, t_i. Equation (6.143) summarizes robust estimates for the variable variance and the projection pursuit can be seen as a simple regression-based approach to obtain a robust objective function, for example

$$\mathrm{PI}\left(t_i\right) = \widehat{\sigma}(t_i). \tag{6.145}$$

Here, PI stands for projection index and represents the robust objective functions. As examples, various objective functions for the data mean $\widehat{\sigma}$ are summarized in (6.143).

6.3.1.3 M-estimator

This is a maximum likelihood estimator for minimizing the residuals, for example the residuals associated with the jth process variable of a PCA model $g_j(k)$

$$\min_{\theta} \sum_{k=1}^{K} \rho\left(g_j(k), \theta\right). \tag{6.146}$$

for which a variety of estimators have been proposed, including

$$\begin{aligned}
\rho_{\mathrm{L}_\theta}(g_j(k), \theta) &= \frac{|g_j(k)|^\theta}{\theta} \\
\rho_{\mathrm{FAIR}}(g_j(k), \theta) &= \theta^2 \left[\frac{|g_j(k)|}{\theta} - \log\left(1 + \frac{|g_j(k)|}{\theta}\right)\right] \\
\rho_{\mathrm{HUBER}}(g_j(k), \theta) &= \begin{cases} \frac{1}{2} g_j^2(k), & \text{if} \quad |g_j(k)| \leq \theta \\ \theta(|g_j(k)| - \frac{\theta}{2}), & \text{if} \quad |g_j(k)| > \theta \end{cases} \\
\rho_{\mathrm{CAUCHY}}(g_j(k), \theta) &= \frac{\theta^2}{2} \log\left(1 + \left(\frac{g_j(k)}{\theta}\right)^2\right) \\
\rho_{\mathrm{WELSCH}}(g_j(k), \theta) &= \frac{\theta^2}{2}\left\{1 - \exp\left[-\left(\frac{g_j(k)}{\theta}\right)^2\right]\right\}.
\end{aligned} \tag{6.147}$$

The parameter θ serves as a tuning parameter.

6.3.1.4 Least median of squares – LMS

This is one of the most popular methods and was developed by Rousseeuw (1984) for robustly estimating variance. This technique replaces the sum of the squared residuals with the robust median

$$\min \operatorname{med}\left(g_i^2(k)\right). \tag{6.148}$$

In other words, the estimator is the smallest value for the median of the squared residuals computed over the complete reference data set. In contrast to the M-estimator, the LMS estimator does not present a weighted least squares problem and the determination of a solution can be computationally demanding.

6.3.2 Trimming approaches

Trimming approaches exclude some *extreme* samples that are considered outliers for determining a robust estimate. Associated methods that the research literature has proposed include the trimmed least squares, multivariate trimming, the minimum volume estimator and the minimum covariance determinant estimator.

6.3.2.1 Least trimmed squares – LTS

This is the simplest approach and relates to the classification of samples based on their residue magnitude (Rousseeuw 1984). Those samples producing the largest residuals are considered outliers and, accordingly, excluded from the computation of the estimate. The LTS method gives rise to the solution of the following minimization problem

$$\min_{\theta} \sum_{k=1}^{K^-} \rho\left(\widetilde{g}_j(k), \theta\right) \tag{6.149}$$

where $\widetilde{g}_j(k)$ is referred to as an ordered residual that is ranked according to the magnitude of the residual (crescent magnitude). Those with the largest magnitude are removed so that $K^- \leq K$ samples remain. With regards to Figure 6.18, it should be noted that the LTS method only tackles samples that produce orthogonal or *bad* leverage outliers.

6.3.2.2 Multivariate trimming – MVT

Instead of the use of residuals, the MVT technique relies on the distance between the data points to produce a robust estimate (Gnanadesikan and Kettenring 1972; Maronna 1976). Assuming the data follow a Gaussian distribution function, the MVT method iteratively discards extreme values which, in turn, generates a PDF that shows significant departures from the theoretical one.

6.3.2.3 Minimum volume estimator – MVE

This approach is similar to the MVT technique in that it assumes that the data can be described by a predefined shape. More precisely, the MVE method determines a multivariate ellipsoid that hugs at least 50% of the samples. Points that fall outside this ellipsoid are not considered for estimating a model.

6.3.2.4 Minimum covariance determinant estimator – MCD

The MCD method is similar in approach to the MVE and MVT techniques in that it relates to the assumed cluster property of uncorrupted observations (Gnanadesikan and Kettenring 1972; Maronna 1976). Utilizing a cross-validation procedure, this technique is able to give a robust estimation of the data location and dispersion. In a univariate case, the MCD approach reduces to a LTS estimator where each data point receives a weight of one if it belongs to the robust confidence interval and zero otherwise. Rocke and Woodruff (1996); Rousseeuw and Driessen (1999) pointed out that MCD is theoretically superior to MVT, and Davies (1992) showed that MCD possesses better statistical properties compared to MVE.

6.4 Small sample sets

Reference data that include relatively few samples compared to the number of process variables present challenges in determining confidence limits/regions for statistical inference. Numerous textbooks on statistics outline that the confidence limits, determining the acceptance region for estimating parameters or hypothesis tests, widens with a reduction in the size of the reference set. This, in turn, can have a profound and undesirable effect upon the number of Type I and II errors.

As an example, the confidence interval for estimating the variable mean for a particular process variable z is given by

$$\widehat{\bar{z}} - a_{\bar{z}} \leq \bar{z} \leq \widehat{\bar{z}} + a_{\bar{z}}. \tag{6.150}$$

The true mean value, $\bar{z}$ lies under the assumption that $E\{(z - \bar{z})^2\} = \sigma^2$ is known within this confidence interval, limited by the parameter $a_{\bar{z}}$, which is given by

$$a_{\bar{z}} = \frac{c\sigma}{\sqrt{K}}. \tag{6.151}$$

Here, c defines the confidence interval for a zero mean Gaussian distribution of unit variance, $\phi(\cdot)$, and is given by $1 - \alpha = \int_{-c}^{c} \phi(u)du$. For example, significances of 0.05 and 0.01 require c to be 1.960 and 2.576, respectively. The relationship in (6.151), however, shows a direct dependence between the length of the confidence interval for estimating $\bar{z}$ and the number of samples, K, since σ and c are constant. Qualitatively, if K is large $a_{\bar{z}}$ will be small and vice versa.

The same problem emerges when determining the upper and lower control limits for Shewhart charts, and the control limits for the Hotelling's T^2 and Q statistics. This section revisits the issue of constructing non-negative quadratic forms and associated control limits using small reference sets. A non-negative quadratic form, such as the Hotelling's T^2 statistic, has the following definition

$$T^2 = \left(\mathbf{z} - \widehat{\bar{\mathbf{z}}}\right)^T \widehat{\mathbf{S}}_{z_0z_0}^{-1} \left(\mathbf{z} - \widehat{\bar{\mathbf{z}}}\right). \tag{6.152}$$

Here,

- $\mathbf{z} \sim \mathcal{N}\left\{\bar{\mathbf{z}}, \mathbf{S}_{z_0z_0}\right\}$ and $\widehat{\bar{\mathbf{z}}} = \frac{1}{K}\sum_{k=1}^{K}\mathbf{z}(k)$ are a data vector and the estimated sample mean, respectively; and
- $\widehat{\mathbf{S}}_{z_0z_0} = \frac{1}{K-1}\sum_{k=1}^{K}\left(\mathbf{z}(k) - \widehat{\bar{\mathbf{z}}}\right)\left(\mathbf{z}(k) - \widehat{\bar{\mathbf{z}}}\right)^T$ is the estimate of $\mathbf{S}_{z_0z_0}$ for a total of K independent samples.

The estimation of the data covariance matrix $\widehat{\mathbf{S}}_{z_0z_0}$ follows a Wishart distribution (Tracey *et al.* 1992). Under the assumption that the estimation of $\mathbf{S}_{z_0z_0}$ is independent of each $\mathbf{z}(k)$, $k = \{1, 2, \ldots, K\}$, the T^2 statistic follows an F-distribution

$$T^2 \sim \frac{n_z(K^2 - 1)}{K(K - n_z)}\mathcal{F}(n_z, K - n_z). \tag{6.153}$$

Here, the estimates of $\bar{\mathbf{z}}$ and $\mathbf{S}_{z_0z_0}$ have the distributions

- $\widehat{\bar{\mathbf{z}}} \sim \mathcal{N}\left\{\bar{\mathbf{z}}, \frac{1}{K}\mathbf{S}_{z_0z_0}\right\}$; and
- $(K - 1)\widehat{\mathbf{S}}_{z_0z_0} \sim \mathcal{W}(K - 1, \mathbf{S}_{z_0z_0})$,

where $\mathcal{W}(.)$ is a Wishart distribution. The often observed high degree of correlation in the recorded variable set, described in Section 1.2 and Chapter 2, is addressed by defining a reduced set of LVs

$$\widehat{\mathbf{t}} = \widehat{\mathbf{P}}^T\mathbf{z}_0 \quad \widehat{\mathbf{t}} \in \mathbb{R}^n, \widehat{\mathbf{P}} \in \mathbb{R}^{n_z \times n} \tag{6.154}$$

Chapter 3 showed that $n < n_z$ yields two non-negative quadratic forms for PCA, and $n < n_x$ gives rise to three non-negative quadratic forms for PLS. Concentrating on PCA, the Hotelling's T^2 and Q statistics are defined as

$$T^2 = \left(\mathbf{z} - \widehat{\bar{\mathbf{z}}}\right)^T \widehat{\mathbf{P}}\widehat{\mathbf{\Lambda}}^{-1}\widehat{\mathbf{P}}^T\left(\mathbf{z} - \widehat{\bar{\mathbf{z}}}\right) = \widehat{\mathbf{t}}^T\widehat{\mathbf{\Lambda}}^{-1}\widehat{\mathbf{t}} \tag{6.155}$$

and

$$Q = \left(\mathbf{z} - \widehat{\bar{\mathbf{z}}}\right)^T\left[\mathbf{I} - \widehat{\mathbf{P}}\widehat{\mathbf{P}}^T\right]\left(\mathbf{z} - \widehat{\bar{\mathbf{z}}}\right), \tag{6.156}$$

respectively. As before, $\widehat{\mathbf{P}}$ and $\widehat{\mathbf{\Lambda}}$ store estimated dominant n eigenvectors and eigenvalues of $\widehat{\mathbf{S}}_{z_0z_0}$, respectively.

Remark 6.4.1 *Assuming that the estimate of* $\mathbf{\Lambda}$ *follows a Wishart distribution and that the samples used to determine this estimate are independent from those used to estimate* $\mathbf{P}$*, the Hotelling's* T^2 *statistic follows a scaled F-distribution with* n *and* $K - n$ *degrees of freedom, that is* $K(K - n)T^2 \sim n(K^2 - 1)F(n, K - n)$.

The above remark relates to the estimation of $\mathbf{\Lambda}$

$$\widehat{\mathbf{\Lambda}} = \frac{1}{K-1}\sum_{k=1}^{K}\widehat{\mathbf{t}}(k)\widehat{\mathbf{t}}^T(k) = \frac{1}{K-1}\sum_{k=1}^{K}\widehat{\mathbf{P}}^T\mathbf{z}_0(k)\mathbf{z}_0^T(k)\widehat{\mathbf{P}} = \widehat{\mathbf{P}}^T\widehat{\mathbf{S}}_{z_0z_0}\widehat{\mathbf{P}}, \tag{6.157}$$

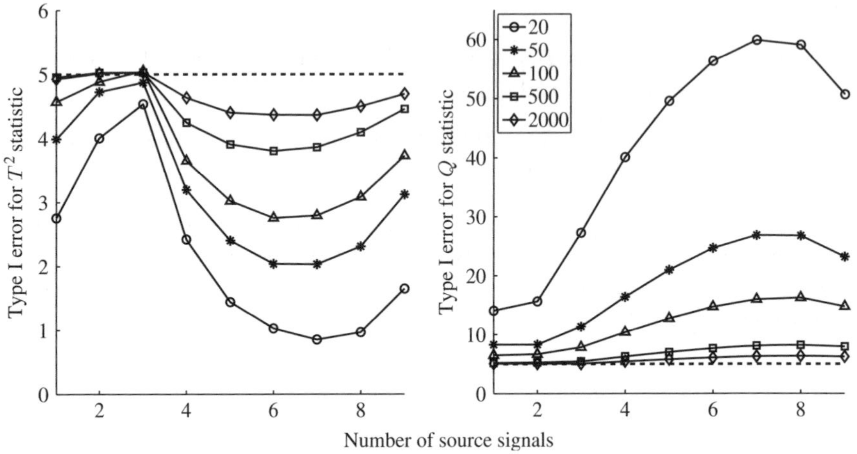

Figure 6.19 Type I error [%] *for* $\alpha = 0.05$.

which follows a Wishart distribution if and only if $\mathbf{t}(k)$ follow a multivariate Gaussian distribution and $\widehat{\mathbf{P}}$ is estimated from a different sample set. If this assumption is not met, the Hotelling's T^2 statistic does not follow an F-distribution. Approximations of the Q statistic have been proposed in Nomikos and MacGregor (1995).

The following Monte Carlo experiment illustrates the problem of determining the control limits for non-negative quadratic forms if K is small. This example is similar to that in Ramaker *et al.* (2004) and includes fifteen process variables, three source variables and an error vector

$$\mathbf{z}_0(k) = \boldsymbol{\xi}_1 s_1(k) + \boldsymbol{\xi}_2 s_2(k) + \boldsymbol{\xi}_3 s_3(k) + \mathbf{g}(k) \tag{6.158}$$

where $\boldsymbol{\xi}_1$, $\boldsymbol{\xi}_2$, and $\boldsymbol{\xi}_3 \in \mathbb{R}^{15}$ are arbitrary unit length vectors and $s_1(k)$, $s_2(k)$ and $s_3(k) \in \mathbb{R}$ are statistically independent Gaussian sequences of zero mean and variances 5, 3 and 1, respectively. The error vector $\mathbf{g}(k) \sim \mathcal{N}\{\mathbf{0}, 0.05\mathbf{I}\}$.

From this process, a number of reference sets was simulated, which form the the basis for determining the Hotelling's T^2 and Q statistics. To determine the Type I error for a significance of α, one additional set of 250 test samples that was not used as a reference set were simulated. The reference sets included $K = 20$, 50, 100, 500 and 1000 samples and were simulated a total of 100 times (Monte Carlo experiment). The control limits were obtained for each experiment with a significance of $\alpha = 0.05$. Figure 6.19 shows the results of these Monte Carlo experiments for a variety of retained components, ranging from 1 to 14.

Given that each point in Figure 6.19 represents the average Type I error, it is interesting to note that the smallest departure of the T^2 statistic arises for $n = 3$. Any other n produced a more significant departure. As expected, the smaller the size of the reference set, the more pronounced the departure from the theoretical 5%. Whilst this example yielded an up to 4% difference in Type I error for the

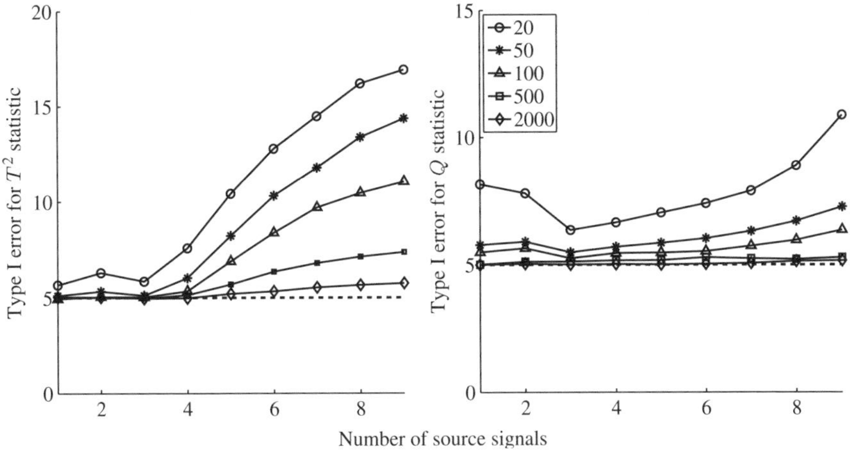

Figure 6.20 Type I error [%] *(LOO CV) for* $\alpha = 0.05$.

T^2 statistic and $K = 20$, a more considerable impact is noticeable for the Q statistic.

To overcome this issue, Ramaker *et al.* (2004) proposed the use of leave-one-out cross validation (LOO CV)[9] to determine the score variables and the mismatch between the original variable, $\mathbf{z}_0$, and its projection onto the model subspace $\mathbf{g} = \mathbf{z}_0 - \widehat{\mathbf{P}}\widehat{\mathbf{t}} = \left[\mathbf{I} - \widehat{\mathbf{P}}\widehat{\mathbf{P}}^T\right]\mathbf{z}_0$. More precisely, the kth data sample is removed from the reference set and the remaining $K - 1$ samples are used to estimate P_{-k}. This is followed by a determination of the retained scores and the residual vector for the kth sample, that is $\widehat{\mathbf{t}}(k) = \widehat{\mathbf{P}}_{-k}^T\mathbf{z}_0(k)$ and $\mathbf{g}(k) = \mathbf{z}_0(k) - \widehat{\mathbf{P}}_{-k}\widehat{\mathbf{t}}(k)$.

The application of LOO CV therefore produces a total of K t-scores and residual vectors which are then used to determine the covariance matrix $\widehat{\mathbf{S}}_{tt}$ and the control limit for the Q statistic. Comparing Figures 6.19 and 6.20 allows a direct comparison between the original approach and LOO CV, respectively. Although the latter technique yielded a significant reduction in the Type I error of the Q statistic, it did not show any improvement for the Hotelling's T^2 statistic. To the contrary, the LOO CV approach produced a very significant increase in the Type I errors.

Although the work in Ramaker *et al.* (2004) highlighted that non-negative squared forms are difficult to establish using small reference sets, it did not identify a theoretical rationale that explains the unwanted increase in number of Type I error (Q statistic) and Type II error (Hotelling's T^2 statistic). Analyzing Remark 6.4.1, however, reveals that the Hotelling's T^2 statistic can only follow an F-distribution when the estimate of $\mathbf{S}_{tt}$ relies on data that were not used for the estimation of $\mathbf{P}$. For conventional PCA, however, $\widehat{\mathbf{P}}$ and $\widehat{\mathbf{S}}_{tt}$ store the eigenvectors and the eigenvalues of $\widehat{\mathbf{S}}_{zz}$ and hence, statistical independence is not guaranteed.

[9] The principle of which is discussed in Subsection 2.4.1.

Moreover, the use of LOO CV produces a total of K different model subspaces and residual subspaces. This, in turn, implies that there is no guarantee that the estimate of $\mathbf{S}_{tt}$ follows a Wishart distribution. Despite the fact that this approach produced a substantial reduction in Type I error for the Q statistic, the sequence of K residual vectors corresponds, therefore, to a total of K different residual subspaces. A more minor issue is the computational demand to implement the LOO CV approach.

Next, we discuss an alternative approach that overcomes the problems of the LOO CV technique. To improve this method, a more rigorous separation of the data is required to guarantee statistical independence for distribution functions of:

- $\widehat{\mathbf{P}}$ and $\mathbf{z}_0$ (to estimate $\mathbf{S}_{tt}$ and T_α^2); and
- $\mathbf{g}$ and (to estimate Q_α).

The proposed division produces two independent reference sets

$$\mathbf{Z}_1^T = \begin{bmatrix} \mathbf{z}(1) & \mathbf{z}(2) & \cdots & \mathbf{z}(K_1) \end{bmatrix} \tag{6.159}$$

and

$$\mathbf{Z}_2^T = \begin{bmatrix} \mathbf{z}(K_1+1) & \mathbf{z}(K_1+2) & \cdots & \mathbf{z}(K) \end{bmatrix} \tag{6.160}$$

of equal length. Next, an estimate of $\mathbf{P}$, $\widehat{\mathbf{P}}_1$, based on

$$\widehat{\mathbf{S}}_{z_0z_0}^{(1)} = \frac{1}{K_1-1}\sum_{k=1}^{K_1}\left(\mathbf{z}(k)-\widehat{\bar{\mathbf{z}}}_1\right)\left(\mathbf{z}(k)-\widehat{\bar{\mathbf{z}}}_1\right)^T \tag{6.161}$$

is obtained, followed by computing

$$\widehat{\mathbf{S}}_{tt}^{(2)} = \frac{1}{K-K_1-1}\sum_{k=K_1+1}^{K}\widehat{\mathbf{P}}_1^T\left(\mathbf{z}(k)-\widehat{\bar{\mathbf{z}}}_1\right)\left(\mathbf{z}(k)-\widehat{\bar{\mathbf{z}}}_1\right)^T\widehat{\mathbf{P}}_1. \tag{6.162}$$

The proposed division of the reference data set guarantees that the distribution functions for $\widehat{\mathbf{P}}_1$ and $\mathbf{Z}_2$ are statistically independent. This, in turn, implies that

$$\widehat{\bar{\mathbf{z}}}_1 \sim \mathcal{N}(\mathbf{0}, \tfrac{1}{K-K_2}\mathbf{S}_{z_0z_0}) \tag{6.163}$$

and

$$\left(K-K_1-1\right)\widehat{\mathbf{S}}_{tt}^{(2)} \sim \mathcal{W}\left(K-K_1-1, \widehat{\mathbf{P}}_1^T\mathbf{S}_{z_0z_0}\widehat{\mathbf{P}}_1\right). \tag{6.164}$$

Moreover, using this data separation, the score variables now have the following distribution for new observations that are not included in $\mathbf{Z}_1$ and $\mathbf{Z}_2$

$$\mathbf{t} = \widehat{\mathbf{P}}_1^T\left(\mathbf{z}-\widehat{\bar{\mathbf{z}}}_1\right) \sim \mathcal{N}\left(\mathbf{0}, \tfrac{K-K_1+1}{K-K_1}\widehat{\mathbf{P}}_1^T\mathbf{S}_{z_0z_0}\widehat{\mathbf{P}}_1\right) \tag{6.165}$$

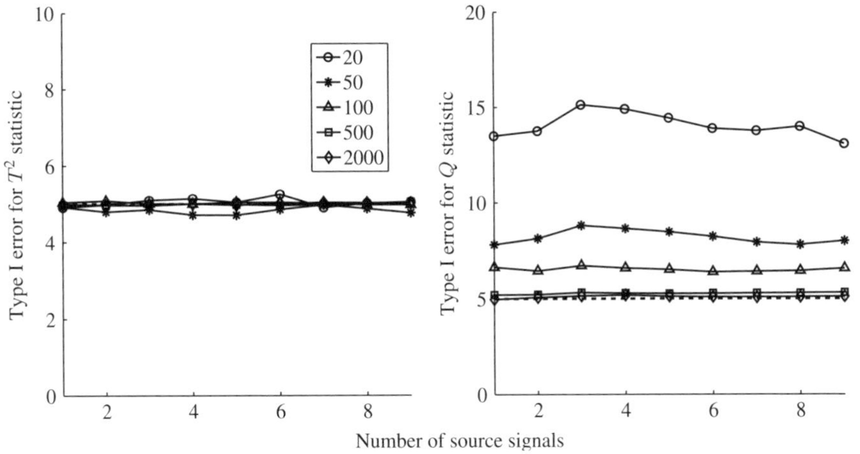

Figure 6.21 *Number of Type I error [%] (two-stage PCA) for a significance of 0.05.*

and consequently, the T^2 statistic follows an F-distribution, that is

$$T^2 = \mathbf{t}^T \widehat{\mathbf{S}}_{tt_2}^{-1} \mathbf{t} \sim \frac{n\left(\left(K - K_1\right)^2 - 1\right)}{\left(K - K_1\right)\left(K - K_1 - n\right)} \mathcal{F}\left(n, K - K_1 - n\right). \tag{6.166}$$

Utilizing the same approach to determine the $\widehat{Q}_\alpha$ statistic, that is, computing the residual vectors from the reference set $\mathbf{Z}_2$ and the estimate of $\mathbf{P}$ from $\mathbf{Z}_1$, Figure 6.21 shows, as expected, that the 50% percentile of the Monte Carlo experiments for $\alpha = 0.05$. Note that an equal separation of the reference data set resulted in the determination of only 10 PCs if 20 samples were available. This, on the other hand, implied that a total of nine discarded PCs could be analyzed. For small sample sets, including less than 100 samples, an increase in the Type I error for the Q statistic arose. This can be attributed to the fact that the distribution function of the Q statistic, used to determine $\widehat{Q}_\alpha$, is an approximation, which requires a larger data set to be accurate.

The main focus of this section, however, is on the T^2 statistic, which the data division showed to outperform the approach by Ramaker *et al.* (2004). More precisely, the 50% percentile is very close to $\alpha = 0.05$ for any combination of the number of retained PCs and size of the reference sets.

6.5 Tutorial session

Question 1. Explain why PCA produces a biased estimation of the orientation of the model and residual subspaces when the error covariance matrix for a set of process variables, $\mathbf{z}_0$, is not of the form $\mathbf{S}_{gg} = \sigma_g \mathbf{I}$. What is the impact of a biased estimation in terms of extracting the source and error variables?

Question 2. For the application of maximum likelihood PCA, What is the reason for constraining the maximum number of estimated parameters according to (6.46)? In case the number of error variances is larger than this maximum number, discuss ways to estimate the error covariance matrix.

Question 3. Discuss the advantage of using PLS over standard ordinary least squares in determining a parametric regression model if the input variable set is highly correlated. What happens if there is a linear dependency among the input variable set?

Question 4. Explain why OLS, PLS and MRPLS produce a biased estimation of the parameter regression matrix between the input and output variables if both variable sets are corrupted by an error term. Explain the impact of this estimation bias upon the extraction of source signals according to the PLS and MRPLS data structures in (2.24) and (2.51)?

Question 5. What are outliers and how can they be categorized? What is the effect of outliers on the determination of PCA and PLS models?

Project 1. Based on a Monte Carlo simulation, use the example described in (6.53) to (6.55) and alter the magnitude of the individual diagonal elements of $\mathbf{S}_{gg}$ as well as their differences in value and compare the accuracy of the model subspace between the MLPCA and the PCA estimates. How does increasing or reducing the number of samples affect the accuracy?

Project 2. Contrast the stopping rule introduced in Subsection 6.1.6 with those discussed in Subsection 2.4.1 using the simulation example in (6.53) to (6.55). For this comparison, change the magnitude of the individual diagonal elements of $\mathbf{S}_{gg}$ as well as their differences in value. How does increasing or reducing the number of samples affect the estimate of n?

Project 3. Repeat the Monte Carlo experiment for the data structure described in (6.87) to (6.89c). Comment on the experimental results in terms of the accuracy of the OLS estimation for a varying sample size $K = 100$, 200, 500, 1000 and 2000. Compare the results with the PLS estimation of the regression matrix for each K by varying the number of retained LV sets $n = 1, 2, \ldots, 10$ and comment on the results obtained. Is there an optimal trade off between the accuracy of the parameter estimation, model prediction error and the number of retained sets of LVs for each K?

Project 4. For PCA, study the impact of outliers model using a Monte Carlo simulation on the basis of the example described in (6.53) to (6.54) by defining $\mathbf{S}_{gg} = \sigma_g \mathbf{I}$, with $\sigma_g^2 = 0.05$. Inject 1%, 2%, 5%, 10%, 20% and 50% of outliers into a simulated reference set of $K = 200, 500, 1000, 2000$ and 5000 samples and comment upon the accuracy of the estimated model and residual subspaces. Next, use the Stahel-Donoho scale estimator to determine a robust estimation of the data covariance matrix and examine how the accuracy of estimating the model and residual subspaces improves when using the robust estimate?

Project 5. Repeat the Monte Carlo simulation described in (6.158) by altering the error variance σ_{g}^2, the variances of the source signals $\sigma_{s_1}^2$, $\sigma_{s_2}^2$ and $\sigma_{s_3}^2$ and the number of retained LV sets.

7

Monitoring multivariate time-varying processes

As outlined in the first three chapters, MSPC relies on linear parametric PCA or PLS models that are time invariant. Such models, however, may become unrepresentative in describing the variable interrelationships some time after they were identified. Gallagher *et al.* (1997) pointed out that most industrial processes are time-varying and that the monitoring of such processes, therefore, requires the adaptation of PCA and PLS models to accommodate this behavior. In addition to the parametric models, the monitoring statistics, including their control limits, may also have to vary with time, as discussed in Wang *et al.* (2003). Another and very important requirement is that the adapted MSPC monitoring model must still be able to detect abnormal process behavior.

Focussing on PCA, this chapter discusses three techniques that allow an adaptation of the PCA model, Recursive PCA (RPCA), Moving Window PCA (MWPCA), and a combination of both. Embedding an adaptive PLS model for constructing the associated monitoring statistics, however, is a straightforward extension and is discussed in Section 7.7. The research literature discussed adaptive PLS algorithms, for example in Dayal and MacGregor (1997c); Helland *et al.* (1991); Qin (1998); Wang *et al.* (2003). For the non-causal data representation in (2.2) and the causal ones in (2.24) and (2.51), two properties are of particular interest:

- the speed of adaptation, describing how fast the monitoring model changes with new events; and
- the speed of computation, that is the time the algorithm takes to complete one iteration for model adaptation.

Statistical Monitoring of Complex Multivariate Processes: With Applications in Industrial Process Control, First Edition. Uwe Kruger and Lei Xie.

The sensitivity issue, that is how sensitive the adaptive MSPC method is in detecting incipient changes, requires the use of a multiple step ahead application of the currently updating model. In other words, the updated model is not applied when the next sample is available but when the next $\mathfrak{K} > 1$ sample becomes available. This includes the application of the adaptive PCA model as well as the computation of the monitoring statistics and the control limits.

This chapter presents a discussion of the relevant literature on monitoring time-varying processes, followed in Sections 7.2 and 7.3 by a discussion on how to adapt PCA models, which also include the adaptation of the univariate monitoring statistics and their associated control limits. To show the working of an adaptive MSPC model, Sections 7.3, 7.5 and 7.6 summarize application studies to a simulation example, data from a simulated industrial process and recorded data from a furnace process, respectively. Finally, Section 7.8 presents a tutorial session including small projects and questions to help familiarization with this material.

7.1 Problem analysis

The literature has proposed two approaches for updating identified models. The first one is related to a moving window that slides along the data and the other is a recursive formulation. The principle behind the moving window approach is well-known. The window progresses along the data as new observations become available and a new process model is generated by including the newest sample and excluding the oldest one. On the other hand, recursive techniques update the model for an ever-increasing data set that includes new samples without discarding the old ones.

For process monitoring, recursive methods offer efficient computation by updating the process model using the previous model rather than completely building it from the original data (Dayal and MacGregor 1997c; Helland *et al.* 1991; Li *et al.* 2000; Qin 1998). Although conceptually simple and successfully employed for process monitoring (Li *et al.* 2000), Recursive Principal Component Aanalysis (RPCA) may be difficult to implement in practice for the following two reasons:

- the data set on which the model is updated is ever-growing, leading to a reduction in the speed of adaptation as the data size increases. This issue is discussed in Wang *et al.* (2005); and
- RPCA includes older data that become increasingly unrepresentative of the time-varying process. If a forgetting factor is introduced to down-weight older samples, the selection of this factor can be difficult without *a priori* knowledge of likely fault conditions.

In comparison, the Moving Window Principal Component Analysis (MWPCA) formulation can overcome some of the above problems by including a sufficient number of data points in the time-window, from which to build the adaptive process model. More precisely, MWPCA allows older samples to be discarded in favor of newer ones that are more representative of the current process operation. Furthermore, the use of a constant size of window results in a constant speed of model adaptation. This, however, may cause a problem when the window has to cover a large number of samples in order to include sufficient process variation for modeling and monitoring purposes since the computational speed of MWPCA then drops significantly. If a smaller window size is attempted to improve computational efficiency, data within the window may then not adequately reveal the underlying relationships between the process variables. An additional danger of a short window is that the resulting model may adapt to process changes so quickly that abnormal behavior may remain undetected.

To improve computational efficiency without compromising the window size, a fast moving window PCA scheme has been proposed by Wang *et al.* (2005). This scheme relies on the adaptation of the RPCA procedure and yields an MWPCA algorithm for updating (inclusion of a new sample) and downdating (removal of the oldest sample). Blending recursive and moving window techniques has proved beneficial in computing the computation of the discrete Fourier transform and in least-squares approximations (Aravena 1990; Fuchs and Donner 1997).

Qin (1998) discussed the integration of a moving window approach into recursive PLS, whereby the process data are grouped into sub-blocks. Individual PLS models are then built for each data block. When the window moves along the blocks, a PLS model for the selected window is calculated using the sub-PLS models rather than the original sub-block data. The fast MWPCA is intuitively adapted, as the window slides along the original data sample by sample. The general computational benefits of the new algorithm are analyzed in terms of the number of floating point operations required to demonstrate a significantly increased computational efficiency.

Another important aspect worth considering when monitoring a time-varying process is how to adapt the control limits as the process moves on. In a similar fashion to the work in Wang *et al* (2005, 2003), this chapter uses a $\mathfrak{K}$-step-ahead horizon in the adaptive PCA monitoring procedure. This implies that the adapted PCA model, including the adapted control limit, is applied after $\mathfrak{K} > 1$ samples are recorded. The advantage of using an older process model for process monitoring is demonstrated by application to simulated data for fault detection. In order to simplify the equations in Sections 7.2 and 7.3 the notation $\widehat{\cdot}$ to denote estimates, for example those of the correlation matrix or the eigenvalues, is omitted.

7.2 Recursive principal component analysis

This section introduces the working of a recursive PCA algorithm, which requires:

- the adaptation of the data covariance or correlation matrix;
- a recalculation of the eigendecomposition; and
- an adjustment of the control limits.

The introduction of the RPCA formulation also provides, in parts, an introduction into MWPCA, as the latter algorithm incorporates some of the steps of RPCA derived below. The next section then develops the MWPCA approach along with an adaptation of the control limits and a discussion into suitable methods for updating the eigendecomposition.

The application study in Chapter 4 showed that using the covariance or correlation matrix makes negligible difference if the process variables have a similar variance. Similar observations have also been reported in Jackson (2003). If this is not the case, a scaling is required as the PCA model may otherwise be dominated by a few variables that have a larger variance. This is discussed extensively in the literature. For example, Section 3.3 in reference Jackson (2003) outlines that the entries in the correlation matrix does not have units[1] whereas the covariance matrix may have different units. The same reference, page 65, makes the following remark.

Remark 7.2.1 *The use of the correlation matrix is so widespread in some fields of application that many practitioners never use the covariance matrix at all and may not be aware that this is a viable option in some instances.*

Given that it is preferred to use the correlation matrix, its adaptation is achieved efficiently by adaptively calculating the current correlation matrix from the previous one and the new observation, as discussed in Li *et al.* (2000). For the original data matrix $\mathbf{Z} \in \mathbb{R}^{K \times n_z}$, which includes n_z process variables collected until time instant K, the mean and standard deviation are given by $\bar{\mathbf{z}}$ and $\mathbf{\Sigma} = \mathrm{diag}\{\sigma_1 \cdots \sigma_{n_z}\}$. The original data matrix $\mathbf{Z}$ is then scaled to produce $\widetilde{\mathbf{Z}}_0 = \left[\mathbf{Z} - \mathbf{1}\bar{\mathbf{z}}^T\right]\mathbf{\Sigma}^{-1}$, such that each variable now has zero mean and unit variance. According to (2.2) and Table 2.1, $\mathbf{z}_0 = \mathbf{z} - \bar{\mathbf{z}}$ and the notation $\sim$ results from the scaling to unit variance. The correlation matrix, $\mathcal{C}_{z_0 z_0}$ obtained from the scaled reference data set is accordingly

$$\mathcal{C}_1 = \tfrac{1}{K-1}\widetilde{\mathbf{Z}}_0^T\widetilde{\mathbf{Z}}_0. \tag{7.1}$$

Note that the change in subscript from $z_0 z_0$ to 1 is to distinguish between the *old* correlation matrix from the adapted one, $\mathcal{C}_2$, when the *new* sample $\mathbf{z}(K+1)$ becomes available. The mean value of the augmented data matrix is given by

$$\left[\begin{array}{c} \left.\begin{matrix} \mathbf{z}^T(1) \\ \mathbf{z}^T(2) \\ \vdots \\ \mathbf{z}^T(K) \end{matrix}\right\} = \mathbf{Z} \\ \hline \mathbf{z}^T(K+1) \end{array}\right] \tag{7.2}$$

[1] To be more precise, the unit of the elements of the correlation matrix is 1.

and can be updated as follows

$$\bar{\mathbf{z}}_2 = \tfrac{K}{K+1}\bar{\mathbf{z}}_1 + \tfrac{1}{K+1}\mathbf{z}(K+1). \tag{7.3}$$

Again, the subscripts 1 and 2 for the mean vector and the standard deviations are to discriminate between the *old* and *adapted* ones. The adapted standard deviation of the ith process variable is

$$\sigma_{i_2}^2 = \tfrac{K-1}{K}\sigma_{i_1}^2 + \Delta\bar{z}_{i_2}^2 + \tfrac{1}{K}\left(z_i(K+1) - \bar{z}_{i_2}\right)^2 \tag{7.4}$$

with $\Delta\bar{\mathbf{z}}_2 = \bar{\mathbf{z}}_2 - \bar{\mathbf{z}}_1$. Given that $\boldsymbol{\Sigma}_2 = \mathrm{diag}\{\sigma_{1_2} \cdots \sigma_{n_{Z_2}}\}$, the centering and scaling of the new sample, $\mathbf{z}(K+1)$, is

$$\widetilde{\mathbf{z}}_0(K+1) = \boldsymbol{\Sigma}_2^{-1}\left(\mathbf{z}(K+1) - \bar{\mathbf{z}}_2\right). \tag{7.5}$$

Utilizing $\Delta\bar{\mathbf{z}}_2$, $\boldsymbol{\Sigma}_2$, $\boldsymbol{\Sigma}_1$, $\widetilde{\mathbf{z}}_0(K+1)$ and the old correlation matrix $\mathcal{C}_1$, the updated correlation matrix $\mathcal{C}_2$ is given by

$$\mathcal{C}_2 = \tfrac{K-1}{K}\boldsymbol{\Sigma}_2^{-1}\boldsymbol{\Sigma}_1\mathcal{C}_1\boldsymbol{\Sigma}_1\boldsymbol{\Sigma}_2^{-1} + \boldsymbol{\Sigma}_2^{-1}\Delta\bar{\mathbf{z}}_2\Delta\bar{\mathbf{z}}_2^T\boldsymbol{\Sigma}_2^{-1} + \tfrac{1}{K}\widetilde{\mathbf{z}}_0(K+1)\widetilde{\mathbf{z}}_0^T(K+1). \tag{7.6}$$

The eigendecomposition of $\mathcal{C}_2$ then provides the required new PCA model.

7.3 Moving window principal component analysis

On the basis of the adaptation procedure for RPCA, it is now shown how to derive an efficient adaptation of $\mathcal{C}_1$ involving an updating stage, as for RPCA, and a downdating stage for removing the contribution of the oldest sample. This adaptation requires a three step procedure outlined in the next subsection. Then, Subsection 7.3.2 shows that the up- and downdating of $\mathcal{C}_1$ is numerically efficient for large window sizes. Subsection 7.3.3 then introduces a $\mathfrak{K}$-step-ahead application of the adapted MWPCA model to improve the sensitivity of the on-line monitoring scheme. The adaptation of the control limits and the monitoring charts is discussed in Subsections 7.3.4 and 7.3.5, respectively. Finally, Subsection 7.3.6 provides a discussion concerning the required minimum size for the moving window.

7.3.1 Adapting the data correlation matrix

RPCA updates $\mathcal{C}_1$ by incorporating the new sample (updating). A conventional moving window approach operates by first discarding the oldest sample (downdating) from the model and then adding the new sample (updating). Figure 7.1 shows details of this two-step procedure for a window size $\mathcal{K}$, with k being a sample index. MWPCA is based on this up- and downdating, but incorporates the adaptation developed for RPCA (Li *et al.* 2000). The three matrices in Figure 7.1 represent the data in the previous window (Matrix I), the result of removing the oldest sample $\mathbf{z}(k)$ (Matrix II), and the current window of selected

Old Window		Intermediate Data		New Window
$\begin{pmatrix} \mathbf{z}(k) \\ \mathbf{z}(k+1) \\ \vdots \\ \mathbf{z}(k+\mathcal{K}-1) \end{pmatrix}_{\mathcal{K}\times n_z}$	$\Longrightarrow$	$\begin{pmatrix} \mathbf{z}(k+1) \\ \vdots \\ \mathbf{z}(k+\mathcal{K}-1) \end{pmatrix}_{(\mathcal{K}-1)\times n_z}$	$\Longrightarrow$	$\begin{pmatrix} \mathbf{z}(k+1) \\ \vdots \\ \mathbf{z}(k+\mathcal{K}-1) \\ \mathbf{z}(k+\mathcal{K}) \end{pmatrix}_{\mathcal{K}\times n_z}$
Matrix I		Matrix II		Matrix III
$(\bar{\mathbf{z}}_k, \mathbf{\Sigma}_k, \mathcal{C}_k)$		$(\bar{\mathbf{z}}^*, \mathbf{\Sigma}^*, \mathcal{C}^*)$		$(\bar{\mathbf{z}}_{k+1}, \mathbf{\Sigma}_{k+1}, \mathcal{C}_{k+1})$

Figure 7.1 Two-step adaptation to construct new data window.

data (Matrix III) produced by adding the new sample $\mathbf{z}(k+\mathcal{K})$ to Matrix II. Next, the adaptations of the mean vectors, the standard deviations and the correlation matrices for Matrix II and III are determined.

Step 1: Matrix I to Matrix II

The downdating of the effect of removing the oldest sample from Matrix I can be computed in a similar way to that shown in (7.3).

$$\bar{\mathbf{z}}^* = \frac{1}{\mathcal{K}-1}\left(\mathcal{K}\bar{\mathbf{z}}_k - \mathbf{z}(k)\right). \tag{7.7}$$

Equation (7.8) describes how to downdate the impact of $\mathbf{z}(k)$ upon the variable mean

$$\Delta\bar{\mathbf{z}}^* = \bar{\mathbf{z}}_k - \bar{\mathbf{z}}^*. \tag{7.8}$$

Using (7.7) and (7.8) the variance of the ith process variable becomes

$$\sigma_i^{*^2} = \frac{\mathcal{K}-1}{\mathcal{K}-2}\sigma_{i_k}^2 - \frac{\mathcal{K}-1}{\mathcal{K}-2}\Delta\bar{z}_i^{*^2} - \frac{1}{\mathcal{K}-2}\left(z_i(k) - \bar{z}_{i_k}\right)^2 \tag{7.9}$$

and (7.10) stores the standard deviations of the n_z process variables

$$\mathbf{\Sigma}^* = \mathrm{diag}\left\{\sigma_1^* \quad \cdots \quad \sigma_{n_z}^*\right\}. \tag{7.10}$$

Finally, the impact of recursively downdating $\mathbf{z}(k)$ from $\mathcal{C}_k$ follows from the above equations. For simplicity, the matrix $\widetilde{\mathcal{C}}$ is now introduced to compute $\mathcal{C}^*$

$$\widetilde{\mathcal{C}} = \frac{\mathcal{K}-2}{\mathcal{K}-1}\mathbf{\Sigma}_k^{-1}\mathbf{\Sigma}^*\mathcal{C}^*\mathbf{\Sigma}^*\mathbf{\Sigma}_k^{-1} \tag{7.11}$$

which can be further divided into

$$\widetilde{\mathcal{C}} = \mathcal{C}_k - \mathbf{\Sigma}_k^{-1}\Delta\bar{\mathbf{z}}^*\Delta\bar{\mathbf{z}}^{*^T}\mathbf{\Sigma}_k^{-1} - \frac{1}{\mathcal{K}-1}\widetilde{\mathbf{z}}_0(k)\widetilde{\mathbf{z}}_0^T(k) \tag{7.12}$$

where

$$\widetilde{\mathbf{z}}_0(k) = \mathbf{\Sigma}_k^{-1}\left(\mathbf{z}(k) - \bar{\mathbf{z}}_k\right). \tag{7.13}$$

The downdating for the correlation matrix after elimination of the oldest sample, Matrix II, can now be expressed in (7.14).

$$\mathcal{C}^* = \tfrac{\mathcal{K}-1}{\mathcal{K}-2}\boldsymbol{\Sigma}^{*^{-1}}\boldsymbol{\Sigma}_k\widetilde{\mathcal{C}}\boldsymbol{\Sigma}_k\boldsymbol{\Sigma}^{*^{-1}}. \tag{7.14}$$

Step 2: Matrix II to Matrix III

This step involves the updating of the PCA model by incorporating the new sample. Based on (7.3) and (7.7) the updated mean vector is

$$\bar{\mathbf{z}}_{k+1} = \tfrac{1}{\mathcal{K}}\left((\mathcal{K}-1)\,\bar{\mathbf{z}}^* + \mathbf{z}(k+\mathcal{K})\right). \tag{7.15}$$

The change in the mean vectors are computed from (7.15) and (7.16)

$$\Delta\bar{\mathbf{z}}_{k+1} = \bar{\mathbf{z}}_{k+1} - \bar{\mathbf{z}}^*, \tag{7.16}$$

and the standard deviation of the ith variable follows from (7.17).

$$\sigma^2_{i_{k+1}} = \tfrac{\mathcal{K}-2}{\mathcal{K}-1}\sigma_i^{*^2} + \Delta\bar{z}^2_{i_{k+1}} - \tfrac{1}{\mathcal{K}-1}\left(z_i(k+\mathcal{K}) - \bar{z}_{i_{k+1}}\right)^2 \tag{7.17}$$

and (7.18)

$$\boldsymbol{\Sigma}_{k+1} = \mathrm{diag}\left\{\sigma_{1_{k+1}} \quad \cdots \quad \sigma_{n_{Z_{k+1}}}\right\}. \tag{7.18}$$

Finally, the scaling of the newest sample, $\mathbf{z}(k+\mathcal{K})$, and the updating of the correlation matrix are described in (7.19)

$$\widetilde{\mathbf{z}}_0(k+\mathcal{K}) = \boldsymbol{\Sigma}^{-1}_{k+1}\left(\mathbf{z}(k+\mathcal{K}) - \bar{\mathbf{z}}_{k+1}\right) \tag{7.19}$$

and (7.20)

$$\mathcal{C}_{k+1} = \tfrac{\mathcal{K}-2}{\mathcal{K}-1}\boldsymbol{\Sigma}^{-1}_{k+1}\boldsymbol{\Sigma}^*\mathcal{C}^*\boldsymbol{\Sigma}^*\boldsymbol{\Sigma}^{-1}_{k+1} + \boldsymbol{\Sigma}^{-1}_{k+1}\Delta\bar{\mathbf{z}}_{k+1}\Delta\bar{\mathbf{z}}^T_{k+1}\boldsymbol{\Sigma}^{-1}_{k+1} + \tfrac{1}{\mathcal{K}-1}\widetilde{\mathbf{z}}_0(k+\mathcal{K})\widetilde{\mathbf{z}}_0^T(k+\mathcal{K}), \tag{7.20}$$

respectively. Combining Steps 1 and 2 allows deriving Matrix III from Matrix I, the adapted mean, standard deviation and correlation matrix, which is shown next.

Step 3: Combination of Step 1 and Step 2

Including downdating, (7.7), and updating, (7.15), adapting the mean vector directly yields

$$\bar{\mathbf{z}}_{k+1} = \bar{\mathbf{z}}_k + \tfrac{1}{\mathcal{K}}\left(\mathbf{z}(k+\mathcal{K}) - \mathbf{z}(k)\right). \tag{7.21}$$

The adapted standard deviations follow from combining (7.9) and (7.17)

$$\sigma^2_{i_{k+1}} = \sigma^2_{i_k} + \Delta\bar{z}^2_{i_{k+1}} - \Delta\bar{z}_i^{*^2} + \frac{\left(z_i(k+\mathcal{K}) - \bar{z}_{i_{k+1}}\right)^2 - \left(z_i(k) - \bar{z}_{i_k}\right)^2}{\mathcal{K}-1} \tag{7.22}$$

Table 7.1 Procedure to update correlation matrix for the MWPCA approach.

Step	Equation	Description
1	$\bar{\mathbf{z}}^* = \frac{1}{\mathcal{K}-1}\left(\mathcal{K}\bar{\mathbf{z}}_k - \mathbf{z}(k)\right)$	Mean of Matrix II
2	$\Delta\bar{\mathbf{z}}^* = \bar{\mathbf{z}}_k - \bar{\mathbf{z}}^*$	Difference between means
3	$\widetilde{\mathbf{z}}_0(k) = \mathbf{\Sigma}_k^{-1}\left(\mathbf{z}(k) - \bar{\mathbf{z}}_k\right)$	Scale the discarded sample
4	$\widetilde{\mathcal{C}} = \mathcal{C}_k - \mathbf{\Sigma}_k^{-1}\Delta\bar{\mathbf{z}}^*\Delta\bar{\mathbf{z}}^{*^T}\mathbf{\Sigma}_k^{-1} - \frac{1}{\mathcal{K}-1}\widetilde{\mathbf{z}}_0(k)\widetilde{\mathbf{z}}_0^T(k)$	Bridge over Matrix I and III
5	$\bar{\mathbf{z}}_{k+1} = \frac{1}{\mathcal{K}}\left((\mathcal{K}-1)\,\bar{\mathbf{z}}^* + \mathbf{z}(k+\mathcal{K})\right)$	Mean of Matrix III
6	$\Delta\bar{\mathbf{z}}_{k+1} = \bar{\mathbf{z}}_{k+1} - \bar{\mathbf{z}}^*$	Difference between means
7	$\sigma_{i_{k+1}}^2 = \sigma_{i_k}^2 + \Delta\bar{z}_{i_{k+1}}^2 - \Delta\bar{z}_i^{*^2} + \frac{\left(z_i(k+\mathcal{K}) - \bar{z}_{i_{k+1}}\right)^2 - \left(z_i(k) - \bar{z}_{i_k}\right)^2}{\mathcal{K}-1}$	Standard deviation of Matrix III
8	$\mathbf{\Sigma}_{k+1} = \mathrm{diag}\left\{\sigma_{1_{k+1}} \cdots \sigma_{n_{Z_{k+1}}}\right\}$	Store standard deviations in Matrix III
9	$\widetilde{\mathbf{z}}_0(k+\mathcal{K}) = \mathbf{\Sigma}_{k+1}^{-1}\left(\mathbf{z}(k+\mathcal{K}) - \bar{\mathbf{z}}_{k+1}\right)$	Scale the new sample
10	$\mathcal{C}_{k+1} = \frac{\mathcal{K}-2}{\mathcal{K}-1}\mathbf{\Sigma}_{k+1}^{-1}\mathbf{\Sigma}_k\widetilde{\mathcal{C}}\mathbf{\Sigma}_k\mathbf{\Sigma}_{k+1}^{-1} + \mathbf{\Sigma}_{k+1}^{-1}\Delta\bar{\mathbf{z}}_{k+1}\Delta\bar{\mathbf{z}}_{k+1}^T\mathbf{\Sigma}_{k+1}^{-1} + \frac{1}{\mathcal{K}-1}\mathbf{z}_0(k+\mathcal{K})\mathbf{z}_0(k+\mathcal{K})^T$	Correlation matrix of Matrix III

where $\Delta\bar{z}_{i_{k+1}} = \frac{1}{\mathcal{K}}\left(z_i\,(k+\mathcal{K}) - z_i\,(k)\right)$ and $\Delta\bar{z}_i^* = \frac{1}{\mathcal{K}-1}\left(z_i\,(k) - \bar{z}_{i_k}\right)$. Substituting (7.12) and (7.14) into (7.20) produces the adapted correlation matrix of Matrix III

$$\begin{aligned}\mathcal{C}_{k+1} = \mathbf{\Sigma}_{k+1}^{-1}\mathbf{\Sigma}_k\widetilde{\mathcal{C}}\mathbf{\Sigma}_k\mathbf{\Sigma}_{k+1}^{-1} + \mathbf{\Sigma}_{k+1}^{-1}\Delta\bar{\mathbf{z}}_{k+1}\Delta\bar{\mathbf{z}}_{k+1}^T\mathbf{\Sigma}_{k+1}^{-1} \\ + \frac{1}{\mathcal{K}-1}\widetilde{\mathbf{z}}_0(k+\mathcal{K})\mathbf{z}_0^T(k+\mathcal{K}).\end{aligned} \tag{7.23}$$

The combination of Steps 1 and 2 constitutes the fast moving window technique, which is summarized in Table 7.1 for convenience.

The MWPCA technique gains part of its computational efficiency by incorporating the efficient update and downdate procedures. This is examined in more detail in the Subsection 7.3.3. Subsection 7.3.2 discusses computational issues regarding the adaptation of the eigendecomposition.

7.3.2 Adapting the eigendecomposition

Methods for updating the eigendecomposition of symmetric positive definite matrices have been extensively studied over the past decades. The following list includes the most commonly proposed methods for recursively adapting such matrices:

- *rank one modification* (Bunch *et al.* 1978; Golub 1973);
- *inverse iteration* (Golub and van Loan 1996; van Huffel and Vandewalle 1991);
- *Lanczos tridiagonalization* (Cullum and Willoughby 2002; Golub and van Loan 1996; Paige 1980; Parlett 1980);
- *first order perturbation* (Champagne 1994; Stewart and Sun 1990; Willink 2008);
- *projection-based adaptation* (Hall *et al.* 1998, 2000, 2002); and
- *data projection method* (Doukopoulos and Moustakides 2008).

Alternative work relies on gradient descent methods (Chatterjee *et al.* 2000) which are, however, not as efficient.

The computational efficiency of the listed algorithms can be evaluated by their number of floating point (flops) operations consumed, which is listed in Table 7.2. Evaluating the number of flops in terms of the order $O(\cdot)$, that is, $O\left(n_z^3\right)$ is of order n_z^3, highlights that the data projection method and the first order perturbation are the most economic methods. Given that $n_z > n_k$, where n_k is the estimated number of source signals for the kth data window, Table 7.2 suggests that the data projection method is more economic than the first order perturbation method. It should be noted, however, that the data projection method:

- adapts the eigenvectors but not the eigenvalues; and
- assumes that the number of source signals, n_k is constant.

If the eigenvectors are known, the eigenvalues can easily be computed as

$$\lambda_{i_{k+1}} = \mathbf{p}_{i_{k+1}}^T \mathcal{C}_{k+1} \mathbf{p}_{i_{k+1}}. \tag{7.24}$$

If the number of source signals is assumed constant, the additional calculation of the eigenvalues renders the first order perturbation method computationally more economic, since the computation of each adapted eigenvalue is $O\left({}^1\!/\!{}_2 n_z^2\right)$. In practice, the number of source signals may vary, for example resulting from

Table 7.2 Efficiency of adaptation methods.

Adaptation Method	Computational Cost
rank one modification	$O\left(2n_z^3\right)$
inverse iteration	$O\left(n_z^3\right)$
Lanczos tridiagonalization	$O\left(2.5n_z^2 n_k\right)$
first order perturbation	$O\left(n_z^2\right)$
projection-based	$O\left(n_z^3\right)$
data projection method	$O\left(n_z n_k^2\right)$

throughput or grade changes which result in transients during which this number of the assumed data model $\mathbf{z}_0 = \Xi\mathbf{s} + \mathbf{g}$ can change. Examples of this are available in Li *et al.* (2000) and Sections 7.6 and 7.7 below. The assumption for the first order perturbation method is that the adapted correlation matrix can be written in the following form

$$\mathcal{C}_{k+1} = \mathcal{C}_k + \mu\left[\widetilde{\mathbf{z}}_0(k+1)\widetilde{\mathbf{z}}_0^T(k+1) - \mathcal{C}_k\right] \tag{7.25}$$

where $\mu \in \mathbb{R}$ is a small positive value. By selecting $\mu = 1/\mathcal{K}+1$ the above equation represents an approximation of the correlation matrix, since

$$\begin{aligned}\mathcal{C}_{k+1} &\approx \tfrac{\mathcal{K}}{\mathcal{K}+1}\mathcal{C}_k + \tfrac{1}{\mathcal{K}+1}\widetilde{\mathbf{z}}_0(k+1)\widetilde{\mathbf{z}}_0^T(k+1)\\ &\approx \mathcal{C}_k + \tfrac{1}{\mathcal{K}+1}\widetilde{\mathbf{z}}_0(k+1)\widetilde{\mathbf{z}}_0^T(k+1) - \tfrac{1}{\mathcal{K}+1}\mathcal{C}_k\\ &\approx \mathcal{C}_k + \mu\left[\widetilde{\mathbf{z}}_0(k+1)\widetilde{\mathbf{z}}_0^T(k+1) - \mathcal{C}_k\right].\end{aligned} \tag{7.26}$$

On the basis of the above discussion, it follows that:

- an updated and downdated version of the data correlation matrix is available and hence, the adaptation does not need to be part of the adaptation of the eigendecomposition;
- the faster first order perturbation and data projection methods are designed for recursive but not moving window formulations;
- the dominant n_{k+1} eigenvectors as well as the eigenvalues need to be adapted;
- the number of retained PCs may change; and
- the algorithm should not be of $O\left(n_z^3\right)$.

Fast methods for adapting the model and residual subspaces rely on orthogonal projections, such as Gram-Schmidt orthogonalization (Champagne 1994; Doukopoulos and Moustakides 2008). Based on an iterative calculation of a QR decomposition in Golub and van Loan (1996, page 353) the following orthonormalization algorithm can be utilized to determine the adapted eigenvectors:

1. select an old basis as initial basis for new model plane, $\mathbf{P}_{k+1_0} = \mathbf{P}_k$;
2. find an orthonormal basis, $\mathbf{P}_{k+1_{i+1}}$, for the plane spanned by $\mathcal{C}_{k+1}\mathbf{P}_{k+1_i}$ using the modified Gram-Schmidt procedure, for example;
3. check for convergence, i.e. if $\left|\left\|\left[\mathbf{P}_{k+1_{i+1}}^T\mathbf{P}_{k+1_i}\right]\right\|_2^2 - n_k\right| < \varepsilon$ then terminate else go to Step 2 by setting $\mathbf{P}_{k+1_i} = \mathbf{P}_{k+1_{i-1}}$.

This algorithm converges exponentially and proportional to $\left(\lambda_{n_{k+1}} \big/ \lambda_{n_{k+1}+1}\right)^{n_z}$ (Doukopoulos and Moustakides 2008). Moreover, the computational efficiency is of $O\left(2n_z n_k^2\right)$ as discussed in Golub and van Loan (1996, page 232).

The underlying assumption of this iterative algorithm is that the number of source signals, n, is time invariant. In practice, however, this number may vary as stated above. Knowing that the computational cost is $O\left(2n_z n_k^2\right)$, it is apparent that any increase in the number of source signals increases the computational burden quadratically. Applying the following pragmatic approach can account for a varying n:

1. Select the initial (and complete) set of eigenvectors and eigenvalues from the first window, for example by applying a *divide and conquer algorithm* (Mastronardi *et al.* 2005).
2. Set counter $j = 0$.
3. Determine the number of source signals as discussed in Subsection 2.4.1, for example by applying the VRE or VPC techniques, and store the first $n_1 + 1$ eigenvectors in an augmented matrix $\mathbf{P}_1^\star = \begin{bmatrix} \mathbf{P}_1 & \mathbf{p}_{n_1+1} \end{bmatrix}$.
4. Adapt mean, variance, and correlation matrix, $\bar{\mathbf{z}}_{k+1}$, $\sigma_{i_{k+1}}^2$ $(i = 1, \cdots, n_z)$ and $\mathcal{C}_{k+1}$, respectively, as outlined in Section 7.2 (RPCA) or Subsection 7.3.1 (MWPCA).
5. Utilize the iterative calculation to determine an adapted set for the first $n_k + 1 + j$ eigenvectors, $\mathbf{P}_k^\star$.
6. Determine n_{k+1} on the basis of the adapted correlation and eigenvector matrix, $\mathcal{C}_{k+1}$ and $\mathbf{P}_k^\star$.
7. Check if n_{k+1} is within the range $1, \cdots, n_k + 1 + j$. If this is the case:
   (a) select n_{k+1} as this minimum;
   (b) adjust the size of $\mathbf{P}_k^\star$ if $n_{k+1} \neq n_k$, which yields $\mathbf{P}_{k+1}^\star$;
   (c) set counter $j = 0$; and
   (d) go to Step 8.

   If this is not the case:
   (a) set $j = j + 1$;
   (b) augment the matrix $\mathbf{P}_k^\star = \begin{bmatrix} \mathbf{P}_k^\star & \mathbf{p}_{n_1+1+j} \end{bmatrix}$; and
   (c) return to Step 5.
8. Compute the adapted eigenvalues $\lambda_{i_{n_{k+1}}}$, $i = 1, \cdots, n_{k+1}$ using (7.24).
9. When the next sample becomes available, set $k = k + 1$ and go to Step 4.

The adaptation of the eigendecomposition is therefore:

- of $O\left(2n_z\left(n_k + 1\right)^2\right)$ if $n_{k+1} \leq n_k$; and
- increases to $O\left(2n_z\left(n_{k+1} + 1\right)^2\right)$ if $n_{k+1} > n_k$.

The adaptation of the eigenvalues in Step 8 is of $O\left({}^1/{}_2 n_z\left(n_z+1\right)n_{k+1}\right)$. Overall, the number of floating point operations therefore is $O\left(n_z^2\right)$. Next, we examine the computational cost for the proposed moving window adaptation procedure.

7.3.3 Computational analysis of the adaptation procedure

After evaluating the computational complexity for the adaptation of the eigendecomposition, this section now compares the adaptation using the up- and downdating approach with a recalculation of the correlation matrix using all samples in the new window. The aim of this comparison is to determine the computational efficiency of this adaptation and involves the numbers of floating point operations (flops) consumed for both methods. For determining the number of flops, we assume that:

- the addition and multiplication of two values requires one flop; and
- that factors such as $\frac{1}{\mathcal{K}}$ or $\frac{1}{\mathcal{K}-1}$ have been determined prior to the adaptation procedure.

Moreover, the number of flops for products of two vectors and matrix products that involve one diagonal matrix, e.g. (7.22), are of $O\left(n_z^2\right)$ flops and scaling operations of vectors using diagonal matrices, such as (7.19), are of $O\left(n_z\right)$.

Table 7.3 presents general algebraic expressions for the number of flops required for updating the correlation matrix in both algorithms. It should be noted that the recalculation of the MWPCA model is of $O\left(\mathcal{K}n_z^2\right)$. In contrast, fast MWPCA is only of $O\left(n_z^2\right)$. The two algorithms can be compared by plotting the ratio of flops consumed by a recalculation over those required by the up- and downdating method. Figure 7.2 shows the results of this comparison for a variety of configurations, that is, varying window length, $\mathcal{K}$, and number of variables, n_z.

Figure 7.2 shows that the computational speed advantage of the fast MWPCA can exceed 100. The larger the window size, the more significant the advantage. However, with an increasing number of variables, the computational advantage is reduced. Using the expressions in Table (7.3), a hypothetical case can be constructed to determine when the up- and downdating procedure is more economic. For a given number of process variables, a window length that is larger than $\mathcal{K}^*$

$$\mathcal{K}^* \geq 12\frac{n_z+2}{n_z+5} \tag{7.27}$$

Table 7.3 Number of flops consumed for adapting correlation matrix.

MWPCA technique	Expression
Recomputing correlation matrix	$\mathcal{K}n_z^2+5\mathcal{K}n_z+n_z^2+3n_z$
Using up- and downdating approach	$13n_z^2+27n_z$

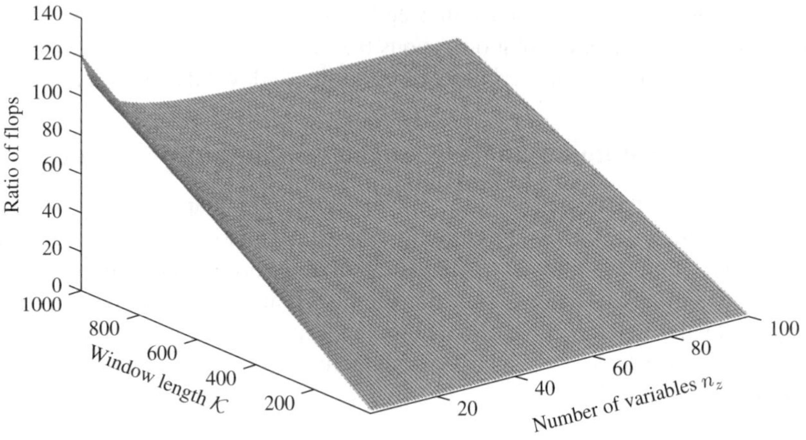

Figure 7.2 Ratio of flops for recalculation of PCA model over up- and downdating approach for various $\mathcal{K}$ and n_z numbers.

results in a computationally faster execution of the introduced up- and downdating approach. A closer inspection of (7.27) reveals that equality requires the window length $\mathcal{K}$ to be smaller than n_z if $n_z \geq 10$. In order to reveal the underlying correlation structure within the process variables, however, it is imperative to guarantee that $\mathcal{K} \gg n_z$. Practically, the proposed up- and downdating method offers a fast adaptation of the correlation matrix that is of $O\left(n_z^2\right)$. Together with the adaptation of the eigendecomposition, one adaptation step is therefore of $O\left(n_z^2\right)$. The required adaptation of the control limits is discussed next.

7.3.4 Adaptation of control limits

Equations (3.5) and (3.16) describe how to compute the control limits for both non-negative quadratic statistics. Given that the following parameters can vary:

- the number of retained components, n_{k+1}; and
- the discarded eigenvalues, $\lambda_{n+1_{k+1}}, \lambda_{n+2_{k+1}}, \cdots, \lambda_{n_{z_{k+1}}}$,

both control limits may need to be recomputed, as n_{k+1} may be different to n_k and the eigenvalues may change too. The number of source signals can be computed by the VRE or VPC criterion for example. The adaptation of the eigendecomposition, however, includes the retained components of the correlation matrix only. Adapted values for the discarded eigenvalues are therefore not available.

The adaptation of Q_α can, alternatively, be carried out by applying (3.29) to (3.31), as proposed by Wang *et al.* (2003). This has also been discussed in Nomikos and MacGregor (1995) for monitoring applications to batch processes. Equation (3.29) outlines that the parameters required to approximate the control limit, $Q_\alpha = g\chi^2_\alpha(h)$ include the first two statistical moments of the Q statistic

$$g = \frac{E\left\{Q^2\right\}}{2E\left\{Q\right\}} \qquad h = 2\frac{\left(E\left\{Q\right\}\right)^2}{E\left\{Q^2\right\}}. \tag{7.28}$$

A moving window adaptation of these are given by:

$$\bar{Q}_{k+1} = \bar{Q}_k + \tfrac{1}{\mathcal{K}}\Delta Q^-_{k+1} \tag{7.29}$$

to estimate the mean and

$$\sigma^2_{Q_{k+1}} = \sigma^2_{Q_k} + \tfrac{1}{\mathcal{K}(\mathcal{K}-1)}\left(\Delta Q^-_{k+1}\right)^2 + \tfrac{1}{\mathcal{K}-1}\left(\Delta Q^-_{k+1}\left(\Delta Q^+_{k+1} - 2\right)\right) \tag{7.30}$$

for the variance. Here

- $\Delta Q^-_{k+1} = Q(k+1) - Q(k-\mathcal{K}+1)$; and
- $\Delta Q^+_{k+1} = Q(k+1) + Q(k-\mathcal{K}+1)$.

After computing $\bar{Q}_{k+1}$ and $\sigma^2_{Q_{k+1}}$, the parameters g_{k+1} and h_{k+1} can be calculated

$$g_{k+1} = \frac{\sigma^2_{Q_{k+1}}}{2\bar{Q}_{k+1}} \qquad h_{k+1} = 2\frac{\bar{Q}^2_{k+1}}{\sigma^2_{Q_{k+1}}}. \tag{7.31}$$

After developing and evaluating the moving window adaptation, the next subsection shows how to delay the application of the adapted PCA model.

7.3.5 Process monitoring using an application delay

The literature on adaptive modeling advocates the application of adapted models for the next available sample before a readaptation is carried out. This has also been applied in earlier work on adaptive MSPC (Lee and Vanrolleghem 2003; Li *et al.* 2000; Wang *et al.* 2003). The adaptation of each variable mean and variance as well as the correlation matrix, however, may allow incipient faults to be adapted. This may render such faults undetectable particulary for small window sizes and gradually developing fault conditions. Increasing the window size would seem to be a straightforward solution to this problem. Larger window sizes, however, result in a slower adaptation speed and changes in the variable interrelationships that the model should adapt may consequently not be adequately adapted. Therefore, the adaptation using a too large window has the potential to produce an increased Type I error.

To prevent the adaptation procedure from adapting incipiently developing faults Wang *et al.* (2005) proposed the incorporation of a delay for applying the adapted PCA model. More precisely, the previously adapted PCA model is not applied to analyze the next recorded sample. Rather, it is used to evaluate the sample to be recorded $\mathfrak{K} \geq 1$ time steps later. Figure 7.3 exemplifies this for an incipient fault described by a ramp. When the $(k+1)$th sample

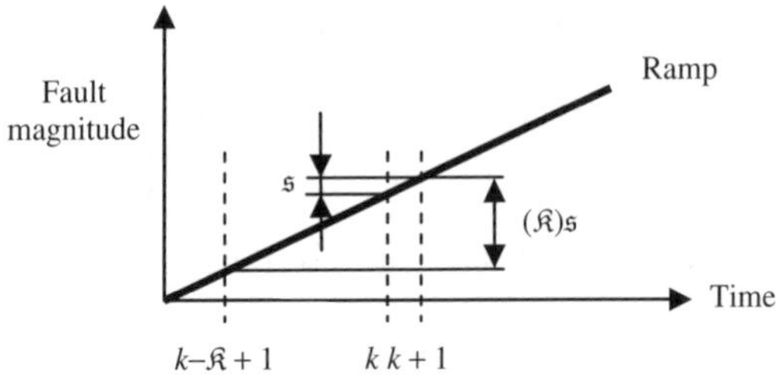

Figure 7.3 Influence of delayed application for detecting a ramp-type fault.

becomes available, the model which is adapted by including the $(k+1-\mathfrak{K})$th and discarding the $(k+1-\mathfrak{K}-\mathcal{K})$th samples is used to monitor the process, rather than the one including the kth and discarding the $(k-\mathcal{K})$th samples. As Figure 7.3 illustrates, the *older* model is more sensitive to this ramp-type fault since the recent model is likely to have been corrupted by the samples describing the impact of the ramp fault.

Incorporating the $\mathfrak{K}$-step-ahead application of the monitoring model results in the following construction of the Hotelling's T^2 statistic

$$T^2(k+1) = \widetilde{\mathbf{z}}_0^T(k+1)\mathbf{P}_{k-\mathfrak{K}}\boldsymbol{\Lambda}_{k-\mathfrak{K}}^{-1}\mathbf{P}_{k-\mathfrak{K}}^T\widetilde{\mathbf{z}}_0^T(k+1). \tag{7.32}$$

Both $\boldsymbol{\Lambda}_{k-\mathfrak{K}}$ and $\mathbf{P}_{k-\mathfrak{K}}$ are obtained from the $(k-\mathfrak{K})$th model, while $\widetilde{\mathbf{z}}_0(k+1)$ is the $(k+1)$th sample scaled using the mean and variance for that model, that is, $\widetilde{\mathbf{z}}_0(k+1) = \boldsymbol{\Sigma}_{k-\mathfrak{K}}^{-1}\left(\mathbf{z}(k+1) - \bar{\mathbf{z}}_{k-\mathfrak{K}}\right)$. The Q statistic for this sample is

$$Q(k+1) = \widetilde{\mathbf{z}}_0^T(k+1)\left[\mathbf{I} - \mathbf{P}_{k-\mathfrak{K}}\mathbf{P}_{k-\mathfrak{K}}^T\right]\widetilde{\mathbf{z}}_0(k+1). \tag{7.33}$$

It should be noted that a one-step-ahead prediction corresponds to $\mathfrak{K} = 1$. Another advantage of the application delay is the removal of samples that lead to Type I errors for both univariate statistics. Such *violating* samples are earmarked and excluded from the adaptation process. This can further prevent samples describing incipient faults to corrupt the monitoring model. The increase in sensitivity for detecting incipient fault conditions for $\mathfrak{K} > 1$ is demonstrated in the next three sections.

7.3.6 Minimum window length

This issue relates to the minimum number of samples required to provide a sufficiently accurate estimate of the data covariance matrix from the data within the sliding window. Following from the discussion in Section 6.4, if the window size is small, the variances for estimating the mean and variance/covariance is significant.

The number of samples required to estimate the variance of a single variable has been extensively discussed in the 1950s and early 1960s (Graybill

1958; Graybill and Connell 1964; Graybill and Morrison 1960; Greenwood and Sandomire 1950; Leone *et al.* 1950; Tate and Klett 1959; Thompson and Endriss 1961). Based on this early work, Gupta and Gupta (1987) derived an algorithmic expression to determine the required sample size for multivariate data sets.

For a Gaussian distributed variable set, if the variable set is independently distributed, that is, the covariance matrix is a diagonal matrix, the minimum number of samples is approximately $\mathcal{K}_{\min} = 1 + 2(z_{\alpha^*}/\varepsilon)^2$, where $\alpha^* = \frac{1}{2}[1 - (1-\alpha)^{1/n_z}]$, α is the significance, ε is the relative error and z_{α^*} defines the confidence interval of a zero mean Gaussian distribution of unity variance. As an example, to obtain the variance of $n_z = 20$ independently distributed (i.d.) variables for $\varepsilon = 0.1$ and $\alpha = 0.05$ requires a total of $\mathcal{K}_{\min} = 1821$ samples. Table 2 in Gupta and Gupta (1987) provides a list of required samples for various configurations of n_z, ε and α.

In most practical cases, it cannot be assumed that the covariance matrix is diagonal, and hence the theoretical analyses in the previous paragraph are only of academic value. However, utilizing the algorithm developed in Russell *et al.* (1985), Gupta and Gupta (1987) showed that writing the elements of the covariance matrix and its estimate in vector form, that is, $\mathrm{vec}\{\mathbf{S}_{z_0z_0}\} = \begin{pmatrix}\sigma_1^2 & \sigma_{21}^2 & \cdots & \sigma_{n_z}^2\end{pmatrix}^T$ and $\mathrm{vec}\{\widehat{\mathbf{S}}_{z_0z_0}\} = \begin{pmatrix}\widehat{\sigma}_1^2 & \widehat{\sigma}_{21}^2 & \cdots & \widehat{\sigma}_{n_z}^2\end{pmatrix}^T$ with $\sigma_{21}^2 = r_{12}\sigma_1\sigma_2$ and $\widehat{\sigma}_{12}^2$ its estimate, allows defining the random vector $\mathfrak{z} = \sqrt{\mathcal{K}-1}(\mathrm{vec}\{\widehat{\mathbf{S}}_{z_0z_0}\} - \mathrm{vec}\{\mathbf{S}_{z_0z_0}\}) \in \mathbb{R}^{n_z\times(n_z+1)/2}$, since a covariance matrix is symmetric.

The vector $\mathfrak{z} \sim \mathcal{N}\left\{\mathbf{0}, \mathbf{S}_{\mathfrak{z}\mathfrak{z}}\right\}$, where the elements of $\mathbf{S}_{\mathfrak{z}\mathfrak{z}}$ are $\mathrm{cov}\{\mathfrak{z}_{ij}, \mathfrak{z}_{kl}\} = \sigma_{ih}\sigma_{jl} + \sigma_{il}\sigma_{jk}$ (Muirhead 1982). For a given ε and α and the definition of $\mathfrak{z}^* = \varepsilon\sqrt{\mathcal{K}-1}\begin{pmatrix}|\sigma_{11}| & |\sigma_{12}| & \cdots & |\sigma_{n_zn_z}|\end{pmatrix}^T$, the probability $1-\alpha$ can be obtained through integration of

$$F(-\mathfrak{z}^* < \mathfrak{z} < \mathfrak{z}^*) = \int\limits_{-\mathfrak{z}^*<\mathfrak{z}<\mathfrak{z}^*} \frac{|\mathbf{S}_{\mathfrak{z}\mathfrak{z}}|^{-1/2}}{(2\pi)^{n_z\times(n_z+1)/4}} \exp(-\tfrac{1}{2}\mathfrak{z}^T\mathbf{S}_{\mathfrak{z}\mathfrak{z}}^{-1}\mathfrak{z})\mathrm{d}\mathfrak{z}.$$

As the limits of the integration depend on the number of samples, the integral can be evaluated using the algorithm proposed by Russell *et al.* (1985).

It is important to note, however, that the resultant size of the reference sample depends on $\mathbf{S}_{\mathfrak{z}\mathfrak{z}}$, which is usually unknown. Despite this, the analysis of the integral allows the following conclusions to be drawn: (i) an increase in the number of recorded variables yields a larger size of $\mathfrak{z}$ as well and (ii) highly correlated variables may require a reduced reference set.

These conclusions follow from the discussion in Gupta and Gupta (1987). Particularly the last point, that a increasing degree of correlation among the process variables may lead qualitatively to a reduction in the number of samples required, is of interest. The preceding discussion therefore highlights that window size does not only depend on the size of the variable set. Given that the variables of industrial processes are expected to possess a high degree of correlation implies that window size may not necessarily increase sharply for large variable sets.

Another, more pragmatic, approach is discussed in Chiang *et al.* (2001), which relies on the estimation of the critical value for the Hotelling's T^2 statistic. As discussed in Subsection 3.1.2, assuming the data covariance matrix is known, the Hotelling's T^2 statistic follows a χ^2 distribution and the critical value is given by $\chi^2_\alpha(n_z)$. On the other hand, if the data covariance matrix needs to be estimated, the Hotelling's T^2 statistic follows an F-distribution, for which the critical value can be obtained as shown in (3.5). Tracey *et al.* (1992) outlined that the critical value of an F-distribution asymptotically converges to that of a χ^2 distribution, that is:

$$\lim_{\mathcal{K}\to\infty} \frac{n_z\left(\mathcal{K}^2-1\right)}{\mathcal{K}\left(\mathcal{K}-n_z\right)}\mathcal{F}_\alpha\left(n_z, \mathcal{K}-n_z\right) \to \chi^2_\alpha\left(n_z\right). \tag{7.34}$$

Defining the relative difference between both critical values

$$\epsilon = \frac{\frac{n_z\left(\mathcal{K}^2-1\right)}{\mathcal{K}(\mathcal{K}-n_z)}\mathcal{F}_\alpha\left(n_z, \mathcal{K}-n_z\right) - \chi^2_\alpha\left(n_z\right)}{\chi^2_\alpha\left(n_z\right)} \tag{7.35}$$

which gives rise to

$$\mathcal{K}_{\min} = \frac{2\sqrt{1+n_z^2-\gamma\left(n_z, \mathcal{K}_{\min}\right)} - n_z}{2\left(1-\gamma\left(n_z, \mathcal{K}_{\min}\right)\right)} \tag{7.36}$$

and can be solved by iteration. Here, $\gamma\left(n_z, K_{\min}\right) = {}^{(1+\varepsilon)\chi^2_\alpha(n_z)}/{}_{\mathcal{F}_\alpha(n_z, K_{\min}-n_z)}$. Table 2.2 in Chiang *et al.* (2001) lists solutions for various values of n_z with ϵ and α being 0.1 and 0.05, respectively. The minimum number of required samples in this table suggests that it should be roughly 10 times the number of recorded variables. Chiang *et al.* (2001) highlighted that this pragmatic approach does not take the correlation among the process variables into account and may yield a minimum number that is, in fact, too small. More precisely, Section 11.6 in Anderson (2003) describes that confidence limits for eigenvalues and eigenvectors depend on $\mathcal{K}$.

As control limits for the Q statistic, however, depend on the discarded eigenvalues, which follows from (3.16), inaccurately estimated discarded eigenvalues may have a significant and undesired impact upon the computation of Q_α. This implies that the number suggested in (7.36) may be sufficient to determine an appropriate minimum number for constructing the control limit for the Hotelling's T^2 statistic. However, a significantly larger number may be required in order to prevent erroneous results for computing the control limit of the Q statistic.

The suggested value can therefore be used as a guideline knowing that it is advisable to opt for a larger $\mathcal{K}_{\min}$. As the above discussion and the analysis in Section 6.4 highlight, the number of samples required to construct accurate estimates of the data covariance/correlation matrix is still an open issue for the research community.

7.4 A simulation example

This section studies the delayed application of an adaptive MSPC monitoring model using a simulated process. The example process is designed to represent slowly changing behavior in the form of a ramp. Such situations are common in industrial practice and include leakages, pipe blockages, catalyst degradations, or performance deteriorations in individual process units. If an adaptive MSPC monitoring approach is applied in this scenario, such gradual and incipient changes may be accommodated by model adaptation and hence remain unnoticed.

The aim of this section is therefore to study whether the proposed adaptation can detect such incipient faults. A description of the simulated process is given first, followed by an application of a standard PCA-based monitoring model in Subsection 7.4.2. Finally, Subsection 7.4.3 then shows the application of MWPCA and studies the impact of an application delay.

7.4.1 Data generation

The process has four process variables and is based on the following data structure

$$\begin{pmatrix} z_1 \\ z_2 \\ z_3 \\ z_4 \end{pmatrix} = \boldsymbol{\Xi}\mathbf{s} + \mathbf{g} = \begin{bmatrix} 1 & 0 \\ 0 & 1 \\ 1.7 & 0.8 \\ -0.6 & 0.02 \end{bmatrix} \begin{pmatrix} s_1 \\ s_2 \end{pmatrix} + \begin{pmatrix} g_1 \\ g_2 \\ g_3 \\ g_4 \end{pmatrix}. \tag{7.37}$$

Each of the above variables follow a Gaussian distribution, with

$$E\left\{\mathbf{s}\right\} = \mathbf{0} \quad E\left\{\mathbf{s}\mathbf{s}^T\right\} = \begin{bmatrix} 1 & -0.3 \\ -0.3 & 1 \end{bmatrix} \tag{7.38}$$

for the source variables and

$$E\left\{\mathbf{g}\right\} = \mathbf{0} \quad E\left\{\mathbf{g}\mathbf{g}^T\right\} = 0.1\mathbf{I} \tag{7.39}$$

for the error variables. Moreover, the source and error variables are independent, that is, $E\left\{\mathbf{s}\mathbf{g}^T\right\} = \mathbf{0}$. From this process, a total of 6000 samples were generated. To simulate an incipient fault condition, a ramp with a slope of 0.0015 between two samples was superimposed on the first source variable from sample 3501 onwards

$$s_{1_f}(k) = \begin{cases} s_1(k) & k < 3500 \\ s_1(k) + (k-3500)\,0.0015 & k \geq 3500 \end{cases}. \tag{7.40}$$

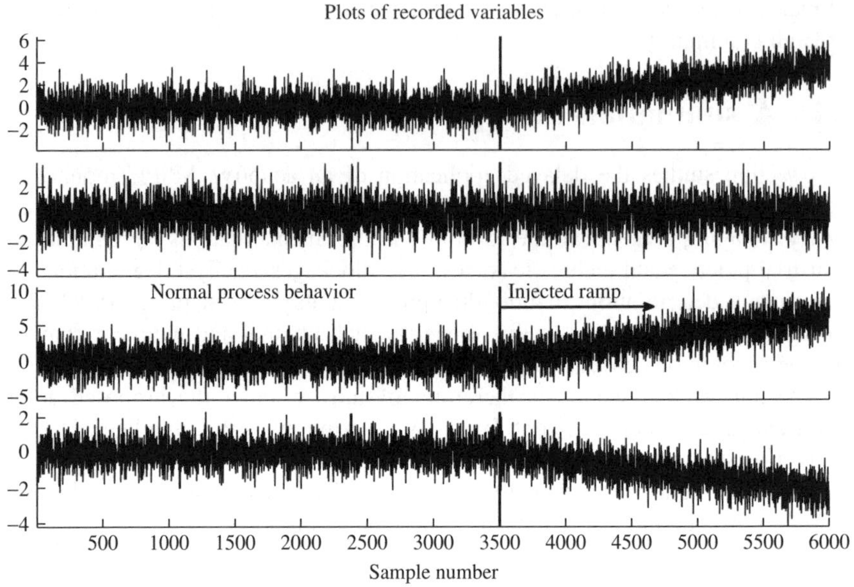

Figure 7.4 Simulated data for process described by Equations (7.37) to (7.40).

The relationships between the variables, i.e. Ξ and $\mathbf{S}_{gg}$, remain unchanged. Thus, the orientation of the model subspace or the residual subspace did not change over time. The simulated process could therefore be regarded as changing to a different, and undesired operating region. Figure 7.4 shows plots of the four process variables for the entire data set. Utilizing this data set for monitoring the simulated process allows us to demonstrate that a varying application delay can:

- accommodate the injected ramp by considering this change as normal; or
- consider this ramp as a process fault that must be detected.

This is discussed in Subsection 7.4.3, following an application of PCA.

7.4.2 Application of PCA

Concentrating on $\mathbf{S}_{z_0 z_0}$, the scaling does not affect the error covariance matrix $\mathbf{S}_{gg} = \sigma_g^2 \mathbf{I}$ with $\sigma_g^2 = 0.1$. Hence the two discarded eigenvalues are 0.1. Moreover, none of the elements in $\Xi = \begin{bmatrix} \xi_1 & \xi_2 \end{bmatrix}$ changed. This implies that the orientation of the model subspace is not affected by the fault condition. Secondly, (3.7) outlines that the Q statistic is not affected either, since the error vector remains unchanged. Hence, neither the orientation of the residual subspace nor the variance of the discarded principal components changed.

In fact, the first n score variables predominantly describe the variation of the source signals, which follows (2.8). With this in mind, the Q statistic describing the fault condition, Q_f, becomes

$$Q_f(k) = \left(\Xi\left(\mathbf{s}(k) + \Delta\mathbf{s}(k)\right) + \mathbf{g}(k)\right)^T \left[\mathbf{I} - \mathbf{P}\mathbf{P}^T\right]\left(\Xi\left(\mathbf{s}(k) + \Delta\mathbf{s}(k)\right) + \mathbf{g}(k)\right). \tag{7.41}$$

and hence

$$Q_f(k) = \left(\mathbf{s}(k) + \Delta\mathbf{s}(k)\right)^T \underbrace{\Xi^T\mathbf{P}_d}_{=\mathbf{0}} \mathbf{P}_d^T \Xi\left(\mathbf{s}(k) + \Delta\mathbf{s}(k)\right) + \mathbf{g}^T(k)\mathbf{P}_d\mathbf{P}_d^T\mathbf{g}(k). \tag{7.42}$$

which implies that $Q_f(k) = Q(k) \leq Q_\alpha$. The above relationship utilized the fact that:

- the source and error variables are uncorrelated;
- the model subspace, spanned by the column space of Ξ, is orthogonal to the residual subspace spanned by the column vectors of $\mathbf{P}_d$; and
- the abnormal condition is described here by

$$\Delta\mathbf{s}(k) = \begin{pmatrix} \Delta s_1(k) \\ \Delta s_2(k) \end{pmatrix} = \begin{cases} \begin{pmatrix} (k - 3500)\,0.0015 \\ 0 \end{pmatrix} & k \geq 3500 \\ \mathbf{0} & k < 3500 \end{cases}.$$

The same analysis for the Hotelling's T^2 statistic yields

$$\begin{aligned} T_f^2(k) &= T^2(k) + T_1^2(k) + T_2^2(k) \\ T_f^2(k) &= T^2(k) + \Delta s_1^2(k)\left(\frac{\left(\boldsymbol{\xi}_1^T\mathbf{p}_1\right)^2}{\lambda_1} + \frac{\left(\boldsymbol{\xi}_1^T\mathbf{p}_2\right)^2}{\lambda_2}\right) \\ &\quad + \Delta s_1(k)\left(2\left(\frac{\boldsymbol{\xi}_1^T\mathbf{p}_1}{\lambda_1}\mathbf{p}_1^T + \frac{\boldsymbol{\xi}_1^T\mathbf{p}_2}{\lambda_2}\mathbf{p}_2^T\right)\mathbf{z}_0(k)\right). \end{aligned} \tag{7.43}$$

The above equation shows that the ramp-type fault has two effects upon the T^2 statistic. The term $T_1^2(k) = \Delta s_1^2(k)\left(\lambda_1^{-1}\left(\boldsymbol{\xi}_1^T\mathbf{p}_1\right)^2 + \lambda_2^{-1}\left(\boldsymbol{\xi}_1^T\mathbf{p}_2\right)^2\right)$ describes a parabola of the form $\mathfrak{A}\,(k - 3500)^2$

$$\underbrace{0.0015^2\left(\frac{\left(\boldsymbol{\xi}_1^T\mathbf{p}_1\right)^2}{\lambda_1} + \frac{\left(\boldsymbol{\xi}_1^T\mathbf{p}_2\right)^2}{\lambda_2}\right)}_{\mathfrak{A}}(k - 3500)^2 \quad k \geq 3500. \tag{7.44}$$

The second term is a Gaussian distributed contribution with a quadratically increasing variance, $E\left\{T_2^2\right\} = \mathfrak{B}\,(k-3500)^2$

$$E\left\{T_2^2\right\} = E\left\{\left(2\left(\frac{\boldsymbol{\xi}_1^T\mathbf{p}_1}{\lambda_1}\mathbf{p}_1\mathbf{z}_0 + \frac{\boldsymbol{\xi}_1^T\mathbf{p}_2}{\lambda_2}\mathbf{p}_2^T\mathbf{z}_0\right)\right)^2\right\}\Delta s_1^2(k)$$

$$E\left\{T_2^2\right\} = E\left\{\left(2\left(\frac{\boldsymbol{\xi}_1^T\mathbf{p}_1}{\lambda_1}t_1 + \frac{\boldsymbol{\xi}_1^T\mathbf{p}_2}{\lambda_2}t_2\right)\right)^2\right\}\Delta s_1^2(k) \tag{7.45}$$

$$E\left\{T_2^2\right\} = \underbrace{0.003^2\left(\frac{\left(\boldsymbol{\xi}_1^T\mathbf{p}_1\right)^2}{\lambda_1} + \frac{\left(\boldsymbol{\xi}_1^T\mathbf{p}_2\right)^2}{\lambda_2}\right)}_{\mathfrak{B}}(k-3500)^2.$$

To determine the parameters $\mathfrak{A}$ and $\mathfrak{B}$, we need the first two eigenvector-eigenvalue pairs of the covariance matrix corresponding to the example process described in (7.37) to (7.40)

$$\mathbf{S}_{z_0z_0} = \boldsymbol{\Xi}\mathbf{S}_{ss}\boldsymbol{\Xi}^T + \sigma_g^2\mathbf{I} = \begin{bmatrix} 1.1000 & -0.3000 & 1.4600 & -0.6060 \\ -0.3000 & 1.1000 & 0.2900 & 0.2000 \\ 1.4600 & 0.2900 & 2.8140 & -0.8702 \\ -0.6060 & 0.2000 & -0.8702 & 0.4676 \end{bmatrix} \tag{7.46}$$

which has the following dominant eigenpairs

$$\mathbf{p}_1 = \begin{pmatrix} -0.4808 \\ -0.0133 \\ -0.8280 \\ 0.2882 \end{pmatrix} \qquad \lambda_1 = 3.9693 \tag{7.47a}$$

$$\mathbf{p}_2 = \begin{pmatrix} 0.2951 \\ -0.9079 \\ -0.2247 \\ -0.1952 \end{pmatrix} \qquad \lambda_2 = 1.3123. \tag{7.47b}$$

With $\boldsymbol{\xi}_1^T = \begin{pmatrix} 1.0 & 0.0 & 1.7 & -0.6 \end{pmatrix}$, the parameters $\mathfrak{A}$ and $\mathfrak{B}$ are equal to

$$\mathfrak{A} = 2.4101 \times 10^{-6} \qquad \mathfrak{B} = 9.6406 \times 10^{-6}. \tag{7.48}$$

When the ramp-type fault is injected for a total of 2500 samples, that is, from sample 3501 to the end of the simulated set, the parabola has a height of $2.4101 \times 10^{-6} \times 2500^2 = 15.0634$ and the standard deviation of T_2^2 is $\sqrt{9.6406 \times 10^{-6} \times 2500^2} = 7.7623$. It is interesting to note that the parameters $\mathfrak{A}$ and $\mathfrak{B}$ are equal up to a scaling factor of 4. In other words the height of the quadratic term T_1^2 is one fourth of the variance of T_2^2.

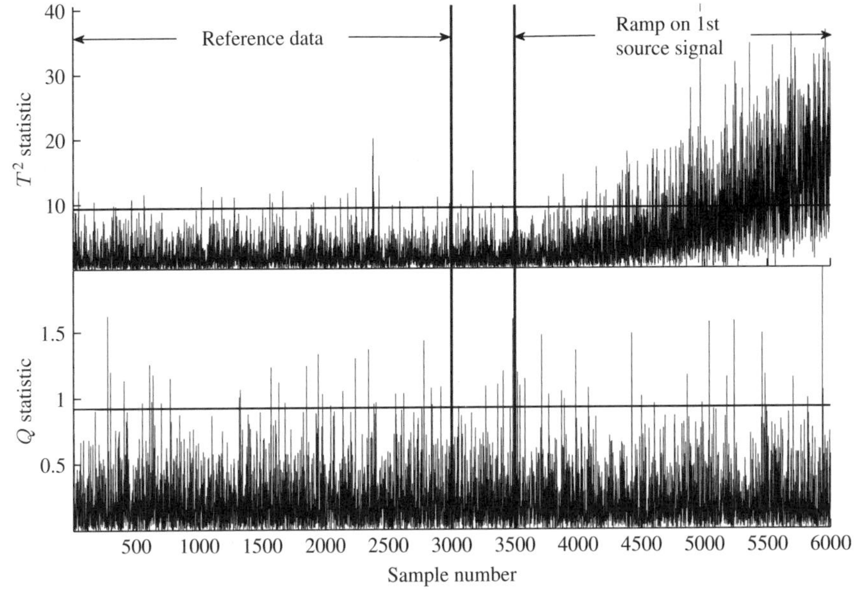

Figure 7.5 Plots of the Hotelling's T^2 *and* Q *statistics for the data set in Figure 7.4.*

Figure 7.5 plots both monitoring statistics for the 6000 simulated samples. Whilst the first 3000 samples served as a reference data set to identify a PCA model including the construction of the monitoring statistics and their control limits, this monitoring model was applied to entire data set. For the first 3500 samples, both statistics described normal process behavior. After the injection of the fault, significant violations of the T^2 statistic arose from around the 4200th sample onwards, whilst the Q statistic remained insensitive to the fault.

This suggests a delay in detecting this event of about 700 samples. In SPC, such a delay is often referred as the *average run length*, that is, the difference in which a process commences to run at a normal operating condition and that at which the monitoring scheme indicates a change from acceptable to rejectable quality level. According to (7.43) to (7.45), the ramp will augment the T^2 statistic by superimposing a quadratic bias term and normally distributed sequence which increases in variance as the fault becomes more severe. This can be confirmed by inspecting the upper plot in Figure 7.5. The next subsection studies the influence of a $\mathfrak{K}$-step-ahead application of an adapted model.

7.4.3 Utilizing MWPCA based on an application delay

To illustrate the effect of applying the adapted MSPC monitoring model with a delay of $\mathfrak{K}$ samples is now studied. Commencing with the traditional $\mathfrak{K} = 1$ approach, Figure 7.6 shows both monitoring statistics for a window length of

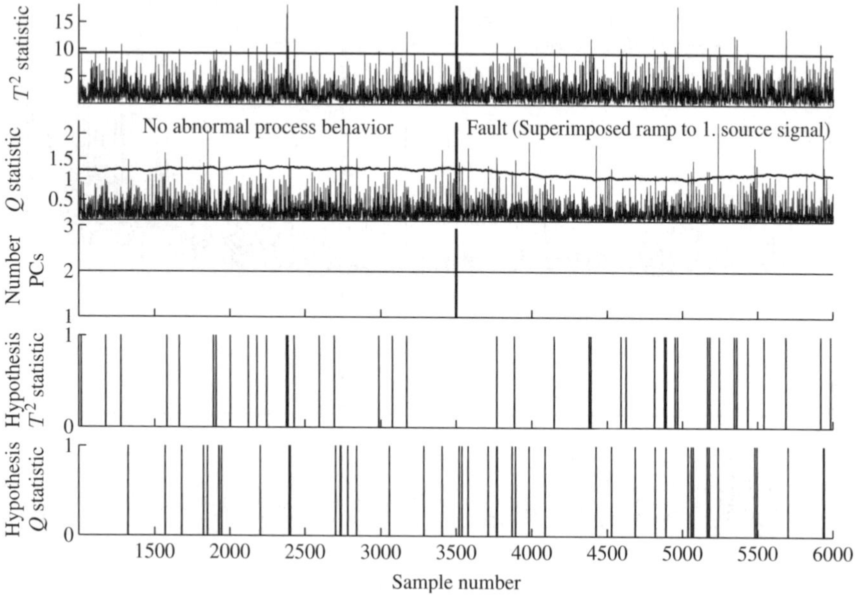

Figure 7.6 Monitoring a ramp change using $\mathcal{K} = 1000$ *and* $\mathfrak{K} = 1$.

$\mathcal{K} = 1000$. By closer inspection, the number of Type I errors does not exceed 1% and hence, the injected ramp-type fault cannot be detected. Moreover, the lower two plots give a clear picture when the null hypothesis, that the process is in-statistical-control, is accepted (value of zero) or when this hypothesis is rejected (value of one).

Besides the Type I error being close to 1%, the hypothesis plots for the T^2 and Q statistics do not indicate a higher density of rejections of the null hypothesis between samples 3500 and 6000. If we assume that such behavior is a *normal* occurrence, for example the performance deterioration of an operating unit, and the adaptive monitoring model should accommodate this behavior, selecting the values for $\mathcal{K}$ and $\mathfrak{K}$ is appropriate.

The middle plot in Figure 7.6 shows how the estimated number of source signals vary over time. This number was estimated using the VRE technique, described in Subsection 2.4.1. Since the injected fault does not affect the geometry of the PCA model subspace nor the residual subspace, the number of source signals does not change. Hence, the adaptation procedure constantly determines, as expected, that two latent component sets are sufficient.

However, if we do not consider this behavior *normal* then the selection of both parameters will render this fault consequently undetectable. The preceding analysis showed that an increase in $\mathcal{K}$ will reduce the impact of new samples upon the covariance/correlation matrix. With reference to (7.11), (7.12), (7.14), (7.20) and (7.23) this follows from the fact that $\frac{\mathcal{K}-2}{\mathcal{K}-1}$ and $\frac{1}{\mathcal{K}-1}$ asymptotically converge to 1 and 0 as $\mathcal{K} \rightarrow \infty$, respectively.

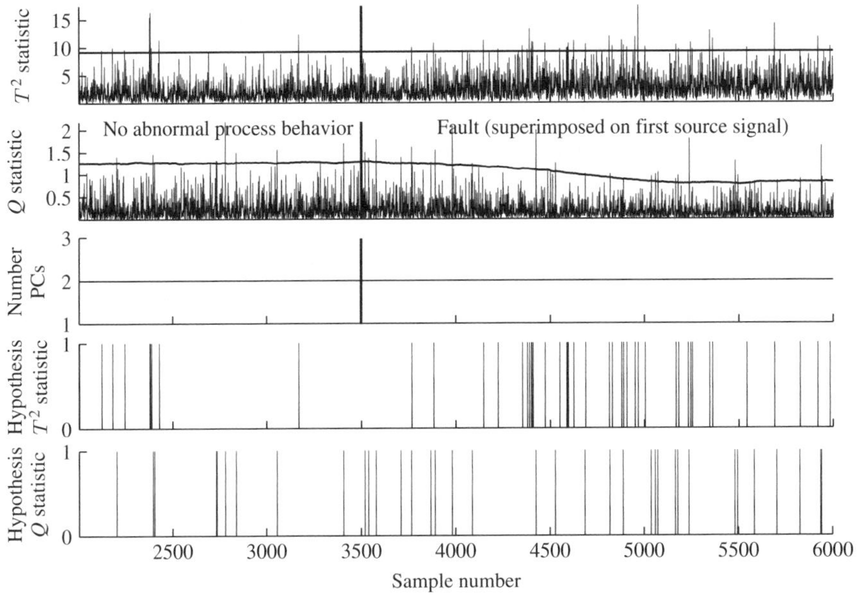

Figure 7.7 Monitoring a ramp change using $\mathcal{K} = 2000$ and $\mathfrak{K} = 100$.

On the other hand, Figure 7.4 shows that a delayed application of the adapted MSPC model may increase the impact of a fault upon the process performance $\mathfrak{K}$ samples earlier. Selecting $\mathcal{K}$ and $\mathfrak{K}$ to be 2000 and 100, Figure 7.7 shows that the ramp-type fault can now be detected. Comparing the lower two plots in Figure 7.7 with those of Figure 7.6 yields a statistically significant number of violations of the Hotelling's T^2 statistic between samples 3500 and 6000.

The empirical significance level is now 1.4% which indicates an out-of-statistical-control performance. In contrast, the number of violations of the Q statistic for the same data section is close to 1% and hence this statistic suggests we should accept the hypothesis that the process is in-statistical-control. Altering the window length $\mathcal{K}$ and the application delay $\mathfrak{K}$ allows studying the influence of both parameters upon the sensitivity in detecting the ramp-type fault.

Table 7.4 presents the result of such an analysis where the empirical significance is determined for the number of violations for the data range 3001 to 6000 and divided by the total number of 3000 samples. By browsing through the columns of this table, it is interesting to note that the empirical significance for the Q statistic in any configuration is very close to the 1%. Following the analysis in (7.42), this is expected since the fault does not affect the Q statistic.

For a better visualization of the results in Table 7.4, Figure 7.8 shows the constant number of Type I errors for the Q statistic in any configuration. With regards to the T^2 statistic, a different picture emerges. As expected, the larger the window length the less new samples affect the adaptation of the monitoring

Table 7.4 Results of $\mathfrak{K}$-step-ahead application for various window lengths $\mathcal{K}$.

	$\mathcal{K} = 1000$		$\mathcal{K} = 2000$		$\mathcal{K} = 3000$	
$\mathfrak{K}$	α_{T^2}	α_Q	α_{T^2}	α_Q	α_{T^2}	α_Q
1	0.0083	0.0097	0.0093	0.0100	0.0100	0.0097
2	0.0087	0.0093	0.0093	0.0100	0.0100	0.0097
3	0.0090	0.0097	0.0093	0.0100	0.0103	0.0097
4	0.0097	0.0093	0.0093	0.0100	0.0103	0.0097
6	0.0093	0.0097	0.0093	0.0100	0.0100	0.0097
8	0.0093	0.0097	0.0093	0.0097	0.0100	0.0097
10	0.0093	0.0097	0.0093	0.0097	0.0103	0.0097
15	0.0100	0.0097	0.0083	0.0097	0.0103	0.0093
20	0.0103	0.0097	0.0090	0.0097	0.0103	0.0097
25	0.0100	0.0093	0.0090	0.0097	0.0107	0.0097
30	0.0107	0.0090	0.0100	0.0100	0.0110	0.0100
35	0.0107	0.0093	0.0100	0.0100	0.0113	0.0100
40	0.0110	0.0093	0.0103	0.0100	0.0113	0.0100
45	0.0113	0.0087	0.0103	0.0100	0.0117	0.0100
50	0.0113	0.0090	0.0103	0.0100	0.0117	0.0100
60	0.0113	0.0090	0.0107	0.0097	0.0120	0.0097
70	0.0117	0.0100	0.0113	0.0097	0.0130	0.0097
80	0.0123	0.0093	0.0117	0.0097	0.0130	0.0097
90	0.0123	0.0093	0.0123	0.0097	0.0130	0.0097
100	0.0127	0.0087	0.0123	0.0097	0.0130	0.0097
120	0.0130	0.0093	0.0127	0.0097	0.0143	0.0093
140	0.0140	0.0097	0.0130	0.0097	0.0147	0.0100
160	0.0150	0.0100	0.0140	0.0097	0.0153	0.0100
180	0.0157	0.0097	0.0147	0.0097	0.0173	0.0100
200	0.0163	0.0100	0.0160	0.0097	0.0183	0.0097
220	0.0170	0.0093	0.0170	0.0097	0.0200	0.0100
240	0.0183	0.0100	0.0177	0.0100	0.0213	0.0097
260	0.0190	0.0097	0.0203	0.0097	0.0227	0.0100
280	0.0207	0.0103	0.0217	0.0093	0.0243	0.0100
300	0.0210	0.0103	0.0237	0.0097	0.0260	0.0100
350	0.0213	0.0107	0.0267	0.0100	0.0307	0.0093
400	0.0230	0.0100	0.0297	0.0103	0.0340	0.0097
450	0.0267	0.0103	0.0320	0.0110	0.0413	0.0103
500	0.0313	0.0110	0.0373	0.0110	0.0450	0.0103
600	0.0383	0.0113	0.0480	0.0107	0.0573	0.0107
700	0.0463	0.0113	0.0617	0.0103	0.0737	0.0107
800	0.0583	0.0110	0.0720	0.0110	0.0920	0.0117
900	0.0710	0.0110	0.0920	0.0110	0.1123	0.0117
1000	0.0867	0.0107	0.1113	0.0107	0.1297	0.0120

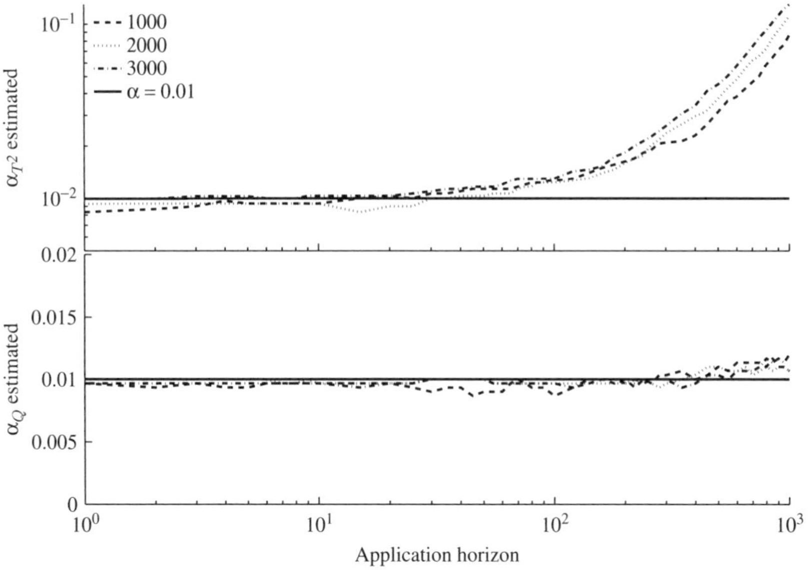

Figure 7.8 Plots of estimated significance for T^2 and Q statistics for various application horizons $\mathfrak{K}$ and window lengths $\mathcal{K}$.

model. In Figure 7.4, the dash-dot line represents a window length of $\mathcal{K} = 3000$, which confirms this.

The figure also shows that increased empirical significance levels for $\mathcal{K} = 3000$ emerged for an application horizon of $\mathfrak{K} \approx 20$. For $\mathcal{K} = 1000$ and $\mathcal{K} = 2000$, the sensitivity in detecting this fault condition decreases. A clear and increasing empirical significance level can be noticed for $\mathfrak{K} \approx 30$. The increases for the latter two configurations are also not as pronounced as for the window length of $\mathcal{K} = 3000$.

7.5 Application to a Fluid Catalytic Cracking Unit

This section applies the adaptive monitoring scheme to a realistic simulation of a Fluid Catalytic Cracking Unit (FCCU) that is described in McFarlane *et al.* (1993). This application is intended to include incipient time-varying behavior that represents a normal operational change and a second more pronounced process fault. Both conditions take the shape of a ramp, where the adaptive monitoring approach must incorporate the first change in order to prevent false alarms. In contrast, the adaptive monitoring approach must be able to detect the second change. A detailed description of this process is given next, prior to a discussion of how the data was generated and how the adaptive monitoring model was established in Subsection 7.5.2. Then, Subsection 7.5.3 presents a pre-analysis of the simulated data set. This is followed by describing the monitoring results

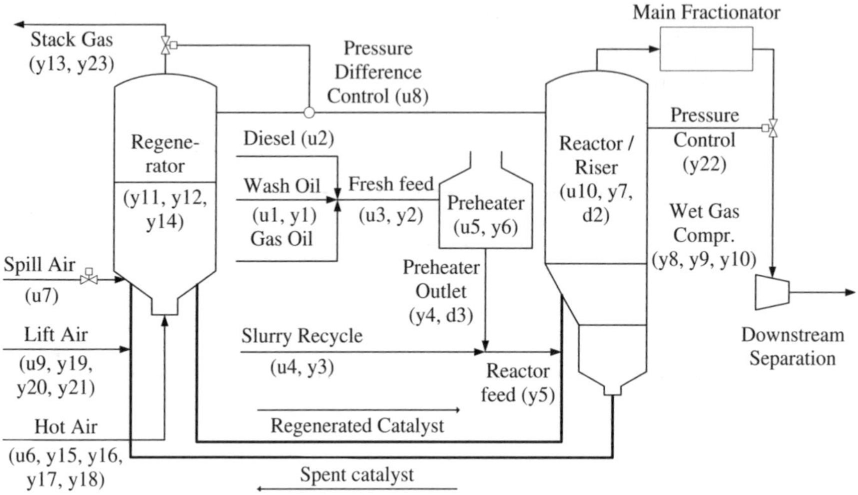

Figure 7.9 Schematic diagram of a fluid catalytic cracking unit.

using PCA and MWPCA with an application delay of one instance in Subsections 7.5.4 and 7.5.5, respectively.

7.5.1 Process description

An FCCU is an important economic unit in oil refining operations. It typically receives several heavy feedstocks from other refinery operations and cracks these streams to produce lighter, more valuable components that are eventually blended into gasoline and other products. Figure 7.9 presents a schematic diagram of this particular Model IV FCCU, which is similar to that in McFarlane *et al.* (1993).

The principal feed to the FCCU is gas oil, but heavier diesel and wash oil streams also contribute to the total feed stream. This fresh feed is preheated in a furnace and then passed to the riser, where it is mixed with hot, regenerated catalyst from the regenerator. In addition to the feed stream, slurry from the main fractionator bottoms is also recycled to the riser. The hot catalyst from the regenerator provides the heat necessary for the endothermic cracking reactions. These produce gaseous products which are passed to the main fractionator for separation. Wet gas off the top of the main fractionator is elevated to the pressure of the light ends plant by the wet gas compressor. Further separation of light components occurs in this light ends separation section that are not included in this simulation model.

As a result of the cracking process inside the reactor, a carbonaceous material, known as coke, is deposited on the surface of the catalyst. Since the deposited coke depletes the catalyst property, spent catalyst is recycled to the regenerator where it is mixed with air in a fluidized bed for regeneration of its catalytic properties. The regeneration occurs when oxygen reacts with the deposited coke to produce carbon monoxide and carbon dioxide. The air is provided by a high-capacity combustion air blower and a smaller lift air blower. In addition to contributing to the combustion process, air from the lift air blower assists with the catalyst circulation between the reactor and the regenerator. Complete details of the mechanistic simulation model for this particular model IV FCCU can be found in McFarlane *et al.* (1993) including a complete list of recorded variables.

The input variables of the FCCU simulator are listed in McFarlane *et al.* (1993, page 288, Table 3). Table 7.5 summarizes the construction of the input sequences to generate the data that were used in this study. In addition to a number of regulatory controllers, the riser temperature in the reactor was controlled to a setpoint of 985°F using a PI controller. This controller determines the

Table 7.5 Input sequences applied to FCCU simulator.

Manipulated variable[1]	Description	Unit	Mean value	Standard deviation
$u1$	Wash oil flow setpoint	$^{lb}/_{s}$	13.8	0.005
$u2$	Diesel oil flow setpoint	$^{lb}/_{s}$	0.0	0.0
$u3$	Fresh feed flow setpoint	$^{lb}/_{s}$	Controller output	Controller output
$u4$	Slurry recycle setpoint	$^{lb}/_{s}$	5.25	0.0025
$u5$	Furnace fuel flow setpoint	$^{scf}/_{s}$	34.0	0.0
$u6$	Combustion air blower suction valve		1.0	0.0
$u7$	Spill air valve		0.0	0.0
$u8$	Reactor/regenerator differential pressure setpoint	psi	−3.37715	0.0
$u9$	Lift air flowrate setpoint	$^{lb}/_{s}$	75.46545	0.0
$u10$	Reactor pressure setpoint	$psia$	33.01966	0.0
Disturbance variable[1]	**Description**	**Unit**	**Mean value**	**Standard deviation**
$d1$	Ambient air temperature	°F	75.0	0.0
$d2$	Effective coking factor		1.0	0.025[2]
$d3$	Preheater outlet temperature	°F	460.9	0.2

[1]The location of $u1, \ldots, u10$ and $d1$, $d2$, $d3$ is shown in Figure 7.9

[2]Drawn from a uniform distribution within range $\begin{bmatrix} 0 & 0.025 \end{bmatrix}$ and subtracted from 1.0.

setpoint value for the total fresh feed. For the kth sample, the controller output is determined by the setpoint error $e_{Riser}(k)$, which is the difference between the setpoint value of 985° F and the actual measurement of the riser temperature, the integral over the setpoint error, here approximated using a numerical integration based on the trapezoidal rule, and an offset of 126.0 lb/s

$$\begin{matrix}\text{Fresh feed} \\ \text{flow setpoint}\end{matrix} (k) = \left[126.0 + K_P e_{Riser}(k) + K_I T_S \sum_{l=0}^{k} e_{Riser}(l) \right] \text{lb/s} \quad (7.49)$$

Applying a variety of standard tuning rules, suitable values for the controller parameter K_P and K_I were found to be −0.105 and −0.01, respectively, and T_S is the sampling time of 1 minute.

7.5.2 Data generation

The FCCU simulator provides readings for a total of 36 variables, listed on page 289 in McFarlane *et al.* (1993) (Table 4). From these, 23 variables, listed in Table 7.6, were included in the subsequent analysis and form the data vector **z**.

Table 7.6 Process variables included in the analysis of the FCCU.

Variable[1]	Description	Unit
$y1 = z_1$	Flow of wash oil to reactor riser	lb/s
$y2 = z_2$	Flow of fresh feed to reactor riser	lb/s
$y3 = z_3$	Flow of slurry to reactor riser	lb/s
$y4 = z_4$	Temperature of fresh feed entering furnace	$^\circ F$
$y5 = z_5$	Temperature of fresh feed entering reactor riser	$^\circ F$
$y6 = z_6$	Furnace firebox temperature	$^\circ F$
$y7 = z_7$	Temperature of reactor riser	$^\circ F$
$y8 = z_8$	Wet gas compressor suction pressure	$psia$
$y9 = z_9$	Wet gas compressor inlet suction flow	$ICFM$
$y10 = z_{10}$	Wet gas flow to the vapor recovery unit	mol/s
$y11 = z_{11}$	Temperature of regenerator bed	$^\circ F$
$y12 = z_{12}$	Regenerator pressure	$psia$
$y13 = z_{13}$	Concentration of oxygen in regenerator stack gas	$mole\%$
$y14 = z_{14}$	Level of catalyst in standpipe	ft
$y15 = z_{15}$	Combustion air blower inlet suction flow	$ICFM$
$y16 = z_{16}$	Combustion air blower throughput	lb/s
$y17 = z_{17}$	Combustion air flow to the regenerator	lb/s
$y18 = z_{18}$	Combustion air blower discharge pressure	$psia$
$y19 = z_{19}$	Lift air blower inlet suction flow	$ICFM$
$y20 = z_{20}$	Actual speed of the lift air blower	RPM
$y21 = z_{21}$	Lift air blower throughput	lb/s
$y22 = z_{22}$	Wet gas compressor suction valve position	
$y23 = z_{23}$	Stack gas valve position	

[1]The location of the recorded variables $y1, \ldots, y23$ is shown in Figure 7.10

The excluded 13 variables were constant and hence did not offer any information for monitoring the unit. The FCCU system was simulated for a sampling frequency of once per minute. The controller interaction to maintain the riser temperature at $985°F$ also occurred at a sampling interval of 1 minute. In order to simulate measurement noise, each of the recorded variables was superimposed by independently distributed noise sequences that followed a Gaussian distribution. These sequences had a mean of zero and a variance of 5% to that of the uncorrupted variable.

In this configuration, 15 000 samples were recorded. The two abnormal conditions were a deteriorating performance of the furnace and a fault in the combustion air blower. The next two subsections summarize the effects of these conditions by analyzing the mechanistic model in McFarlane *et al.* (1993).

7.5.2.1 Injecting a performance deterioration of the furnace

This is a naturally occurring phenomena that is practically addressed through a routine maintenance of the unit. The effect of a performance deterioration can be felt in the enthalpy balance within the furnace. The main variables affected are the furnace firebox temperature and the fresh feed temperature to the riser. This behavior describes a performance deterioration in heat exchangers and translated into a decrease in the furnace overall heat transfer coefficient UA_f. According to McFarlane *et al.* (1993, page 294), the change in UA_f affects the temperature within the firebox, T_3, through the following enthalpy balance

$$\frac{dT_3}{dt} = \frac{1}{\tau_{fb}}\left(F_5\Delta H_{fu} - UA_f T_{lm} - Q_{loss}\right), \tag{7.50}$$

where:

- $\tau_{fb} = 200s$ is the furnace firebox time constant;
- F_5 is the fuel gas flow to the furnace in $\frac{scf}{s}$;
- $\Delta H_{fu} = 1000\frac{B.t.u.}{scf}$ is heat of combusting the furnace fuel;
- $T_{lm} = \frac{(T_3-T_1)(T_3-T_2)}{\ln\left(\frac{T_3-T_1}{T_3-T_2}\right)}$ is the log mean temperature difference;
- $T_1 = 460.9°F$ is the fresh feed temperature entering the furnace;
- T_2 is the fresh feed temperature entering the reactor in $°F$;
- T_3 is the furnace firebox temperature in $°F$; and
- Q_{loss} is the heat loss from the furnace in $\frac{B.t.u.}{s}$.

The analysis in McFarlane *et al.* (1993) also yields that T_2 is affected by alterations of the parameter UA_f

$$\frac{dT_2}{dt} = \frac{1}{\tau_{fo}}\left(T_{2_{ss}} - T_2\right), \tag{7.51}$$

where:

- $\tau_{fo} = 60s$ is the furnace time constant;
- $T_{2_{ss}} = T_1 + \frac{UA_f T_{lm}}{F_3}$; and
- F_3 is the flow of fresh feed to the reactor.

This naturally occurring deterioration was injected after the first 5000 samples were recorded. The change in the parameter UA_f was as follows

$$UA_f(k) = \begin{cases} 25\frac{B.t.u.}{s} & k < 5000 \\ 25\frac{B.t.u.}{s} - 10^{-5}(k - 5000)\frac{B.t.u.}{s} & k \geq 5000 \end{cases}. \tag{7.52}$$

It is important to note that the deteriorating performance of units is dealt with by routine inspections and scheduled maintenance programs. For process monitoring, this implies that the on-line monitoring scheme must adapt to performance deterioration like this one unless this deterioration directly affects the product quality or has an adverse effect upon other operation units. On the other hand, the monitoring scheme must be sensitive in detecting fault conditions, for example in individual units, and correctly reveal their progression through the process so that experienced plant operators are able to identify the root causes of such events and respond appropriately. The generated data set included the injection of a fault located in the combustion air blower that is discussed next.

7.5.2.2 Injecting a loss in the combustion air blower

This fault was a gradual loss in the air blower capacity for any number of reasons. The discussion in McFarlane *et al.* (1993) outlines that this fault affects the combustion air blower throughput as follows

$$F_6 = \frac{\widetilde{F}_6 p_1 F_{surca}}{T_{atm} - 459.6^\circ F}. \tag{7.53}$$

Here

- F_6 is the combustion air blower throughput in $\frac{lb}{s}$;
- p_1 is the combustion air blower suction pressure in *psia*;
- $F_{surca} = 45,100ICFM$ is the combustion air blower inlet suction flow; and
- T_{atm} is the atmospheric temperature of $75^\circ F$.

The fault condition was injected after 10 000 samples were simulate by altering the coefficient $\widetilde{F}_6$ as follows:

$$\widetilde{F}_6(k) = \begin{cases} 0.04511\frac{^\circ F}{ft} & k < 10\,000 \\ 0.04511\frac{^\circ F}{ft} - 5 \times 10^{-8}(k - 10\,000)\frac{^\circ F}{ft} & k \geq 10\,000 \end{cases}. \tag{7.54}$$

Note that the constant $\widetilde{F}_6$ includes conversions from $\frac{ft^3}{min}$ to $\frac{ft^3}{s}$ (*ICFM*) and from $\frac{lb}{in^2}$ to $\frac{lb}{ft^2}$ (*psia*). A change in F_6 affects an alteration of the combustion air blower suction pressure, p_1

$$\frac{dp_1}{dt} = \kappa_1 \left(F_{V_6} - F_6\right) \tag{7.55}$$

and the combustion air blower discharge pressure, p_2

$$\frac{dp_2}{dt} = \kappa_2 \left(F_6 - F_{V_7} - F_7\right) \tag{7.56}$$

where κ_1 and κ_2 are constants (McFarlane *et al.* 1993) if the atmospheric temperature is assumed to be constant and F_{V_6}, F_{V_7} and F_7 are the flows through the combustion air blower suction valve and combustion air blower vent valve, and the combustion air flow to the regenerator in $\frac{lb}{s}$, respectively, and are given by

$$F_{V_6} = \kappa_3 \sqrt{p_{atm} - p_1} \quad F_{V_7} = \kappa_4 \sqrt{p_2 - p_{atm}} \quad F_7 = \kappa_5 \sqrt{p_2 - p_{rgb}}. \tag{7.57}$$

Here, $p_{rbg} = p_6 + \kappa_6 W_{reg}$ is the pressure at the bottom of the regenerator, κ_3 to κ_6 are constants, p_{atm} is the atmospheric pressure (assumed constant), p_6 is the regenerator pressure and W_{reg} is the inventory of catalyst in the regenerator. It is important to note that (7.55) to (7.57) are interconnected. For example, p_1 is dependent upon F_{V_6} and vice versa. In fact, most of the variables related to the combustion air blower are affected by this fault, including the combustion air blower suction flowrate F_{succa}

$$F_{succa} = \left(45,000 + \sqrt{1.581 \times 10^9 - 1.249 \times 10^6 \left(\frac{14.7 p_2}{p_1}\right)^2}\right) ICFM. \tag{7.58}$$

Within the regenerator, a change in the combustion air flow to the regenerator affects the operation of the smaller lift air blower, including the lift air blower speed, s_a

$$s_a = \left(s_{a_{min}} + 1100 V_{lift}\right) RPM \quad V_{lift} = V_{lift}\left(F_7, F_9, F_{10}\right) \tag{7.59}$$

where V_{lift} is the lift air blower steam valve, which regulates the total air flow to the regenerator, $F_T = F_7 + F_9 + F_{10}$. This, in turn, implies that this minor fault in the combustion air blower does not affect the reacting conditions in the regenerator. Other variables of the lift air blower that are affected by a change in the lift air blower speed include the lift air blower suction flowrate, F_{sucla}

$$F_{sucla} = \left(F_{base} \frac{s_a}{s_b}\right) ICFM \tag{7.60}$$

where s_b is the base speed of the lift air blower and F_{base} is the air lift compressor inlet suction flow at base conditions and lift air blower throughput, F_8

$$F_8 = \frac{\left(0.04511\frac{^\circ F}{ft}\right) p_{atm} F_{sucla}}{T_{atm} - 459.6^\circ F}. \tag{7.61}$$

7.5.3 Pre-analysis of simulated data

The analysis of this unit involved a total of 23 variables, shown in Table 7.6. Figure 7.10 shows the recorded data set including the performance deterioration of the furnace and the fault in the combustion air blower. In this figure, the variables are plotted in the order they are listed in Table 7.6: the upper left plot depicts the first six variables, the upper right plot shows variables 7 to 12, the lower left plot presents variables 13 to 18 and the lower right plot charts the remaining variables.

Following the analysis of the fault conditions in Subsection 7.5.2, the performance deterioration of the furnace affected the temperature of fresh feed entering reactor riser, variable 5, and the furnace firebox temperature, variable 6. Inspecting the middle section of the plots in Figure 7.10, data points 5001 to 10 000, these are indeed the only two variables that showed an effect on the performance

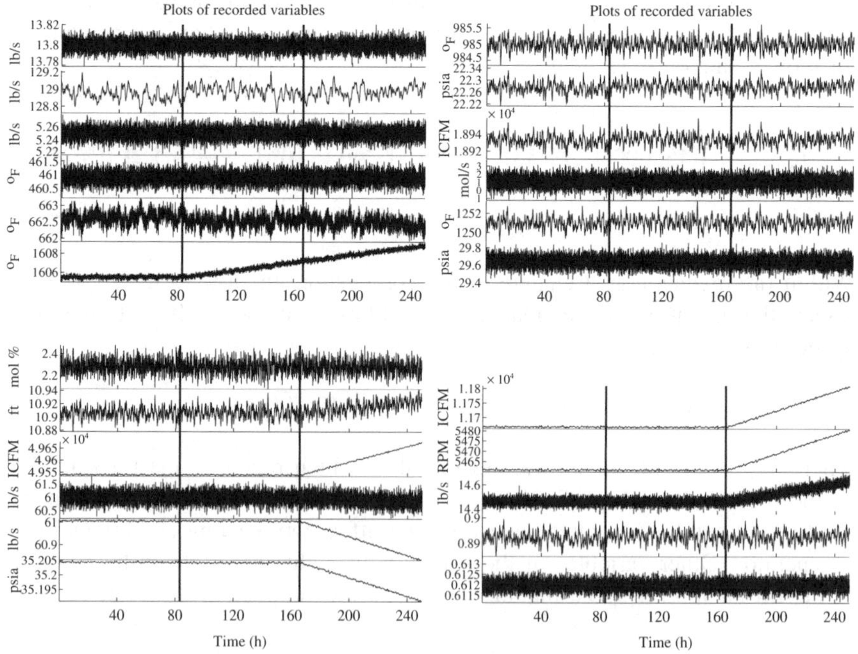

Figure 7.10 Simulated data sequences for FCCU process.

deterioration. However, the fresh feed temperature to the riser was hardly affected although a minor negative trend can be noticed in Figure 7.10. Concentrating on the right section of the simulated data, data points 10 001 to 15 000, affected by the fault in the combustion air blower are variables 14 to 21.

With regards to the underlying mechanistic model of the FCCU in the previous subsection, the recorded variables confirmed that the effect of this fault was mainly felt in both air blowers but did not have a noticeable impact in the regenerator and hence the reactor riser. The impact upon the catalyst level in the standpipe, variable 16 can be attributed to the increase in spill air, and lift air had a minor effect on the catalyst circulation between the reactor and the regenerator. That a loss in combustion air blower capacity resulted in a reduction in combustion air blower output, including the throughput, F_6, the air flow to the regenerator, F_7, and the discharge pressure, p_2, makes sense physically.

The increase in the combustion air blower inlet suction flow, however, is more difficult to explain. A close inspection of (7.53), (7.55), (7.56) and (7.58) yields that the alterations in the parameter $\widetilde{F}_6$ led to a constantly changing operating point. More precisely, (7.58) suggests that the suction flowrate could only increase if p_2 decreased and p_1 increased or remained constant. In fact, the pressure p_1 remained constant and was, as discussed above, therefore not included in this analysis. Consequently, F_{succa} increased slightly, as p_2 reduced in value.

According to the model-based analysis in this and the previous subsections, the information encapsulated in the recorded variables revealed a correct signature of the combustion air blower fault as well as the naturally occurring performance deterioration of the furnace. The next two subsections present the application of PCA and the discussed MWPCA approach to detect and diagnose both events.

7.5.4 Application of PCA

The first step is the identification of a PCA model and involves the estimation of $\mathbf{S}_{z_0 z_0}$ and the number of source signals. The first 5000 samples of the 23 recorded variables described normal process operation and were divided into two sets of 2500 samples. The first 2500 samples were used to obtain the eigendecomposition of the $\widehat{\mathbf{S}}_{z_0 z_0}$. Figure 7.11 summarizes the results of applying the VRE criterion, detailed in Section 2.4.1, and highlights that the minimum is for three source signals. This implies that recorded variables possess a high degree of correlation.

The second half of the reference set was used to estimate the covariance matrix of the score variables, $\widehat{\mathbf{\Lambda}}$. It should be noted that an independent estimation of the PCA model and the score covariance matrix is required, which follows from the discussion in Tracey *et al.* (1992). Figure 7.12 shows the performance of the Hotelling's T^2 and Q statistics for the entire data set of 15 000 samples. As expected, the first 5000 samples (83 h and 20 min) show the process in statistical control. However, from around 100 h into the data set, excessive violations of the confidence limits arose for both statistics indicating an *out-of-statistical-control* situation.

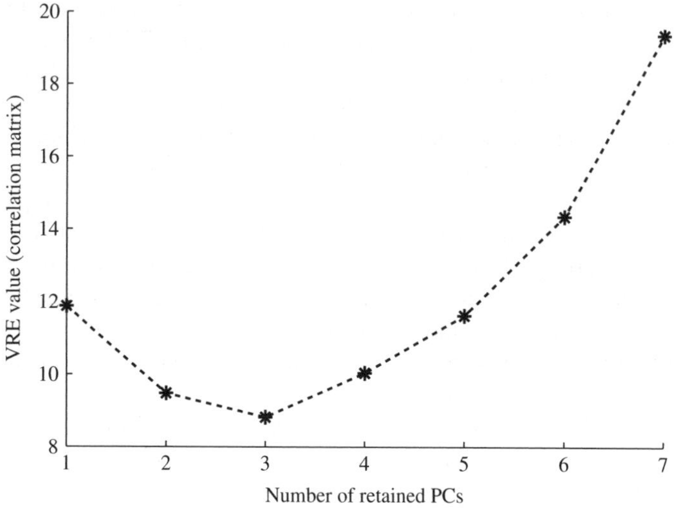

Figure 7.11 Selection of the number of source signals using the VRE criterion.

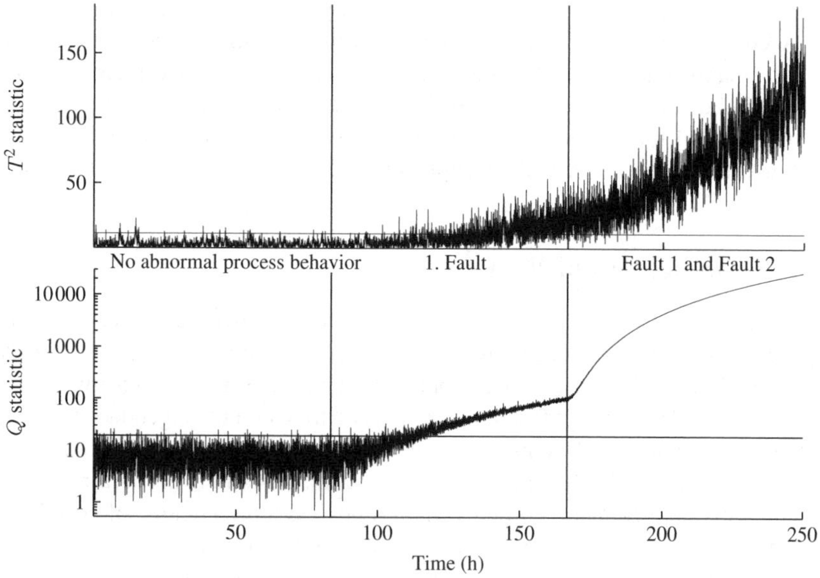

Figure 7.12 Application of PCA to data set shown in Figure 7.10.

As stated above, however, the data portion representing the middle section of the data (Fault 1), describes a performance deterioration of the furnace which naturally occurs over time. Consequently, it is desirable if the on-line monitoring approach is capable of masking this behavior. Inspecting the performance of the PCA model for the third portion of the data outlines that it can detect both conditions, the performance deterioration in the furnace and the loss in combustion air blower capacity. The application of PCA, therefore, showed an on-line monitoring approach requires to be adaptive in order to accommodate the performance deterioration. The adaptive algorithm, however, must still be able to detect the loss in combustion air blower capacity. Subsection 7.5.4 applies the MWPCA approach to the generated data.

7.5.5 Application of MWPCA

The time-invariant PCA model could detect the presence of both simulated events, which is undesired. The MWPCA method has been designed to adapt the model if the relationship between the recorded process variables is time-variant. The aim of this subsection is to examine whether the performance deterioration of the furnace can be adapted and whether the loss of combustion air blower capacity can be detected.

7.5.5.1 Determining an adaptive MWPCA model

The first step for establishing an adaptive MWPCA model is the selection of window size. For this, 2000 samples were selected to ensure that the data set within the window is large enough to reveal the underlying relationships between the recorded variables. This selection, however, is difficult and presents a tradeoff between the speed of adaptation and the requirement to extract the variable interrelationships of the 23 variables listed in Table 7.6. Table 2.2 in Chiang *et al.* (2001) suggests that a minimum number of samples is 284 for a total of 25 variables. The discussion in Subsection 7.3.6, however, showed that this number may be too small in the presence of a high degree and estimation of the control limit for the Q statistic. Following the discussion in Subsection 7.3.6, the significantly larger selection for the window size over the suggested one using (7.35) and (7.36) is therefore required.

Figure 7.13 shows the results of applying a MWPCA model for an application delay of $\mathfrak{K} = 1$. Contrasting Figures 7.13 and 7.12 reveals that utilizing MWPCA removes the excessive number of violations that the PCA model showed in response to Fault 1. The last two plots of Figure 7.13 show the number of violations of the Hotelling's T^2 and Q statistics. These plots confirm that the number of violations in the first and second portion of the data show an average number of violation of 1% for the Q and around 0.65% of violations for the Hotelling's T^2 statistic. It can therefore be concluded that MWPCA was able to accommodate the slow performance deterioration in the furnace. However, comparing the last portion of the data, the application of MWPCA could not detect Fault 2, the gradual loss in combustion air blower capacity. This was different with the use of PCA, as Figure 7.12 confirms. It is therefore imperative to rely on the $\mathfrak{K}$-step ahead application of the currently adapted monitoring model, which is examined next.

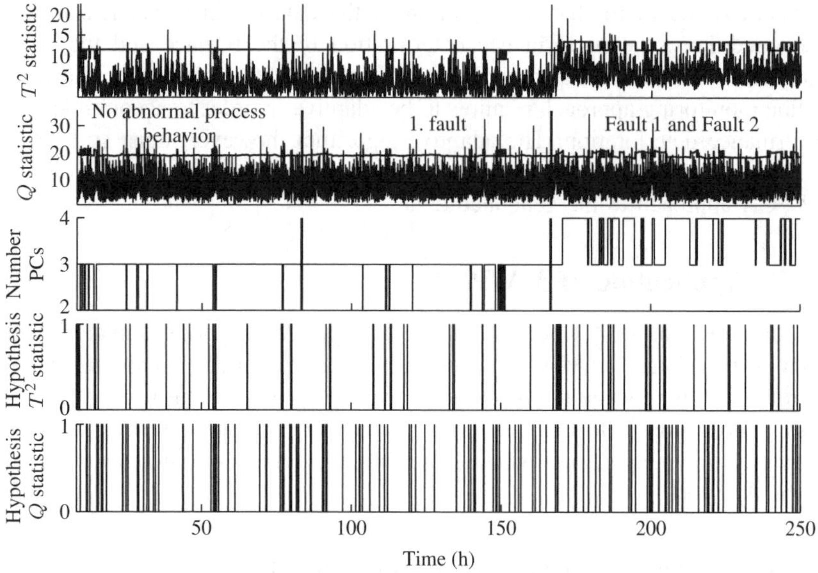

Figure 7.13 Application of MWPCA for $\mathcal{K} = 2000$ *to data set shown in Figure 5.7.*

7.5.5.2 Utilizing MWPCA based on an application delay

In order to determine the application delay, Subsection 7.4.3 discussed that this parameter $\mathfrak{K}$ can be determined empirically. By selecting $\mathfrak{K} = 1, 2, \ldots$ the empirical significance for the Hotelling's T^2 and Q statistics can determined for each integer value and listed. Determining from which selected $\mathfrak{K}$ the empirical significance exceeds the selected significance α then provides a threshold. Figure 7.14 summarizes the results of applying a MWPCA model for the selected window size of $\mathcal{K} = 2000$ and a varying number for $\mathfrak{K}$ ranging from 1 to 100.

The empirical significance was determined for the first 10 000 samples, which included the first and the middle portion of the data describing the performance deterioration of the furnace. Figure 7.14 highlights that Q statistic produces empirical significance levels between 0.009 and 0.01 for $\mathfrak{K}$-values below 60. For the Hotelling's T^2 statistic, $\mathfrak{K}$-values above ten yield empirical significance values exceeding $\alpha = 0.01$. Consequently, the application delay was selected to be $\mathfrak{K} = 10$.

Figure 7.15 shows the performance of the delayed application of the adaptive MWPCA monitoring model. A direct comparison between Figures 7.13 and 7.15 shows that the application delay has no noticeable difference with regards to the number of violations for the first two portions of the data. A difference, though, is the number of source signals. For the first two portions of the data, the VRE criterion determines this number for $\mathfrak{K} = 1$ to be between two and three. In contrast, a constant number of three PCs is suggested for the MWPCA model based on $\mathfrak{K} = 10$. Different from the utilization of an application delay, each

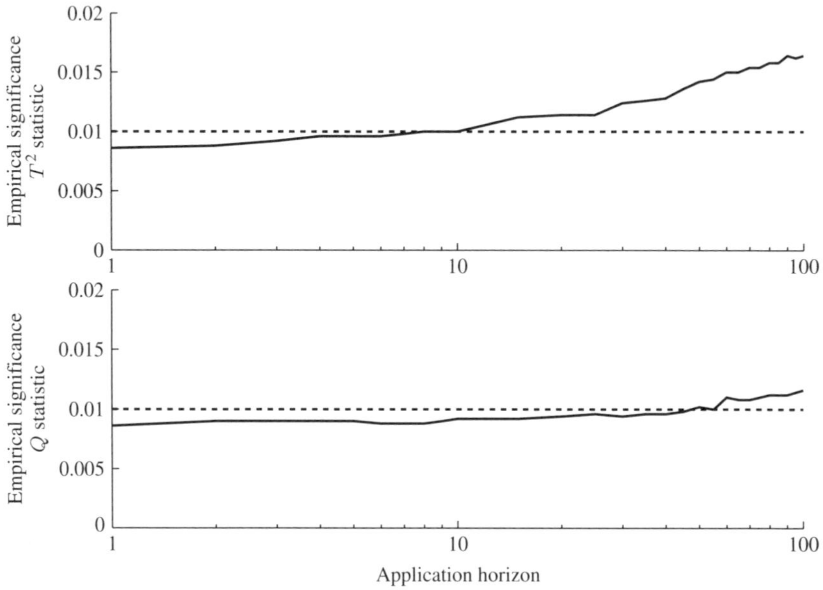

Figure 7.14 Determining application delay $\mathfrak{K}$.

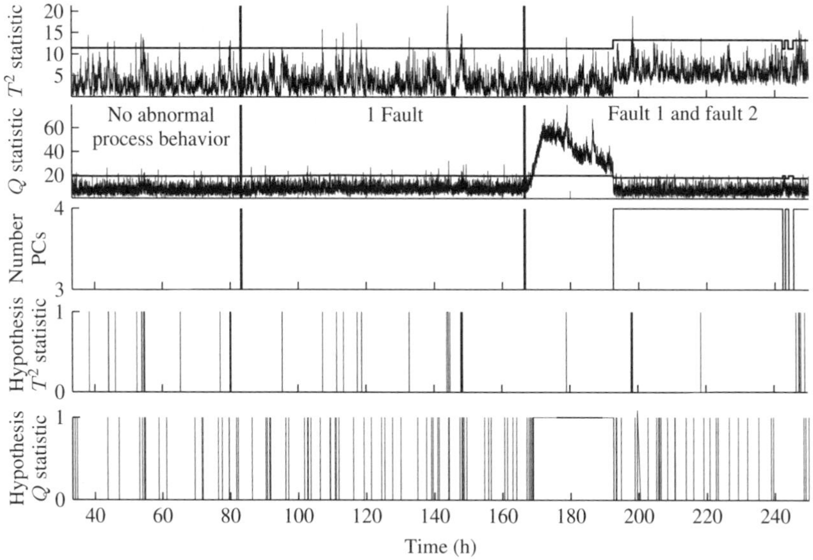

Figure 7.15 Application of MWPCA for $\mathcal{K} = 2000$ and $\mathfrak{K} = 10$ to data set shown in Figure 7.10.

sample was utilized to adapt the MWPCA model including those that produced violations of both univariate monitoring statistics. When adapting the MWPCA model in conjunction with the application delay, samples that produced violations for either statistic were not included.

Analyzing the performance of the delayed application of the MWPCA monitoring model to the third section of the data (describing the performance deterioration of the furnace and the loss in combustion air blower capacity) shows that the fault condition can now be detected by the Q statistic. After the first 500 samples violating the Q statistic, $\mathfrak{K}$ was reduced from 10 to 1 and samples that generate violations were again included in the adaptation. This was just after 170 hours into the data set and showed that the MWPCA model could again adapt to both conditions.

After 190 hours, the number of source signals increased to four and the Q statistic did not show significant violations. This implies that both conditions were no longer detectable and is in line with the results of applying MWPCA for a $\mathfrak{K}$ value of one in Figure 7.13. This confirms that setting $\mathfrak{K} = 10$ and excluding samples that produced violations of either statistic made it possible to adapt the naturally occurring performance deterioration. Moreover, the MWPCA monitoring model detected the superimposed process fault describing the loss in combustion air blower capacity.

7.6 Application to a furnace process

This section summarizes the application of PCA and MWPCA based on an application delay to recorded data from a furnace process. The process is briefly described, prior to a summary of the application of PCA to recorded reference data. The section concludes with the application of MWPCA to the reference set and a second data set describing a sensor bias in one of the temperature measurements.

7.6.1 Process description

This process represents an intermediate heating furnace which is part of a Powerforming process (Powerforming is a process developed by ExxonMobile: W. S. Kmak, A kinetic simulation model of the Powerforming process, 68th National AIChE Meeting, Houston TX, 1971). It receives light naphtha and produces a high-octane liquid in a number of fixed-bed reactors, which are supported by a catalyst, at elevated temperatures and high hydrogen pressures.

In general, furnaces are important in chemical processes, as they elevate the temperature of raw materials or intermediate products to the high temperature levels required to enhance the performance of downstream units, for example reaction sections. As exemplified in this section, the harsh environments inside a furnace may be challenging for obtaining accurate sensor readings. For control engineering application, however, accurate sensor readings are important,

Table 7.7 Recorded process variables.

Variable	Description	Unit
z_1	Tube skin temperature 1	$^\circ C$
z_2	Tube skin temperature 2	$^\circ C$
z_3	Tube skin temperature 3	$^\circ C$
z_4	Tube skin temperature 4	$^\circ C$
z_5	Tube skin temperature 5	$^\circ C$
z_6	Tube skin temperature 6	$^\circ C$
z_7	Tube skin temperature 7	$^\circ C$
z_8	Tube skin temperature 8	$^\circ C$
z_9	Tube skin temperature 9	$^\circ C$
z_{10}	Naphtha outlet temperature	$^\circ C$

particularly for robust control and advanced process control, for example model predictive control.

This particular furnace operates at different fuel gas pressure levels. The fuel gas flow is uncontrolled and depends on the current operating point. The temperature of the upstream naphtha-feed of the first reactor varies when entering the main furnace, where it is elevated to the specification of the second reactor. Further information concerning the catalytic reforming processes may be found in Pujadó and Moser (2006). A case study similar to that analyzed here is given in Fortuna *et al.* (2005).

Table 7.7 lists the recorded temperature variables of the furnace, which were sampled at a sampling rate of 30 seconds. A data set was recorded over a period of two weeks that included normal operating data that served as reference data here and the occurrences of sensor biases in a number of temperature sensors.

Figure 7.16 shows a section of the recorded data describing around 51 hours (6200 samples) of normal operation. The mean value of the temperature variables changes significantly over time and confirms the need for an adaptive monitoring approach. Moreover, the data show some irregular patterns that are encircled.

7.6.2 Description of sensor bias

Thermocouples for measuring skin temperatures are prone to measurement biases which usually recover after some hours. Should the temperature readings form part of a feedback control structure, it is necessary to detect such sensor faults as early as possible and to take appropriate action. During the recording period, several such events arose. Figure 7.17 shows one occurrence of a sensor bias in the thermocouple measuring Skin Temperature 5. It is important to note that the actual bias is just around 25°C. Comparing this with the range of temperature values in Figure 7.16, such a small bias can easily be overlooked by plant personnel. The next two subsections show the application of PCA and MWPCA to the normal operating data and the application of MWPCA to the data set describing the sensor bias.

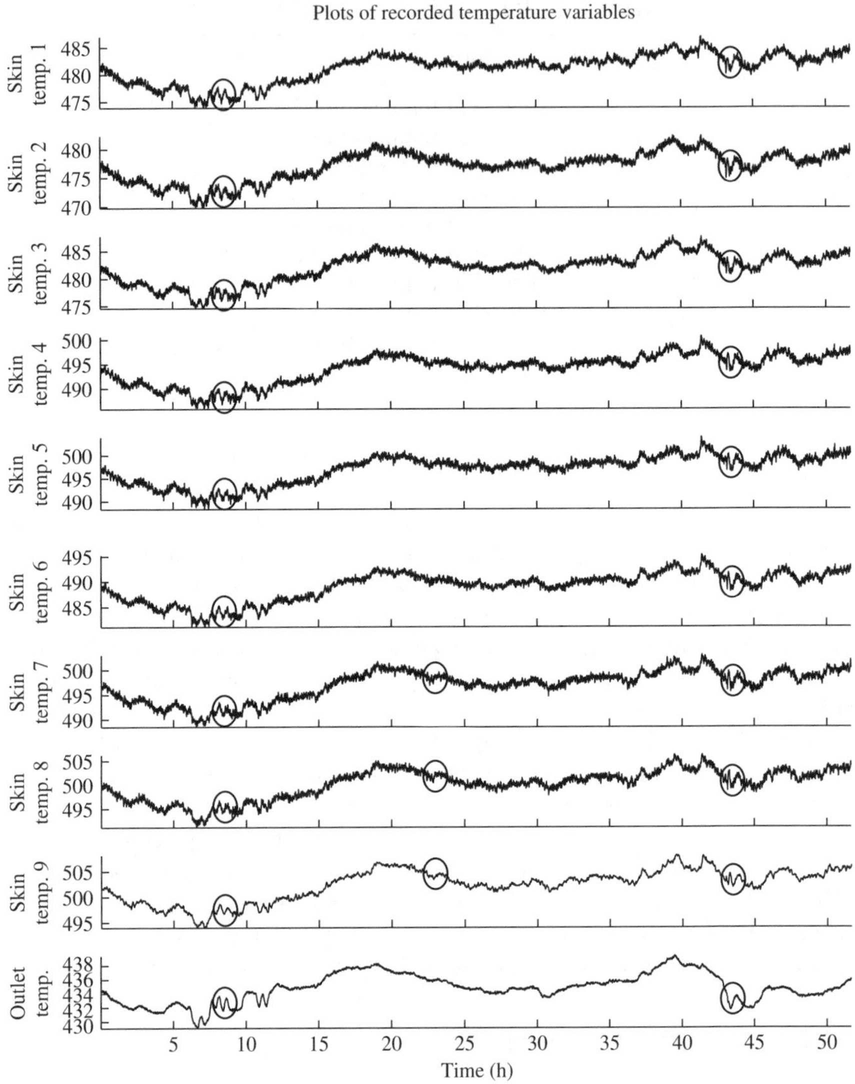

Figure 7.16 Reference data of furnace process.

7.6.3 Application of PCA

To demonstrate that a fixed PCA model may run into difficulties if the mean and/or variance of the process variables changes significantly, this subsection applies a PCA monitoring model that is determined from the first half of the data and applied to the entire data set. The identification of a PCA model commenced by subtracting the mean of each variable and dividing it by the standard deviation, estimated from the samples in the first half of the data, and

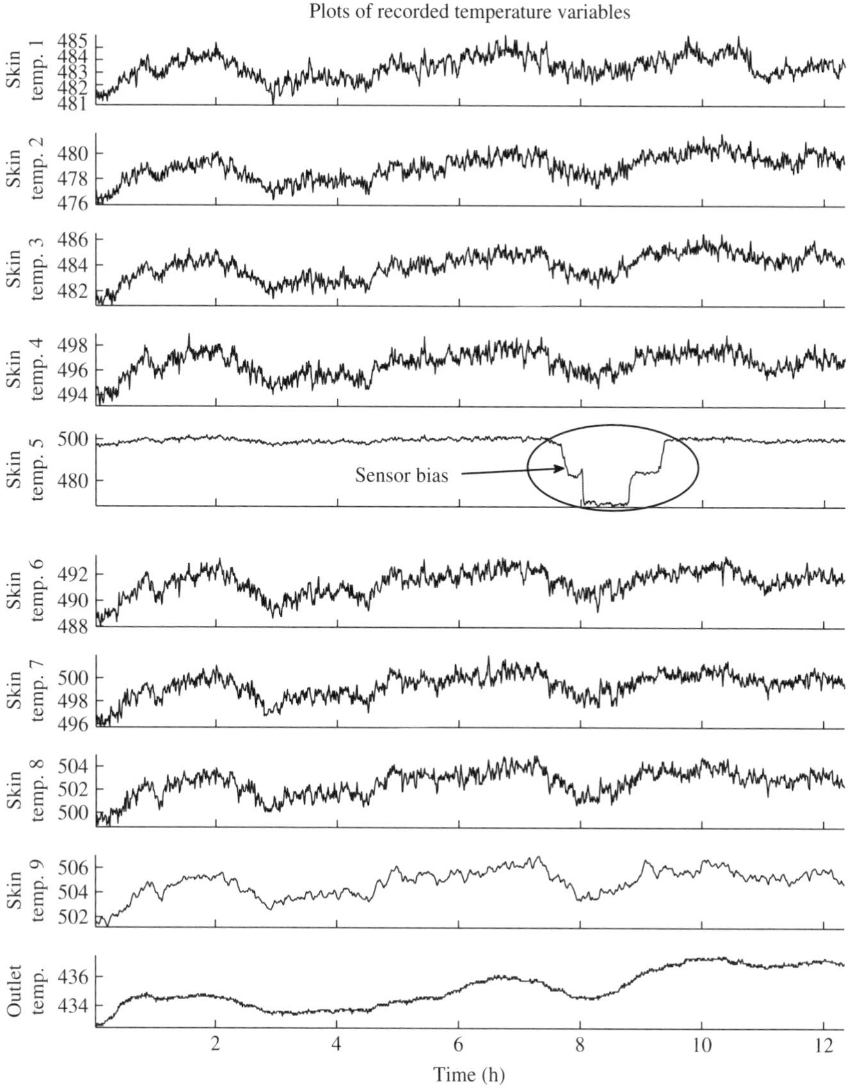

Figure 7.17 Data of furnace process describing sensor bias.

the estimation of the data correlation matrix. The eigendecomposition of the estimated covariance matrix then provided the required information to establish a PCA monitoring model.

The upper plot in Figure 7.18 shows the eigenvalues of the estimated data correlation matrix in descending order. The results of applying the VRE criterion is shown in the lower plot of Figure 7.18 and suggests three source signals. The control limits for the Hotelling's T^2 and Q statistics had to be determined. For

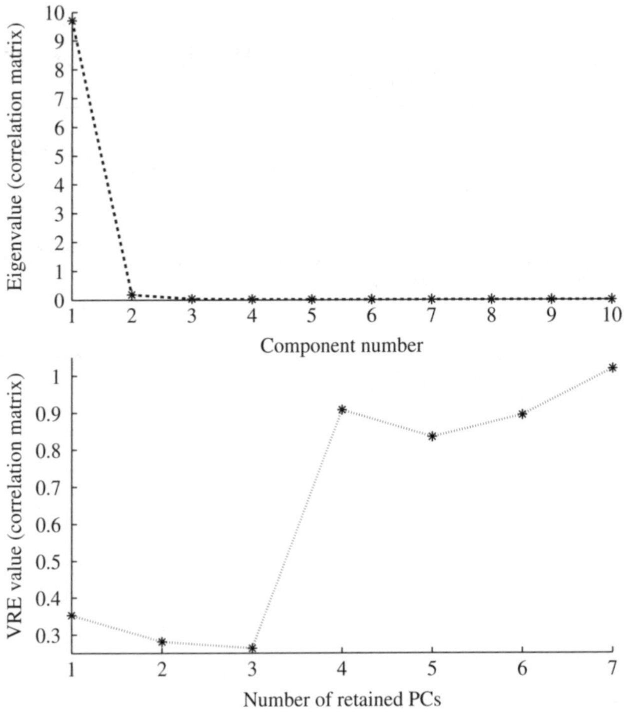

Figure 7.18 Eigenvalues of $\widehat{\mathfrak{C}}_{z_0 z_0}$ *(upper plot) and selection of* n *(lower plot).*

$\alpha = 0.01$, the control limits of the Hotelling's T^2 and Q statistics were 11.3684 and 0.2252, respectively.

Figure 7.19 shows the performance of the PCA monitoring model over the entire data set. The Q statistic does not yield a significant number of violations. In contrast, the Hotelling's T^2 statistic presents a different picture. Whilst the first half shows few violations, there are massive violations in the second half of the data set. This is indicative of excessive common cause variation.

By comparing Figures 7.16 and 7.19, the last third of the data incorporates more variation in all of the temperature readings compared to the remaining data set, particulary the middle portion between 12 to 35 hours into the data set. Although the behavior of the furnace is not abnormal in the second half of the data, the fixed PCA model shows significant violations. This example, therefore, demonstrates that an adaptive model is needed in order adapt to the changes in variable mean, which is discussed in the next subsection.

7.6.4 Utilizing MWPCA based on an application delay

In order to establish a MWPCA model, the window size and the application delay need to be determined. Given that the number of variables is significantly smaller

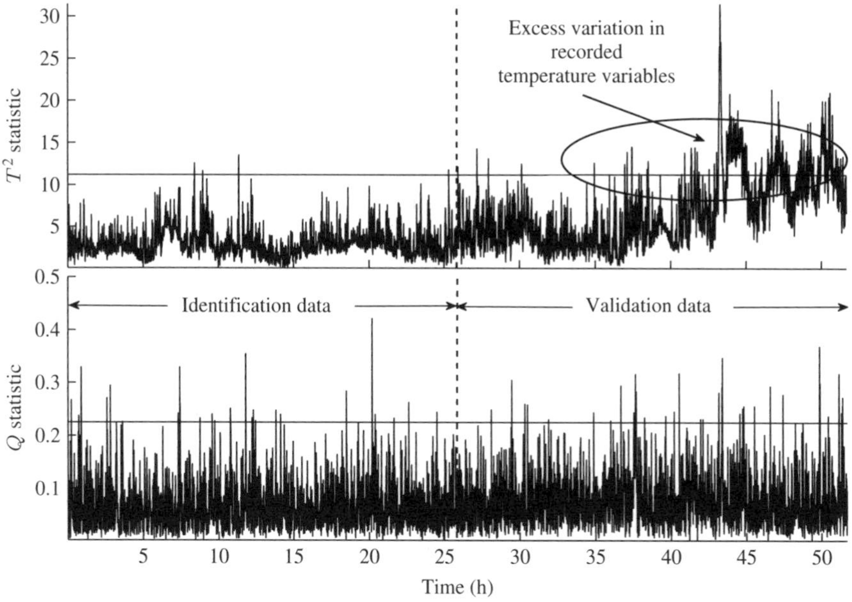

Figure 7.19 Application of PCA to data set shown in Figure 7.16.

compared to the FCCU application study in Section 7.5, the window size may be made smaller. For a total of ten recorded process variables, Table 2.2 in Chiang *et al.* (2001) suggests that the minimum number of samples to populate the data covariance/correlation matrix is 118. Following the discussion in Subsection 7.3.6, this number does not take the high degrees of correlation into account.

Inspecting Figure 7.16 highlights that the ten temperature readings follow a very similar pattern and are, consequently, highly correlated, which the upper plot in Figure 7.18 confirms. More precisely, the last 7 eigenvalues are close to zero. Regarding the analysis in Subsection 7.3.6, given that these eigenvalues determine the control limit of the Q statistic, any significant estimation error would have a profound impact for determining Q_α, as the estimation error depends reciprocally on the window size. To ensure that the window size is significantly larger then the suggestion of 118, $\mathcal{K}$ was selected to be 900.

The next step involved choosing a value of the application delay $\mathfrak{K}$. As illustrated in Subsections 7.4.3 and 7.5.5, this delay can be determined empirically. The empirical significance for various values of $\mathfrak{K}$ was obtained for the reference data. Table 7.8 summarizes the results and outlines that significant increases of the empirical significance arise for $\mathfrak{K}$ values over 20 for the Hotelling's T^2 statistic and for $\mathfrak{K}$ values over 30 for the Q statistic. This suggests a selection of $\mathfrak{K} = 20$.

After selecting $\mathcal{K} = 900$ and $\mathfrak{K} = 20$, Figure 7.20 confirms that MWPCA model adapts this behavior. Moreover, the MWPCA monitoring model is still able to detect the erratic, abrupt and unsteady glitches, encircled in Figure 7.16.

Table 7.8 Empirical significance.

Application horizon	$\widehat{\alpha}_{T^2}$	$\widehat{\alpha}_Q$
1	0.008107	0.003506
2	0.008764	0.004163
3	0.008326	0.004382
4	0.008326	0.004601
5	0.008326	0.004601
6	0.008545	0.004820
7	0.008545	0.004820
8	0.009202	0.005039
9	0.008764	0.006135
10	0.008983	0.006354
15	0.009422	0.006573
20	0.010955	0.007011
25	0.011394	0.008545
30	0.012489	0.011174
35	0.014461	0.014899
40	0.017528	0.017090
45	0.018843	0.017748
50	0.020377	0.019500

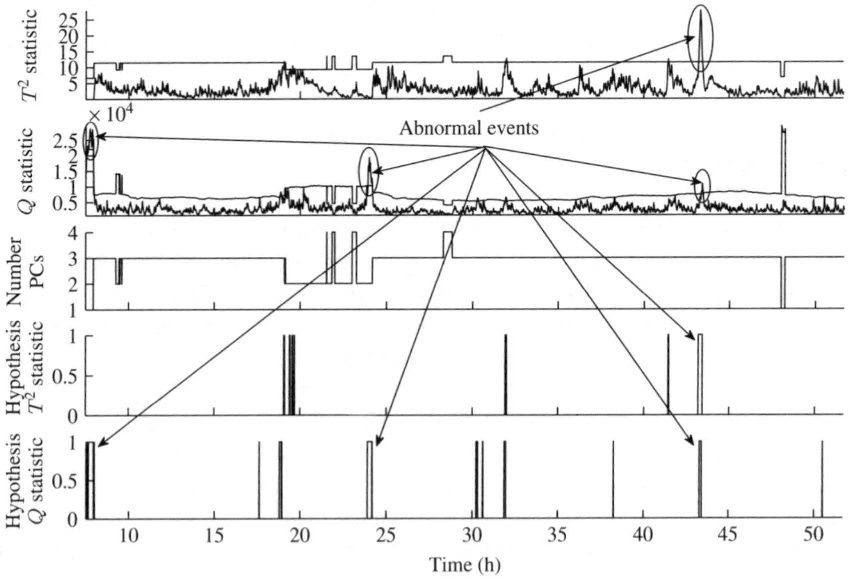

Figure 7.20 Applying MWPCA to data shown in Figure 7.16, $\mathcal{K} = 900$ *and* $\mathfrak{K} = 20$.

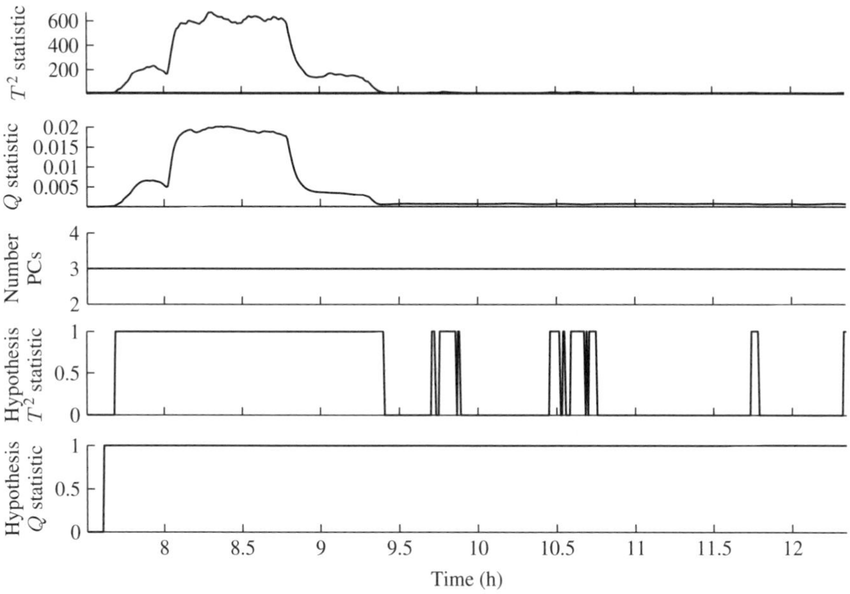

Figure 7.21 Applying MWPCA to data shown in Figure 7.17, $\mathcal{K} = 900$ *and* $\mathfrak{K} = 20$.

Samples corresponding to these short violations are removed from adapting the MWPCA model to ensure that no corruption of the monitoring model arises. Overall, the number of violations does not exceed the significance of 0.01, which implies that the reference data described the process in-statistical-control. It is also interesting to note that the number of source signals varies between one and four. Changes in this number, however, occur infrequently and three PCs are retained most of the time.

Finally, Figure 7.21 shows that the MWPCA model can detect the sensor bias in Skin Temperature 5. The Q statistic is sensitive to this event just after 7 hours and 37 minutes into the data set. The first 7 hours and 30 minutes of data cover the initial moving window Hence, the abscissa in Figures 7.20 and 7.21 does not start from zero. The Hotelling's T^2 statistic provided constant violations of its control limit starting from 7 hours and 40 minutes into the data. After detecting the sensor bias, the adaptation of the MWPCA model was suppressed.

According to Figure 7.17, the sensor bias arose just after 7 hours and 35 minutes and remained up until 9 hours and 20 minutes into the data set. By comparison, the sensor fault could be detected almost instantly. The bottom two plots in Figure 7.21 highlight that the Hotelling's T^2 statistic violates the control limit from 9 hours and 25 minutes and thereafter with noticeable and sporadic violations just before 10 hours and after 10 hours and 30 minutes and after 11 hours and 40 minutes. In contrast, the Q statistic remains violating its control limit.

That the Hotelling's T^2 statistic does not show significant violations after 9 hours and 25 minutes implies that the three source signals showed a state of in-statistical control. However, the underlying geometry of the PCA model has changed compared to the now time-invariant model that was last updated before detecting the sensor bias. If a plant operator examines the situation at that point and determines that the sensor readings are correct, the adaptation of the MWPCA model can continue.

The analysis of both data sets, therefore, confirmed that the application MWPCA can adapt to changes in the variable mean and data correlation matrix and subsequently the PCA monitoring model. The adaptive monitoring model is also able to detect the sensor bias. To verify which of the sensors is faulty, contribution charts or variable reconstruction can be used as discussed in Subsection 3.2.1.

7.7 Adaptive partial least squares

In a similar fashion to PCA, the adaptation of a PLS model can be carried out on the basis of a recursive or a moving window formulation. The first step involves the adaptation of the estimated mean vectors for the input and output variables. This can be accomplished by applying (7.3) for the recursive and (7.21) for the moving window formulation by replacing the process with the input and output variables.

The next step is the adaptation of the covariance and cross-covariance matrices. For the reasons outlined in Jackson (2003) and Remark 7.2.1 above, it is advisable to scale the process variables to unity variance. The covariance and cross-covariance matrix consequently become the correlation and cross-correlation matrix. This entails the division of the input and output variable sets by the estimated standard deviation, which must be adapted too. Equations (7.4) and (7.22) show the recursive and moving window adaptation of the standard deviation, respectively.

This section first outlines how to adapt the correlation and cross-correlation matrices recursively and by applying a moving window formulation in Subsections 7.7.1 and 7.7.2, respectively. Subsection 7.7.3 then discusses how to adapt the number of source signals n and finally, Subsection 7.7.4 summarizes the adaptation of the PLS model. It should be noted that the adaptation of the control limits follows from the discussion in Subsection 7.3.4 and is therefore not covered in this section.

It is also advisable to consider the use of an application delay (Subsection 7.3.5) to improve the sensitivity in detecting incipient fault conditions. This follows from the benefits outlined by the simulation examples in Sections 7.4 and 7.5. Finally, another important aspect is the minimum size of the initial reference set (RPLS) or the size of the moving window (MWPLS). As outlined in Subsection 7.3.6, this is still an issue that has not been exhaustively studied and still requires further attention by the research community. As before, the $\widehat{\cdot}$ notation to denote estimates of the weight vectors is omitted in this section to simplify the presentation of the equations.

7.7.1 Recursive adaptation of $\mathbf{S}_{x_0x_0}$ and $\mathbf{S}_{x_0y_0}$

The adaptation of the correlation matrix of the input variables $\mathcal{C}_{x_0x_0}$ is identical to adaptation of $\mathcal{C}_{z_0z_0}$ for recursive PCA in Section 7.2. Including the $(k+1)$th sample of the input variables, (7.62) shows the recursive update of $\mathcal{C}^{(k)}_{x_0x_0}$ to become $\mathcal{C}^{(k+1)}_{x_0x_0}$

$$\begin{aligned}\mathcal{C}^{(k+1)}_{x_0x_0} =& \tfrac{k-1}{k}\left[\boldsymbol{\Sigma}^{(k+1)}_x\right]^{-1}\boldsymbol{\Sigma}^{(k)}_x\mathcal{C}^{(k)}_{x_0x_0}\boldsymbol{\Sigma}^{(k)}_x\left[\boldsymbol{\Sigma}^{(k+1)}_x\right]^{-1}\\&+\left[\boldsymbol{\Sigma}^{(k+1)}_x\right]^{-1}\Delta\bar{\mathbf{x}}_{k+1}\Delta\bar{\mathbf{x}}_{k+1}\left[\boldsymbol{\Sigma}^{(k+1)}_x\right]^{-1}\\&+\tfrac{1}{k}\widetilde{\mathbf{x}}_0(k+1)\widetilde{\mathbf{x}}^T_0(k+1).\end{aligned} \tag{7.62}$$

Here, $\boldsymbol{\Sigma}^{(k+1)}_x$, $\Delta\bar{\mathbf{x}}_{k+1}$ and $\widetilde{\mathbf{x}}_0(k+1)$ are obtained in the same way as shown in (7.4) and (7.5) for the data vector $\mathbf{z}_0(k+1)$.

The recursive adaptation of the cross-correlation matrix $\mathcal{C}_{x_0y_0}$ requires the adaptation of the mean and variance of both variable sets and is given by

$$\begin{aligned}\mathcal{C}^{(k+1)}_{x_0y_0} =& \tfrac{k-1}{k}\left[\boldsymbol{\Sigma}^{(k+1)}_x\right]^{-1}\boldsymbol{\Sigma}^{(k)}_x\mathcal{C}^{(k)}_{x_0y_0}\boldsymbol{\Sigma}^{(k)}_y\left[\boldsymbol{\Sigma}^{(k+1)}_x\right]^{-1}\\&+\left[\boldsymbol{\Sigma}^{(k+1)}_x\right]^{-1}\Delta\bar{\mathbf{x}}_{k+1}\Delta\bar{\mathbf{y}}_{k+1}\left[\boldsymbol{\Sigma}^{(k+1)}_y\right]^{-1}\\&+\tfrac{1}{k}\widetilde{\mathbf{x}}_0(k+1)\widetilde{\mathbf{y}}^T_0(k+1).\end{aligned} \tag{7.63}$$

Table 7.9 summarizes the steps for determining the auxiliary variables involved in (7.62) and (7.63) and the complete recursive adaptation of both matrices.

7.7.2 Moving window adaptation of $\mathbf{S}_{x_0x_0}$ and $\mathbf{S}_{x_0y_0}$

As shown in Section 7.3, the recursive adaptation of for $\mathcal{C}^{(k)}_{x_0x_0}$ and $\mathcal{C}^{(k)}_{x_0y_0}$ can be developed further to yield a moving window adaptation. This is based on the three-step procedure, which is outlined in Figure 7.1. By reformulating (7.7) to (7.23), Table 7.10 summarizes the steps involved in the moving window adaptation of $\mathcal{C}^{(k)}_{x_0x_0}$ and $\mathcal{C}^{(k)}_{x_0y_0}$.

7.7.3 Adapting the number of source signals

Subsection 2.4.2 provides a list of stopping rules to determine the number of source signals. From these, techniques that rely on reference sets, such as cross validation or bootstrapping methods may not suitable, given that the relationship among and between the input and output variables are assumed to be time varying.

An alternative is to evaluate the accuracy of predicting the output variables. Equation (2.147) shows how to describe the accuracy in terms of the PRESS statistic for the initial model. With a new sample becoming available, the accuracy for predicting the output variables of this sample can be determined and compared with the prediction accuracy of the initial model.

Table 7.9 Recursive adaptation of covariance and cross-covariance matrices.

No.	Variables	Description
1	$\bar{\mathbf{x}}_{k+1} = \frac{k}{k+1}\left(k\bar{\mathbf{x}}_k + \mathbf{x}(k+1)\right)$, $\bar{\mathbf{y}}_{k+1} = \frac{k}{k+1}\left(k\bar{\mathbf{y}}_k + \mathbf{y}(k+1)\right)$	Adapt the mean vectors
2	$\Delta\bar{\mathbf{x}}_{k+1} = \bar{\mathbf{x}}_{k+1} - \bar{\mathbf{x}}_k$, $\Delta\bar{\mathbf{y}}_{k+1} = \bar{\mathbf{y}}_{k+1} - \bar{\mathbf{y}}_k$	Difference between consecutive mean vectors
3	$\left(\sigma_{x,i}^{(k+1)}\right)^2 = \frac{k-1}{k}\left(\sigma_{x,i}^{(k)}\right)^2 + \Delta\bar{x}_{k+1,i}^2 + \frac{1}{k}\left(x_i(k+1) - \bar{x}_{k+1,i}\right)^2$ $\boldsymbol{\Sigma}_x^{(k+1)} = \mathrm{diag}\left\{\sigma_{x,1}^{(k+1)} \cdots \sigma_{x,n_x}^{(k+1)}\right\}$	Adapt diagonal matrix storing the standard deviations of the input variables
4	$\left(\sigma_{y,i}^{(k+1)}\right)^2 = \frac{k-1}{k}\left(\sigma_{y,i}^{(k)}\right)^2 + \Delta\bar{y}_{k+1,i}^2 + \frac{1}{k}\left(y_i(k+1) - \bar{y}_{k+1,i}\right)^2$ $\boldsymbol{\Sigma}_y^{(k+1)} = \mathrm{diag}\left\{\sigma_{y,1}^{(k+1)} \cdots \sigma_{y,n_y}^{(k+1)}\right\}$	Adapt diagonal matrix storing the standard deviations of the output variables
5	$\widetilde{\mathbf{x}}_0(k+1) = \left[\boldsymbol{\Sigma}_x^{(k+1)}\right]^{-1}\left(\mathbf{x}(k+1) - \bar{\mathbf{x}}_{k+1}\right)$, $\widetilde{\mathbf{y}}_0(k+1) = \left[\boldsymbol{\Sigma}_y^{(k+1)}\right]^{-1}\left(\mathbf{y}(k+1) - \bar{\mathbf{y}}_{k+1}\right)$	Centering and scaling the new sample of the input and output variables
6	$\boldsymbol{\mathcal{C}}_{x_0x_0}^{(k+1)} = \frac{k-1}{k}\left[\boldsymbol{\Sigma}_x^{(k+1)}\right]^{-1}\boldsymbol{\Sigma}_x^{(k)}\boldsymbol{\mathcal{C}}_{x_0x_0}^{(k)}\boldsymbol{\Sigma}_x^{(k)} \left[\boldsymbol{\Sigma}_x^{(k+1)}\right]^{-1} + \left[\boldsymbol{\Sigma}_x^{(k+1)}\right]^{-1}\Delta\bar{\mathbf{x}}_{k+1}\Delta\bar{\mathbf{x}}_{k+1}\left[\boldsymbol{\Sigma}_x^{(k+1)}\right]^{-1} + \frac{1}{k}\widetilde{\mathbf{x}}_0(k+1)\widetilde{\mathbf{x}}_0^T(k+1)$ $\boldsymbol{\mathcal{C}}_{x_0y_0}^{(k+1)} = \frac{k-1}{k}\left[\boldsymbol{\Sigma}_x^{(k+1)}\right]^{-1}\boldsymbol{\Sigma}_x^{(k)}\boldsymbol{\mathcal{C}}_{x_0y_0}^{(k)}\boldsymbol{\Sigma}_y^{(k)} \left[\boldsymbol{\Sigma}_y^{(k+1)}\right]^{-1} + \left[\boldsymbol{\Sigma}_x^{(k+1)}\right]^{-1}\Delta\bar{\mathbf{x}}_{k+1}\Delta\bar{\mathbf{y}}_{k+1}\left[\boldsymbol{\Sigma}_y^{(k+1)}\right]^{-1} + \frac{1}{k}\widetilde{\mathbf{x}}_0(k+1)\widetilde{\mathbf{y}}_0^T(k+1)$	Recursive adaptation of $\boldsymbol{\mathcal{C}}_{x_0x_0}$ and $\boldsymbol{\mathcal{C}}_{x_0y_0}$

For a moving window formulation, the selection of n is then as follows

$$\begin{aligned}\mathrm{PRESS}_{n_{k+1}} &= \frac{1}{n_y}\sum_{i=1}^{n_y}\left\|\widetilde{\mathbf{y}}_0(k+\mathcal{K}) - \widehat{\mathbf{Q}}_{k+1}\left[\widehat{\mathbf{R}}_{k+1}\right]^T\widetilde{\mathbf{x}}_0\right\|^2 \\ \mathrm{PRESS}_{n_{k+1}} &\leq \mathrm{PRESS}_{n_1},\end{aligned} \tag{7.64}$$

Table 7.10 Moving window adaptation of $\mathcal{C}_{x_0x_0}$ and $\mathcal{C}_{x_0y_0}$.

Step	Equations	Description
1	$\overline{\mathbf{x}}^* = \frac{1}{\mathcal{K}-1}\left(\mathcal{K}\overline{\mathbf{x}}_k - \mathbf{x}(k)\right),$ $\overline{\mathbf{y}}^* = \frac{1}{\mathcal{K}-1}\left(\mathcal{K}\overline{\mathbf{y}}_k - \mathbf{y}(k)\right)$	Mean of Matrix II
2	$\Delta\overline{\mathbf{x}}^* = \overline{\mathbf{x}}_k - \overline{\mathbf{x}}^*,\ \Delta\overline{\mathbf{y}}^* = \overline{\mathbf{y}}_k - \overline{\mathbf{y}}^*$	Difference in mean vectors
3	$\widetilde{\mathbf{x}}_0(k) = \left[\mathbf{\Sigma}_x^{(k)}\right]^{-1}\left(\mathbf{x}(k) - \overline{\mathbf{x}}_k\right),$ $\widetilde{\mathbf{y}}_0(k) = \left[\mathbf{\Sigma}_y^{(k)}\right]^{-1}\left(\mathbf{y}(k) - \overline{\mathbf{y}}_k\right)$	Scale discarded sample for input and output variables
4	$\widetilde{\mathcal{C}}_{x_0x_0} = \mathcal{C}_{x_0x_0}^{(k)} - \left[\mathbf{\Sigma}_x^{(k)}\right]^{-1}\Delta\overline{\mathbf{x}}^*\Delta\overline{\mathbf{x}}^{*^T}\left[\mathbf{\Sigma}_x^{(k)}\right]^{-1}$ $- \frac{1}{\mathcal{K}-1}\widetilde{\mathbf{x}}_0(k)\widetilde{\mathbf{x}}_0^T(k),$ $\widetilde{\mathcal{C}}_{x_0y_0} = \mathcal{C}_{x_0y_0}^{(k)} - \left[\mathbf{\Sigma}_x^{(k)}\right]^{-1}\Delta\overline{\mathbf{x}}^*\Delta\overline{\mathbf{y}}^{*^T}\left[\mathbf{\Sigma}_y^{(k)}\right]^{-1}$ $- \frac{1}{\mathcal{K}-1}\widetilde{\mathbf{x}}_0(k)\widetilde{\mathbf{y}}_0^T(k)$	Bridge old and new data window
5	$\overline{\mathbf{x}}_{k+1} = \frac{1}{\mathcal{K}}\left((\mathcal{K}-1)\,\overline{\mathbf{v}}^* + \mathbf{x}(k+\mathcal{K})\right),$ $\overline{\mathbf{y}}_{k+1} = \frac{1}{\mathcal{K}}\left((\mathcal{K}-1)\,\overline{\mathbf{y}}^* + \mathbf{y}(k+\mathcal{K})\right)$	Mean of new data window
6	$\Delta\overline{\mathbf{v}}_{k+1} = \overline{\mathbf{v}}_{k+1} - \overline{\mathbf{v}}^*,\ \Delta\overline{\mathbf{y}}_{k+1} = \overline{\mathbf{y}}_{k+1} - \overline{\mathbf{y}}^*$	Mean vector difference
7	$\left(\sigma_{x,i}^{(k+1)}\right)^2 = \left(\sigma_{x,i}^{(k)}\right)^2 + \left(\Delta\bar{x}_{k+1,i}\right)^2$ $- \left(\Delta\bar{x}_{k+1,i}^*\right)^2 + \frac{\left(x_i(k+K)-\bar{x}_{k+1,i}\right)^2 - \left(x_i(k)-\bar{x}_{k,i}\right)^2}{K-1},$ $\left(\sigma_{y,i}^{(k+1)}\right)^2 = \left(\sigma_{y,i}^{(k)}\right)^2 + \left(\Delta\bar{y}_{k+1,i}\right)^2$ $- \left(\Delta\bar{y}_{k+1,i}^*\right)^2 + \frac{\left(y_i(k+K)-\bar{y}_{k+1,i}\right)^2 - \left(y_i(k)-\bar{y}_{k,i}\right)^2}{K-1}$	Standard deviations for variables in new data window
8	$\mathbf{\Sigma}_x^{(k+1)} = \mathrm{diag}\left\{\sigma_{x,1}^{(k+1)} \cdots \sigma_{x,n_x}^{(k+1)}\right\},$ $\mathbf{\Sigma}_y^{(k+1)} = \mathrm{diag}\left\{\sigma_{y,1}^{(k+1)} \cdots \sigma_{y,n_y}^{(k+1)}\right\}$	Storing standard deviations in diagonal scaling matrices
9	$\widetilde{\mathbf{x}}_0(k+\mathcal{K}) = \left[\mathbf{\Sigma}_x^{(k+1)}\right]^{-1}\left(\mathbf{x}(k+\mathcal{K}) - \overline{\mathbf{x}}_{k+1}\right),$ $\widetilde{\mathbf{y}}_0(k+\mathcal{K}) = \left[\mathbf{\Sigma}_y^{(k+1)}\right]^{-1}\left(\mathbf{y}(k+\mathcal{K}) - \overline{\mathbf{y}}_{k+1}\right)$	Scale input and output vector of new sample

(*continued overleaf*)

Table 7.10 *(continued)*

Step	Equations	Description
10	$\mathcal{C}_{x_0x_0}^{(k+1)} = \frac{\mathcal{K}-2}{\mathcal{K}-1}\left[\mathbf{\Sigma}_x^{(k+1)}\right]^{-1}\mathbf{\Sigma}_x^{(k)}\widetilde{\mathcal{C}}_{x_0x_0}\mathbf{\Sigma}_x^{(k)}$ $\left[\mathbf{\Sigma}_x^{(k+1)}\right]^{-1}+\left[\mathbf{\Sigma}_x^{(k+1)}\right]^{-1}\Delta\bar{\mathbf{x}}_{k+1}\Delta\bar{\mathbf{x}}_{k+1}^T\left[\mathbf{\Sigma}_x^{(k+1)}\right]^{-1}$ $+\frac{1}{\mathcal{K}-1}\widetilde{\mathbf{x}}_0(k+\mathcal{K})\widetilde{\mathbf{x}}_0(k+\mathcal{K}),$ $\mathcal{C}_{x_0y_0}^{(k+1)} = \frac{\mathcal{K}-2}{\mathcal{K}-1}\left[\mathbf{\Sigma}_x^{(k+1)}\right]^{-1}\mathbf{\Sigma}_x^{(k)}\widetilde{\mathcal{C}}_{x_0y_0}\mathbf{\Sigma}_y^{(k)}$ $\left[\mathbf{\Sigma}_y^{(k+1)}\right]^{-1}+\left[\mathbf{\Sigma}_x^{(k+1)}\right]^{-1}\Delta\bar{\mathbf{x}}_{k+1}\Delta\bar{\mathbf{y}}_{k+1}^T\left[\mathbf{\Sigma}_y^{(k+1)}\right]^{-1}$ $+\frac{1}{\mathcal{K}-1}\widetilde{\mathbf{x}}_0(k+\mathcal{K})\widetilde{\mathbf{y}}_0(k+\mathcal{K})$	Update of correlation and cross-correlation matrix

where $\text{PRESS}_{n_{k+1}}$ is the resultant PRESS statistic of the $(k+\mathcal{K})$th sample for the retention of n sets of LVs, $\widehat{\mathbf{Q}}_{k+1} \in \mathbb{R}^{n_y \times n}$ and $\widehat{\mathbf{R}}_{k+1} \in \mathbb{R}^{n_x \times n}$ are updated matrices, $k+1$ referring to the new sample, storing the first n_{k+1} q́-loading and r-weight vectors, respectively, and PRESS_{n_1} describes the accuracy of the initial PLS model, retaining the initial selection of n_1 sets of LVs. The next subsection describes the complete adaptation procedure for a PLS model.

7.7.4 Adaptation of the PLS model

As for adaptive PCA models, the number of source signals may vary over time, resulting from throughput, grade changes or operator interventions that may yield short-term transients behavior between different operating conditions for example. On the other hand, the difference between the initial and adapted weight vectors, that is, $\mathbf{w}_i^{(k+1)} - \mathbf{w}_i^{(k)}$, is expected to be small. This implies that the iterative PLS routine converges after the first few iteration steps for each set of LVs unless significant change in the process behavior has arisen.

To maintain this efficiency for a varying number of source signals, it is possible to employ the approach introduced in Subsection 7.3.2. This entails the storage and adaptation of a number of sets of weight vectors that is larger than n_k. If $n_{k+1} > n_k$, this this approach ensures that more sets of LVs are available to test whether $\text{PRESS}_{n_{k+1}} \leq \text{PRESS}_{n_1}$. The number of pairs of weight vectors that are temporarily stored are $n_k + 1 + j$, where j is an integer that is also adaptively computed. The following list of steps summarizes the adaptive PLS algorithm, based on the steps discussed above:

1. Obtain an initial PLS model.

2. Set counter $j = 0$.

3. Determine initial number of source signals as as discussed in Subsection 2.4.2 and store initial q́-weight vectors in the matrix $\mathbf{Q}_1^* = \begin{bmatrix}\mathbf{Q}_1 & \mathbf{q}_{n_1+1}\end{bmatrix}$.

4. Adapt mean, scaling matrices as well as correlation and cross-correlation matrices using a recursive (Table 7.9) or moving window (Table 7.10) formulation.

5. Update the $1 \leq i \leq n_k + 1 + j$ q- and w-weight vectors:
   (a) initiate iteration $i = 1$, $\mathbf{M}^{(1)}_{x_0 y_0} = \mathcal{C}_{x_0 y_0}$;
   (b) set ${}_0\mathbf{q}_i^{(k+1)} = \mathbf{q}_i^{(k)}$;
   (c) compute $\mathbf{w}_i^{(k+1)} = \mathbf{M}^{(i)}_{x_0 y_0} {}_0\mathbf{q}_i^{(k+1)}$;
   (d) scale w-weight vector $\mathbf{w}_i^{(k+1)} = \mathbf{w}_i^{(k+1)} / \left\| \mathbf{w}_i^{(k+1)} \right\|$;
   (e) calculate ${}_1\mathbf{q}_i^{(k+1)} = \mathbf{M}^{(i)}_{y_0 x_0} \mathbf{w}_i^{(k+1)}$;
   (f) scale q-weight vector ${}_1\mathbf{q}_i^{(k+1)} = {}_1\mathbf{q}_i^{(k+1)} / \left\| {}_1\mathbf{q}_i^{(k+1)} \right\|$;
   (g) if $\left\| {}_1\mathbf{q}_i^{(k+1)} - {}_0\mathbf{q}_i^{(k+1)} \right\| > \epsilon$ set ${}_0\mathbf{q}_i^{(k+1)} = {}_1\mathbf{q}_i^{(k+1)}$ and go to Step (c) else go to Step (h);
   (h) determine r-weight vector:
      - $\mathbf{r}_i^{(k+1)} = \mathbf{w}_i^{(k+1)}$ if $i = 1$; and
      - $\mathbf{r}_i^{(k+1)} = \mathbf{w}_i^{(k+1)} - \sum_{m=1}^{i-1} \mathbf{p}_m^{(k+1)^T} \mathbf{w}_i^{(k+1)} \mathbf{r}_m^{(k+1)}$ if $1 < i \leq n_k + 1 + j$;
   (i) calculate p- and q́-loading vectors:
      - $\mathbf{p}_i^{(k+1)} = \mathcal{C}^{(k+1)}_{x_0 x_0} \mathbf{r}_i^{(k+1)} / \mathbf{r}_i^{(k+1)^T} \mathcal{C}^{(k+1)}_{x_0 x_0} \mathbf{r}_i^{(k+1)}$;
      - $\acute{\mathbf{q}}_i^{(k+1)} = \mathbf{M}^{(i)}_{y_0 x_0} \mathbf{r}_i^{(k+1)} / \mathbf{r}_i^{(k+1)^T} \mathcal{C}^{(k+1)}_{x_0 x_0} \mathbf{r}_i^{(k+1)}$;
   (j) deflate cross-covariance matrix
      - $\mathbf{M}^{(i+1)}_{x_0 y_0} = \mathbf{M}^{(i)}_{x_0 y_0} - \mathbf{p}_i^{(k+1)} \left(\mathbf{r}_i^{(k+1)^T} \mathcal{C}^{(k+1)}_{x_0 x_0} \mathbf{r}_i^{(k+1)} \right) \acute{\mathbf{q}}_i^{(k+1)^T}$;
   (k) set $i = i + 1$ and return to Step (b) unless $i = n_k + 1 + j$.

6. Determine n_{k+1} such that $\text{PRESS}_{n_{k+1}} \leq \text{PRESS}_{n_1}$. If $n_{k+1} \leq n_k + 1 + j$:
   (a) Select n_{k+1} as the smallest integer for which $\text{PRESS}_{n_{k+1}} \leq \text{PRESS}_{n_1}$;
   (b) Define $\mathbf{Q}^*_{k+1} = \left[\mathbf{Q}_{k+1} \quad \mathbf{q}_{n_{k+1}} \right]$;
   (c) Set $j = 0$;
   (d) Go to Step 7.

   If $\text{PRESS}_{n_{k+1}} > \text{PRESS}_{n_1}$ for $n_{k+1} = n_k + 1 + j$:
   (a) Set $j = j + 1$;
   (b) Augment the matrix $\mathbf{Q}^*_k = \left[\mathbf{Q}^*_k \quad \mathbf{q}_{n_1+j+1} \right]$;
   (c) Return to Step 5(b) after setting $i = n_k + 1 + j$.

7. When the next sample becomes available, set $k = k + 1$ and return to Step 4.

7.8 Tutorial Session

Question 1: Explain how window size affects the accuracy of the adapted model and how does it affect the detectability of faults? Summarize and comment upon the recommendations in the existing work on determining the window length.

Question 2: What are the advantages and disadvantages of using a recursive and a moving window adaptation of PCA and PLS models?

Question 3: What is the reason behind the introduction of an application delay?

Question 4: What are the steps for adapting a PCA- and PLS-based monitoring model?

Project: Using a Monte Carlo simulation on the basis of the example in (7.37) to (7.40), determine the effect of a varying window length $\mathcal{K}$, a varying application horizon $\mathfrak{K}$ and a varying slope for the ramp describing time-varying process behavior upon the empirical significance level. If the ramp is considered as a fault condition, how does the window length and the application horizon affect the average run length?

8

Monitoring changes in covariance structure

Over the past decades, many successful MSPC application studies have been reported in the literature, for example Al-Ghazzawi and Lennox (2008); Aparisi (1998); Duchesne *et al.* (2002); Knutson (1988), Kourti and MacGregor (1995, 1996), Kruger *et al.* (2001), MacGregor *et al.* (1991), Marcon *et al.* (2005), Piovoso and Kosanovich (1992), Raich and Çinar (1996), Sohn *et al.* (2005), Tates *et al.* (1999), Veltkamp (1993), Wilson (2001). This chapter shows that the conventional MSPC framework, however, may be insensitive to certain fault conditions that affect the underlying geometric relationships of the LV sets. Section 8.1 demonstrates that even substantial alterations in the geometry of the sample projections may not yield acceptance of the alternative hypothesis that the process is out-of-statistical-control.

As the construction of the model and residuals subspaces as well as the control ellipses/ellipsoid for PCA/PLS models originate from data covariance and cross-covariance matrices, this problem is referred to as a change in covariance structure. Any change in these matrices consequently affects the orientation of these subspaces. Thus, in order to detect such alterations, it is imperative to monitor changes in the underlying data covariance structure, which Section 8.2 highlights. This section also presents preliminaries of the *statistical local approach* that allows constructing non-negative squared statistics that directly relate to the orientation of the model and residual subspaces and the control ellipses/ellipsoid.

This problem has been addressed by Ge *et al.* (2010, 2011), Kruger and Dimitriadis (2008); Kruger *et al.* (2007), and Kumar *et al.* (2002) by introducing a different paradigm to the MSPC-based framework. Blending the determination of the LV sets into the statistical local approach gives rise to the construction

---

Statistical Monitoring of Complex Multivariate Processes: With Applications in Industrial Process Control, First Edition. Uwe Kruger and Lei Xie.

of statistics, which Section 8.3 introduces for PCA. These statistics are referred to as *primary residuals* that follow an unknown but non-Gaussian distribution.

It follows from the *central limit theorem* that a sum of random variables follow asymptotically a Gaussian distribution. This is taken advantage of in defining *improved residuals* that are based on the *primary residuals*. Section 8.4 revisits the simulation examples in Section 8.1 and shows that the deficiency of conventional MSPC can be overcome by deriving monitoring charts from the improved residuals.

Sections 8.5 introduces a fault diagnosis scheme to extract fault signatures for determining potential root causes of abnormal events. Section 8.6 applies the introduced monitoring approach to experimental data from a gearbox system. As in Section 8.4, the application study of the gearbox system highlights that the improved residuals are more sensitive in detecting abnormal process behavior when compared to conventional score variables.

Section 8.7 then discusses some theoretical aspects that stem from blending the statistical local approach into the conventional MSPC framework. This includes a direct comparison between the monitoring functions derived in Sections 8.3 and the score variables obtained by the PCA models and provides a detailed analysis the Hotelling's T^2 and Q statistics derived from the improved residuals. The chapter concludes in Section 8.8 with a tutorial session concerning the material covered, including questions as well as homework and project assignments.

8.1 Problem analysis

This section presents examples demonstrating that conventional MSPC-based process monitoring maybe insensitive to changes in the covariance structure of the process variables. A statistic, developed here, describes under which conditions traditional fault detection charts are insensitive to such changes. All stochastic variables in this section are assumed to be of zero mean, which, according to (2.2), implies that $\mathbf{z} = \mathbf{z}_0$. For simplicity, this section uses the data vector $\mathbf{z}$ instead of $\mathbf{z}_0$.

8.1.1 First intuitive example

This example involves two process variables constructed from two i.d. source variables of zero mean, s_1 and s_2, which have a variance of $\sigma_1^2 = 10$ and $\sigma_2^2 = 2$. The following transformation describes the construction of the process variables

$$\begin{pmatrix} z_1^{(0)} \\ z_2^{(0)} \end{pmatrix} = \underbrace{\begin{bmatrix} \sqrt{3}/2 & -1/2 \\ 1/2 & \sqrt{3}/2 \end{bmatrix}}_{\mathbf{T}^{(0)}} \begin{pmatrix} s_1 \\ s_2 \end{pmatrix}. \tag{8.1}$$

Here, $\mathbf{T}^{(0)}$ is a transformation matrix and the index (0) refers to the reference covariance structure. Equation (8.1) is an anticlockwise rotation of the original axes by 30°. Thus, $z_1^{(0)}$ and $z_2^{(0)}$ are coordinates of the rotated base, while s_1 and s_2 are coordinates of the original base. The covariance matrix of $\mathbf{z}^{(0)^T} = \begin{pmatrix} z_1^{(0)} & z_2^{(0)} \end{pmatrix}$ is

$$E\left\{\mathbf{z}^{(0)}\mathbf{z}^{(0)^T}\right\} = \begin{bmatrix} \sqrt{3}/2 & -1/2 \\ 1/2 & \sqrt{3}/2 \end{bmatrix} \begin{bmatrix} 10 & 0 \\ 0 & 2 \end{bmatrix} \begin{bmatrix} \sqrt{3}/2 & 1/2 \\ -1/2 & \sqrt{3}/2 \end{bmatrix} = \begin{bmatrix} 8 & 2\sqrt{3} \\ 2\sqrt{3} & 4 \end{bmatrix}. \quad (8.2)$$

From (8.1), a total of 100 samples for $z_1^{(0)}$ and $z_2^{(0)}$ are generated. The plots in column (a) of Figure 8.1 show the corresponding scatter diagram (upper plot) and the Hotelling's T^2 statistic. The anticlockwise rotation can be noticed from the orientation of the ellipse. Moreover, the rotation does not affect the length of the semimajor and semiminor. For $\alpha = 0.01$, $T_\alpha^2 = 9.21034$, and the values of semimajor and semiminor are $\sqrt{T_\alpha^2 \sigma_1^2} = \sqrt{92.1034} = 9.5971$ and $\sqrt{T_\alpha^2 \sigma_2^2} = \sqrt{18.4207} = 4.2919$, respectively. A detailed discussion on how to construct control ellipses is given in Subsection 1.2.3. Specifically designed changes in the covariance structure of $z_1^{(0)}$ and $z_2^{(0)}$ are carried out next in order to demonstrate that conventional MSPC may not be able to detect them.

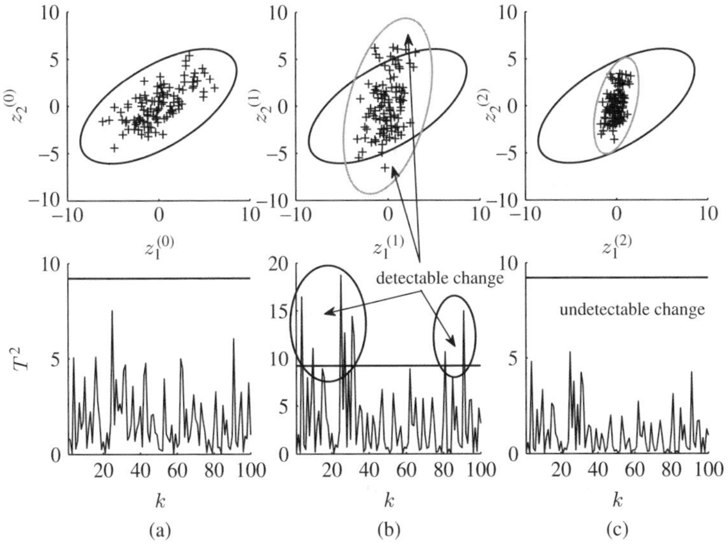

Figure 8.1 Detectable and undetectable changes in covariance structure.

8.1.1.1 First change in covariance structure

The following transformation changes the covariance structure between $z_1^{(0)}$ and $z_2^{(0)}$

$$\begin{pmatrix} z_1^{(1)} \\ z_2^{(1)} \end{pmatrix} = \underbrace{\begin{bmatrix} \sqrt{2}/2 & -\sqrt{2}/2 \\ \sqrt{2}/2 & \sqrt{2}/2 \end{bmatrix}}_{\mathbf{T}^{(1)}} \begin{pmatrix} z_1^{(0)} \\ z_2^{(0)} \end{pmatrix} = \underbrace{\begin{bmatrix} (\sqrt{3}-1)/2\sqrt{2} & -(\sqrt{3}+1)/2\sqrt{2} \\ (\sqrt{3}+1)/2\sqrt{2} & (\sqrt{3}-1)/2\sqrt{2} \end{bmatrix}}_{\mathbf{T}^{(1)}\mathbf{T}^{(0)}} \begin{pmatrix} s_1 \\ s_2 \end{pmatrix} \quad (8.3)$$

where $\mathbf{T}^{(1)}$ describes an anticlockwise rotation by 45° and the index (1) refers to the first change. Consequently, $\mathbf{T}^{(1)}\mathbf{T}^{(0)}$ first represents an anticlockwise rotation by 30° to produce $\mathbf{z}^{(0)}$ and a subsequent rotation by 45° to determine $\mathbf{z}^{(1)}$. The variables $\mathbf{z}^{(1)}$ are the coordinates to a base that is rotated by 75° relative to the original cartesian base. The covariance matrix for $\mathbf{z}^{(1)}$, $E\{\mathbf{z}^{(1)}\mathbf{z}^{(1)^T}\}$, is

$$E\left\{\mathbf{z}^{(1)}\mathbf{z}^{(1)^T}\right\} = \begin{bmatrix} (\sqrt{3}-1)/2\sqrt{2} & -(\sqrt{3}+1)/2\sqrt{2} \\ (\sqrt{3}+1)/2\sqrt{2} & (\sqrt{3}-1)/2\sqrt{2} \end{bmatrix} \begin{bmatrix} 10 & 0 \\ 0 & 2 \end{bmatrix}$$
$$\times \begin{bmatrix} (\sqrt{3}-1)/2\sqrt{2} & (\sqrt{3}+1)/2\sqrt{2} \\ -(\sqrt{3}+1)/2\sqrt{2} & (\sqrt{3}-1)/2\sqrt{2} \end{bmatrix} \quad (8.4)$$
$$E\left\{\mathbf{z}^{(1)}\mathbf{z}^{(1)^T}\right\} = \begin{bmatrix} 6-2\sqrt{3} & 2 \\ 2 & 6+2\sqrt{3} \end{bmatrix}.$$

Using (8.3), a total of 100 samples are generated for $\mathbf{z}^{(1)}$. From this set, the column plots associated with (b) in Figure 8.1 show the scatter diagram (upper plot) and the Hotelling's T^2 statistic (lower plot). For the scatter diagram, the dashed and solid lines represent the control ellipse for the variable sets $\mathbf{z}^{(1)}$ and $\mathbf{z}^{(0)}$, respectively. Furthermore, the Hotelling's T^2 statistic for each sample is computed with respect to $E\{\mathbf{z}^{(0)}\mathbf{z}^{(0)^T}\}$. Since eight points fall outside the confidence regions for the scatter diagram and the control limit of the Hotelling's T^2 statistic, the charts correctly indicate an out-of-statistical-control situation. Consequently, this change in covariance structure between $z_1^{(0)}$ and $z_2^{(0)}$ is identifiable.

8.1.1.2 Second change in covariance structure

The same experiment is now repeated, but this time the variance of the i.d. sequences s_1 and s_2 is $\sigma_1 = 3$ and $\sigma_2 = 1/2$, respectively. Applying (8.3) to first produce $z_1^{(0)}$ and $z_2^{(0)}$ and subsequently $z_1^{(2)}$ and $z_2^{(2)}$ gives rise to the covariance matrix

$$E\left\{\mathbf{z}^{(2)}\mathbf{z}^{(2)^T}\right\} = \begin{bmatrix} \left(\sqrt{3}-1\right)/2\sqrt{2} & -\left(\sqrt{3}+1\right)/2\sqrt{2} \\ \left(\sqrt{3}+1\right)/2\sqrt{2} & \left(\sqrt{3}-1\right)/2\sqrt{2} \end{bmatrix} \begin{bmatrix} 3 & 0 \\ 0 & 1/2 \end{bmatrix}$$

$$\times \begin{bmatrix} \left(\sqrt{3}-1\right)/2\sqrt{2} & \left(\sqrt{3}+1\right)/2\sqrt{2} \\ -\left(\sqrt{3}+1\right)/2\sqrt{2} & \left(\sqrt{3}-1\right)/2\sqrt{2} \end{bmatrix} \tag{8.5}$$

$$E\left\{\mathbf{z}^{(2)}\mathbf{z}^{(2)^T}\right\} = \begin{bmatrix} \left(7-2\frac{1}{2}\sqrt{3}\right)/4 & 5/8 \\ 5/8 & \left(7+2\frac{1}{2}\sqrt{3}\right)/4 \end{bmatrix}.$$

With the reduced variance for s_1 and s_2, 100 samples are generated using (8.3). The plots in column (c) of Figure 8.1 show the scatter diagram of $z_1^{(2)}$ and $z_2^{(2)}$ and the Hotelling's T^2 statistic based on $E\left\{\mathbf{z}^{(0)}\mathbf{z}^{(0)^T}\right\}$. The dashed control ellipse corresponds to $z_1^{(2)}$ and $z_2^{(2)}$ and the solid one refers to $z_1^{(0)}$ and $z_2^{(0)}$. Despite significant alterations to the covariance structure of $z_1^{(0)}$ and $z_2^{(0)}$ these changes are undetected since the dashed control ellipse is inside the solid one. Therefore, the alteration renders the scatter diagrams and the Hotelling's T^2 statistic blind.

In essence, if changes to the covariance structure arise that lead to small alterations in the geometry of statistical confidence regions and limits, such events may not be detectable. Next, a more detailed statistical analysis is presented to formulate conditions which render conventional multivariate analysis insensitive.

8.1.2 Generic statistical analysis

The intuitive analysis in the previous subsection suggested that changes in the covariance structure manifest themselves in alterations of the eigenvalues and eigenvectors of the covariance matrix. This follows from (8.2), (8.4) and (8.5). However, this analysis was restricted to rotations of the control ellipse and is therefore limited in a multivariate context. More precisely, since MSPC techniques decompose the data space(s) into model and residual subspaces, a more generic condition must to be developed to investigate whether the above insensitivity can generally arise.

Concentrating on the non-negative quadratic Hotelling's T^2 and Q statistics, violations of their control limits are indicative of such changes. This postulates the following condition for changes in the covariance structure to be undetectable.

Condition 8.1.1 *A change in the covariance structure of the process variables is undetectable if and only if the Type I error with respect to the control limits or region of the original covariance structure does not exceed the significance* α.

This represents a condition that can be satisfied by examining the control limits of the non-negative quadratic statistics. Subsection 3.1.2 showed that the control limit of the Hotelling's T^2 statistic is, asymptotically, the critical value of a χ^2 distribution for the significance α. On the other hand, the control limit of the Q statistic can be approximated by a χ^2 distribution (Box 1954; Jackson and Mudholkar 1979; MacGregor and Kourti 1995; Satterthwaite 1941). With this in mind, it follows that

$$F\left(\sum_{i=m_0}^{m_1} \frac{t_i^2}{\sigma_{t_i}^2}\right) \approx \eta\chi^2(\theta) \tag{8.6}$$

where η and θ are a weight factor and the number of degrees of freedom of a χ^2 distribution, respectively. It should be noted that the approximation in (8.6) is also applicable to the Hotelling's T^2 statistic.

In the case of PCA,

- $\sigma_{t_i}^2$ is the ith largest eigenvalue of $\mathbf{S}_{zz}$, and m_0 and m_1 are 1 and n, respectively, for the Hotelling's T^2 statistic; and
- $\sigma_{t_i}^2 = 1$, and m_0 and m_1 are $n+1$ and n_z, respectively, for the Q statistic.

For PLS,

- $\sigma_{t_i}^2$ is $\sum_{j=1}^{n_x} r_{ji}^2\sigma_j + 2\sum_{j=1}^{n_x-1}\sum_{l=j+1}^{n_x} r_{ji}r_{li}\sigma_{jl}^2$, and m_0 and m_1 are 1 and n, respectively, for the Hotelling's T^2 statistic; and
- $\sigma_{t_i}^2 = 1$, and m_0 and m_1 are $n+1$ and n_x, respectively, for the Q_e statistic.

Although the relationship below is also applicable to the Q_f statistic for PLS, this analysis is not considered here.

Estimating the sample mean and variance of the sequence $\sum_{i=m_0}^{m_1}\left({t_i^{(0)}}^2(1)/\sigma_i\right)$, $\sum_{i=m_0}^{m_1}\left({t_i^{(0)}}^2(2)/\sigma_i\right), \ldots, \sum_{i=m_0}^{m_1}\left({t_i^{(0)}}^2(K_0)/\sigma_i\right)$, (3.30) and (3.31) show that

$$\eta_0 = \frac{\widehat{\sigma}_0^2}{2\widehat{\mu}_0} \qquad \theta_0 = \frac{2\widehat{\mu}_0^2}{\widehat{\sigma}_0^2}, \tag{8.7}$$

if K_0 is sufficiently large. Here, the sub- and superscript (0) refer, as before, to the reference condition, and $\widehat{\mu}_0$ and $\widehat{\sigma}_0^2$ are the estimated mean and variance, respectively. For a second sequence, $\sum_{i=m_0}^{m_1}\left({t_i^{(1)}}^2/\sigma_i\right)$, which contains a total of K_1 samples $\sum_{i=m_0}^{m_1}\left({t_i^{(1)}}^2(1)/\sigma_i\right)$, $\sum_{i=m_0}^{m_1}\left({t_i^{(1)}}^2(2)/\sigma_i\right), \ldots, \sum_{i=m_0}^{m_1}\left({t_i^{(1)}}^2(K_1)/\sigma_i\right)$, describing a change in the variable covariance structure, the parameters $\widehat{\mu}_1$ and $\widehat{\sigma}_1^2$ can be obtained. Here, the sub- and superscript (1) refer to the second operating condition. Using the estimates $\widehat{\mu}_0$, $\widehat{\mu}_1$, $\widehat{\sigma}_0^2$ and $\widehat{\sigma}_1^2$ allows formulating the following condition for detecting the second and abnormal operating condition.

Condition 8.1.2 *If the control limit for* $\sum_{i=m_0}^{m_1}\left({t_i^{(1)}}^2/\sigma_i\right)$*, obtained for a significance* α*, is approximated by* $\eta_1\chi_\alpha^2(\theta_1)$*, where* η_1 *and* θ_1 *are values* *for* η *and* θ *estimated from* $\sum_{i=m_0}^{m_1}\left({t_i^{(1)}}^2(1)/\sigma_i\right)$, $\sum_{i=m_0}^{m_1}\left({t_k^{(1)}}^2(2)/\sigma_i\right)$, $\cdots$, $\sum_{i=m_0}^{m_1}\left({t_k^{(1)}}^2(K_1)/\sigma_i\right)$ *for a sufficiently large* K_1, *is smaller or equal to the control limit for* $\sum_{i=m_0}^{m_1}\left({t_k^{(0)}}^2/\sigma_i\right)$, *approximated by* $\eta_0\chi^2_\alpha(\theta_0)$, *this change is undetectable.*

Under the application of above condition, score-based process monitoring using conventional MSPC may be insensitive to changes in the variable covariance structure, which the next subsection illustrates using a three-variable example.

8.1.3 Second intuitive example

The three variables are defined by a linear combination of the two zero mean i.d. source signals $s_1^{(0)}$ and $s_2^{(0)}$, which have a variance of $\sigma_1^2 = 15^1/_2$ and $\sigma_2^2 = 7^1/_2$. As before, the superscript $^{(0)}$ refers to the original covariance structure. According to (2.2), the zero mean error vector $\mathbf{g}$, augmented to the common cause variation $\Xi\mathbf{s}^{(0)}$, has an error covariance matrix $\mathbf{S}_{gg} = {}^1/_2\mathbf{I}$. Furthermore, (8.8) defines the score and loading vectors for the data vector $\mathbf{z} = \Xi\mathbf{s} + \mathbf{g}$.

$$\begin{pmatrix} z_1^{(0)} \\ z_2^{(0)} \\ z_3^{(0)} \end{pmatrix} = \underbrace{\begin{bmatrix} 1/\sqrt{6} & 1/\sqrt{2} & 1/\sqrt{3} \\ 1/\sqrt{6} & -1/\sqrt{2} & 1/\sqrt{3} \\ 2/\sqrt{6} & 0 & -1/\sqrt{3} \end{bmatrix}}_{\mathbf{T}^{(0)}} \begin{pmatrix} t_1^{(0)} \\ t_2^{(0)} \\ t_3^{(0)} \end{pmatrix}. \tag{8.8}$$

The matrix $\mathbf{T}^{(0)}$ stores the eigenvectors of $\mathbf{S}_{zz}^{(0)}$ and ${\mathbf{t}^{(0)}}^T = \left(t_1^{(0)}\ t_2^{(0)}\ t_3^{(0)}\right)$ is a vector storing the score variables. Under the assumption that $E\left\{\mathbf{s}\mathbf{g}^T\right\} = \mathbf{0}$, the covariance matrix of $\mathbf{z}^{(0)}$, $\mathbf{S}_{zz}^{(0)}$, is equal to

$$\begin{aligned} \mathbf{S}_{zz}^{(0)} &= \begin{bmatrix} 1/\sqrt{6} & 1/\sqrt{2} & 1/\sqrt{3} \\ 1/\sqrt{6} & -1/\sqrt{2} & 1/\sqrt{3} \\ 2/\sqrt{6} & 0 & -1/\sqrt{3} \end{bmatrix} \begin{bmatrix} 16 & 0 & 0 \\ 0 & 8 & 0 \\ 0 & 0 & 1/2 \end{bmatrix} \begin{bmatrix} 1/\sqrt{6} & 1/\sqrt{6} & 2/\sqrt{6} \\ 1/\sqrt{2} & -1/\sqrt{2} & 0 \\ 1/\sqrt{3} & 1/\sqrt{3} & -1/\sqrt{3} \end{bmatrix} \\ \mathbf{S}_{zz}^{(0)} &= \begin{bmatrix} 41/6 & -7/6 & 31/6 \\ -7/6 & 41/6 & 31/6 \\ 31/6 & 31/6 & 65/6 \end{bmatrix} \end{aligned} \tag{8.9}$$

which follows from (6.5). Moreover, the column space of Ξ is equal to the first two column vectors of $\mathbf{T}^{(0)}$. For simplicity, is assumed here that Ξ contains these column vectors, implying that the orthogonal complement, $\Xi^\perp$, is the transpose of the third column vector and the generalized inverse, $\Xi^\dagger$, is the transpose of Ξ.

The contribution of the first, second and third principal components to the sum of the variances of the three process variables are $^{32}/_{49}\cdot 100\% = 65.31\%$, $^{16}/_{49}\cdot 100\% = 32.65\%$ and $^{1}/_{49}\cdot 100\% = 2.04\%$, which follows from (2.116) to (2.122). Equation (6.73) highlights that the first two score variables mainly describe the two source variables, which contribute 97.94% to this sum of variances, whilst the contribution of the third score variable is 2.04% and, according to (3.7), relates to $\mathbf{g}$.

The eigenvectors $\mathbf{p}_1^T = \left({}^1/\sqrt{6} \quad {}^1/\sqrt{6} \quad {}^2/\sqrt{6}\right)$ and $\mathbf{p}_2^T = \left({}^1/\sqrt{2} \quad -{}^1/\sqrt{2} \quad 0\right)$ span the model subspace and $\mathbf{p}_3^T = \left({}^1/\sqrt{3} \quad {}^1/\sqrt{3} \quad -{}^1/\sqrt{3}\right)$ spans the residual subspace. As the data space corresponding to z_1, z_2 and z_3 is a Cartesian space, the minimum angles of the axes z_1, z_2 and z_3 to the third eigenvector are 54.74°, 54.74° and 125.26°, respectively. The critical value of a χ^2 distribution for two degrees of freedom and $\alpha = 0.01$ is $\chi_\alpha^2 = 9.21034$. The lengths of the semimajor and semiminor of the control ellipse (first two score variables) are, therefore, $\sqrt{\chi_\alpha^2 \cdot \lambda_1} = \sqrt{147.3654} = 12.1394$ and $\sqrt{\chi_\alpha^2 \cdot \lambda_2} = \sqrt{73.6827} = 8.5839$, respectively, $\lambda_1 = 16$ and $\lambda_2 = 8$.

To introduce alterations to this data covariance structure and to examine whether these alterations are detectable, a total of four changes are considered. Each of these changes relates to an anticlockwise rotation of the original variable set by 30°. Equation (8.10) shows the corresponding rotation matrix $\mathbf{T}^{(1)}$

$$\mathbf{T}^{(1)} = \begin{bmatrix} \sqrt{3}/2 & -1/2 & 0 \\ 1/2 & \sqrt{3}/2 & 0 \\ 0 & 0 & 1 \end{bmatrix}. \tag{8.10}$$

The first change is a simple rotation of the first two variables

$$\mathbf{z}^{(1)} = \mathbf{T}^{(1)}\mathbf{T}^{(0)}\mathbf{t}^{(1)}, \tag{8.11}$$

where $\mathbf{t}^{(1)} = \mathbf{t}^{(0)}$. The remaining three changes also alter the variance of the score variables, listed in Table 8.1, which produces the data vectors $\mathbf{z}^{(2)}$, $\mathbf{z}^{(3)}$ and $\mathbf{z}^{(4)}$

$$\mathbf{z}^{(2)} = \mathbf{T}^{(1)}\mathbf{T}^{(0)}\mathbf{t}^{(2)} \quad \mathbf{z}^{(3)} = \mathbf{T}^{(1)}\mathbf{T}^{(0)}\mathbf{t}^{(3)} \quad \mathbf{z}^{(4)} = \mathbf{T}^{(1)}\mathbf{T}^{(0)}\mathbf{t}^{(4)}. \tag{8.12}$$

There are now the following five variable sets:

1. the reference set $\mathbf{z}^{(0)}$ yielding the loading vectors stored in $\mathbf{T}^{(0)}$ and score variances of 16, 8 and $^1/_2$;

2. the variable set $\mathbf{z}^{(1)}$ representing the loading vectors $\mathbf{T}^{(1)}\mathbf{T}^{(0)}$ and score variance of 16, 8 and $^1/_2$;

3. the variable set $\mathbf{z}^{(2)}$ producing the same loading vectors as $\mathbf{z}^{(1)}$ but yields score variance of 4, 2 and 0.15;

Table 8.1 Variance of score variables $t_1^{(m)}$, $t_2^{(m)}$ and $t_3^{(m)}$.

Condition/ Variable	$\lambda_1 = E\left\{t_1^{(m)^2}\right\}$	$\lambda_2 = E\left\{t_2^{(m)^2}\right\}$	$\lambda_3 = E\left\{t_3^{(m)^2}\right\}$
$m = 0$	16	8	0.5
$m = 1$	16	8	0.5
$m = 2$	4	2	0.15
$m = 3$	4	2	0.125
$m = 4$	4	2	0.1

4. variable set $\mathbf{z}^{(3)}$ which produces the same loading vectors as $\mathbf{z}^{(2)}$ but the variance of the third score variable is 0.125; and

5. finally variable set $\mathbf{z}^{(4)}$ which again yields the same loading vectors as $\mathbf{z}^{(2)}$ but the variance of the third score variable is 0.1.

To demonstrate how different these five variable sets are requires the inspection of the corresponding covariance matrices for $\mathbf{z}^{(0)}$, $\mathbf{z}^{(1)}$, $\mathbf{z}^{(2)}$, $\mathbf{z}^{(3)}$ and $\mathbf{z}^{(4)}$

$$\mathbf{S}_{zz}^{(0)} = 1/6 \begin{bmatrix} 41 & -7 & 31 \\ -7 & 41 & 31 \\ 31 & 31 & 65 \end{bmatrix}$$

$$\mathbf{S}_{zz}^{(1)} = 1/12 \begin{bmatrix} 82+7\sqrt{3} & -7 & 31\left(\sqrt{3}-1\right) \\ -7 & 82-7\sqrt{3} & 31\left(\sqrt{3}+1\right) \\ 31\left(\sqrt{3}-1\right) & 31\left(\sqrt{3}+1\right) & 130 \end{bmatrix}$$

$$\mathbf{S}_{zz}^{(2)} = 1/120 \begin{bmatrix} 206+17\sqrt{3} & -17 & 77\left(\sqrt{3}-1\right) \\ -17 & 206-17\sqrt{3} & 77\left(\sqrt{3}+1\right) \\ 77\left(\sqrt{3}-1\right) & 77\left(\sqrt{3}+1\right) & 326 \end{bmatrix} \quad (8.13)$$

$$\mathbf{S}_{zz}^{(3)} = 1/64 \begin{bmatrix} 82+7\sqrt{3} & -7 & 31\left(\sqrt{3}-1\right) \\ -7 & 82-7\sqrt{3} & 31\left(\sqrt{3}+1\right) \\ 31\left(\sqrt{3}-1\right) & 31\left(\sqrt{3}+1\right) & 130 \end{bmatrix}$$

$$\mathbf{S}_{zz}^{(4)} = 1/60 \begin{bmatrix} 102+9\sqrt{3} & -9 & 39\left(\sqrt{3}-1\right) \\ -9 & 102-9\sqrt{3} & 39\left(\sqrt{3}+1\right) \\ 39\left(\sqrt{3}-1\right) & 39\left(\sqrt{3}+1\right) & 162 \end{bmatrix}.$$

The next step is to perform a total of 1000 Monte Carlo simulations for each of the five variable sets, $\mathbf{z}^{(0)}, \cdots, \mathbf{z}^{(4)}$. According to Condition 8.1.2, the changes in the covariance structure cannot be detected if the control limits associated with the variable sets representing $\mathbf{z}^{(1)}$, $\mathbf{z}^{(2)}$, $\mathbf{z}^{(3)}$ and $\mathbf{z}^{(4)}$ are smaller or equal to the control limit corresponding to $\mathbf{z}^{(0)}$. It is important to note, however, that the non-negative quadratic statistics must be constructed from the PCA model related to the variable set $\mathbf{z}^{(0)}$. The calculation of the score variables for each of the five variable sets is

$$\begin{aligned} \widetilde{\mathbf{t}}^{(0)} &= \mathbf{T}^{(0)^T}\mathbf{z}^{(0)} = \mathbf{T}^{(0)^T}\mathbf{T}^{(0)}\mathbf{t}^{(0)} = \mathbf{t}^{(0)} \\ \widetilde{\mathbf{t}}^{(1)} &= \mathbf{T}^{(0)^T}\mathbf{z}^{(1)} = \mathbf{T}^{(0)^T}\mathbf{T}^{(1)}\mathbf{T}^{(0)}\mathbf{t}^{(1)} \neq \mathbf{t}^{(1)} \\ \widetilde{\mathbf{t}}^{(2)} &= \mathbf{T}^{(0)^T}\mathbf{z}^{(2)} = \mathbf{T}^{(0)^T}\mathbf{T}^{(1)}\mathbf{T}^{(0)}\mathbf{t}^{(2)} \neq \mathbf{t}^{(2)} \end{aligned} \quad (8.14)$$

$$\widetilde{\mathbf{t}}^{(3)} = \mathbf{T}^{(0)^T}\mathbf{z}^{(3)} = \mathbf{T}^{(0)^T}\mathbf{T}^{(1)}\mathbf{T}^{(0)}\mathbf{t}^{(3)} \neq \mathbf{t}^{(3)}$$

$$\widetilde{\mathbf{t}}^{(4)} = \mathbf{T}^{(0)^T}\mathbf{z}^{(4)} = \mathbf{T}^{(0)^T}\mathbf{T}^{(1)}\mathbf{T}^{(0)}\mathbf{t}^{(4)} \neq \mathbf{t}^{(4)}.$$

Based on (8.14), the five Hotelling's T^2 statistics are now constructed from the first two elements of the score vectors $\widetilde{\mathbf{t}}^{(0)}, \cdots, \widetilde{\mathbf{t}}^{(4)}$ and the score covariance matrix $\mathbf{\Lambda} = \text{diag}\left\{16 \quad 8\right\}$. The Q statistics are simply the squared values of the third elements of $\mathbf{t}^{(0)}, \cdots, \mathbf{t}^{(4)}$. Each of the 1000 Monte-Carlo simulation experiments include a total of $K = 100$ samples. This gives rise to a total of 1000 estimates for the control limits of the Hotelling's T^2 and Q statistics for $\mathbf{z}^{(0)}, \ldots, \mathbf{z}^{(4)}$. To assess the sensitivity in detecting each of the four changes, the 2.5 and the 97.5 percentiles as well as the median can be utilized.

Figure 8.2 (a) shows the range limited by the 2.5 and 97.5 percentiles of the control limit for each of the five Hotelling's T^2 statistics $\widehat{T}_\alpha^{(0)^2}, \cdots, \widehat{T}_\alpha^{(4)^2}$. Plot (b) in this figure shows the ranges for the control limits of $\widehat{Q}_\alpha^{(0)}, \cdots, \widehat{Q}_\alpha^{(4)}$. The circle inside each of the ranges represents the median. Examining the range for the Hotelling's T^2 statistic in relation to Condition 8.1.2, it is clear that the Hotelling's T^2 statistic is insensitive to any of the changes introduced to the original covariance structure.

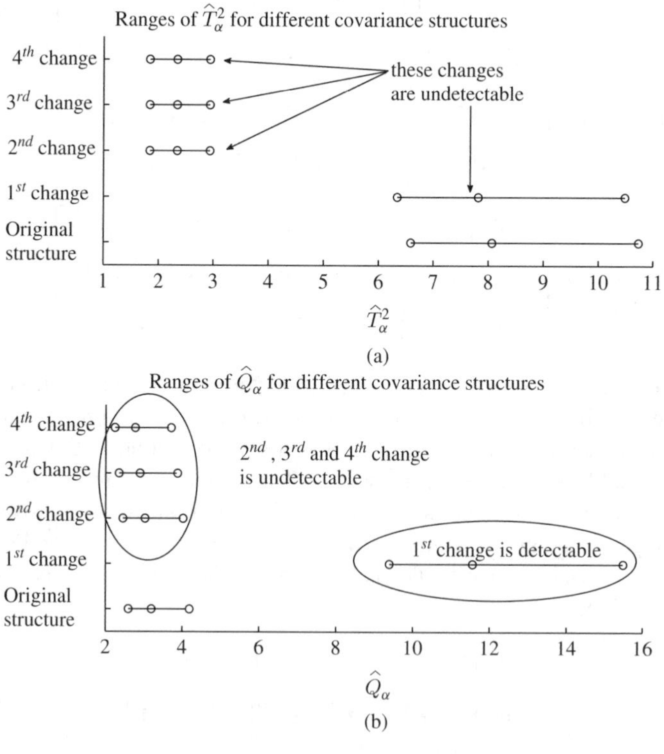

Figure 8.2 Analysis of detectability for different covariance structures.

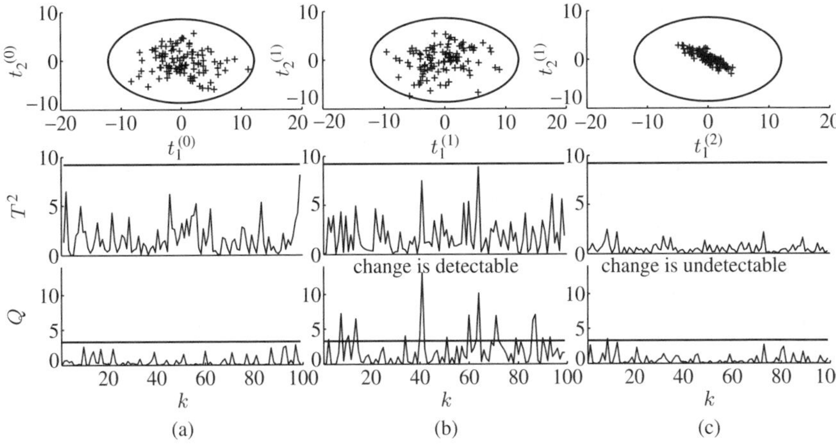

Figure 8.3 Detectable and undetectable changes in covariance structure.

A different picture emerges when making the same comparison for the Q statistic. While the range for $Q_\alpha^{(0)}$ covers values between 2.2 and 4 (roughly), the values for $Q_\alpha^{(1)}$ range between around 9.5 and 15.5.[1] According to Condition 8.1.2, this implies that this first alteration is detectable by the Q statistic. In contrast, the remaining three changes may not be detectable as the ranges for $\widehat{Q}_\alpha^{(2)}$, $\widehat{Q}_\alpha^{(3)}$ and $\widehat{Q}_\alpha^{(4)}$ have a significant overlap with the range for $\widehat{Q}_\alpha^{(0)}$. More precisely, the 2.5 and 97.5 percentiles for $\widehat{Q}_\alpha^{(0)}$ are larger than those for $\widehat{Q}_\alpha^{(2)}$, $\widehat{Q}_\alpha^{(3)}$ and $\widehat{Q}_\alpha^{(4)}$. Consequently, the second to fourth alterations are not detectable by the Hotelling's T^2 and may not be detectable by the Q statistic either.

To graphically illustrate the above findings, a total of 100 samples are generated for variable sets $\mathbf{z}^{(0)}$, $\mathbf{z}^{(1)}$ and $\mathbf{z}^{(2)}$. Referring to these sets as *data set 1*, *data set 2* and *data set 3*, corresponding to $\mathbf{z}^{(0)}$, $\mathbf{z}^{(1)}$ and $\mathbf{z}^{(2)}$, respectively, Figure 8.3 shows the results of analyzing them using a PCA model established from data set 1. In this figure, the column in rows (a), (b) and (c) represent the analysis of data set 1, data set 2 and data set 3, respectively. The upper plots show the control ellipse and the scatter plots of data sets 1 to 3. The plots in the middle and lower row of Figure 8.3 present the Hotelling's T^2 and the Q statistics, respectively.

The plots associated with index (a) indicate that the projection of each of the 100 samples of data set 1 onto the model subspace fall inside the control ellipse. This, in turn, implies that none of the samples results in a violation of the control limit of the Hotelling's T^2 statistic. Also the residual Q statistic has not violated its control limit, $Q_{0.01} = 3.2929$, for any of the 100 samples of data set 1. Hence, the hypothesis that the process is in-statistical-control must be accepted.

A different result emerges when inspecting the plots associated with data set 2, representing an anticlockwise rotation of $z_1 - z_2$ axis by 30°. Although

[1] This relatively large range outlines, again, the problem of selecting an appropriate size for the reference data set, discussed in Section 6.4 and Subsection 7.3.6.

projecting the samples onto the model subspace shows no projections outside the control ellipse, the Q statistic highlights that the squared distance of a total of 16 samples from the model subspace is larger than 3.2929. This change is therefore detectable.

Finally, the plots corresponding to data set 3 point out that the projected samples onto the model subspace fall inside the control ellipse and that the squared distance of each sample from the model subspace is less than 3.2929. Consequently, this change remains undetected, which is undesirable. The remainder of this chapter describes the incorporation of the statistical local approach into the MSPC framework to detect such changes.

With regards to the second, third and fourth alterations, one could justifiably argue that if the third eigenvalue is not changed from 0.5 to 0.15, 0.125 and 0.1, respectively, any of these changes are detectable by the Q statistic. This follows from (6.4) and (6.5), which highlight that λ_3 corresponds to the noise variance. According to Figure 8.3, the rotation of the control ellipse changes its orientation relative to the original model subspace. Thus, samples that are further away from the center of the ellipse but still inside produce a larger distance to the original model subspace.

If the axes of the rotated control ellipse are linear combinations of the eigenvectors spanning the model subspace, the rotated ellipse remains inside the model subspace. Hence, such an alteration of the covariance structure has no effect on the residual subspace and hence the Q statistic. Revisiting the geometric analysis in Figure 8.1, a change in the orientation and dimension may yield a control ellipse that lies within the original ellipse and is on the model subspace. Equations (6.7) to (6.11) outline that such an alteration results from a change in the covariance matrix of the source signals and may, consequently, remain undetected.

8.2 Preliminary discussion of related techniques

After outlining that the basic MSPC monitoring framework may not detect certain changes in the data covariance structure, a different paradigm is required to address this issue. Revisiting the analysis in Figure 8.1, the exact shape and orientation of a control ellipse is defined by the eigenvectors and eigenvalues of $\mathbf{S}_{z_0 z_0}$. In other words, if the orientation of the eigenvectors and the eigenvalues could be monitored on-line, any change in the covariance structure can consequently not go unnoticed. It is therefore required to formulate monitoring functions that directly relate to the eigen decomposition of $\mathbf{S}_{z_0 z_0}$.

Basseville (1988) described a statistical theory, known as the *statistical local approach*, that can be readily utilized to define vector-valued monitoring functions, referred to as *primary residual* vectors $\boldsymbol{\phi}$, of the form

$$\boldsymbol{\phi} = \boldsymbol{\phi}\left(\mathfrak{p}, \mathbf{z}_0\right) \tag{8.15}$$

where $\mathfrak{p}$ is a vector of model parameters and $\mathbf{z}_0 \sim \mathcal{N}\left\{\mathbf{0}, \mathbf{S}_{z_0z_0}\right\}$. For simplicity, the distribution function of $\boldsymbol{\phi}(\cdot)$ is assumed to be unknown at this point.

The parameter vectors includes the eigendecomposition of $\mathbf{S}_{z_0z_0}$ for PCA and $\mathbf{S}_{x_0x_0}$ and $\mathbf{S}_{x_0y_0}$ for PLS. The construction of the primary residuals for PCA is discussed in Sections 8.3. For a statistical inference based on $\boldsymbol{\phi}(\cdot)$, however, the following problem arises. How to construct a monitoring framework if $F(\boldsymbol{\phi}(\cdot))$ cannot be assumed to be Gaussian or is unknown, as assumed thus far?

This question can be answered by assuming that $\mathbf{z}_0$ stores i.i.d. sequences, that is, $E\left\{\mathbf{z}_0(k)\mathbf{z}_0^T(l)\right\} = \delta_{kl}\mathbf{S}_{z_0z_0}$, where k and l are sample indices. As the distribution function of $\boldsymbol{\phi}(\cdot)$ depends on the distribution function of $\mathbf{z}_0$, instances of $\boldsymbol{\phi}\left(\mathfrak{p}, \mathbf{z}_0\right)$ are also i.i.d. Under these conditions, the following sum of the primary residual vectors

$$\boldsymbol{\theta} = \frac{1}{\sqrt{K}} \sum_{k=1}^{K} \boldsymbol{\phi}\left(\mathfrak{p}, \mathbf{z}_0(k)\right) \tag{8.16}$$

follows, asymptotically, a Gaussian distribution function, which is a result of the CLT. Subsection 8.7.1 provides a detailed discussion and a proof of the CLT. The sum in (8.16) is defined as the *improved residual* vector and is, asymptotically, Gaussian distribution. If $E\left\{\boldsymbol{\phi}\left(\mathfrak{p}, \mathbf{z}_0\right)\right\} = \mathbf{0}$ and $E\left\{\boldsymbol{\phi}\left(\mathfrak{p}, \mathbf{z}_0\right)\boldsymbol{\phi}^T\left(\mathfrak{p}, \mathbf{z}_0\right)\right\} = \mathbf{S}_{\phi\phi}$, $\boldsymbol{\theta} \sim \mathcal{N}\left\{\mathbf{0}, \mathbf{S}_{\phi\phi}\right\}$ and can be utilized to construct scatter diagrams as well as a Hotelling's T^2 statistic as discussed in Subsection 3.1.2.

For PCA, it is sufficient to develop primary residuals related to the eigenvalues and the eigenvectors of $\mathbf{S}_{z_0z_0}$, as they determine the orientation of the model and the residual subspaces, and the size and orientation of the control ellipse. For PLS, however, there are two interrelated data spaces. Project 2 in the tutorial session of this chapter extends the development of improved and primary residuals for PLS.

For PCA, the next section discuss the construction of primary and improved residuals describing changes in the geometry of the model and residual subspaces and summarizes their basic statistical properties.

8.3 Definition of primary and improved residuals

Sections 2.1 and 9.3 outline that a PCA monitoring model is completely described by the eigendecomposition of $\mathbf{S}_{z_0z_0}$. This includes the orientation of the model and residual subspaces as well as the orientation and size of the n dimensional control ellipsoid. Consequently, the primary residuals rely on the eigendecomposition of $\mathbf{S}_{z_0z_0}$, and are derived in Subsection 8.3.1 using the definition of the ith eigenvector $\mathbf{S}_{z_0z_0}\mathbf{p}_i - \lambda_i\mathbf{p}_i = \mathbf{0}$. Subsection 8.3.2 shows that primary residuals can also be obtained from $\mathbf{p}_i^T\mathbf{S}_{z_0z_0}\mathbf{p}_i - \lambda_i = 0$. Subsections 8.3.3 and 8.3.4 contrast both types of primary residuals and determine their statistical properties. Finally, Subsection 8.3.5 shows the construction of improved residuals.

8.3.1 Primary residuals for eigenvectors

Starting with the definition of the objective function for obtaining the ith eigenvector

$$\mathbf{p}_i \triangleq \arg\max_{\mathbf{p}} \left\{ E\left\{\mathbf{p}^T \mathbf{z}_0 \mathbf{z}_0^T \mathbf{p}\right\} - \lambda_i \left(\mathbf{p}^T \mathbf{p} - 1\right)\right\}, \tag{8.17}$$

the partial derivative of (8.17) allows determining the optimal solution

$$\mathbf{p}_i = \arg \frac{\partial}{\partial \mathbf{p}} \left\{ E\left\{\mathbf{p}^T \mathbf{z}_0 \mathbf{z}_0^T \mathbf{p}\right\} - \lambda_i \left(\mathbf{p}^T \mathbf{p} - 1\right)\right\} = \mathbf{0} \tag{8.18}$$

which is given by

$$\mathbf{p}_i = \arg \left\{ E\left\{2 t_i \mathbf{z}_0\right\} - 2\lambda_i \mathbf{p}_i \right\} = \mathbf{0}. \tag{8.19}$$

The above equation relies on the fact that $\mathbf{z}_0^T \mathbf{p}_i = t_i$. Now, defining

$$\boldsymbol{\phi}_i \triangleq 2 t_i \mathbf{z}_0 - 2\lambda_i \mathbf{p}_i \tag{8.20}$$

allows simplifying Equation (8.19) to become

$$\mathbf{p}_i \triangleq \arg \left\{ E\left\{\boldsymbol{\phi}_i\right\}\right\} = \mathbf{0} \tag{8.21}$$

and consequently

$$E\left\{\boldsymbol{\phi}_i\right\}\big|_{\mathbf{p}=\mathbf{p}_i} = \mathbf{0}. \tag{8.22}$$

It follows from (8.22) that in the vicinity of $\mathbf{p}_i$, defined by $\boldsymbol{\omega}\left(\mathbf{p}_i\right)$ for which $\mathbf{p}_i \notin \boldsymbol{\omega}\left(\mathbf{p}_i\right)$, the following holds true

$$E\left\{\boldsymbol{\phi}_i\right\}\big|_{\mathbf{p}\in\boldsymbol{\omega}(\mathbf{p}_i)} \neq \mathbf{0}. \tag{8.23}$$

Equations (8.22) and (8.23) imply that each loading vector $\mathbf{p}_i$ produces a corresponding statistic $\boldsymbol{\phi}_i$ such that $E\left\{\boldsymbol{\phi}_i\right\} = \mathbf{0}$, when $\mathbf{p}$ is the equal to the ith eigenvector of $\mathbf{S}_{z_0 z_0}$. In contrast, any deviation from zero indicates that $\mathbf{p}_i$ is no longer the eigenvector associated with the ith eigenvalue.

The next step is to define two parameter vectors that store the eigenvectors spanning the model and residual subspaces. The vector for the model subspace, $\mathfrak{p}$, is

$$\mathfrak{p} = \begin{pmatrix} \mathbf{p}_1^T & \mathbf{p}_2^T & \cdots & \mathbf{p}_n^T \end{pmatrix}^T \in \mathbb{R}^{n_z n} \tag{8.24}$$

and that of the residual subspace, $\mathfrak{p}_d$, is defined as

$$\mathfrak{p}_d = \begin{pmatrix} \mathbf{p}_{n+1}^T & \mathbf{p}_{n+2}^T & \cdots & \mathbf{p}_{n_z}^T \end{pmatrix}^T \in \mathbb{R}^{n_z(n_z-n)}. \tag{8.25}$$

This gives rise to the following two primary residual vectors for the model subspace

$$\boldsymbol{\phi} = \begin{pmatrix} \boldsymbol{\phi}_1^T & \boldsymbol{\phi}_2^T & \cdots & \boldsymbol{\phi}_n^T \end{pmatrix}^T \in \mathbb{R}^{n_z n} \tag{8.26}$$

and the residual subspace

$$\boldsymbol{\phi}_d = \begin{pmatrix} \boldsymbol{\phi}_{n+1}^T & \boldsymbol{\phi}_{n+2}^T & \cdots & \boldsymbol{\phi}_{n_z}^T \end{pmatrix}^T \in \mathbb{R}^{n_z(n_z-n)}. \tag{8.27}$$

The next subsection develops primary residual vectors for the eigenvalues of $\mathbf{S}_{z_0 z_0}$.

8.3.2 Primary residuals for eigenvalues

Pre-multiplying (8.20) by $\mathbf{p}_i^T$ gives rise to

$$\mathbf{p}_i^T \boldsymbol{\phi}_i = 2t_i \mathbf{p}_i^T \mathbf{z}_0 - 2\lambda_i \mathbf{p}_i^T \mathbf{p}_i = 2\left(t_i^2 - \lambda_i\right) \triangleq \widetilde{\phi}_i. \tag{8.28}$$

The expectation of $\widetilde{\phi}_i\left(\lambda_i, \mathbf{z}_0\right)$ directly follows from (8.22)

$$E\left\{\widetilde{\phi}_i\right\}\Big|_{\lambda=\lambda_i} = 0 \qquad E\left\{\widetilde{\phi}_i\right\}\Big|_{\lambda\in\widetilde{\omega}(\lambda_i)} \neq 0. \tag{8.29}$$

As before, $\widetilde{\omega}\left(\lambda_i\right)$ defines the neighborhood of λ_i, where $\lambda_i \notin \widetilde{\omega}\left(\lambda_i\right)$. This implies that $E\left\{\widetilde{\phi}_i\right\} = 0$ holds true if and only if λ is the ith largest eigenvalue of $\mathbf{S}_{z_0 z_0}$. In a similar fashion to the $\mathfrak{p}$ and $\mathfrak{p}_d$, and $\boldsymbol{\phi}$ and $\boldsymbol{\phi}_d$, $\widetilde{\mathfrak{p}}$ and $\widetilde{\mathfrak{p}}_d$, and $\widetilde{\boldsymbol{\phi}}$ and $\widetilde{\boldsymbol{\phi}}_d$, for the retained and discarded eigenvalues can be defined as follows

$$\begin{aligned}
\widetilde{\mathfrak{p}} &= \begin{pmatrix} \lambda_1 & \lambda_2 & \ldots & \lambda_n \end{pmatrix}^T \in \mathbb{R}^n \\
\widetilde{\mathfrak{p}}_d &= \begin{pmatrix} \lambda_{n+1} & \lambda_{n+2} & \cdots & \lambda_{n_z} \end{pmatrix}^T \in \mathbb{R}^{n_z-n} \\
\widetilde{\boldsymbol{\phi}} &= \begin{pmatrix} \widetilde{\phi}_1 & \widetilde{\phi}_2 & \ldots & \widetilde{\phi}_n \end{pmatrix}^T \in \mathbb{R}^n \\
\widetilde{\boldsymbol{\phi}}_d &= \begin{pmatrix} \widetilde{\phi}_{n+1} & \widetilde{\phi}_{n+2} & \cdots & \widetilde{\phi}_{n_z} \end{pmatrix}^T \in \mathbb{R}^{n_z-n}.
\end{aligned} \tag{8.30}$$

The next subsection provides a detailed examination of the primary residuals.

8.3.3 Comparing both types of primary residuals

The analysis concentrates first on the primary residual vectors $\boldsymbol{\phi}$ and $\boldsymbol{\phi}_d$, which have the dimension $n_z n$ and $n_z\left(n_z - n\right)$, respectively. These dimensions, therefore, depend on the ratio $^n/_{n_z}$. If n is close to n_z or if n is small compared to n_z, the size of $\boldsymbol{\phi}$ or $\boldsymbol{\phi}_d$ can be substantial. This subsection then compares the sensitivity of $\boldsymbol{\phi}$ and $\boldsymbol{\phi}_d$, with $\widetilde{\boldsymbol{\phi}}$ and $\widetilde{\boldsymbol{\phi}}_d$ for detecting changes in the covariance structure.

8.3.3.1 Degrees of freedom for primary residuals $\boldsymbol{\phi}$ and $\boldsymbol{\phi}_d$

A closer inspection of the primary residuals $\boldsymbol{\phi}$ and $\boldsymbol{\phi}_d$ reveals that its elements may be linearly dependent. This is best demonstrated by a joint analysis

$$\begin{pmatrix} \boldsymbol{\phi}_1 \\ \vdots \\ \boldsymbol{\phi}_n \\ \boldsymbol{\phi}_{n+1} \\ \vdots \\ \boldsymbol{\phi}_{n_z} \end{pmatrix} = \begin{pmatrix} 2\mathbf{z}_0 t_1 \\ \vdots \\ 2\mathbf{z}_0 t_n \\ 2\mathbf{z}_0 t_{n+1} \\ \vdots \\ 2\mathbf{z}_0 t_{n_z} \end{pmatrix} - \begin{pmatrix} 2\lambda_1 \mathbf{p}_1 \\ \vdots \\ 2\lambda_n \mathbf{p}_n \\ 2\lambda_{n+1} \mathbf{p}_{n+1} \\ \vdots \\ 2\lambda_{n_z} \mathbf{p}_{n_z} \end{pmatrix} \tag{8.31}$$

which can alternatively be written as

$$\begin{pmatrix} \boldsymbol{\phi}_1 \\ \vdots \\ \boldsymbol{\phi}_{n_z} \end{pmatrix} = \begin{pmatrix} 2\left(\sum_{i=1}^{n_z} \mathbf{p}_i t_i\right) t_1 - 2\lambda_1 \mathbf{p}_1 \\ \vdots \\ 2\left(\sum_{i=1}^{n_z} \mathbf{p}_i t_i\right) t_{n_z} - 2\lambda_{n_z} \mathbf{p}_{n_z} \end{pmatrix}. \tag{8.32}$$

In matrix-vector form, (8.32) becomes

$$\begin{pmatrix} \boldsymbol{\phi}_1 \\ \boldsymbol{\phi}_2 \\ \boldsymbol{\phi}_3 \\ \vdots \\ \boldsymbol{\phi}_{n_z} \end{pmatrix} = 2 \underbrace{\begin{bmatrix} \mathbf{p}_1 & \mathbf{p}_2 & \mathbf{p}_3 & \cdots & \mathbf{p}_{n_z} & \mathbf{0} & \mathbf{0} & \cdots & \mathbf{0} \\ \mathbf{0} & \mathbf{p}_1 & \mathbf{0} & \cdots & \mathbf{0} & \mathbf{p}_2 & \mathbf{p}_3 & \cdots & \mathbf{0} \\ \mathbf{0} & \mathbf{0} & \mathbf{p}_1 & \cdots & \mathbf{0} & \mathbf{0} & \mathbf{p}_2 & \cdots & \mathbf{0} \\ \vdots & \vdots & \vdots & & \vdots & \vdots & \vdots & & \vdots \\ \mathbf{0} & \mathbf{0} & \mathbf{0} & \cdots & \mathbf{p}_1 & \mathbf{0} & \mathbf{0} & \cdots & \mathbf{p}_{n_z} \end{bmatrix}}_{\boldsymbol{\Psi}} \begin{pmatrix} t_1^2 - \lambda_1 \\ t_1 t_2 \\ t_1 t_3 \\ \vdots \\ t_1 t_{n_z} \\ t_2^2 - \lambda_2 \\ t_2 t_3 \\ \vdots \\ t_{n_z}^2 - \lambda_{n_z}^2 \end{pmatrix} \tag{8.33}$$

Since $\boldsymbol{\Psi} \in \mathbb{R}^{n_z^2 \times n_z(n_z+1)/2}$ has full column rank, its rank is equal to $n_z(n_z+1)/2$. More precisely, a total of $n_z(n_z-1)/2$ elements in the combined primary residual vector are linearly dependent upon the remaining $n_z(n_z+1)/2$ ones.

For the primary residual vectors $\boldsymbol{\phi}$ and $\boldsymbol{\phi}_d$, this has the following consequence: *if the number of the elements in*:

- $\boldsymbol{\phi} \in \mathbb{R}^{n_z n}$ and
- $\boldsymbol{\phi}_d \in \mathbb{R}^{n_z(n_z-n)}$

is larger than or equal to

$$\boldsymbol{\phi} \quad : \quad n_z n \geq \frac{n_z\left(n_z+1\right)}{2}$$

$$\boldsymbol{\phi}_d \quad : \quad n_z\left(n_z-n\right) \geq \frac{n_z\left(n_z+1\right)}{2}$$

there is a linear dependency between these primary residuals. This gives rise to linear dependency among the elements in $\boldsymbol{\phi}$ and $\boldsymbol{\phi}_d$ under the following conditions

$$\begin{array}{llll}
\boldsymbol{\phi} & : & n_z n \leq {}^{n_z\left(n_z+1\right)}/_{2} & \rightarrow \text{ no linear dependency} \\
\boldsymbol{\phi} & : & n_z n > {}^{n_z\left(n_z+1\right)}/_{2} & \rightarrow \text{ linear dependency} \\
\boldsymbol{\phi}_d & : & n_z\left(n_z-n\right) \leq {}^{n_z\left(n_z+1\right)}/_{2} & \rightarrow \text{ no linear dependency} \\
\boldsymbol{\phi}_d & : & n_z\left(n_z-n\right) > {}^{n_z\left(n_z+1\right)}/_{2} & \rightarrow \text{ linear dependency}
\end{array}$$

and leads to the following criteria for $\boldsymbol{\phi}$

$$\begin{aligned}
2nn_z &\leq n_z^2 + n_z \\
2n &\leq n_z + 1 \\
n &\geq {}^{\left(n_z-1\right)}/_{2}
\end{aligned} \tag{8.34}$$

and $\boldsymbol{\phi}_d$

$$\begin{aligned}
2n_z\left(n_z-n\right) &\leq n_z^2 + n_z \\
2\left(n_z-n\right) &\leq n_z + 1 \\
n &\geq {}^{\left(n_z-1\right)}/_{2}.
\end{aligned} \tag{8.35}$$

From the above relationships, it follows that

$$n_z - 1 \leq 2n \leq n_z + 1 \tag{8.36}$$

which can only be satisfied if $n = {}^{n_z}/_{2}$ if n_z is even and ${}^{n_z-1}/_{2} \leq n \leq {}^{n_z+1}/_{2}$ if n_z is odd. Figure 8.4 summarizes the above findings and shows graphically which condition leads to a linear dependency of the primary residuals in $\boldsymbol{\phi}$ and $\boldsymbol{\phi}_d$.

The importance of these findings relates to the construction of the primary residual vectors, since the number of source signals is determined as part of the identification of a principal component model. In other words, the original size of $\boldsymbol{\phi}$ and $\boldsymbol{\phi}_d$ is $n_z n$ and $n_z\left(n_z-n\right)$, respectively, and known *a priori*. If the analysis summarized in Figure 8.4 reveals that elements stored in the primary residual vectors are linearly dependent, the redundancy can be removed by eliminating redundant elements in $\boldsymbol{\phi}$ or $\boldsymbol{\phi}_d$, such that this number is smaller or equal to ${}^{n_z\left(n_z+1\right)}/_{2}$ in both vectors.

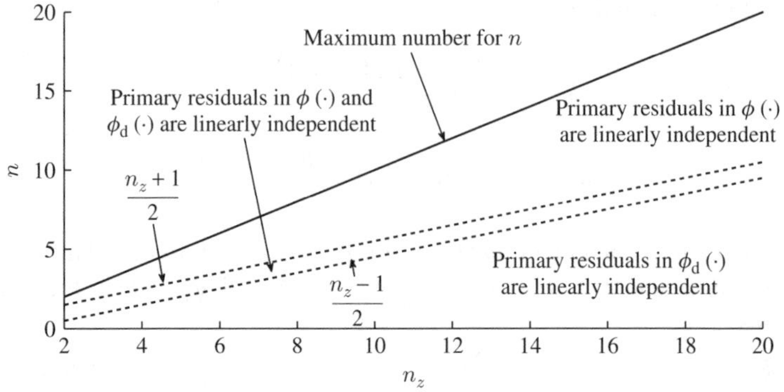

Figure 8.4 Linear dependency among elements in $\boldsymbol{\phi}(\mathbf{p}, \mathbf{z}_0)$ and $\boldsymbol{\phi}_d(\mathbf{p}_d, \mathbf{z}_0)$.

8.3.3.2 Sensitivity analysis for $\boldsymbol{\phi}$, $\boldsymbol{\phi}_d$, $\widetilde{\boldsymbol{\phi}}$ and $\widetilde{\boldsymbol{\phi}}_d$

To investigate whether the primary residuals $\boldsymbol{\phi}_i$ and $\widetilde{\phi}_i$ can both detect changes in the eigenvalues and the eigenvectors of $\mathbf{S}_{z_0z_0}$, the examination focuses on:

- the primary residuals $\boldsymbol{\phi}$ and $\widetilde{\boldsymbol{\phi}}$ to evaluate their sensitivity in detecting changes in the eigenvectors and eigenvalues associated with the orientation of the model subspace and the orientation and size of the control ellipsoid; and
- the primary residuals $\boldsymbol{\phi}_d$ and $\widetilde{\boldsymbol{\phi}}_d$ to examine their sensitivity in detecting changes in the eigenvectors and eigenvalues related to the orientation of the residual subspace and, according to (3.16), the approximation of the distribution function of the sum of squared residuals.

The resultant analysis yields the following two lemmas, which are proved below.

Lemma 8.3.1 *For a change in the orientation of the model subspace and/or the orientation/size of the control ellipsoid, the primary residual vectors $\boldsymbol{\phi}$ and $\widetilde{\boldsymbol{\phi}}$ are sensitive in detecting this change, as their expectation differs from zero.*

Lemma 8.3.2 *For a change associated with the orientation of the residual subspace and/or, the approximation of the distribution function of the sum of the squared residuals, the primary residual vectors $\boldsymbol{\phi}_d$ and $\widetilde{\boldsymbol{\phi}}_d$ can both detect this change by producing an expectation that is different from zero.*

Proof. The proof commences by rewriting (8.17) as follows

$$\begin{aligned} \mathbf{p}_i &= \arg\max_{\mathbf{p}} \left\{ \mathbf{p}^T E\left\{\mathbf{z}_0\mathbf{z}_0^T\right\}\mathbf{p} - \lambda_i\left(\mathbf{p}^T\mathbf{p} - 1\right)\right\} \\ \mathbf{p}_i &= \arg\max_{\mathbf{p}} \left\{ \mathbf{p}^T \mathbf{S}_{z_0z_0}\mathbf{p} - \lambda_i\left(\mathbf{p}^T\mathbf{p} - 1\right)\right\} \end{aligned} \quad (8.37)$$

and investigating the impact of a change in $\mathbf{p}_i$, that is, $\mathbf{p}_i^* = \mathbf{p}_i + \Delta\mathbf{p}_i$, $\|\Delta\mathbf{p}_i\| \ll \|\mathbf{p}_i\|$, and λ_i, i.e. $\lambda_i^* = \lambda_i + \Delta\lambda_i$, $|\Delta\lambda_i| \ll \lambda_i$.

Directional changes of the *i*th eigenvector. Assuming that λ_i remains unchanged, (8.19) can be rewritten on the basis of (8.37)

$$2\mathbf{S}_{z_0z_0}\mathbf{p}_i - 2\lambda_i\mathbf{p}_i = \mathbf{0}. \tag{8.38}$$

Knowing that a change in the covariance structure between the recorded process variables produces a different $\mathbf{S}_{z_0z_0}$, denoted here by $\mathbf{S}^*_{z_0z_0}$, (8.38) becomes

$$2\mathbf{S}^*_{z_0z_0}\mathbf{p}_i - 2\lambda_i\mathbf{p}_i = \boldsymbol{\epsilon}_i \neq \mathbf{0}. \tag{8.39}$$

The expectation of the primary residual vector $E\left\{\boldsymbol{\phi}_i\right\} = \boldsymbol{\epsilon}_i$ and given by

$$\begin{aligned} E\left\{\boldsymbol{\phi}_i\right\} &= 2\left[\mathbf{S}^*_{z_0z_0} - \lambda_i\mathbf{I}\right]\mathbf{p}_i \\ E\left\{\boldsymbol{\phi}_i\right\} &= 2\left(\mathbf{S}^*_{z_0z_0}\mathbf{p}_i - \frac{\lambda_i\mathbf{S}_{z_0z_0}\mathbf{p}_i}{\lambda_i}\right) \\ E\left\{\boldsymbol{\phi}_i\right\} &= 2\left[\mathbf{S}^*_{z_0z_0} - \mathbf{S}_{z_0z_0}\right]\mathbf{p}_i = \boldsymbol{\epsilon}_i. \end{aligned} \tag{8.40}$$

It follows that $\boldsymbol{\epsilon}_i$ depends on the changes of the elements in $\mathbf{S}_{z_0z_0}$. Equation (8.40) shows that the condition $E\{\boldsymbol{\phi}_i\} = \mathbf{0}$ only arises if and only if $\mathbf{p}_i$ is also an eigenvector of $\mathbf{S}^*_{z_0z_0}$ associated with λ_i. This situation, however, cannot arise for all $1 \le i \le n_{\underset{\sim}{z}}$ unless $\mathbf{S}^*_{z_0z_0} = \mathbf{S}_{z_0z_0}$. An important question is whether the primary residual ϕ_i also reflect a directional changes of $\mathbf{p}_i$. This can be examined by subtracting $2\mathbf{S}^*_{z_0z_0}\mathbf{p}_i^* - 2\lambda_i\mathbf{p}_i^* = \mathbf{0}$ from (8.39), where $\mathbf{p}_i^*$ is the eigenvector of $\mathbf{S}^*_{z_0z_0}$ associated with λ_i, which yields

$$2\left(\mathbf{S}^*_{z_0z_0}\Delta\mathbf{p}_i - \lambda_i\Delta\mathbf{p}_i\right) = \boldsymbol{\epsilon}_i. \tag{8.41}$$

Pre-multiplying the above equation by $\mathbf{p}_i^T$ produces

$$2\underbrace{\left(\mathbf{p}_i^T\mathbf{S}^*_{z_0z_0} - \lambda_i\mathbf{p}_i^T\right)}_{\neq\mathbf{0}^T}\underbrace{\Delta\mathbf{p}_i}_{\neq\mathbf{0}} = \mathbf{p}_i^T\boldsymbol{\epsilon}_i = \epsilon_i \neq 0. \tag{8.42}$$

It is important to note that if the pre-multiplication is carried out by the transpose of $\mathbf{p}_i^* \neq \mathbf{p}_i$, (8.42) becomes zero, since $\mathbf{p}_i^{*^T}\mathbf{S}^*_{z_0z_0} - \lambda_i\mathbf{p}_i^{*^T} = \mathbf{0}^T$. Consequently, any directional change of $\mathbf{p}_i$ manifests itself in $E\{\widetilde{\phi}_i\} = \epsilon_i \neq 0$. This, in turn, implies that both primary residual vectors, $\boldsymbol{\phi}_i$ and $\widetilde{\phi}_i$, are sufficient in detecting any directional change in $\mathbf{p}_i$ by a mean different from zero. It should also be noted that if $\mathbf{p}_i^T\boldsymbol{\epsilon}_i = 0$ if both vectors are orthogonal to each other. A closer inspection of (8.42), however, yields that only the trivial case of $\mathbf{S}^*_{z_0z_0} = \mathbf{S}_{z_0z_0}$ can produce $\epsilon_i = 0$.

Changes in the ith eigenvalue. Now, λ_i changes under the assumption that $\mathbf{p}_i$ remains constant. For this change, (8.39) becomes

$$2\mathbf{S}^*_{z_0z_0}\mathbf{p}_i - 2\lambda_i\mathbf{p}_i = \boldsymbol{\epsilon}_i. \tag{8.43}$$

Subtracting $\mathbf{S}^*_{z_0z_0}\mathbf{p}_i = \lambda_i^*\mathbf{p}_i$, based on the correct eigenvalue λ_i^*, from Equation (8.43) gives rise to

$$\boldsymbol{\epsilon}_i = 2\left(\lambda_i - \lambda_i^*\right)\mathbf{p}_i = 2\underbrace{\Delta\lambda_i}_{\neq 0}\mathbf{p}_i, \tag{8.44}$$

and hence, $E\{\boldsymbol{\phi}_i\} = 2\Delta\lambda_i\mathbf{p}_i$, which implies that $\boldsymbol{\phi}_i$ is sensitive to the change in λ_i. Finally, pre-multiplication of (8.44) by $\mathbf{p}_i^T$ yields

$$2\Delta\lambda_i = \epsilon_i, \tag{8.45}$$

where $\epsilon_i = \mathbf{p}_i^T\boldsymbol{\epsilon}_i$. Thus, $E\{\widetilde{\phi}_i\} = 2\Delta\lambda_i$. This analysis highlights that both primary residual vectors, $\boldsymbol{\phi}_i$ and ϕ_i, can detect the change in λ_i.

The above lemmas outline that any change in the covariance structure of $\mathbf{z}_0$ can be detected by $\boldsymbol{\phi}_i$ and ϕ_i. Given that:

- the dimensions of the primary residuals $\widetilde{\boldsymbol{\phi}}$ and $\widetilde{\boldsymbol{\phi}}_d$ are significantly smaller than those of $\boldsymbol{\phi}$ and $\boldsymbol{\phi}_d$, respectively;
- the primary residuals for the eigenvectors and the eigenvalues, $\boldsymbol{\phi}_i$ and $\widetilde{\phi}_i$, can detect a change in the covariance structure of $\mathbf{z}_0$; and
- the elements in the primary residual vectors $\boldsymbol{\phi}$ and $\boldsymbol{\phi}_d$ cannot generally be assumed to be linearly independent,

it is advisable to utilize the primary residual vectors $\widetilde{\boldsymbol{\phi}}$ and $\widetilde{\boldsymbol{\phi}}_d$ for process monitoring. For simplicity, the parameter vectors are now denoted as follows $\boldsymbol{\lambda} = \widetilde{\mathbf{p}}$ and $\boldsymbol{\lambda}_d = \widetilde{\mathbf{p}}_d$. Moreover, the tilde used to discriminate between $\boldsymbol{\phi}_i$ and its scaled sum $\mathbf{p}_i^T\boldsymbol{\phi}_i = \widetilde{\phi}_i$ is no longer required and can be omitted. The next subsection analyzes the statistical properties of $\boldsymbol{\phi}$ and $\boldsymbol{\phi}_d$.

8.3.4 Statistical properties of primary residuals

According to (8.29), the expectation of both primary residual vectors, $\boldsymbol{\phi}$ and $\boldsymbol{\phi}_d$, is equal to zero. The remaining statistical properties of ϕ_i include its variance, the covariance of ϕ_i and ϕ_j, the distribution function of ϕ_i and the central moments of ϕ_i. This allows constructing the covariance matrices for $\boldsymbol{\phi}$ and $\boldsymbol{\phi}_d$, $\mathbf{S}_{\phi\phi} \in \mathbb{R}^{n\times n}$ and $\mathbf{S}_{\phi_d\phi_d} \in \mathbb{R}^{(n_z-n)\times(n_z-n)}$, respectively.

Variance of $\boldsymbol{\phi}_i$. The variance of ϕ_i can be obtained as follows:

$$E\left\{\phi_i^2\right\} = E\left\{\left(2\left(t_i^2 - \lambda_i\right)\right)^2\right\} \tag{8.46}$$

which can be simplified to

$$E\left\{\phi_i^2\right\} = 4\left(E\left\{t_i^4\right\} - 2\lambda_i E\left\{t_i^2\right\} + \lambda_i^2\right). \tag{8.47}$$

Given that:

- $t_i = \mathbf{p}_i^T \mathbf{z}_0$; and
- $\mathbf{z}_0 \sim \mathcal{N}\left\{\mathbf{0}, \mathbf{S}_{z_0 z_0}\right\}$

it follows that $t_i \sim \mathcal{N}\left\{0, \lambda_i\right\}$. As t_i is Gaussian distributed, central moments of $E\left\{t_i^m\right\}$ are 0 if m is odd and $\lambda^{m/2}(m-1)!!$. If m is even.[2] For $m = 2$, $E\left\{t_i^m\right\} = \lambda_i$ and for $m = 4$, $E\left\{t_i^m\right\} = 3\lambda_i^2$. Substituting this into (8.47) gives rise to

$$E\left\{\phi_i^2\right\} = 4\left(3\lambda_i^4 - 2\lambda_i^2 + \lambda_i^2\right) = 8\lambda_i^2. \tag{8.48}$$

Covariance of $\boldsymbol{\phi}_i$ and $\boldsymbol{\phi}_j$. The covariance between two primary residuals is

$$E\left\{\phi_i \phi_j\right\} = E\left\{4\left(t_i^2 - \lambda_i\right)\left(t_j^2 - \lambda_j\right)\right\} \tag{8.49}$$

and can be simplified to

$$E\left\{\phi_i \phi_j\right\} = 4\left(E\left\{t_i^2 t_j^2\right\} - \lambda_j E\left\{t_i^2\right\} - \lambda_i E\left\{t_j^2\right\} + \lambda_i \lambda_j\right). \tag{8.50}$$

Now, substituting $E\left\{t_i^2\right\} = \lambda_i$, $E\left\{t_j^2\right\} = \lambda_j$ and $E\left\{t_i^2 t_j^2\right\} = \lambda_i \lambda_j$, which follows from the Isserlis theorem (Isserlis 1918) and the fact that t_i and t_j are statistically independent and Gaussian distributed, (8.50) reduces to

$$E\left\{\phi_i \phi_j\right\} = 4\left(2\lambda_i \lambda_j - 2\lambda_i \lambda_j\right) = 0. \tag{8.51}$$

Consequently, there is no covariance between ϕ_i and ϕ_j, implying that the covariance matrices for $\boldsymbol{\phi}$ and $\boldsymbol{\phi}_d$ reduce to diagonal matrices.

Distribution function of $\boldsymbol{\phi}_i$. The random variable

$$\widetilde{\phi}_i = \frac{\phi_i + 2\lambda_i}{2\lambda_i} \tag{8.52}$$

yields the following distribution function

$$\widetilde{\phi}_i = \left(\frac{t_i}{\sqrt{\lambda_i}}\right)^2 \sim \chi^2(1) \tag{8.53}$$

[2] !! is the double factorial and the product of the odd numbers only, e.g. $9!! = 1 \cdot 3 \cdot 5 \cdot 7 \cdot 9 = 945$.

since ${}^{t_i}/_{\sqrt{\lambda_i}} \sim \mathcal{N}\{0, 1\}$. In other words, the distribution function of ϕ_i can be obtained by substituting the transformation in (8.52) into the distribution function of a χ^2 distribution with one degree of freedom

$$F\left(\widetilde{\phi}_i\right) = \frac{1}{\sqrt{2}\Gamma\left(1/2\right)} \int_0^{\widetilde{\phi}_i} \widetilde{\psi}_i^{-1/2} e^{-\widetilde{\psi}_i/2} \mathrm{d}\widetilde{\psi}_i \tag{8.54}$$

which gives rise to

$$F\left(\phi_i\right) = \frac{1}{2\sqrt{2\lambda_i}\Gamma\left(1/2\right)} \int_{-2\lambda_i}^{\frac{\phi_i+2\lambda_i}{2\lambda_i}} \left(\frac{\psi_i+2\lambda_i}{2\lambda_i}\right)^{-1/2} e^{-\frac{\psi_i+2\lambda_i}{4\lambda_i}} \mathrm{d}\psi_i. \tag{8.55}$$

With respect to (8.55), the PDF $f\left(\phi_i\right) > 0$ within the interval $\left(-2\lambda_i, \infty\right)$, which follows from the fact that $t_i^2 \geq 0$. In (8.54) and (8.55), $\Gamma\left(1/2\right)$ is the gamma function, defined by the improper integral $\Gamma\left(1/2\right) = \int_0^\infty t^{-1/2} \exp\left(-t\right) \mathrm{d}t$. Figure 8.5 shows the probability density function of the primary residuals for various values of λ_i. The vertical lines in this figure represent the asymptotes at $-2\lambda_i$.

Central moments of $\boldsymbol{\phi_i}$. The determination of the central moments of ϕ_i relies on evaluating the definition for central moments, which is given by

$$E\left\{\phi_i^m\right\} = E\left\{2^m\left(t_i^2 - \lambda_i\right)^m\right\} = \int_{-2\lambda_i}^{\infty} \phi_i^m f\left(\phi_i\right) \mathrm{d}\phi_i. \tag{8.56}$$

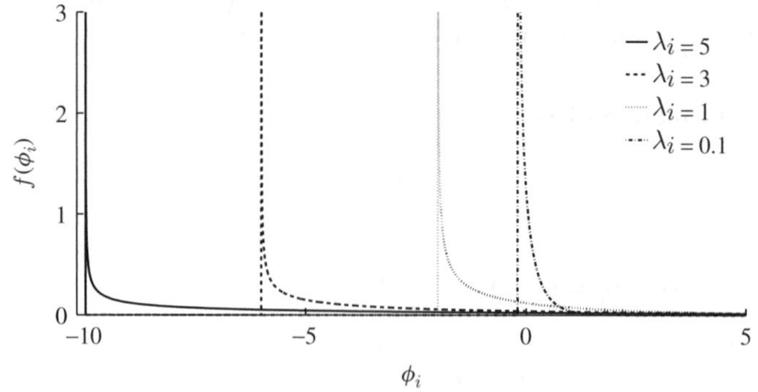

Figure 8.5 Probability density function of ϕ_i for different values of λ_i.

According to (8.56), the central moments can be obtained directly by evaluating the expectation $E\{(t_i^2 - \lambda_i)^m\}$, which gives rise to

$$2^m E\left\{\left(t_i^2 - \lambda_i\right)^m\right\} = 2^m E\left\{\sum_{j=0}^{m} (-1)^j \binom{m}{j} t_i^{2(m-j)} \lambda_i^j\right\}. \tag{8.57}$$

Isolating the terms in (8.57) that are associated with t_i^{2j} and substituting the central moments for $E\{t_i^{2j}\}$ yields

$$2^m E\left\{\left(t_i^2 - \lambda_i\right)^m\right\} = \left(2\lambda_i\right)^m \left(\sum_{j=0}^{m} (-1)^j \binom{m}{j} (2(m-j)-1)!!\right), \tag{8.58}$$

where

$$\binom{m}{j} = \frac{m!}{j!\,(m-j)!} \tag{8.59}$$

are binomial coefficients and $m! = 1 \cdot 2 \cdot 3 \cdots (m-1) \cdot m$. Table 8.2 summarizes the first seven central moments of ϕ_i.

8.3.5 Improved residuals for eigenvalues

Equation (8.16) shows that the improved residuals are time-based sums of the primary residuals and asymptotically Gaussian distributed, given that the primary residuals are i.i.d. sequences. Following from the geometric analysis of the data structure $\mathbf{z}_0 = \Xi\mathbf{s} + \mathbf{g}$ and its assumptions, discussed in Subsection 2.1.1, the model and residual subspaces are spanned by the n dominant and the remaining $n_z - n$ eigenvectors of $E\left\{\mathbf{z}_0\mathbf{z}_0^T\right\} = \mathbf{S}_{z_0z_0}$, respectively.

Table 8.2 First seven central moments of ϕ_i.

Order m	Central moment $E\left(\phi_i^m\right)$
1	0
2	$2^2 2\lambda_i^2 = 8\lambda_i^2$
3	$2^3 8\lambda_i^3 = 64\lambda_i^3$
4	$2^4 60\lambda_i^4 = 960\lambda_i^4$
5	$2^5 544\lambda_i^5 = 17,408\lambda_i^5$
6	$2^6 6040\lambda_i^6 = 386,560\lambda_i^6$
7	$2^7 79008\lambda_i^7 = 10,113,024\lambda_i^7$

Using the definition of the primary residuals for the eigenvalues, the improved residuals become

$$\theta_i\left(\mathbf{z}_0, K\right) = \frac{1}{\sqrt{K}}\sum_{k=1}^{K}\phi_i(\mathbf{z}_0(k)) = \frac{1}{\sqrt{K}}\sum_{k=1}^{K}\left(\mathbf{p}_i^T\mathbf{z}_0(k)\mathbf{z}_0^T(k)\mathbf{p}_i - \lambda_i\right). \tag{8.60}$$

As the eigenvectors and eigenvalues are functions of $\mathbf{z}_0 \sim \mathcal{N}\left\{\mathbf{0}, \mathbf{S}_{z_0z_0}\right\}$, the dependencies on these parameters can be removed from (8.16) and hence, $\theta_i = \theta_i\left(\mathbf{z}_0, K\right)$ with K being the number of samples and $\phi_i = \phi_i(\mathbf{z}_0(k))$. The first and second order moments of $\theta_i\left(\mathbf{z}_0, K\right)$ are as follows

$$\begin{aligned}
E\left\{\theta_i\left(\mathbf{z}_0, K\right)\right\} &= E\left\{\frac{1}{\sqrt{K}}\sum_{k=1}^{K}\left(\mathbf{p}_i^T\mathbf{z}_0(k)\mathbf{z}_0^T(k)\mathbf{p}_i - \lambda_i\right)\right\}\\
E\left\{\theta_i\left(\mathbf{z}_0, K\right)\right\} &= \frac{1}{\sqrt{K}}\mathbf{p}_i^T\sum_{k=1}^{K}E\left\{\mathbf{z}_0(k)\mathbf{z}_0^T(k)\right\}\mathbf{p}_i - \sqrt{K}\lambda_i\\
E\left\{\theta_i\left(\mathbf{z}_0, K\right)\right\} &= \sqrt{K}\left(\mathbf{p}_i^T\mathbf{S}_{z_0z_0}\mathbf{p}_i - \lambda_i\right) = 0
\end{aligned} \tag{8.61}$$

and

$$\begin{aligned}
E\left\{\theta_i^2\left(\mathbf{z}_0, K\right)\right\} &= \tfrac{1}{K}E\left\{\sum_{k=1}^{K}\phi_i^2\left(\mathbf{z}_0(k)\right)\right\}\\
E\left\{\theta_i^2\left(\mathbf{z}_0, K\right)\right\} &= \tfrac{1}{K}\sum_{k=1}^{K}E\left\{\phi_i^2\left(\mathbf{z}_0(k)\right)\right\} -\\
&\quad \tfrac{2}{K}\sum_{l=1}^{K-1}\sum_{k=l+1}^{K}E\left\{\phi_i\left(\mathbf{z}_0(l)\right)\phi_i\left(\mathbf{z}_0(k)\right)\right\}\\
E\left\{\theta_i^2\left(\mathbf{z}_0, K\right)\right\} &= 2\lambda_i^2 - \tfrac{2}{K}\sum_{l=1}^{K-1}\sum_{k=l+1}^{K}E\left\{t_i^2(l)t_i^2(k) - \lambda_i t_i^2(l) - \lambda_i^2 t_i^2(k) + \lambda_i^2\right\}\\
E\left\{\theta_i^2\left(\mathbf{z}_0, K\right)\right\} &= 2\lambda_i^2 - \tfrac{2}{K}\sum_{l=1}^{K-1}\sum_{k=l+1}^{K}\left(\lambda_i^2 - \lambda_i^2 - \lambda_i^2 + \lambda_i^2\right) = 2\lambda_i^2,
\end{aligned} \tag{8.62}$$

respectively. Note that the factor 2 in (8.28) has been removed, as it is only a scaling factor. The variance of ϕ_i is therefore $2\lambda_i^2$. That $E\left\{t_i^2(l)t_i^2(k)\right\} = \lambda_i^2$ follows from the Isserlis theorem (Isserlis 1918). The improved residuals can now be utilized in defining non-negative quadratic statistics.

The separation of the data space into the model and residual subspaces yielded two non-negative quadratic statistics. These describe the variation of the sample projections onto the model subspace (Hotelling's T^2 statistic) and onto the residual subspace (Q statistic). With this in mind, the primary residuals associated with the n largest eigenvalues and remaining $n_z - n$ identical eigenvalues can be used to construct the Hotelling's T^2 and residual Q statistics, respectively.

Intuitively, the definition of these statistics is given by

$$
\begin{aligned}
T_\theta^2(K) &= \sum_{i=1}^{n} \frac{\theta_i^2\left(\mathbf{z}_0, K\right)}{2\lambda_i^2} \\
T_\theta^2(K) &= \begin{pmatrix}\theta_1(\mathbf{z}_0, K) & \cdots & \theta_n(\mathbf{z}_0, K)\end{pmatrix} \begin{bmatrix} 1/2\lambda_1^2 & \cdots & 0 \\ \vdots & \ddots & \vdots \\ 0 & \cdots & 1/2\lambda_n^2 \end{bmatrix} \begin{pmatrix}\theta_1(\mathbf{z}_0, K) \\ \vdots \\ \theta_n(\mathbf{z}_0, K)\end{pmatrix} \\
T_\theta^2(K) &= \tfrac{1}{2}\boldsymbol{\theta}^T(\mathbf{z}_0, K)\boldsymbol{\Lambda}^{-2}\boldsymbol{\theta}(\mathbf{z}_0, K) \\
Q_\theta(K) &= \sum_{i=n+1}^{n_z} \theta_i^2\left(\mathbf{z}_0, K\right) = \begin{pmatrix}\theta_1(\mathbf{z}_0, K) & \cdots & \theta_1(\mathbf{z}_0, K)\end{pmatrix} \begin{pmatrix}\theta_1(\mathbf{z}_0, K) \\ \vdots \\ \theta_1(\mathbf{z}_0, K)\end{pmatrix} \\
Q_\theta(K) &= \boldsymbol{\theta}_d^T(\mathbf{z}_0, K)\boldsymbol{\theta}_d(\mathbf{z}_0, K)
\end{aligned} \tag{8.63}
$$

and follows the definition of the conventional Hotelling's T^2 and Q statistics in (3.8) and (3.15), respectively.

As the number of recorded samples, K, grows so does the upper summation index in (8.60). This, however, presents the following problem. A large K may dilute the impact of a fault upon the sum in (8.60) if only the last few samples describe the abnormal condition. As advocated in Chapter 7, however, this issue can be addressed by considering samples that are inside a sliding window only. Defining the window size by k_0, the incorporation of a moving window yields the following formulation of (8.60)

$$
\theta_i(\mathbf{z}_0, k) = \frac{1}{\sqrt{k_0}} \sum_{l=k-k_0+1}^{k} \phi_i\left(\mathbf{z}_0(l)\right). \tag{8.64}
$$

The selection of k_0 is a trade-off between accuracy and sensitivity. The improved residuals converge asymptotically to a Gaussian distribution, which demands larger values for k_0. On the other hand, a large k_0 value may dilute the impact of a fault condition and yield a larger average run length, which is the time it takes to detect a fault from its first occurrence. The selection of k_0 is discussed in the next section, which revisits the simulation examples in Section 8.1.

8.4 Revisiting the simulation examples of Section 8.1

This section revisits both examples in Section 8.1, which were used to demonstrate that the conventional MSPC framework may not detect changes in the underlying covariance structure.

8.4.1 First simulation example

Figure 8.1 showed that the scatter diagram and the Hotelling's T^2 statistic only detected the first change but not the second one. Recall that both changes resulted in a rotation of the control ellipse for $z_1^{(0)}$ and $z_2^{(0)}$ by 45°. Whilst the variance of both score variables remained unchanged, the variances for the second change were significantly reduced such that the rotated control ellipse was inside the original one.

Given that both changes yield a different eigendecomposition for the variable pairs $z_1^{(1)}$, $z_2^{(1)}$ and $z_1^{(2)}$, $z_2^{(2)}$, the primary residuals are expected to have a mean different from zero. Before determining improved residuals, however, k_0 needs to be determined. If k_0 is too small the improved residuals may not follow a Gaussian distribution accurately, and a too large k_0 may compromise the sensitivity in detecting slowly developing faults (Kruger and Dimitriadis 2008; Kruger *et al.* 2007).

Although the transformation matrix $\mathbf{T}^{(0)}$ and the variances of the i.d. score variables $\sigma_{s_1}^2$ and $\sigma_{s_2}^2$ are known here, the covariance matrix $\mathbf{S}_{z_0 z_0}$ and its eigendecomposition would need to be estimated in practice. Table 8.3 summarizes the results of estimating the covariance of both improved residual variables for a variety of sample sizes and window lengths.

As per their definition, the improved residuals asymptotically follow a Gaussian distribution of zero mean and variance $2\lambda_i^2$ if the constant term in (8.28) is not considered. The mean and variance for θ_1 and θ_2 are $2 \times 10^2 = 200$ and $2 \times 2^2 = 8$, respectively. The covariance $E\left\{\theta_1\theta_2\right\} = 0$ is also estimated in Table 8.3.

The entries in this table are averaged values for 1000 Monte Carlo simulations. In other words, for each combination of K and k_0 a total of 1000 data sets are simulated and the mean, variance and covariance values for each set are the averaged estimates. The averages of each combination indicate that the main effect for an accurate estimation is K, the number of reference samples of θ_1 and θ_2. Particularly window sizes above 50 require sample sizes of 2000 or above to be accurate.

This is in line with expectation, following the discussion in Section 6.4. The entries in Table 8.3 suggest that the number of reference samples for θ_1 and θ_2, K, need to be at least 50 times larger then the window size k_0. Another important issue is to determine how large k_0 needs to be to accurately follow a Gaussian distribution. Figure 8.6 shows Gaussian distribution functions in comparison with the estimated distribution functions of ϕ_1 and ϕ_2, and θ_1 and θ_2 for $k_0 = 10$, 50 and 200.

As expected, the upper plot in this figure shows that the distribution function of primary residuals depart substantially from a Gaussian distribution (straight line). In fact, (8.55) and Figure 8.5 outline that they follow a central χ^2 distribution. The plots in the second, third and bottom row, however, confirm that the sum of the primary residuals converge to a Gaussian distribution.

Table 8.3 Estimated means and variances of improved residuals.

K/k_0	Variable	10	20	50	100	200	500
100	$\widehat{\bar{\theta}}_1$	0.077	−0.268	0.152	−0.624	0.482	0.137
	$\widehat{\bar{\theta}}_2$	−0.031	−0.005	−0.042	−0.049	0.003	0.029
	$\widehat{\bar{\sigma}}^2_{\theta_1}$	180.96	161.56	117.68	64.57	33.70	12.98
	$\widehat{\bar{\sigma}}^2_{\theta_{12}}$	0.200	−0.156	0.512	−0.077	−0.264	−0.022
	$\widehat{\bar{\sigma}}^2_{\theta_2}$	7.245	6.554	4.652	2.672	1.354	0.525
200	$\widehat{\bar{\theta}}_1$	−0.001	−0.132	0.062	−0.128	−0.255	−0.652
	$\widehat{\bar{\theta}}_2$	−0.011	0.010	−0.002	−0.076	−0.008	0.047
	$\widehat{\bar{\sigma}}^2_{\theta_1}$	192.81	180.26	151.28	113.99	69.15	26.59
	$\widehat{\bar{\sigma}}^2_{\theta_{12}}$	0.211	0.039	−0.370	0.015	0.066	0.114
	$\widehat{\bar{\sigma}}^2_{\theta_2}$	7.538	7.334	6.133	4.874	2.592	1.070
500	$\widehat{\bar{\theta}}_1$	0.143	−0.145	−0.049	−0.137	−0.927	0.088
	$\widehat{\bar{\theta}}_2$	0.008	−0.023	−0.038	0.007	0.039	0.082
	$\widehat{\bar{\sigma}}^2_{\theta_1}$	200.88	191.93	179.00	157.26	129.33	69.32
	$\widehat{\bar{\sigma}}^2_{\theta_{12}}$	0.139	0.059	−0.323	0.328	−0.640	−0.379
	$\widehat{\bar{\sigma}}^2_{\theta_2}$	7.966	7.709	7.268	6.322	5.135	2.822
1000	$\widehat{\bar{\theta}}_1$	0.051	0.036	−0.071	0.040	0.114	0.315
	$\widehat{\bar{\theta}}_2$	0.007	0.004	0.014	−0.028	−0.012	0.018
	$\widehat{\bar{\sigma}}^2_{\theta_1}$	199.20	196.82	191.55	182.48	160.39	114.959
	$\widehat{\bar{\sigma}}^2_{\theta_{12}}$	0.159	−0.060	−0.215	0.008	0.134	−0.372
	$\widehat{\bar{\sigma}}^2_{\theta_2}$	7.899	7.866	7.657	7.114	6.511	4.758
2000	$\widehat{\bar{\theta}}_1$	0.011	0.003	−0.052	0.194	−0.128	0.017
	$\widehat{\bar{\theta}}_2$	0.009	−0.003	0.010	−0.000	−0.020	0.019
	$\widehat{\bar{\sigma}}^2_{\theta_1}$	198.96	198.26	196.34	191.06	178.17	148.24
	$\widehat{\bar{\sigma}}^2_{\theta_{12}}$	−0.120	−0.033	0.007	−0.134	0.481	−0.250
	$\widehat{\bar{\sigma}}^2_{\theta_2}$	7.984	7.887	7.870	7.622	7.158	6.209
5000	$\widehat{\bar{\theta}}_1$	−0.012	0.031	0.074	0.070	0.068	0.139
	$\widehat{\bar{\theta}}_2$	−0.002	−0.012	−0.000	−0.004	−0.005	0.045
	$\widehat{\bar{\sigma}}^2_{\theta_1}$	199.85	200.10	197.85	197.44	192.99	184.62
	$\widehat{\bar{\sigma}}^2_{\theta_{12}}$	−0.041	−0.007	0.124	−0.034	−0.179	0.005
	$\widehat{\bar{\sigma}}^2_{\theta_2}$	7.989	7.948	7.934	7.846	7.706	7.091

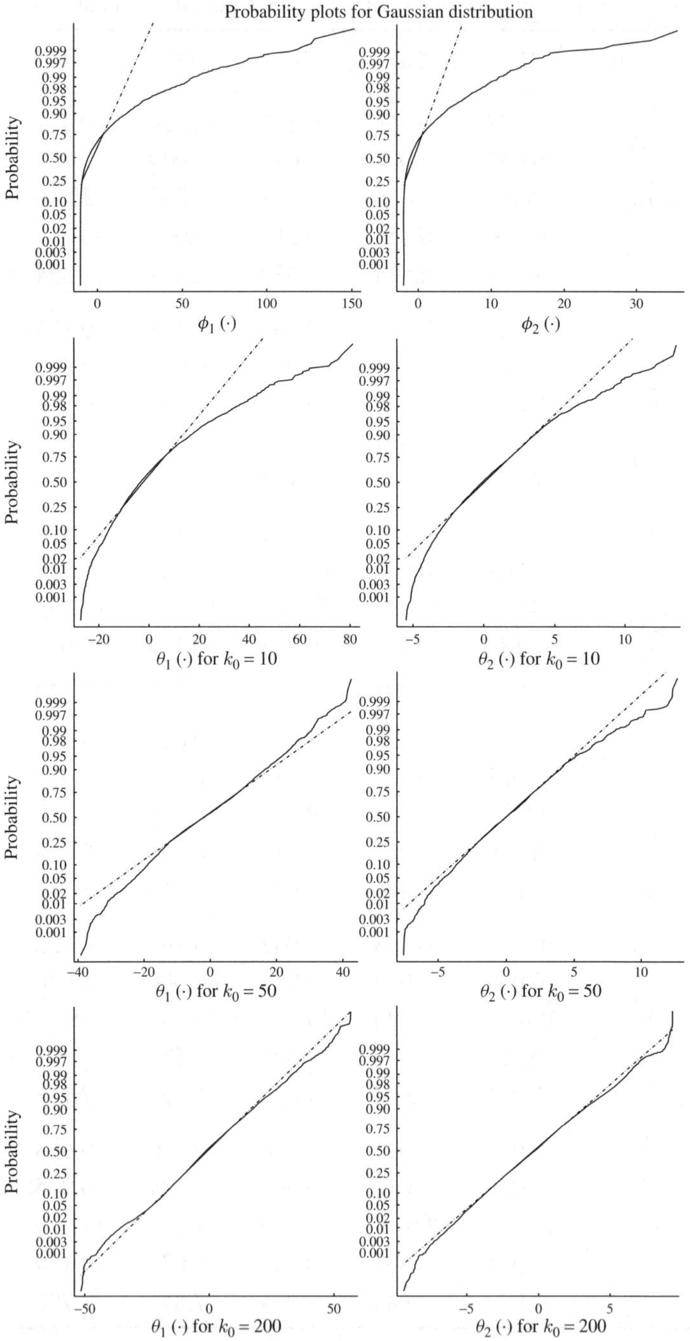

Figure 8.6 Distribution functions of primary and improved residuals.

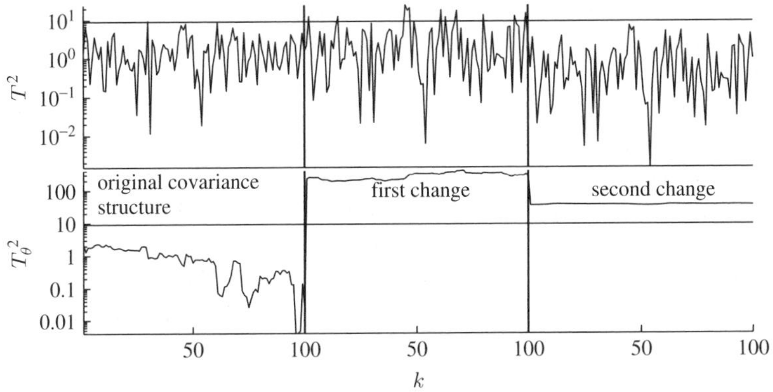

Figure 8.7 First simulation example revisited.

Whilst the smaller window sizes of $k_0 = 10$ and $k_0 = 50$ still resulted in significant departures from the Gaussian distribution, $k_0 = 200$ produced a close approximation of the Gaussian distribution. Together with the analysis of Table 8.3, a window size of $k_0 = 200$ would require a total of $K = 200 \times 50 = 10\,000$ reference samples to ensure that the variance of θ_1 and θ_2 are close to $2\lambda_1^2$ and $2\lambda_2^2$, respectively.

Using the same 1000 Monte Carlo simulations to obtain the average values in Table 8.3 yields an average of 200.28 and 7.865 for $E\{\theta_1^2\}$ and $E\{\theta_2^2\}$, respectively, and -0.243 for $E\{\theta_1\theta_2\}$. After determining an appropriate value for k_0, the Hotelling's T_θ^2 statistics can now be be computed as shown in (8.60).

Figure 8.7 compares the conventional Hotelling's T^2 statistic with the one generated by the statistical local approach. For $k_0 = 200$, both plots in this figure show a total of 100 samples obtained from the original covariance structure (left portion), the first change (middle portion) and the second change (right portion of the plots).

As observed in Figure 8.1, the conventional Hotelling's T^2 statistic could only detect the first change but not the second one. In contrast, the non-negative quadratic statistic based on the statistical local approach is capable of detecting both changes. More precisely, the change in the direction of both eigenvectors (first change) and both eigenvectors and eigenvalues (second change) yields an expectation for both primary residual function that is different from 0.

8.4.2 Second simulation example

Figures 8.2 and 8.3 highlight that conventional MSPC can only detect one out of the four changes of the original covariance structure. The remaining ones, although major, may not be detectable. Each of these changes alter the orientation of the model and residual subspaces as well as the orientation of the control ellipse. This, in turn, also yields a different eigendecomposition in each of the four cases compared to the eigendecomposition of the original covariance structure.

The primary residuals are therefore expected to have mean values that differ from zero. The first step is to determine an appropriate value for k_0. Assuming that the variances for each of the improved residuals, $E\left\{\theta_1^2\right\} = 2\lambda_1^2 = 512$, $E\left\{\theta_2^2\right\} = 2\lambda_2^2 = 128$ and $E\left\{\theta_3^2\right\} = 2\lambda_3^2 = 0.5$, need to be estimated, the same analysis as in Table 8.3 yields that K should be 100 times larger than k_0.

Figure 8.8 compares the estimated distribution function of the improved residuals with a Gaussian distribution function (straight lines) for different values of k_0. The estimation of each distribution function was based on $K = 100 \times 200 = 20\,000$ samples. As the primary residuals are χ^2 distributed the approximated distribution function, consequently, showed no resemblance to a Gaussian one. For $k_0 = 10$ and $k_0 = 50$, the estimated distribution function still showed significant departures from a Gaussian distribution. Selection $k_0 = 200$, however, produced a distribution function that is close to a Gaussian one.

This is expected, as the improved residuals are asymptotically Gaussian distributed. In other words, the larger k_0 the closer the distribution function is to a Gaussian one. It is important to note, however, that if k_0 is selected too large it may dilute the impact of a fault condition and render it more difficult to detect. With this in mind, the selection of $k_0 = 200$ presents a compromise between accuracy of the improved residuals and the average run length for detecting an incipient fault condition.

Figure 8.9 contrasts the conventional non-negative quadratic statistics (upper plots) with those based on the statistical local approach (lower plots) for a total of 100 simulated samples. This comparison confirms that the Hotelling's T^2 and Q statistics can only detect the first change but are insensitive to the remaining three alterations.

The non-negative quadratic statistics relating to the statistical local approach, however, detect each change. It is interesting to note that the first change only affected the Q_θ statistic, whilst the impact of the remaining three changes manifested themselves in the Hotelling's T_θ^2 statistic. This is not surprising, however, given that the primary residuals are a centered measure of variance, which follows from (8.28).

To explain this, the variance of the three score variables can be estimated for each covariance structure. Determining the score variables as $\mathbf{t}^{(m)} = \mathbf{P}^T\mathbf{z}^{(m)}$, where $\mathbf{P}$ stores the eigenvectors of $\mathbf{S}_{zz}^{(0)}$, allows us to estimate these variances. Using a Monte Carlo simulation including 1000 runs, Table 8.4 lists the average values of the estimated variances. The Monte Carlo simulations for each of the five covariance structures were based on a sample size of $K = 1000$.

The sensitivity of the Hotelling's T_θ^2 and Q_θ statistics for each alternation follows from the estimated averages in this table. The initial 30° rotation produces slightly similar variances for the first and second principal component. The variance of the third principal component, however, is about three and a half times larger after the rotation. Consequently, the Hotelling's T_θ^2 statistic is only marginally affected by the rotation, whereas a very significant significant impact arises for the Q_θ statistic.

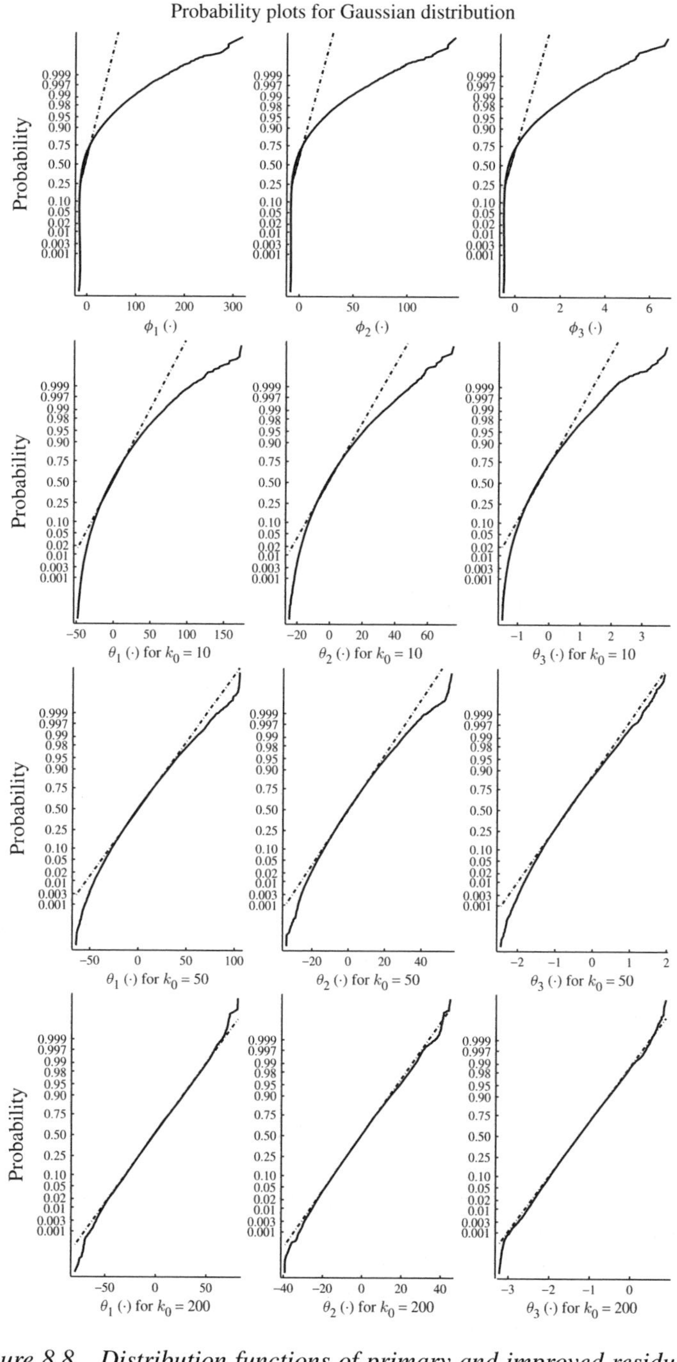

Figure 8.8 Distribution functions of primary and improved residuals.

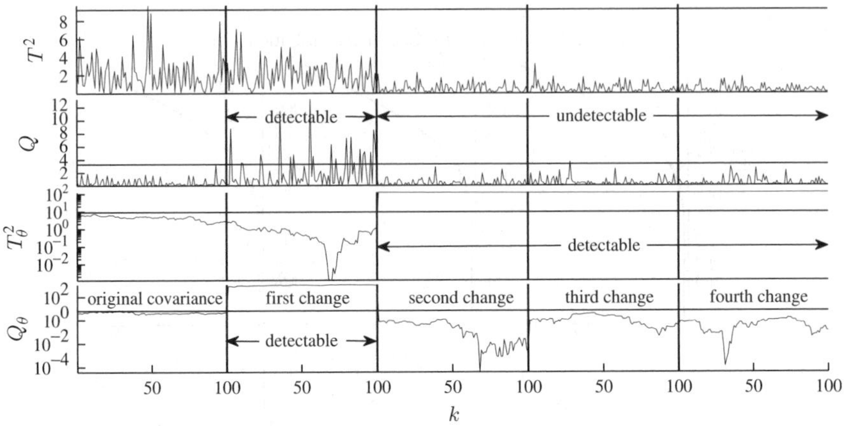

Figure 8.9 Second simulation example revisited.

Table 8.4 Estimated variances of $t_1^{(m)}$, $t_2^{(m)}$ and $t_3^{(m)}$.

$\widehat{\lambda}_j/m$	$m=0$	$m=1$	$m=2$	$m=3$	$m=4$
$\widehat{\lambda}_1$	15.9934	15.2373	3.8177	3.8221	3.8191
$\widehat{\lambda}_2$	7.9907	7.4069	1.8567	1.8558	1.8518
$\widehat{\lambda}_3$	0.4998	1.8104	0.4734	0.4531	0.4326

In contrast, the average eigenvalue for the second, third and fourth alteration produced averaged first and second eigenvalues that are around one quarter of the original ones. The averaged third eigenvalue, however, is very similar to the original one. This explains why these alterations are detectable by the Hotelling's T_θ^2 statistic, while the Q_θ statistic does not show any significant response.

Plotting the improved residuals for each covariance structure and $K = 1000$, which Figure 8.10 shows, also confirms these findings. For a significance of 0.01, the control limits for each improved residual are $\pm 2.58\sqrt{2\lambda_i}$. The larger variance of the third score variable yielded a positive primary residual for the first alteration. Moreover, the smaller variances of the first and second score variables produced negative primary residuals for the remaining changes.

8.5 Fault isolation and identification

For describing a fault condition, Kruger and Dimitriadis (2008) introduced a fault diagnosis approach that extracts the fault signature from the primary residuals. The fault signature can take the form of a simple step-type fault, such as a sensor bias that produces a constant offset, or can have a general deterministic function. For simplicity, the relationship of this diagnosis scheme concentrate first on step-type faults in Subsection 8.5.1. Subsection 8.5.2 then expands this concept to approximate a general deterministic fault signature.

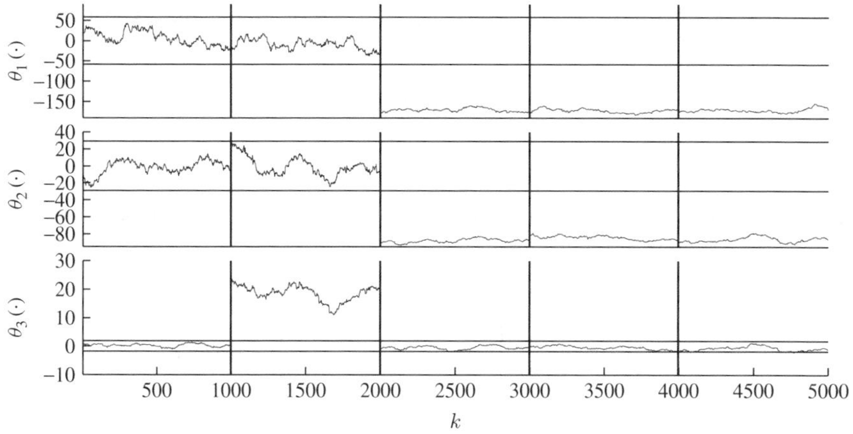

Figure 8.10 Plots of the three improved residuals for each of the five covariance structures.

8.5.1 Diagnosis of step-type fault conditions

The augmented data structure to describe a step-type follows from (3.68)

$$\mathbf{z}_{0_f} = \boldsymbol{\Xi}\mathbf{s} + \mathbf{g} + \Delta\mathbf{z}_0 = \mathbf{z}_0 + \Delta\mathbf{z}_0 \tag{8.65}$$

where $\Delta\mathbf{z}_0 \in \mathbb{R}^{n_z}$ represents an offset term that describes the fault condition. In analogy to the projection-based variable reconstruction approach, the offset can be expressed as follows

$$\Delta\mathbf{z}_0 = \boldsymbol{\upsilon}\mu. \tag{8.66}$$

Here, $\boldsymbol{\upsilon}$ is the *fault direction* and μ is the *fault magnitude*. With respect to the convention introduced by Isermann and Ballé (1997), the detection of a fault condition and the estimation of $\boldsymbol{\upsilon}$ refers to *fault isolation*. As μ describes the size of the fault, the estimation of the fault magnitude represents the *fault identification* step.

Equation (8.67) describes the impact of the offset term upon the primary residual vector for the ith eigenvector

$$\boldsymbol{\phi}_{i_f} = t_f\mathbf{z}_{0_f} - \lambda_i\mathbf{p}_i = \mathbf{z}_{0_f}\mathbf{z}_{0_f}^T\mathbf{p}_i - \lambda_i\mathbf{p}_i \tag{8.67}$$

for omitting the constant of 2 in (8.20). Substituting (8.65) into (8.67) yields

$$\begin{aligned}
\boldsymbol{\phi}_{i_f} &= \left(\mathbf{z}_0 + \Delta\mathbf{z}_0\right)\left(\mathbf{z}_0 + \Delta\mathbf{z}_0\right)^T \mathbf{p}_i - \lambda_i\mathbf{p}_i \\
\boldsymbol{\phi}_{i_f} &= \underbrace{\mathbf{z}_0\mathbf{z}_0^T\mathbf{p}_i - \lambda_i\mathbf{p}_i}_{=\boldsymbol{\phi}_i} + \mathbf{z}_0 \underbrace{\Delta\mathbf{z}_0^T\mathbf{p}_i}_{\mu_i} + \Delta\mathbf{z}_0 \underbrace{\mathbf{z}_0^T\mathbf{p}_i}_{=t_i} + \Delta\mathbf{z}_0\Delta\mathbf{z}_0^T\mathbf{p}_i \\
\boldsymbol{\phi}_{i_f} &= \boldsymbol{\phi}_i + \mathbf{z}_0\mu_i + \Delta\mathbf{z}_0^T t_i + \Delta\mathbf{z}_0\mu_i.
\end{aligned} \tag{8.68}$$

Given that $E\left\{\boldsymbol{\phi}_i\right\} = \mathbf{0}$, $E\left\{\mathbf{z}_0\right\} = \mathbf{0}$ and $E\left\{t_i\right\} = 0$, taking the expectation of (8.86) gives rise to

$$E\left\{\boldsymbol{\phi}_{i_f}\right\} = \Delta\mathbf{z}_0\Delta\mathbf{z}_0^T\mathbf{p}_i$$
$$E\left\{\boldsymbol{\phi}_{i_f}\right\} = \left[\mathbf{I}\otimes\mathbf{p}_i\right]^T\left(\Delta\mathbf{z}_0\otimes\Delta\mathbf{z}_0\right) \tag{8.69}$$

Here $\otimes$ refers to the Kronecker product of two matrices. The results of the two Kronecker products are as follows

$$\left[\mathbf{I}\otimes\mathbf{p}_i\right]^T = \begin{bmatrix} \mathbf{p}_i^T & \mathbf{0}^T & \cdots & \mathbf{0}^T \\ \mathbf{0}^T & \mathbf{p}_i^T & \cdots & \mathbf{0}^T \\ \vdots & \vdots & \ddots & \vdots \\ \mathbf{0}^T & \mathbf{0}^T & \cdots & \mathbf{p}_i^T \end{bmatrix} \in \mathbb{R}^{n_z\times n_z^2} \tag{8.70a}$$

$$\left(\Delta\mathbf{z}_0\otimes\Delta\mathbf{z}_0\right) = \begin{pmatrix} \Delta z_{0_1}^2 \\ \Delta z_{0_1}\Delta z_{0_2} \\ \vdots \\ \Delta z_{0_1}\Delta z_{0_{n_z}} \\ \vdots \\ \Delta z_{0_{n_z}}\Delta z_{0_1} \\ \Delta z_{0_{n_z}}\Delta z_{0_2} \\ \vdots \\ \Delta z_{0_{n_z}}^2 \end{pmatrix} \in \mathbb{R}^{n_z^2}. \tag{8.70b}$$

With $\Delta z_{0_i}\Delta z_{0_j} = \Delta z_{0_j}\Delta z_{0_i}$, (8.69) has a total of $^{n_z(n_z+1)}/_2$ unknowns but only n_z linearly independent equations and is hence an underdetermined system. However, there are a total of n_z equations for $1 \le i \le n_z$. Hence, (8.69) in augmented form becomes

$$\underbrace{\begin{pmatrix} E\left\{\boldsymbol{\phi}_{1_f}\right\} \\ E\left\{\boldsymbol{\phi}_{2_f}\right\} \\ \vdots \\ E\left\{\boldsymbol{\phi}_{n_{z}f}\right\} \end{pmatrix}}_{E\left\{\boldsymbol{\Phi}_f\right\}\in\mathbb{R}^{n_z^2}} = \underbrace{\begin{bmatrix} \left[\mathbf{I}\otimes\mathbf{p}_1\right]^T \\ \left[\mathbf{I}\otimes\mathbf{p}_2\right]^T \\ \vdots \\ \left[\mathbf{I}\otimes\mathbf{p}_{n_z}\right]^T \end{bmatrix}}_{\boldsymbol{\Psi}\in\mathbb{R}^{n_z^2\times n_z^2}} \underbrace{\left(\Delta\mathbf{z}_0\otimes\Delta\mathbf{z}_0\right)}_{\boldsymbol{\zeta}\in\mathbb{R}^{n_z^2}}. \tag{8.71}$$

It is interesting to note that the linear dependency in (8.69) and (8.71) follows from the analysis in Subsection 8.3.3 and particularly (8.33). It is therefore possible to remove the redundant $^{n_z(n_z-1)}/_2$ column vectors of $\boldsymbol{\Psi}$ and $^{n_z(n_z-1)}/_2$ elements of the vector $\boldsymbol{\zeta}$, which gives rise to

$$E\left\{\boldsymbol{\Phi}_f\right\} = \boldsymbol{\Psi}_{\text{red}}\boldsymbol{\zeta}_{\text{red}} \tag{8.72}$$

where $\boldsymbol{\Psi}_{\text{red}} \in \mathbb{R}^{n_z^2 \times \frac{n_z(n_z+1)}{2}}$ and $\boldsymbol{\zeta}_{\text{red}} \in \mathbb{R}^{\frac{n_z(n_z+1)}{2}}$. The expectation on the left hand side of (8.72) can be estimated from the recorded data and the matrix $\boldsymbol{\Psi}_{\text{red}}$ is made up of the elements of loading vectors and hence known. The elements of the vector $\boldsymbol{\zeta}_{\text{red}}$ are consequently the only unknown and can be estimated by the generalized inverse of $\boldsymbol{\Psi}_{\text{red}}$, i.e. $\boldsymbol{\Psi}_{\text{red}}^{\dagger}$

$$\widehat{\boldsymbol{\zeta}}_{\text{red}} = \widehat{\boldsymbol{\Psi}}_{\text{red}}^{\dagger} \widehat{\bar{\boldsymbol{\Phi}}}_f. \tag{8.73}$$

For estimating $\widehat{\bar{\boldsymbol{\Phi}}}_f$, however, it is possible to rely on the improved residuals, since

$$\widehat{\bar{\boldsymbol{\Phi}}}_f = \frac{1}{k_0} \sum_{l=k-k_0+1}^{k} \boldsymbol{\Phi}\left(\mathbf{z}_0(l) + \Delta\mathbf{z}_0\right) = \frac{1}{\sqrt{k_0}} \boldsymbol{\Theta}\left(\mathbf{z}_0 + \Delta\mathbf{z}_0, k\right). \tag{8.74}$$

Here, $\boldsymbol{\Theta}^T\left(\mathbf{z}_0 + \Delta\mathbf{z}_0, k\right) = \left(\boldsymbol{\theta}_1^T(\mathbf{z}_0 + \Delta\mathbf{z}_0, k) \quad \cdots \quad \boldsymbol{\theta}_{n_z}^T(\mathbf{z}_0 + \Delta\mathbf{z}_0, k)\right)$ and $\boldsymbol{\Phi}_f(l) = \boldsymbol{\Phi}\left(\mathbf{z}_0(l) + \Delta\mathbf{z}_0\right)$. In other words, the fault condition can be obtained directly from the improved residuals.

From the estimation of $\widehat{\boldsymbol{\zeta}}_{\text{red}}$, only the terms $\widehat{\Delta z_{0_1}^2}, \widehat{\Delta z_{0_2}^2}, \ldots, \widehat{\Delta z_{0_{n_z}}^2}$ are of interest, as these allow estimation of $\boldsymbol{\upsilon}$ and μ. The estimate of the fault magnitude is given by

$$\widehat{\mu} = \sqrt{\sum_{i=1}^{n_z} \widehat{\Delta z_{0_i}^2}}. \tag{8.75}$$

For estimating the fault direction, however, only the absolute value for each element of $\widehat{\boldsymbol{\upsilon}}$ is available. For determining the sign for each element, the data model of the fault condition can be revisited, which yields

$$E\left\{\mathbf{z}_{0_f} - \Delta\mathbf{z}_0\right\} = \mathbf{0} \tag{8.76}$$

and leads to the following test

$$\bar{z}_{0_{i_f}} \mp \sqrt{\widehat{\Delta z_{0_i}^2}} = \begin{cases} \approx 0 & \text{if the sign is correct} \\ \approx 2\Delta z_{0_{i_f}} & \text{if the sign is incorrect} \end{cases} \quad \text{for all } 1 \leq i \leq n_z. \tag{8.77}$$

After determining all signs using (8.77), the estimation of the fault direction, $\widehat{\boldsymbol{\upsilon}}$, is completed.

It should be noted that the above fault diagnosis scheme is beneficial, as the traditional MSPC approach may be unable to detect changes in the data covariance structure. Moreover, the primary residuals are readily available and the matrix $\widehat{\boldsymbol{\Psi}}_{\text{red}}^{\dagger}$ is predetermined, thus allowing us to estimate the fault signature in a simple and straightforward manner. It should also be noted that $\widehat{\boldsymbol{\zeta}}_{\text{red}}$

provides a visual aid to demonstrate how the fault signature affects different variable combinations. For this, the individual elements in $\widehat{\boldsymbol{\zeta}}_{\text{red}}$ can be plotted in a bar chart. The next subsection discusses how to utilize this scheme for general deterministic fault conditions.

8.5.2 Diagnosis of general deterministic fault conditions

The data structure for a general deterministic fault condition is the following extension of (8.65)

$$\mathbf{z}_{0_f}(k) = \boldsymbol{\Xi}\mathbf{s}(k) + \mathbf{g}(k) + \Delta\mathbf{z}_0(k), \tag{8.78}$$

where $\Delta\mathbf{z}_0(k)$ is some deterministic function representing the impact of a fault condition. Utilizing the fault diagnosis scheme derived in (8.67) to (8.73), the fault signature can be estimated, or to be more precise, approximated by a following moving window implementation of (8.73)

$$\widehat{\bar{\boldsymbol{\Phi}}}_f(k) = \frac{1}{\mathcal{K}} \sum_{l=k-\mathcal{K}+1}^{k} \boldsymbol{\Phi}\left(\mathbf{z}_0\left(l\right) + \Delta\mathbf{z}_0\left(l\right)\right). \tag{8.79}$$

As in Chapter 7, $\mathcal{K}$ is the size of the moving window. The accuracy of approximating the fault signature depends on the selection of $\mathcal{K}$ but also the nature of the deterministic function. Significant gradients or perhaps abrupt changes require smaller window sizes in order to produce accurate approximations. A small sample set, however, has the tendency to produce a less accurate estimation of a parameter, which follows from the discussion in Sections 6.4. To guarantee an accurate estimation of the fault signature, it must be assumed that the deterministic function is smooth and does not contain significant gradients or high frequency oscillation. The fault diagnosis scheme can therefore be applied in the presence of gradual drifts, for example unexpected performance deteriorations as simulated for the FCCU application study in Section 7.5 or unmeasured disturbances that have a gradual and undesired impact upon the process behavior.

One could argue that the average of the recorded process variables within a moving window can also be displayed, which is conceptually simpler than extracting the fault signature from the primary or improved residual vectors. The use of the proposed approach, however, offers one significant advantage. The extracted fault signature approximates the fault signature as a squared curve. In other words, it suppresses values that are close to zero and magnifies values that are larger than one. Hence, the proposed fault diagnosis scheme allows a better discrimination between normal operating conditions and the presence of a fault condition. This is exemplified by a simulation example in the next subsection.

8.5.3 A simulation example

This simulation example follows from the data model of the first intuitive example in Subsection 8.1.1. The two variables have the data and covariance structure described in (8.1) and (8.2), respectively. To construct a suitable deterministic fault condition, the three different covariance structures that were initially used to demonstrate that changes in the covariance structure may not be detectable using conventional MSPC have been revisited as follows. Each of the three covariance structures are identical and equal to that of (8.2). The three variable sets containing a total of 5000 samples each are generated as follows

$$\begin{pmatrix} z_{0_1}^{(0)}(k) \\ z_{0_2}^{(0)}(k) \end{pmatrix} = \begin{bmatrix} \sqrt{3}/2 & -2/2 \\ 2/2 & \sqrt{3}/2 \end{bmatrix} \begin{pmatrix} s_1^{(0)}(k) \\ s_2^{(0)}(k) \end{pmatrix} \tag{8.80a}$$

$$\begin{pmatrix} z_{0_1}^{(1)}(k) \\ z_{0_2}^{(1)}(k) \end{pmatrix} = \begin{bmatrix} \sqrt{3}/2 & -2/2 \\ 2/2 & \sqrt{3}/2 \end{bmatrix} \begin{pmatrix} s_1^{(1)}(k) \\ s_2^{(1)}(k) \end{pmatrix} \tag{8.80b}$$

$$\begin{pmatrix} z_{0_{1_f}}^{(2)}(k) \\ z_{0_{2_f}}^{(2)}(k) \end{pmatrix} = \begin{bmatrix} \sqrt{3}/2 & -2/2 \\ 2/2 & \sqrt{3}/2 \end{bmatrix} \begin{pmatrix} s_1^{(2)}(k) \\ s_2^{(2)}(k) \end{pmatrix} + \begin{pmatrix} 5\sin\left(2\pi k/5000\right) \\ 5\sin\left(4\pi k/5000\right) \end{pmatrix} \tag{8.80c}$$

where $1 \leq k \leq 5000$ is the sample index. It should also be noted that the samples for $\begin{pmatrix} s_1^{(0)}(k) & s_2^{(0)}(k) \end{pmatrix}^T$, $\begin{pmatrix} s_1^{(1)}(k) & s_2^{(1)}(k) \end{pmatrix}^T$ and $\begin{pmatrix} s_1^{(2)}(k) & s_2^{(2)}(k) \end{pmatrix}^T$ are statistically independent of each other. Moreover, each of the source variables has a mean of zero. The properties of the source signals for each of the data sets are therefore

$$E\left\{\begin{pmatrix} s_1^{(0)}(k) \\ s_2^{(0)}(k) \\ s_1^{(1)}(k) \\ s_2^{(1)}(k) \\ s_1^{(2)}(k) \\ s_2^{(2)}(k) \end{pmatrix}\right\} = \begin{pmatrix} 0 \\ 0 \\ 0 \\ 0 \\ 0 \\ 0 \end{pmatrix} \tag{8.81}$$

$$E\left\{\begin{pmatrix} s_1^{(0)}(k) \\ s_2^{(0)}(k) \\ s_1^{(1)}(k) \\ s_2^{(1)}(k) \\ s_1^{(2)}(k) \\ s_2^{(2)}(k) \end{pmatrix} \begin{pmatrix} s_1^{(0)}(l) & s_2^{(0)}(l) & s_1^{(1)}(l) & s_2^{(1)}(l) & s_1^{(2)}(l) & s_2^{(2)}(l) \end{pmatrix}\right\} \tag{8.82}$$

$$= \delta_{kl} \begin{bmatrix} 10 & 0 & 0 & 0 & 0 & 0 \\ 0 & 2 & 0 & 0 & 0 & 0 \\ 0 & 0 & 10 & 0 & 0 & 0 \\ 0 & 0 & 0 & 2 & 0 & 0 \\ 0 & 0 & 0 & 0 & 10 & 0 \\ 0 & 0 & 0 & 0 & 0 & 2 \end{bmatrix}.$$

Concatenating the three data sets then produced a combined data set of 15 000 samples. The fault diagnosis scheme introduced in Subsections 8.5.1 and 8.5.2, was now applied to the combined data set for a window size of $\mathcal{K} = 150$. Figure 8.11 shows the approximated fault signature each of the data sets. As expected, the estimated fault signature for $\Delta\widehat{z^2_{0_1}}$, $\Delta\widehat{z}_{0_1}\Delta\widehat{z}_{0_2}$ and $\Delta\widehat{z^2_{0_2}}$ show negligible departures from zero for the first two data sets. For the third data set, an accurate approximation of the squared fault signature $25\sin^2\left({}^{2\pi k}/_{5000}\right)$ and $25\sin^2\left({}^{4\pi k}/_{5000}\right)$ as well as the cross-product term $25\sin\left({}^{2\pi k}/_{5000}\right)\sin\left({}^{4\pi k}/_{5000}\right)$ (dashed line) can be seen at first glance.

A closer inspection, however, shows a slight delay with which the original fault signature is approximated, particularly for higher frequency fault signatures in the middle and lower plots in Figure 8.11. According to (8.79), this follows from the moving window approach, which produces an average value for the window. Consequently, for sharply increasing or reducing slopes, like in the case of the sinusoidal signal, the use of the moving window compromises the accuracy of the approximation. The accuracy, however, can be improved by reducing in the window size. This, in turn, has a detrimental effect on the smoothness of the approximation.

The last paragraph in Subsection 8.5.2 raises the question concerning the benefit of the proposed fault diagnosis scheme over a simple moving window average of the process variables. To substantiate the advantage of extracting the squared fault signature from the primary residuals instead of the moving window

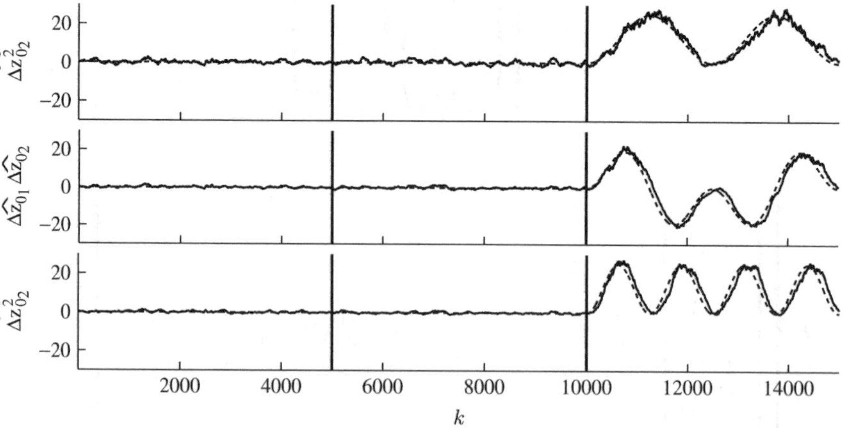

Figure 8.11 Approximated fault signature for $\widehat{\Delta z^2_{0_1}}$, $\widehat{\Delta z_{0_1} z_{0_2}}$ and $\widehat{\Delta z^2_{0_2}}$.

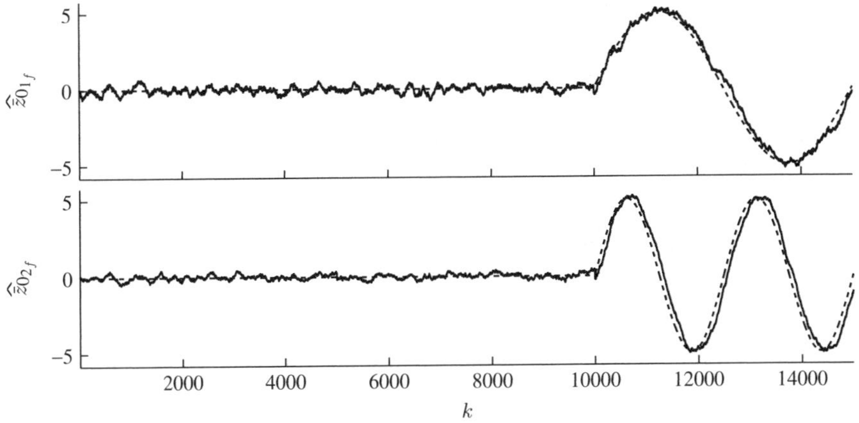

Figure 8.12 Approximated fault signatures for $\hat{\bar{z}}_{0_{1_f}}$ *and* $\hat{\bar{z}}_{0_{1_f}}$.

average of the process variables, Figure 8.12 shows the approximation of the fault signature using a moving window average of the process variables. In order to conduct a fair comparison, the window size for producing the resultant fault signatures in Figure 8.12 was also set to be $\mathcal{K} = 150$.

It is interesting to note that the variance of the estimated fault signature for the first two data sets appears to be significantly larger relative to the variance of the estimated fault signature when directly comparing Figures 8.11 and 8.12. In fact, the amplitude of the sinusoidal signals is squared when using the proposed approach compared to the moving window average of the recorded process variables. Secondly, the accuracy of estimating the fault signature in both cases is comparable.

Based on the results of this comparison, the benefit of the proposed fault diagnosis scheme over a simple moving window average of the process variables becomes clear if the amplitude of the sinusoidal is reduced from five to three for example. It can be expected in this case that the variance of the estimated fault signature for the first 10 000 samples increases more substantial relative to the reduced fault signature. This, however, may compromise a clear and distinctive discrimination between the fault signature and normal operating condition, particularly for smaller window sizes.

8.6 Application study of a gearbox system

This section extends the comparison between the non-negative quadratic statistics constructed from the improved residuals with those based on the score variables using an application study of a gearbox system. This system is mounted on an experimental test rig to record normal operating conditions as well as a number of fault conditions.

The next subsection gives a detailed description of the gearbox system and Subsection 8.6.2 explains how the fault condition was injected into the system. Subsection 8.6.3 then summarizes the identification of a PCA-based monitoring model and the construction of improved residuals. Subsection 8.6.4 finally contrasts the performance of the non-negative quadratic statistics based on the improved residuals with those relying on the score variables.

8.6.1 Process description

Given the widespread use of gearbox systems, the performance monitoring of such systems is an important research area in a general engineering context, for example in mechanical and power engineering applications. A gearbox is an arrangement involving a train of gears that transmit power and regulate rotational speed, for example, from an engine to the axle of a car.

Figure 8.13 presents a schematic diagram of the two-stage helical gearbox system (upper plot) and a similar gearbox to that used to generate the recorded vibration data (lower plot). Table 8.5 provides details of the gearbox, which was operated under full-load conditions of 260Nm.

Figure 8.13 shows that a total of four accelerometers are mounted on this gearbox system, which record the vibration signals simultaneously at a frequency of 6.4 kHz. Each recorded data set includes a total of 32 768 samples. Two data sets were recorded that describe a normal operating condition and a further six data sets that represent a gradually increasing fault condition. Figure 8.14 shows the first 5000 samples of each of the four vibration sensors, z_1 to z_4, for one of the reference sets.

8.6.2 Fault description

For a gearbox system, a tooth breakage is a serious localized fault. Such a fault was simulated here by removing a certain percentage of one tooth in the pinion gear. This tooth removal enabled an experimental representation of a gradual fault advancement under predefined conditions.

The simulated tooth breakage represents the chipping of small parts of one tooth, which is one of the common fault conditions in gearbox systems. For this, the total length of one tooth was gradually removed by increments of 10%. The recorded fault conditions here included a 10% to 50% and a 100% removal of the tooth, that is, a total of six additional data sets that describe the removal of one tooth in various stages. These data sets are referred to here as *10% Fault*, *20% Fault* to *100% Fault*.

Consequences of being unable to detect such faults at early stages include productivity decreases in manufacturing processes, reduced efficiency of engines, equipment damage or even failure. An early detection can provide significant improvements in the reduction of operational and maintenance costs, system down-time, and lead to increased levels of safety, which is an ever-growing concern.

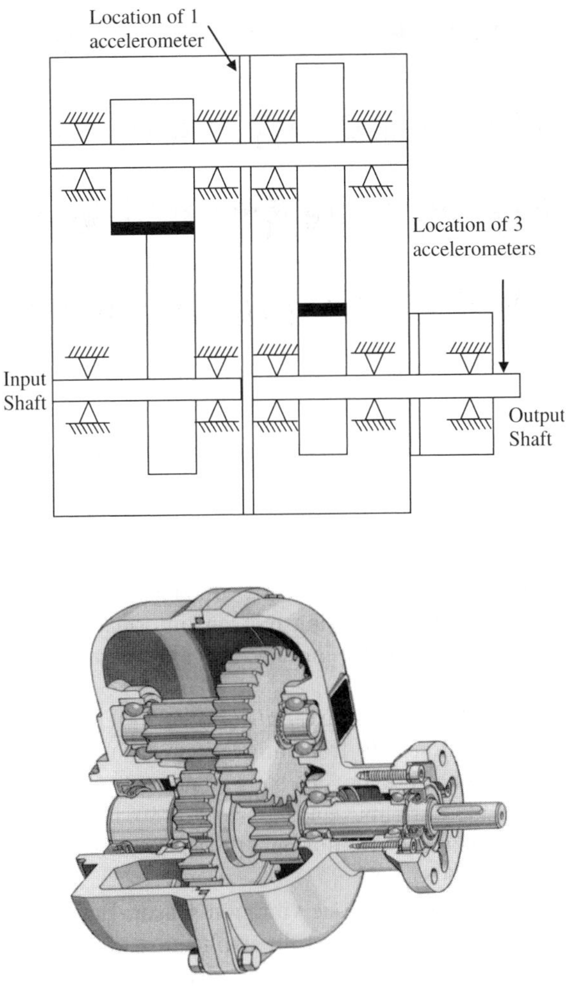

Figure 8.13 Schematic diagram of the gearbox system.

Table 8.5 Specification of gearbox system under study.

Detail	Number of teeth	Speed of stage	Mashing frequency	Contact ratio	Overlap ratio
1st stage	34/70	$24.33\frac{rev}{s}$	$827.73Hz$	1.359	2.890
2nd stage	29/52	$6.59\frac{rev}{s}$	$342.73Hz$	1.479	1.479

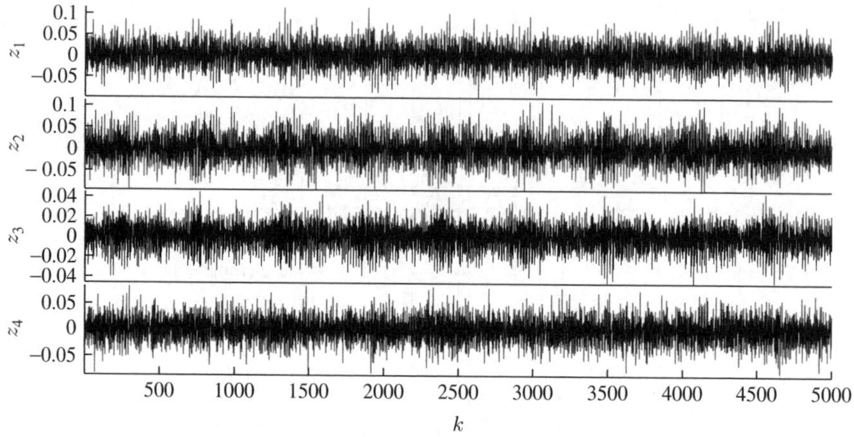

Figure 8.14 First 5000 samples of reference data.

An incipient fault in a mechanical system usually affects certain parameters, such as vibration, noise and temperature. Analyzing these *external variables* allows the performance monitoring of gears, which are usually inaccessible without dismantling the system. Extracting relevant information from the recorded signals is hence essential for detecting any irregularity that could be caused by tooth breakage or wear.

Baydar and Ball (2001), Baydar *et al.* (1999, 2001) and Ge *et al.* (2010) analyzed recorded vibration data from this system using a variety of different methods. Other research work on detecting abnormal operating conditions in gearbox systems include Bartelmus and Zimroz (2009), Hu *et al.* (2007), Stander *et al.* (2002), Staszewski and Tomlinson (1997), Tan and Mba (2005), Wang (2008) and Wu *et al.* (2008).

Since there may be more than one vibration sensor, Baydar and Ball (2001); Baydar *et al.* (1999, 2001) introduced the application of MSPC to successfully detect tooth defects. In a comprehensive comparison, Baydar *et al.* (2001) demonstrated that MSPC is equally as sensitive in detecting tooth defects as signal-based approaches but easier to implement in practice. More recent application of MSPC methods in monitoring mechanical systems are given by He *et al.* (2007, 2009) and Malhi and Gao (2004).

8.6.3 Identification of a monitoring model

This section utilizes PCA to identify a data model according to (2.2). Using one of the reference data sets describing a fault-free operating condition, where no portion of the tooth has been removed, the first steps include the centering and scaling of the data and the estimation of the data correlation matrix.

It follows from Figure 8.14 that the mean value for each vibration signal is close to zero. Estimating the mean yielded values of $\widehat{\bar{z}}_1 = 0.0008$, $\widehat{\bar{z}}_2 = 0.0005$,

$\widehat{\bar{z}}_3 = 0.0005$ and $\widehat{\bar{z}}_4 = 0.0013$. The estimated variances for each sensor are $\widehat{\sigma}_1 = 0.0008$, $\widehat{\sigma}_2 = 0.0010$, $\widehat{\sigma}_3 = 0.0002$ and $\widehat{\sigma}_4 = 0.0006$.

By inspecting the variances, it is apparent that $\widehat{\sigma}_2$ is five times larger than $\widehat{\sigma}_3$ and also significantly larger than $\widehat{\sigma}_1$ and $\widehat{\sigma}_4$. Chapter 4 highlighted that significant differences in variance may lead to dominant contributions of process variables with larger variances than the computed score variables.

Jackson (2003) advocated to use of the correlation matrix (i) to ensure that the variables are dimensionless, that is, their dimension for example $°C$ or *bar* reduces to one and (ii) each process variable has unity variance, which circumvents dominant contributions of variables with large variances. Using the estimated mean and variance, the estimated correlation matrix of the four sensor readings is given by

$$\widehat{\mathcal{C}}_{z_0 z_0} = \begin{bmatrix} 1.0000 & 0.3570 & -0.0759 & 0.0719 \\ 0.3570 & 1.0000 & -0.0791 & 0.3383 \\ -0.0759 & -0.0791 & 1.0000 & -0.0071 \\ 0.0719 & 0.3383 & -0.0071 & 1.0000 \end{bmatrix}. \quad (8.83)$$

The elements of $\widehat{\mathcal{C}}_{z_0 z_0}$ suggest that there is some correlation among sensor readings z_1 and z_2, between variables z_2 and z_4 but variable z_3 shows hardly any correlation with the other sensor readings. Different from the previous application studies in this book, the process variables cannot be seen as highly correlated. Equations (8.84a) and (8.84b) shows the eigendecomposition of $\widehat{\mathcal{C}}_{z_0 z_0}$

$$\widehat{\mathbf{P}} = \begin{bmatrix} -0.5279 & 0.2409 & -0.6480 & 0.4934 \\ -0.6723 & -0.0827 & -0.0422 & -0.7344 \\ 0.1769 & -0.8527 & -0.4900 & -0.0378 \\ -0.4879 & -0.4560 & 0.5816 & 0.4645 \end{bmatrix} \quad (8.84a)$$

$$\widehat{\mathbf{\Lambda}} = \begin{bmatrix} 1.5466 & 0 & 0 & 0 \\ 0 & 1.0099 & 0 & 0 \\ 0 & 0 & 0.9013 & 0 \\ 0 & 0 & 0 & 0.5421 \end{bmatrix}. \quad (8.84b)$$

At first glance, the first two eigenvalues are above one whilst the fourth one is significantly below one and also significantly smaller than the third one. Utilizing the stopping rules for PCA models in Subsection 2.3.1, those that assume a high degree of correlation and a significant contribution of the source signals to the process variables are not applicable here. This, for example, eliminates the VPC and VRE techniques.

An alternative is based on (2.122), which states that the sum of the eigenvalues is equal to the sum of the variances of each process variable. In percentage, the inclusion of $n = 1$, 2, 3 and 4 latent components yields a variance contribution of 38.66%, 63.91%, 86.45% and 100%, respectively. This suggests the retention of three latent components in the PCA model, as 86% of the variance of the

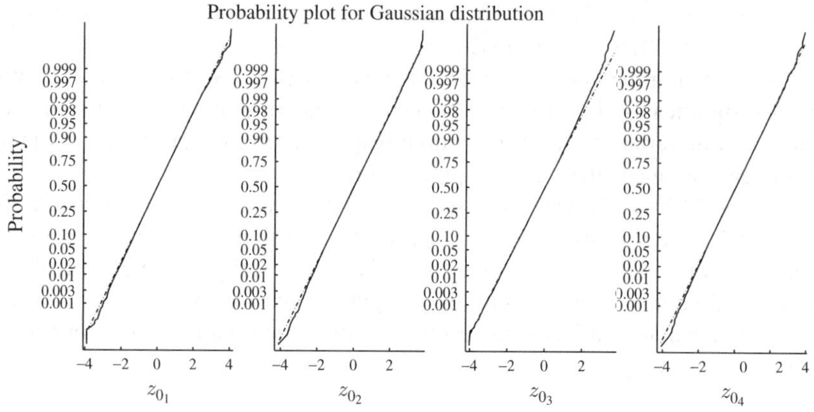

Figure 8.15 Distribution function of sensor readings.

scaled sensor readings can be recovered, and yields the following data model

$$\mathbf{z} = \text{diag}\left\{\sigma_1 \quad \sigma_2 \quad \sigma_3 \quad \sigma_4\right\}(\Xi\mathbf{s} + \mathbf{g}) + \bar{\mathbf{z}}. \tag{8.85}$$

Here, $\mathbf{s} \sim \mathcal{N}\left\{\mathbf{0}, \mathbf{S}_{ss}\right\} \in \mathbb{R}^3$, $\widehat{\mathbf{g}} \sim \mathcal{N}\{\mathbf{0}, 0.5421\mathbf{I}\} \in \mathbb{R}^4$. Estimates of the model and residual subspaces are the first three eigenvectors and the last eigenvector of $\widehat{\mathbf{P}}$, respectively. Up to a similarity transformation, $\widehat{\Xi} = \left[\widehat{\mathbf{p}}_1\ \widehat{\mathbf{p}}_2\ \widehat{\mathbf{p}}_3\right]$, $\widehat{\Xi}^{\dagger} = \left[\widehat{\mathbf{p}}_1\ \widehat{\mathbf{p}}_2\ \widehat{\mathbf{p}}_3\right]^T$ and $\widehat{\Xi}^{\perp} = \widehat{\mathbf{p}}_4^T$. Moreover, Ξ and $\mathbf{S}_{ss}$ are assumed to be unknown.

Figure 8.15 compares the estimated distribution function (solid line) with that of a Gaussian distribution of the same mean and variance (dashed-dot line). This comparison shows a good agreement that confirms the validity of the underlying assumptions for the data model in (8.85).

After establishing a PCA model from one of the reference sets, the next step is to determine the window size k_0. To guarantee statistical independence, the performance of a variety of different sizes was tested using the second reference set. This entails the computation of the score variables, based on the PCA model established from the first reference set, the calculation of the four primary and improved residuals and the estimation of the distribution function for each improved residual.

Figure 8.16 contrasts the four estimated distribution functions for the primary residuals and the improved residuals for $k_0 = 10$, $k_0 = 100$ and $k_0 = 400$ with Gaussian ones of the same mean and variance. As expected, very substantial departures from a Gaussian distribution arise for the primary residuals and the improved residuals for $k_0 = 10$. A closer but still inaccurate approximation emerges for $k_0 = 100$.

Increasing k_0 to 400 gives rise to a substantially more accurate approximation of a Gaussian distribution. As increasing this number further showed insignificant differences and reducing the number decreased the accuracy, the selection of $k_0 = 400$ presents a suitable trade-off between accuracy and sensitivity.

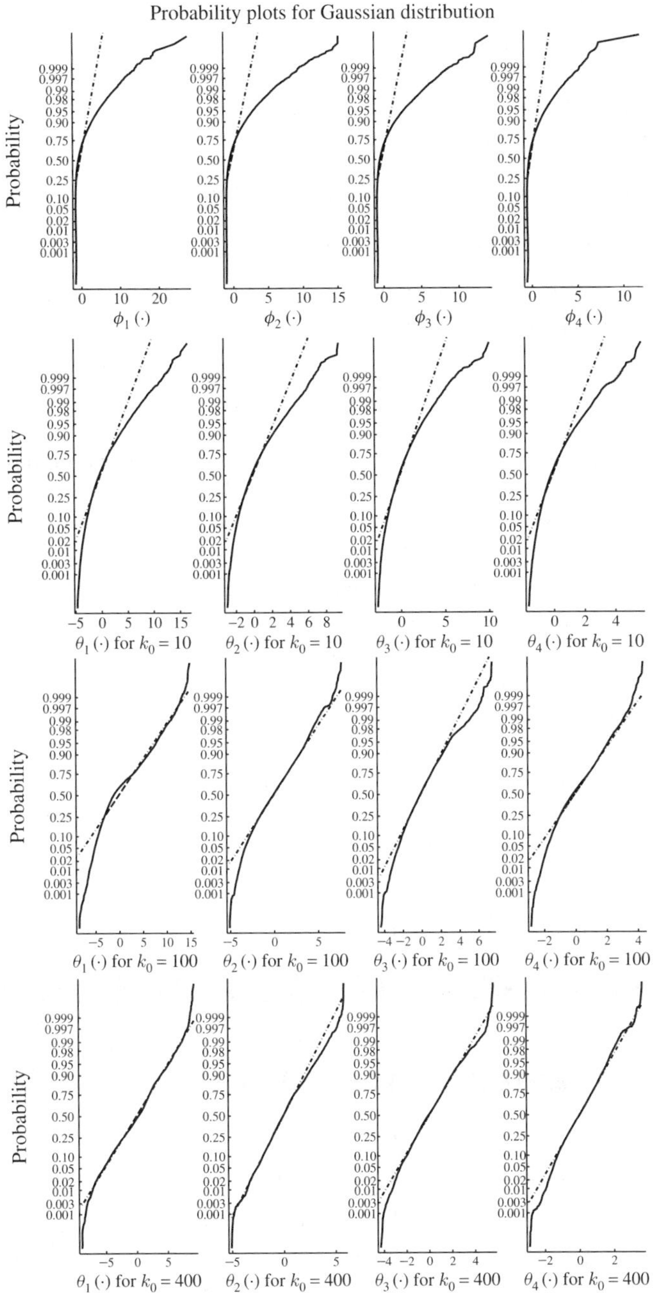

Figure 8.16 Distribution function of primary and improved residuals.

8.6.4 Detecting a fault condition

The PCA monitoring model is now applied to each of the recorded data. These include the two reference sets and the six data sets describing the fault condition. The PCA monitoring model determined in the previous subsection allows establishing a total of four non-negative quadratic monitoring statistics:

1. the conventional Hotelling's T^2 described in (3.8);
2. the residual Q statistic defined in (3.15);
3. the Hotelling's T_θ^2 statistic based on the first three improved residual variables and defined in (8.63); and
4. the residual Q_θ statistic that is constructed from the remaining improved residual variable and also defined in (8.63).

Given that the correlation matrix and the mean vector of the vibration signals are estimates, the control limits for the Hotelling's T^2 statistics are obtained by applying (3.5). Applying (3.16) and (3.29) produced the control limits for the Q and Q_θ statistics, respectively. The significance for each control limit is 0.05.

Figure 8.17 shows the performance of the monitoring statistics for each of the eight conditions. As expected, the first two reference conditions did not yield a statistically significant number of violations of the control limit. For the remaining data sets, referring to the removal of 10%, 20%, 30%, 40%, 50% and 100% for one of the tooth in the pinion gear, a different picture emerged. Each of these conditions led to a significant number of violations for each statistic.

Table 8.6 lists the calculated percentage number of violations of each statistic. This analysis confirms that percentages for the two reference conditions are 5% or below, indicating an in-statistical-control behavior of the gearbox system.

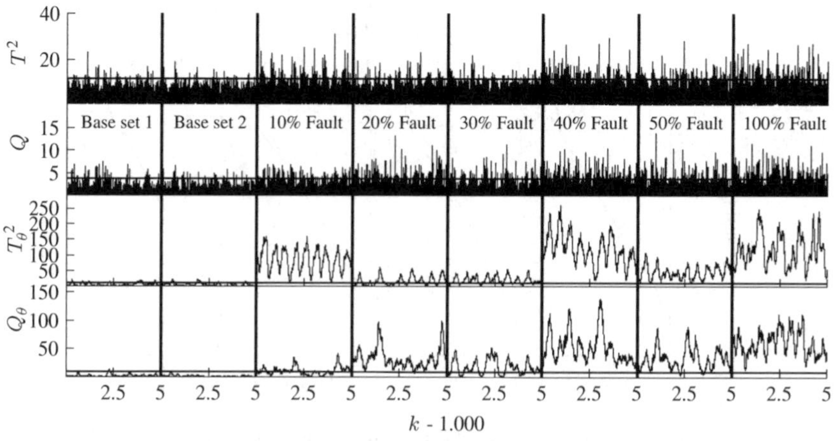

Figure 8.17 Non-negative quadratic statistics for conventional PCA (upper two plots) and the statistical local approach (lower two plots).

Table 8.6 Violations of control limits by monitoring statistic in [%].

Statistic	Percentage violations							
	Base 1	Base 2	10%	20%	30%	40%	50%	100%
T_2	2.05	2.05	6.05	3.45	3.48	8.06	5.39	7.57
Q	4.96	5.00	8.18	10.98	9.39	13.70	11.70	13.68
T_θ^2	2.89	2.88	100	99.52	99.65	100	100	100
Q_θ	3.67	2.18	99.57	100	100	100	100	100

For each of the fault conditions, however, the percentage number of violations exceeds 5% which hence concludes the performance of the gearbox is out-of-statistical-control.

Despite the fact that each monitoring statistic correctly rejected the null hypothesis for each of the fault conditions, it is important to note that the Hotelling's T_θ^2 and Q_θ statistic showed a significantly stronger response to the recorded data involving the manipulated pinion gear. This is in line with the observations in Section 8.5 and confirm the sensitivity of the improved residuals in detecting small alterations in the orientation of the model and residual subspaces and the control ellipsoid.

Identifying PCA models on the basis of each of the eight data sets allows to examine the sensitivity issue in more detail. This relies on benchmarking the second reference set and the six data sets describing the fault condition in various stages against the first reference set. More precisely, the departures of the eigenvectors and the differences for each of the four eigenvalues enables assessing the changes in the orientation of the model and residual subspaces and the orientation of the control ellipsoid. These changes can be described as follows

$$\Delta\varphi_{im} = \arccos\left(\widehat{\mathbf{p}}_{i_0}^T\widehat{\mathbf{p}}_{i_m}\right)\frac{180}{\pi} \qquad \Delta\lambda_{im} = \frac{\widehat{\lambda}_{i_m} - \widehat{\lambda}_{i_0}}{\widehat{\lambda}_{i_0}}100\%. \tag{8.86}$$

Here, the indices $i = 1, \ldots, 4$ and $m = 1, \ldots, 7$ represent the latent component and the data set, respectively, where $m = 1$ symbolizes the second reference data set and $m = 2, \cdots, 7$ corresponds to the data sets 10% Fault, 20% Fault, 30% Fault, 40% Fault, 50% Fault and 100% Fault, respectively. Figure 8.18 summarizes the results of applying (8.86). Whilst the eigenvectors and eigenvalues for the correlation matrices of both reference sets are very close to each other, very significant differences can be noticed for data sets 10% Fault to 100% Fault.

It is interesting to note that the first alteration of the tooth, although only 10% of the length of this tooth was removed, rotated the first eigenvector by around 45°. Apart from the impact of a complete removal of the tooth upon the orientation of the fourth eigenvector, this is the strongest single response.

Similar to the application studies in Subsections 2.1.3 and 6.1.2, the orientation of the model subspace can be assessed by the last eigenvector, which spans the residual subspace. The lower right plot shows a trend that the greater

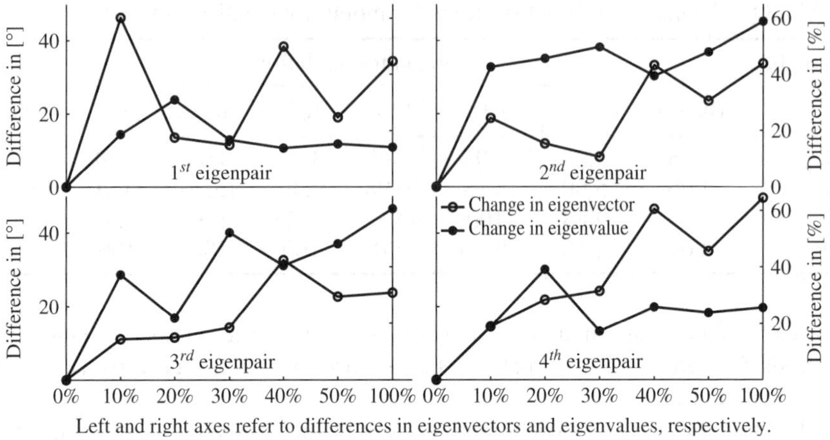

Figure 8.18 Departures of eigenvectors (circles) and eigenvectors (dots) for each set.

the percentage of the tooth is removed, the larger the rotation of the fourth eigenvector and hence the residual subspace. Given that the model and residual subspaces are complementary subspaces, any rotation of the residual subspace will affect the model subspace too.

With this in mind, it can be concluded that the more severe the fault condition the more significant the impact upon the orientation of the model and residual subspaces. A similar trend, however, cannot be observed for the variance of the fourth eigenvalue. Whilst an initial increase can be noticed, this increase levels off at around 20% if larger portions of the tooth are removed.

Apart from the first eigenpair, the second and third pairs show, approximately, a proportionate response to the severity of the fault condition. The more of the tooth is removed, the larger the variance of the third score variable becomes. The removal of 20% and 30% produce a reduction in the variance of the second score variable, relative to the 10% removal. An increasing trend, however, can also be noticed for the variance of the second score variables with the removal of larger portions.

Based on the experimental data obtained, it can be concluded that the presence of a tooth defect increases the variance of the score variables and with it the variance of the vibration signals, which follows from (2.122). It is interesting to note that the relatively small 10% removal of the tooth has a significantly more pronounced effect on the orientation of the control ellipsoid than the removal of 20% and 30%. It is also interesting to note that the removal of 40% has a similar effect upon the monitoring statistics compared to a complete removal of the tooth.

In contrast, removing 50% of the tooth has a less significant effect on the monitoring model than removing only 40%. As stated above, the only direct relationship between the severity of the fault condition and the impact on the geometry of the monitoring model is the orientation of the residual subspace.

Table 8.7 Changes in variance of score variables.

Score	Percentage change for mth condition						
$\Delta\widehat{\sigma}^2_{1m}$	0.002	54.55	9.29	12.46	47.00	22.64	42.44
$\Delta\widehat{\sigma}^2_{2m}$	0.005	23.24	11.80	8.65	36.35	23.98	27.88
$\Delta\widehat{\sigma}^2_{3m}$	0.001	21.16	20.21	22.31	47.55	34.83	47.94
$\Delta\widehat{\sigma}^2_{4m}$	0.001	26.04	50.24	36.17	72.68	54.96	71.94

Finally, Table 8.7 summarizes the percentage changes of the variance of the score variables resulting from the fault condition. The score variances are computed with respect to the PCA model established from the first reference set, producing the following percentage changes

$$\Delta\widehat{\sigma}^2_{im} = \frac{\widehat{\sigma}^2_{im} - \widehat{\sigma}^2_{i0}}{\widehat{\sigma}^2_{i0}} 100\%. \tag{8.87}$$

The entries in Table 8.7 confirm the analysis of the individual conditions in Figure 8.18. The most significant impact upon the variance of the first score variable is the 10% removal of the tooth. For the remaining score variables, the most significant effects are the 40% and 100% removal of the tooth. Moreover, the 10% removal has a considerably stronger impact upon the first three score variances than the 20% and 30% removals. The results in Table 8.7 therefore reflect the observations in the upper two plots in Figure 8.17.

8.7 Analysis of primary and improved residuals

As Subsection 8.3.4 gives a detailed analysis of the statistical properties of the primary residuals, this section presents more theoretical aspects of the primary and improved residuals. The section investigates the first and second moments of the primary residuals of the eigenvectors $\boldsymbol{\phi}_i$, after presenting a detailed proof of the CLT for a sum of random variables that are i.i.d. This is followed by an examination of the covariance matrices for ϕ_i and $\boldsymbol{\phi}_i$ to fault conditions. Finally, Subsection 8.7.3 outlines and proves that the non-negative quadratic statistics obtained from the improved residuals are more sensitive than those computed from the score variables.

8.7.1 Central limit theorem

According to (8.16), the statistical local approach relies on a vector-valued function of improved residuals which are, theoretically, an infinite sum of primary residuals. This subsection shows that

$$\boldsymbol{\theta} = \frac{1}{\sqrt{K}} \sum_{k=1}^{K} \boldsymbol{\phi}\left(\mathfrak{p}, \mathbf{z}_0(k)\right) \tag{8.88}$$

converges in distribution to $\mathcal{N}\left\{\mathbf{0}, \mathbf{S}_{\phi\phi}\right\}$ when $K \to \infty$. To simplify the analysis here, we examine the jth element of $\boldsymbol{\phi}$, ϕ_j. The definition of the primary residuals in the preceding sections outlined that they have a mean of zero, so we can safely assume here that $E\left\{\phi_j\right\} = 0$. Moreover, for the variance of ϕ_j, $E\left\{\phi_j^2\right\}$, we write $\sigma_{\phi_j}^2$. In summary, the first and second moments of ϕ_j are

$$E\left\{\phi_j\right\} = 0 \qquad E\left\{\phi_j^2\right\} = \sigma_{\phi_j}^2 = 2\lambda_j^2. \tag{8.89}$$

With this in mind, the jth element in (8.88) can be reformulated as follows

$$\frac{\theta_j}{\sigma_{\phi_j}} = \frac{1}{\sqrt{K}} \sum_{k=1}^{K} \frac{\phi_j\left(\mathbf{p}, \mathbf{z}_0(k)\right)}{\sigma_{\phi_j}} = \frac{1}{\sigma_{\phi_j}\sqrt{K}} \sum_{k=1}^{K} \phi_j(k). \tag{8.90}$$

Given that the samples, $\mathbf{z}_0(k)$, are drawn independently from the distribution function $\mathcal{N}\left\{\mathbf{0}, \mathbf{S}_{z_0 z_0}\right\}$, the instances of the primary residuals $\phi_j(k)$ are also drawn independently from the distribution function $F\left(\phi_j\right)$ in (8.55). Moreover, as pointed out before, each instance is drawn from the same distribution function $F\left(\phi_j\right)$ and each of these instances are i.i.d. With respect to the preceding discussion, we can now formulate a simplified version of the Lindeberg-Lévy theorem, defining the CLT for the infinite i.i.d. sequence of ϕ_j.

Theorem 8.7.1 *The scaled sum of $\phi_j(1)$, $\phi_j(2)$, $\ldots$, $\phi_j(K)$, which have the same distribution function $F\left(\phi_j\right)$, is asymptotically Gaussian distributed, that is, ${}^{\theta_j}/_{\sigma_{\phi_j}} \sim \mathcal{N}\{0, 1\}$, given that $E\left\{\phi_j(k)\right\} = 0$ and $E\left\{\phi_j^2(k)\right\} = \sigma_{\phi_j}$.*

Proof. The proof of Theorem 8.7.1 commences by defining the characteristic function of the improved residuals

$$\gamma_j(c) = E\left\{e^{\mathrm{i}c\frac{\theta_j}{\sigma_{\phi_j}}}\right\} = \int_{-\infty}^{\infty} e^{\mathrm{i}c\frac{\theta_j}{\sigma_{\phi_j}}}\,\mathrm{d}F\left(\theta_j\right), \tag{8.91}$$

where $\mathrm{i} = \sqrt{-1}$ and $c \in \mathbb{R}$. It is important to note that the characteristic function provides an equal basis for defining and describing the statistical properties of a random variables compared to the cumulative distribution function. For example, if two cumulative distribution functions are identical, so are their characteristic functions. This is taken advantage of here in order to prove that the infinite sum in (8.90) asymptotically follows a Gaussian distribution.

The first step is to substitute the definition of θ_j into (8.91). With respect to the definition of the characteristic function, this gives rise to

$$\gamma_j(c) = E\left\{e^{\mathrm{i}c\frac{\sum_{k=1}^{K}\phi_j(k)}{\sigma_{\phi_j}\sqrt{K}}}\right\} = \prod_{k=1}^{K} E\left\{e^{\mathrm{i}c\frac{\phi_j(k)}{\sigma_{\phi_j}\sqrt{K}}}\right\} = \prod_{k=1}^{K} \gamma_{j_k}(c). \tag{8.92}$$

The fact that the random variables $\phi_j(1), \phi_j(2), \ldots, \phi_j(K)$ are i.i.d. implies that their distribution and characteristic functions are identical. The product in (8.92) can therefore be simplified to

$$\gamma_j(c) = \left(\gamma_j^*(c)\right)^K. \tag{8.93}$$

Here, $\gamma_{j^*}(c) = \sqrt[K]{\gamma_j(c)}$, that is, the characteristic function of ϕ_j is the Kth root of the characteristic function of θ_j. The next step is to develop a Taylor series of γ_{j^*} for $c = 0$

$$\gamma_{j^*} = \gamma_{j^*}(0) + \gamma_{j^*}'(0)c + \tfrac{1}{2}\gamma_{j^*}''(0)c^2 + R(\tau)c^3 \tag{8.94}$$

where the last term, $R(\tau)c^3 = \frac{1}{6}\gamma_{j^*}'''(\tau)c^3$, $\tau \in \mathbb{R}$, is the Lagrangian remainder. The relationships can be obtained from the definition of the characteristic function

$$\begin{aligned} \gamma_{j^*} &= \int_{-\infty}^{\infty} e^{\mathrm{i}c\frac{\phi_j}{\sigma_{\phi_j}\sqrt{K}}} \mathrm{d}F(\phi_j) \\ \Rightarrow \left.\frac{\mathrm{d}^m}{\mathrm{d}c^m}\gamma_{j^*}\right|_{c=0} &= \mathrm{i}^m \int_{-\infty}^{\infty} \left(\frac{\phi_j}{\sigma_{\phi_j}\sqrt{K}}\right)^m \mathrm{d}F(\phi_j) \\ \left.\frac{\mathrm{d}^m}{\mathrm{d}c^m}\gamma_{j^*}\right|_{c=0} &= \mathrm{i}^m E\left\{\left(\frac{\phi_j}{\sigma_{\phi_j}\sqrt{K}}\right)^m\right\}. \end{aligned} \tag{8.95}$$

According to Equation (8.89), the relationships up to order three are therefore:

- $m = 0 \;:\; \gamma_{j^*}(0) = \int_{-\infty}^{\infty} \mathrm{d}F(\phi_j) = 1$;
- $m = 1 \;:\; \gamma_{j^*}'(0) = \mathrm{i}E\left\{\frac{\phi_j}{\sigma_{\phi_j}\sqrt{K}}\right\} = 0$;
- $m = 2 \;:\; \gamma_{j^*}''(0) = (-1)E\left\{\left(\frac{\phi_j}{\sigma_{\phi_j}\sqrt{K}}\right)^2\right\} = -\frac{1}{K}$

$$\gamma_{j^*}'''(\tau) = -\mathrm{i}\int_{-\infty}^{\infty}\left(\frac{\phi_j}{\sigma_{\phi_j}\sqrt{K}}\right)^3 e^{i\tau\frac{\phi_j}{\sigma_{\phi_j}\sqrt{K}}} \mathrm{d}F(\phi_j)\text{; and}$$

- $m = 3 \;:\; \gamma_{j^*}'''(\tau) = \delta(\tau)E\left\{\left(\frac{\phi_j}{\sigma_{\phi_j}\sqrt{K}}\right)^3\right\}$

$$\gamma_{j^*}'''(\tau) = \delta(\tau)\frac{\varrho_{\phi_{j^*}}^3}{\sigma_{\phi_j}^3 K^{3/2}}.$$

Here, $\delta(\tau) \in \mathbb{C}$ is a small correction term and $\varrho^3_{\phi_{j*}}$ is the third moment of ϕ_j. Substituting these relationships in (8.94) yields

$$\gamma_j^* = 1 - \tfrac{1}{2}\tfrac{1}{K}c^2 + \underbrace{\tfrac{1}{6}\delta(\tau)\frac{\varrho^3_{\phi_j^*}}{\sigma^3_{\phi_j} K^{3/2}}}_{R(\tau)} c^3. \tag{8.96}$$

Substituting (8.96) into (8.93) gives rise to

$$\gamma_j\left(c\right) = \left(1 + \tfrac{1}{K}\left(-\frac{c^2}{2}\right) + R(\tau)c^3\right)^K. \tag{8.97}$$

Note that the characteristic function of the standard Gaussian distribution function is $e^{-c^2/2}$ and that $\lim_{K\to\infty}\left(1 + {}^{\nu}/_{K}\right)^K \to e^{\nu}$. Therefore, the expression in (8.97) asymptotically converges to

$$\boxed{\lim_{K\to\infty} \gamma_j\left(c\right) \to e^{-c^2/2}}. \tag{8.98}$$

This follows from $\lim_{K\to\infty} R(\tau) = \left({}^{\delta(\tau)\varrho^3_{\phi_j^*}}/_{6\sigma^3_{\phi_{j*}}}\right) \lim_{K\to\infty} {}^{1}/_{K^{3/2}} \to 0$.

8.7.2 Further statistical properties of primary residuals

After proving the CLT, we now focus on discussing further properties of the primary residuals. Subsection 8.3.4 shows the first two moments for the primary residuals $t_i^2 - \lambda_i$. This subsection determines the covariance and cross-covariance matrices for the primary residuals $t_i\mathbf{z}_0 - \lambda_i\mathbf{p}_i$. It also discusses how the covariance matrices of $t_i^2 - \lambda_i$ and $t_i\mathbf{z}_0 - \lambda_i\mathbf{p}_i$ are affected by a change in the eigenvalues and eigenvectors.

8.7.2.1 Covariance matrices of primary residuals in equation (8.20)

The determination of the covariance and cross-covariance matrices for the primary residual vectors $\boldsymbol{\phi}_i$, and $\boldsymbol{\phi}_i$ and $\boldsymbol{\phi}_j$, $i \neq j$, requires the substitution of $\boldsymbol{\phi}_i = t_i\mathbf{z}_0 - \lambda_i\mathbf{p}_i$ and $\boldsymbol{\phi}_j = t_j\mathbf{z}_0 - \lambda_j\mathbf{p}_j$ into $E\left\{\boldsymbol{\phi}_i\boldsymbol{\phi}_i^T\right\}$ and $E\left\{\boldsymbol{\phi}_i\boldsymbol{\phi}_j^T\right\}$, respectively.

Covariance matrix. Starting with the covariance matrix $E\left\{\boldsymbol{\phi}_i\boldsymbol{\phi}_i^T\right\}$

$$E\left\{\boldsymbol{\phi}_i\boldsymbol{\phi}_i^T\right\} = E\left\{t_i^2\mathbf{z}_0\mathbf{z}_0^T - \lambda_i t_i\mathbf{z}_0\mathbf{p}_i^T - \lambda_i t_i\mathbf{p}_i\mathbf{z}_0^T\right\} + \lambda_i^2\mathbf{p}_i\mathbf{p}_i^T, \tag{8.99}$$

the expectation on the right hand side of (8.99) can be analyzed separately by substituting $\mathbf{z}_0 = t_1\mathbf{p}_1 + t_2\mathbf{p}_2 + \cdots + t_{n_z}\mathbf{p}_{n_z}$, which yields

$$E\left\{t_i^2\mathbf{z}_0\mathbf{z}_0^T\right\} = E\left\{t_i^2\left(\sum_{j=1}^{n_z} t_j\mathbf{p}_j\right)\left(\sum_{m=1}^{n_z} t_m\mathbf{p}_m\right)^T\right\} \tag{8.100a}$$

$$E\left\{\lambda_i t_i \mathbf{z}_0 \mathbf{p}_i^T\right\} = \lambda_i E\left\{t_i\left(\sum_{j=1}^{n_z} t_j \mathbf{p}_j\right)\mathbf{p}_i^T\right\} = \lambda_i^2 \mathbf{p}_i \mathbf{p}_i^T \tag{8.100b}$$

$$E\left\{\lambda_i t_i \mathbf{p}_i \mathbf{z}_0^T\right\} = \lambda_i E\left\{t_i \mathbf{p}_i\left(\sum_{j=1}^{n_z} t_j \mathbf{p}_j\right)\right\} = \lambda_i^2 \mathbf{p}_i \mathbf{p}_i^T. \tag{8.100c}$$

The above simplifications follow from $E\left\{t_i t_j\right\} = 0$ for all $i \neq j$. The fact that there are n_z^2 terms in (8.100a) gives rise to the following separation.

- For $i \neq j \neq m$, there are a total of $n_z^2 - n_z$ terms in this case, which produces

$$E\left\{t_i^2 t_j t_m \mathbf{p}_j \mathbf{p}_m^T\right\} = 0,$$

  which follows from the Isserlis theorem (Isserlis 1918).

- There are $n_z - 1$ cases of $i \neq j = m$, which yield the general expression

$$E\left\{t_i^2 t_j^2\right\} \mathbf{p}_j \mathbf{p}_j^T = \lambda_i \lambda_j \mathbf{p}_j \mathbf{p}_j^T.$$

- The remaining term, $i = j = m$, is equal to

$$E\left\{t_i^4\right\} \mathbf{p}_i \mathbf{p}_i^T = 3\lambda_i^2 \mathbf{p}_i \mathbf{p}_i^T.$$

Putting this all together, (8.100a) reduces to

$$E\left\{t_i^2 \mathbf{z}_0 \mathbf{z}_0^T\right\} = \lambda_i \sum_{j=1 \neq i}^{n_z} \lambda_j \mathbf{p}_j \mathbf{p}_j^T + 3\lambda_i^2 \mathbf{p}_i \mathbf{p}_i^T. \tag{8.101}$$

Substituting (8.100a) to (8.101) into (8.99) finally yields

$$E\left\{\boldsymbol{\phi}_i \boldsymbol{\phi}_i^T\right\} = \lambda_i \sum_{j=1}^{n_z} \lambda_j \mathbf{p}_j \mathbf{p}_j^T + \lambda_i^2 \mathbf{p}_i \mathbf{p}_i^T. \tag{8.102}$$

Cross-covariance matrix. The matrix $E\left\{\boldsymbol{\phi}_i \boldsymbol{\phi}_j^T\right\}, i \neq j$ and $i, j = 1, 2, \ldots, n_z$, is similar to that in (8.99) and is given by

$$E\left\{\boldsymbol{\phi}_i \boldsymbol{\phi}_j^T\right\} = E\left\{t_i t_j \mathbf{z}_0 \mathbf{z}_0^T - \lambda_j t_i \mathbf{z}_0 \mathbf{p}_j^T - \lambda_i t_j \mathbf{p}_i \mathbf{z}_0^T\right\} + \lambda_i \lambda_j \mathbf{p}_i \mathbf{p}_j^T. \tag{8.103}$$

Using the simplifications applied to (8.99), (8.103) reduces to

$$E\left\{t_i t_j \mathbf{z}_0 \mathbf{z}_0^T\right\} = E\left\{t_i t_j\left(\sum_{m=1}^{n_z} t_m \mathbf{p}_m\right)\left(\sum_{\mathrm{m}=1}^{n_z} t_{\mathrm{m}} \mathbf{p}_{\mathrm{m}}\right)^T\right\} \tag{8.104a}$$

$$E\left\{\lambda_j t_i \mathbf{z}_0 \mathbf{p}_j^T\right\} = \lambda_j E\left\{t_i\left(\sum_{m=1}^{n_z} t_m \mathbf{p}_m^T\right)\right\}\mathbf{p}_j^T = \lambda_i \lambda_j \mathbf{p}_i \mathbf{p}_j^T \tag{8.104b}$$

$$E\left\{\lambda_i t_j \mathbf{p}_i \mathbf{z}_0^T\right\} = \lambda_i \mathbf{p}_i E\left\{t_j\left(\sum_{m=1}^{n_z} t_m \mathbf{p}_m\right)^T\right\} = \lambda_i \lambda_j \mathbf{p}_i \mathbf{p}_j^T. \tag{8.104c}$$

Given that $E\left\{t_i t_j t_m t_\mathrm{m}\right\} = 0$ for all $m \neq i, \mathrm{m} \neq j$ and $m \neq j, \mathrm{m} \neq i$, (8.104a) reduces to

$$E\left\{t_i t_j \mathbf{z}_0 \mathbf{z}_0^T\right\} = \sum_{m=1}^{n_z} \sum_{\mathrm{m}=1}^{n_z} \left[E\left\{t_i t_j t_m t_\mathrm{m}\right\} \mathbf{p}_m \mathbf{p}_\mathrm{m}^T\right] = \lambda_i \lambda_j \mathbf{p}_i \mathbf{p}_j^T + \lambda_j \lambda_i \mathbf{p}_j \mathbf{p}_i^T \tag{8.105}$$

Putting it all together, (8.103) finally yields

$$E\left\{\boldsymbol{\phi}_i \boldsymbol{\phi}_j^T\right\} = \lambda_i \lambda_j \mathbf{p}_j \mathbf{p}_i^T \tag{8.106}$$

8.7.2.2 Covariance matrix $E\left\{\boldsymbol{\phi}_i \boldsymbol{\phi}_i^T\right\}$ for a change in λ_i

Under the assumption that the eigenvectors remain constant, changing the covariance of the ith score variable to be $\lambda_i + \Delta\lambda_i$ results in the following alteration of $\boldsymbol{\phi}_i$

$$\boldsymbol{\phi}_i^* = t_i \mathbf{z}_0^* - \lambda_i \mathbf{p}_i = \sum_{m=1}^{n_z} t_i t_m^* \mathbf{p}_m - \lambda_i \mathbf{p}_i. \tag{8.107}$$

The expectation of $\boldsymbol{\phi}_i^*$ is

$$E\left\{\boldsymbol{\phi}_i^*\right\} = E\left\{\sum_{m=1}^{n_z} t_i t_m^* \mathbf{p}_m - \lambda_i \mathbf{p}_i\right\} = \left(\lambda_i + \Delta\lambda_i - \lambda_i\right)\mathbf{p}_i = \Delta\lambda_i \mathbf{p}_i, \tag{8.108}$$

which implies that the covariance matrix matrix becomes

$$E\left\{\left(\boldsymbol{\phi}_i^* - \Delta\lambda_i \mathbf{p}_k\right)\left(\boldsymbol{\phi}_i^* - \Delta\lambda_i \mathbf{p}_k\right)^T\right\} = E\left\{\boldsymbol{\phi}_i \boldsymbol{\phi}_i^T\right\}. \tag{8.109}$$

Thus, a change in the eigenvalues yield the same covariance matrix for $\boldsymbol{\phi}_i$ and $\boldsymbol{\phi}_i^*$.

8.7.2.3 Covariance matrix $E\left\{\boldsymbol{\phi}_i \boldsymbol{\phi}_i^T\right\}$ for change in $\mathbf{p}_i$

An alteration of the eigenvector, $\mathbf{p}_i^* = \mathbf{p}_i + \Delta\mathbf{p}_i$, does not have the same isolated impact upon $\boldsymbol{\phi}_i$, as is the case for a change in the eigenvalue. This is because a rotation of ith eigenvector affects more then just this eigenvector, since the eigenvectors are mutually orthonormal. If we restrict this examination by assuming that only the ith eigenvector is altered and the remaining $n_z - 1$ eigenvectors and the score covariance matrix remain unchanged, $\boldsymbol{\phi}_i$ becomes

$$\boldsymbol{\phi}_i^* = t_i \mathbf{z}_0^* - \lambda_i \mathbf{p}_i = \sum_{j=1}^{n_z} t_i t_j \mathbf{p}_j^* - \lambda_i \mathbf{p}_i. \tag{8.110}$$

Note that $\mathbf{p}_j^* = \mathbf{p}_j$ for all $j \neq i$. The expectation of $\boldsymbol{\phi}_i^*$ is

$$E\left\{\boldsymbol{\phi}_i^*\right\} = \sum_{j=1}^{n_z} E\left\{t_i t_j\right\} \mathbf{p}_j^* - \lambda_i \mathbf{p}_i = \lambda_i \left(\mathbf{p}_i + \Delta\mathbf{p}_i - \mathbf{p}_i\right) = \lambda_i \Delta\mathbf{p}_i, \tag{8.111}$$

which gives rise to the following covariance matrix

$$\begin{aligned}
&E\left\{\left(\boldsymbol{\phi}_i^* - \lambda_i \Delta\mathbf{p}_i\right)\left(\boldsymbol{\phi}_i^* - \lambda_i \Delta\mathbf{p}_i\right)^T\right\} \\
&= E\left\{\left(t_i\left(\sum_{m=1}^{n_z} t_m \mathbf{p}_m^*\right) - \lambda_i\left(\mathbf{p}_i + \Delta\mathbf{p}_i\right)\right)\right. \\
&\qquad \left.\left(t_i\left(\sum_{m=1}^{n_z} t_m \mathbf{p}_m^*\right) - \lambda_i\left(\mathbf{p}_i + \Delta\mathbf{p}_i\right)\right)^T\right\} \\
&= E\left\{t_i^2\left(\sum_{m=1}^{n_z} t_m \mathbf{p}_m^*\right)\left(\sum_{\mathrm{m}=1}^{n_z} t_{\mathrm{m}} \mathbf{p}_{\mathrm{m}}^*\right) - \lambda_i\left(\sum_{m=1}^{n_z} t_i t_m \mathbf{p}_m^*\right)\mathbf{p}_i^{*^T}\right. \\
&\qquad \left. - \lambda_i \mathbf{p}_i^*\left(\sum_{m=1}^{n_z} t_i t_m \mathbf{p}_m^*\right)^T + \lambda_i^2 \mathbf{p}_i^* \mathbf{p}_i^{*^T}\right\} \\
&= \quad \lambda_i \sum_{j=1}^{n_z} \lambda_j \mathbf{p}_j^* \mathbf{p}_j^{*^T} + \lambda_i^2 \mathbf{p}_i^* \mathbf{p}_i^{*^T} = \lambda_i \sum_{j=1 \neq i}^{n_z} \lambda_j \mathbf{p}_j \mathbf{p}_j^T + 2\lambda_i^2 \mathbf{p}_i^* \mathbf{p}_i^{*^T}.
\end{aligned} \tag{8.112}$$

The difference between the covariance matrices of $\boldsymbol{\phi}_i$ and $\boldsymbol{\phi}_i^*$ is therefore

$$\begin{aligned}
&E\left\{\left(\boldsymbol{\phi}_i^* - \lambda_i \Delta\mathbf{p}_i\right)\left(\boldsymbol{\phi}_i^* - \lambda_i \Delta\mathbf{p}_i\right)^T\right\} - E\left\{\boldsymbol{\phi}_i \boldsymbol{\phi}_i^T\right\} \\
&= \; 2\lambda_i^2\left(\mathbf{p}_i \Delta\mathbf{p}_i^T + \Delta\mathbf{p}_i \mathbf{p}_i^T + \Delta\mathbf{p}_i \Delta\mathbf{p}_i^T\right).
\end{aligned} \tag{8.113}$$

8.7.2.4 Covariance of $\widetilde{\boldsymbol{\phi}}_i$ for a change in $\boldsymbol{\lambda}_i$

$$\widetilde{\phi}_i^* = t_i^{*^2} - \lambda_i \tag{8.114}$$

and has the following expectation

$$E\left\{\widetilde{\phi}_i^*\right\} = \sum_{m=1}^{n_z}\sum_{\mathrm{m}=1}^{n_z} \left(\mathbf{p}_i^T \mathbf{p}_m\right) E\left\{t_m^* t_{\mathrm{m}}^*\right\} \left(\mathbf{p}_{\mathrm{m}}^T \mathbf{p}_i\right) - \lambda_i = \Delta\lambda_i, \tag{8.115}$$

which gives rise to the following covariance

$$\begin{aligned} E\left\{\left(\widetilde{\phi}_i^* - \Delta\lambda_i\right)^2\right\} &= E\left\{\left(t_i^{*^2} - \left(\lambda_i + \Delta\lambda_i\right)\right)^2\right\} \\ E\left\{\left(\widetilde{\phi}_i^* - \Delta\lambda_i\right)^2\right\} &= E\left\{t_i^{*^4} - 2t_i^{*^2}\left(\lambda_i + \Delta\lambda_i\right)\right\} + \left(\lambda_i + \Delta\lambda_i\right)^2 \\ E\left\{\left(\widetilde{\phi}_i^* - \Delta\lambda_i\right)^2\right\} &= 2\lambda_i^2 + 4\lambda_i\Delta\lambda_i + 2\Delta\lambda_i^2, \end{aligned} \tag{8.116}$$

which follows from (8.48). The difference between $E\left\{\left(\widetilde{\phi}_i^* - \Delta\lambda_i\right)^2\right\}$ and $E\left\{\widetilde{\phi}_i^2\right\}$ is therefore

$$E\left\{\left(\widetilde{\phi}_i^* - \Delta\lambda_i\right)^2\right\} - E\left\{\widetilde{\phi}_i^2\right\} = 2\Delta\lambda_i\left(2\lambda_i + \Delta\lambda_i\right). \tag{8.117}$$

8.7.2.5 Covariance of $\widetilde{\phi}_i$ for a change in $\mathbf{p}_i$

Changing $\mathbf{p}_i$ to $\mathbf{p}_i + \Delta\mathbf{p}_i$ implies that it is no longer orthogonal to all of the remaining $n_z - 1$ eigenvectors. Assuming that $\left\|\mathbf{p}_i + \Delta\mathbf{p}_i\right\| = 1$ and defining the n_z loading vectors by $\mathbf{p}_j^*$, $\mathbf{p}_j^* = \mathbf{p}_j$ for all $j \neq i$, the expectation of $\widetilde{\phi}_i$ becomes

$$E\left\{\widetilde{\phi}_i^*\right\} = E\left\{t_i^{*^2} - \lambda_i\right\} = E\left\{\mathbf{p}_i^T \mathbf{z}_0^* \mathbf{z}_0^{*^T} \mathbf{p}_i - \lambda_i\right\}. \tag{8.118}$$

Next, substituting $\mathbf{z}_0^* = \sum_{k=1}^{n_z} \mathbf{p}_k^* t_k$ into (8.118) gives rise to

$$\begin{aligned} E\left\{\widetilde{\phi}_i^*\right\} &= E\left\{\mathbf{p}_i^T\left(\sum_{m=1}^{n_z}\mathbf{p}_m^* t_m\right)\left(\sum_{\mathrm{m}=1}^{n_z}\mathbf{p}_{\mathrm{m}}^* t_{\mathrm{m}}\right)^T \mathbf{p}_i - \lambda_i\right\} \\ E\left\{\widetilde{\phi}_i^*\right\} &= \mathbf{p}_i^T\left[\sum_{m=1}^{n_z}\sum_{\mathrm{m}=1}^{n_z}\mathbf{p}_m^* E\left\{t_m t_{\mathrm{m}}\right\}\mathbf{p}_{\mathrm{m}}^{*^T}\right]\mathbf{p}_i - \lambda_i \\ E\left\{\widetilde{\phi}_i^*\right\} &= \sum_{m=1}^{n_z}\left(\mathbf{p}_i^T\mathbf{p}_m^*\right)^2\lambda_k - \lambda_i = \lambda_i\left(\left(\mathbf{p}_i^T\mathbf{p}_i^*\right)^2 - 1\right) < 0. \end{aligned} \tag{8.119}$$

That $E\left\{\widetilde{\phi}_i^*\right\} < 0$ is interesting and follows from the assumption that the length of $\left\|\mathbf{p}_1\right\| = \left\|\mathbf{p}_i + \Delta\mathbf{p}_i\right\| = 1$, that is, $\mathbf{p}_i^T\left(\mathbf{p}_i + \Delta\mathbf{p}_i\right) = \cos\left(\varphi_{(\mathbf{p}_1,\mathbf{p}_i+\Delta\mathbf{p}_i)}\right)$. Using the mean value $\lambda_i\left(\left(\mathbf{p}_i^T\mathbf{p}_i^*\right)^2-1\right)$, the covariance of $\widetilde{\phi}_i^*$ is equal to

$$E\left\{\left(t_i^{*^T}-\lambda_i-\lambda_i\left(\left(\mathbf{p}_i^T\mathbf{p}_i^*\right)^2-1\right)\right)^2\right\}$$

$$=E\left\{\left(\mathbf{p}_i^T\left[\sum_{m=1}^{n_z}\sum_{\mathrm{m}=1}^{n_z}\mathbf{p}_m^*t_mt_{\mathrm{m}}\mathbf{p}_{\mathrm{m}}^{*^T}\right]\mathbf{p}_i\lambda_i\left(\mathbf{p}_i^T\mathbf{p}_i^*\right)^2\right)^2\right\} \tag{8.120}$$

$$=E\left\{\left(\left(\mathbf{p}_i^T\mathbf{p}_i^*\right)^2t_i^2-\lambda_i\left(\mathbf{p}_i^T\mathbf{p}_i^*\right)^2\right)^2\right\}=2\lambda_i^2\left(\mathbf{p}_i^T\mathbf{p}_i^*\right)^4.$$

It follows that $E\left\{\left(\widetilde{\phi}_i^*-\lambda_i\left(\left(\mathbf{p}_i^T\mathbf{p}_i^*\right)^2-1\right)\right)\right\}<E\left\{\widetilde{\phi}_i^2\right\}$, since $\left(\mathbf{p}_i^T\mathbf{p}_i^*\right)^4<1$.

8.7.3 Sensitivity of statistics based on improved residuals

The previous sections showed that the primary residuals $\phi_i=t_i^2-\lambda_i$ are sufficient for detecting changes in the underlying geometry of the data model in (2.2). Furthermore, the mean and variance of these residuals is 0 and $2\lambda_i^2$, respectively, and there is no covariance between the ith and jth primary residuals, that is, $E\left\{\phi_i\phi_j\right\}=0$.

Furthermore, the primary residuals are i.i.d. implying that the improved residuals are asymptotically Gaussian distributed and have the same mean, variance and covariance as the primary residuals. The covariance matrix for the vector containing the improved residuals of the first n eigenpairs of $\mathbf{S}_{z_0z_0}$ is equal to $E\left\{\boldsymbol{\theta}\boldsymbol{\theta}^T\right\}=\frac{1}{2}\boldsymbol{\Lambda}^2$, where $\boldsymbol{\Lambda}$ is the diagonal matrix storing the n dominant eigenvalues of $\mathbf{S}_{z_0z_0}$.

The sensitivity in detecting various fault conditions is now examined for a simple sensor or actuator bias. The scope of Project 3 in the tutorial section covers more complex process faults that alter the eigenvectors and eigenvalues by contrasting the sensitivity of the non-negative quadratic statistics constructed from the improved residuals with those based on the score variables.

8.7.3.1 Sensitivity for detecting sensor or actuator bias

The data model describing a sensor fault is as follows

$$\mathbf{z}^*=\Xi\mathbf{s}+(\mathfrak{g}+\Delta\mathfrak{g})+(\bar{\mathbf{z}}+\Delta\mathbf{z})=\mathbf{z}+\Delta\mathfrak{g}+\Delta\mathbf{z}_0. \tag{8.121}$$

According to Table 2.1, the effect of a sensor or actuator fault can be described by an offset term or a *bias* and a stochastic term or a *precision degradation*. Besides bias and precision degradation, both terms can also describe a complete failure or a drift if the offset term is assumed to be time varying and deterministic. For a sensor bias, described by a constant bias term $\Delta\mathbf{z}_0$, the ith improved residual becomes

$$\theta_i^*\left(k_0\right) = \frac{1}{\sqrt{k_0}} \sum_{k=1}^{k_0} \left(\left(\mathbf{p}_i^T \mathbf{z}_0(k) + \mathbf{p}_i^T \Delta \mathbf{z}_0 \right)^2 - \lambda_i \right). \tag{8.122}$$

Expectation of $\theta_i^*\left(k_0\right)$. The expectation follows from

$$\begin{aligned} E\left\{\theta_i^*\right\} &= E\left\{ \frac{1}{\sqrt{k_0}} \sum_{k=1}^{k_0} \left(t_i^2(k) + 2t_i \mathbf{p}_i^T \Delta \mathbf{z}_0 + \left(\mathbf{p}_i^T \Delta \mathbf{z}_0 \right)^2 \right) - \lambda_i \right\} \\ E\left\{\theta_i^*\right\} &= \frac{1}{\sqrt{k_0}} \sum_{k=1}^{k_0} \left(\lambda_i + \left(\mathbf{p}_i^T \Delta \mathbf{z}_0 \right)^2 - \lambda_i \right) = \sqrt{k_0} \left(\mathbf{p}_i^T \Delta \mathbf{z}_0 \right)^2. \end{aligned} \tag{8.123}$$

Variance of $\theta_i^*\left(k_0\right)$. Defining the mean-centered θ_i^* by

$$\theta_i^* - \sqrt{k_0} \left(\mathbf{p}_i^T \Delta \mathbf{z}_0 \right)^2 = \frac{1}{\sqrt{k_0}} \left(\sum_{k=1}^{k_0} t_i^2(k) + 2t_i(k) \mathbf{p}_i^T \Delta \mathbf{z}_0 + k_0 \lambda_i \right) = \widetilde{\theta}_i^* \tag{8.124}$$

simplifies the determination of the variance of $\theta_i^*\left(k_0\right)$

$$\begin{aligned} E\left\{\widetilde{\theta}_i^{*2}\right\} &= \frac{1}{k_0} E\left\{ \left(\sum_{k=1}^{k_0} \left(t_i^2(k) + 2t_i(k) \mathbf{p}_i^T \Delta \mathbf{z}_0 \right) + k_0 \lambda_i \right)^2 \right\} \\ E\left\{\widetilde{\theta}_i^{*2}\right\} &= \frac{1}{k_0} E\left\{ \left(\sum_{k=1}^{k_0} t_i^2(k) \right)^2 + 4 \left(\mathbf{p}_i^T \Delta \mathbf{z}_0 \right)^2 \left(\sum_{k=1}^{k_0} t_i(k) \right)^2 \right. \\ &\quad + k_0^2 \lambda_i^2 + 4 \mathbf{p}_i^T \Delta \mathbf{z}_0 \left(\sum_{k=1}^{k_0} t_i^2(k) \right) \left(\sum_{l=1}^{k_0} t_i(l) \right) \\ &\quad \left. + 2k_0 \lambda_i \left(\sum_{k=1}^{k_0} t_i^2(k) \right) + 4k_0 \lambda_i \mathbf{p}_i^T \Delta \mathbf{z}_0 \left(\sum_{k=1}^{k_0} t_i(k) \right) \right\} \\ E\left\{\widetilde{\theta}_i^{*2}\right\} &= \frac{1}{k_0} E\left\{ \sum_{k=1}^{k_0} t_i^4(k) + 2 \sum_{k=1}^{k_0-1} \sum_{l=k+1}^{k_0} t_i^2(k) t_i^2(l) \right. \\ &\quad + 4 \left(\mathbf{p}_i^T \Delta \mathbf{z}_0 \right)^2 \left(\sum_{k=1}^{k_0} t_i^2(k) + 2 \sum_{k=1}^{k_0-1} \sum_{l=k+1}^{k_0} t_i(k) t_i(l) \right) \end{aligned} \tag{8.125}$$

$$+ k_0^2\lambda_i^2 + 4\mathbf{p}_i^T \Delta\mathbf{z}_0 \left(\sum_{k=1}^{k_0} \sum_{l=1}^{k_0} t_i^2(k) t_i(l) \right)$$

$$\left. + 2k_0\lambda_i \left(\sum_{k=1}^{k_0} t_i^2(k) \right) + 4k_0\lambda_i \mathbf{p}_i^T \Delta\mathbf{z}_0 \left(\sum_{k=1}^{k_0} t_i(k) \right) \right\}$$

$$E\left\{ \widetilde{\theta}_i^{*^2} \right\} = \tfrac{1}{k_0} \left(3k_0\lambda_i^2 + (k_0 - 1)\, k_0\lambda_i^2 + 4k_0\lambda_i \left(\mathbf{p}_i^T \Delta\mathbf{z}_0\right)^2 + k_0^2\lambda_i^2 + 2k_0^2\lambda_i^2 \right)$$

$$E\left\{ \widetilde{\theta}_i^{*^2} \right\} = 2\lambda_i \left(1 + 2\left(\mathbf{p}_i^T \Delta\mathbf{z}_0\right)^2 \right)$$

which yields that $E\{\widetilde{\theta}_i^{*^2}\} > E\{\theta_i^2\}$. The next step is to examine the impact of $\Delta\mathbf{z}_0$ upon the non-negative quadratic monitoring statistics.

Effect of $\Delta\mathbf{z}_0$ upon $T^2_{\theta*}$. This impact can be described by $E\{T_{\theta^*}^{*^2}\}$

$$E\left\{T^2_{\theta^*}\right\} = E\left\{ \sum_{i=1}^{n} \frac{\theta_i^{*^2}}{2\lambda_i^2} \right\} = \sum_{i=1}^{n} \frac{E\left\{\theta_i^{*^2}\right\}}{2\lambda_i^2} \tag{8.126}$$

which requires examining $E\{\theta_i^{*^2}\}$

$$E\left\{\theta_i^{*^2}\right\} = E\left\{ \tfrac{1}{k_0} \left(\sum_{k=1}^{k_0} \left(t_i^2(k) + 2\left(\mathbf{p}_i^T \Delta\mathbf{z}_0\right) t_i(k) + \left(\mathbf{p}_i^T \Delta\mathbf{z}_0\right)^2 - \lambda_i \right) \right)^2 \right\}$$

$$= E\left\{ \tfrac{1}{k_0} \left(\sum_{k=1}^{k_0} t_i^4(k) + 2 \sum_{k=1}^{k_0-1} \sum_{l=k+1}^{k_0} t_i^2(k) t_i^2(l) \right.\right.$$

$$+ 4\left(\mathbf{p}_i^T \Delta\mathbf{z}_0\right)^2 \left(\sum_{k=1}^{k_0} t_i^2(k) + 2 \sum_{k=1}^{k_0-1} \sum_{l=k+1}^{k_0} t_i(k) t_i(l) \right)$$

$$+ k_0^2 \left(\mathbf{p}_i^T \Delta\mathbf{z}_0\right)^2 + k_0^2\lambda_i^2 + 4\left(\mathbf{p}_i^T \Delta\mathbf{z}_0\right) \left(\sum_{k=1}^{k_0-1} \sum_{l=k+1}^{k_0} t_i^2(k) t_i(l) \right)$$

$$+ 2k_0 \left(\mathbf{p}_i^T \Delta\mathbf{z}_0\right)^2 \left(\sum_{k=1}^{k_0} t_i^2(k) \right) - 2k_0\lambda_i \left(\sum_{k=1}^{k_0} t_i^2(k) \right)$$

$$+ 4k_0\lambda_i \left(\mathbf{p}_i^T \Delta\mathbf{z}_0\right)^3 \left(\sum_{k=1}^{k_0} t_i(k) \right) - 4k_0\lambda_i \left(\mathbf{p}_i^T \Delta\mathbf{z}_0\right) \left(\sum_{k=1}^{k_0} t_i(k) \right)$$

$$\left.\left. - 2k_0^2\lambda_i \left(\mathbf{p}_i^T \Delta\mathbf{z}_0\right)^2 \right) \right\}$$

$$= \tfrac{1}{k_0}\Big(3k_0\lambda_i^2 + (k_0-1)\,k_0\lambda_i^2 + 4k_0\lambda_i\left(\mathbf{p}_i^T\Delta\mathbf{z}_0\right)^2 + k_0^2\left(\mathbf{p}_i^T\Delta\mathbf{z}_0\right)^4 + k_0^2\lambda_i^2 + 2k_0^2\lambda_i\left(\mathbf{p}_i^T\Delta\mathbf{z}_0\right)^2 - 2k_0^2\lambda_i^2 - 2k_0^2\lambda_i\left(\mathbf{p}_i^T\Delta\mathbf{z}_0\right)^2\Big) \tag{8.127}$$

from which it follows that

$$E\left\{\theta_i^{*^2}\right\} = 2\lambda_i^2 + 4\lambda_i\left(\mathbf{p}_i^T\Delta\mathbf{z}_0\right)^2 + k_0\left(\mathbf{p}_i^T\Delta\mathbf{z}_0\right)^4. \tag{8.128}$$

Equation (8.126) can now be evaluated, which yields

$$\begin{aligned} E\left\{T_{\theta^*}^2\right\} &= \sum_{i=1}^{n}\frac{2\lambda_i^2 + 4\lambda_i\left(\mathbf{p}_i^T\Delta\mathbf{z}_0\right)^2 + k_0\left(\mathbf{p}_i^T\Delta\mathbf{z}_0\right)^4}{2\lambda_i^2} \\ &= \sum_{i=1}^{n}\left(1 + \frac{2\left(\mathbf{p}_i^T\Delta\mathbf{z}_0\right)^2}{\lambda_i} + \frac{k_0\left(\mathbf{p}_i^T\Delta\mathbf{z}_0\right)^4}{2\lambda_i^2}\right) \\ &= n + \sum_{i=1}^{n}\left(2 + {}^1\!/\!_2 k_0\left(\mathbf{p}_i^T\Delta\mathbf{z}_0\right)^2\right)\frac{\left(\mathbf{p}_i^T\Delta\mathbf{z}_0\right)^2}{\lambda_i}. \end{aligned} \tag{8.129}$$

This compares favorably to the sensitivity of the conventional Hotelling's T^2 statistic

$$\begin{aligned} E\left\{T^2\right\} &= \sum_{i=1}^{n}\frac{E\left\{t_i^{*^2}\right\}}{\lambda_i} = \sum_{i=1}^{n}\frac{E\left\{\left(\mathbf{p}_i^T\left(\mathbf{z}_0+\Delta\mathbf{z}_0\right)\right)^2\right\}}{\lambda_i} \\ &= \sum_{i=1}^{n}\frac{E\left\{t_i^2 + 2t_i\left(\mathbf{p}_i^T\Delta\mathbf{z}_0\right) + \left(\mathbf{p}_i^T\Delta\mathbf{z}_0\right)^2\right\}}{\lambda_i} \\ &= n + \sum_{i=1}^{n}\frac{\left(\mathbf{p}_i^T\Delta\mathbf{z}_0\right)^2}{\lambda_i}. \end{aligned} \tag{8.130}$$

The difference between (8.128) and (8.129) is then a measure for sensitivity of the Hotelling's T^2 statistic constructed from the improved residuals and the score variables

$$E\left\{T_{\theta^*}^2 - T^2\right\} = \sum_{i=1}^{n}\left(1 + {}^1\!/\!_2 k_0\left(\mathbf{p}_i^T\Delta\mathbf{z}_0\right)^2\right)\frac{\left(\mathbf{p}_i^T\Delta\mathbf{z}_0\right)^2}{\lambda_i} > 0. \tag{8.131}$$

For the Hotelling's T^2 statistics, (8.131) outlines that, irrespective of the window length k_0, the non-negative quadratic statistics constructed from the improved residuals is more sensitive to a set of sensor or actuator biases. Moreover, this equation also highlights that the larger the value of k_0 the more significant this increase in sensitivity becomes.

Constructing the residual Q statistics on the basis of (3.19) yields the same conclusion. Under the assumption that $\mathbf{S}_{\mathfrak{gg}} = \sigma_{\mathfrak{g}}^2 \mathbf{I}$, the $n_z - n$ required eigenvalues for constructing the $T^2_{\theta_d^*}$ and $T^2_{d^*}$, $\lambda_{n+1} = \cdots = \lambda_{n_z} = \sigma_{\mathfrak{g}}^2$. Since $\lambda_1 \geq \lambda_2 \geq \cdots \geq \lambda_n > \sigma_{\mathfrak{g}}^2$, (8.131) also confirms that the increase in sensitivity is more pronounced for the Q statistic.

The application studies in Sections 8.4 and 8.6 confirm the above findings. More precisely, Figures 8.7, 8.9 and 8.17 illustrate that the non-negative quadratic statistics based on the score variables show sporadic violations of their control limits which, however, exceeded the significance level. In contrast, the statistics constructed from the improved residuals showed a considerably stronger response and produced, in almost each case, a constant violation of their control limits.

8.8 Tutorial session

Question 1: Describe under which conditions conventional scatter diagrams, the Hotelling's T^2 statistic and the Q statistic are insensitive to fault conditions. What is the effect of such changes upon Type II errors?

Question 2: Referring to Question 1, analyze how these changes can be detected.

Question 3: Explain why the primary residuals are difficult to use for constructing monitoring charts. How can the associated problems be overcome?

Question 4: What is the problem of using primary residuals that are based on the eigenvectors of the data covariance matrix? Are the primary residuals related to the eigenvalues of the data covariance matrix also affected by this problem? Are both types of primary residuals sensitive to geometric changes in the model and residual subspaces as well as changes in the variances of the source and error variables?

Question 5: Describe the properties of the primary and improved residuals based on the eigenvalues of the covariance matrix.

Question 6: Discuss the assumptions under which the central limit theorem holds true. What are the conditions under which the central limit theorem does not hold true?

Project 1: Use a Monte Carlo simulation based on the second intuitive example, described in (8.8), augment the stochastic vector $\mathbf{z}_0$ by a time-varying mean vector $\bar{\mathbf{z}} = \bar{\mathbf{z}}(k)$ and comment on the Type I and II errors. For PCA, discuss how to develop an adaptive monitoring approach to construct primary residuals. How can such an adaptive approach be utilized to determine improved residuals?

Project 2: Develop primary and improved residuals on the basis of the PLS objective functions for determining the weight and loading vectors. Can a fault that (i) only affects the input variables, or (ii) only affects the output variables, or (iii) affects the input and the output variables, be detected? Discuss the results and propose a reliable monitoring scheme for detecting geometric changes in the weight and loading vectors and the variance of the score variables.

Project 3: Assume that a fault condition affects the ith eigenvector and/or eigenvalue of the covariance matrix $\mathbf{S}_{z_0 z_0}$. Similar to the analysis in Subsection 8.7.3, develop and contrast the relationships describing the impact of such a change upon the Hotelling's T^2 and T_θ^2 statistics.

PART IV

DESCRIPTION OF MODELING METHODS

9

Principal component analysis

This chapter introduces the PCA algorithm, including a discussion showing that the computed score and loading vectors maximize their contribution to the column and row space of a data matrix. A summary of PCA and the introduction of a preliminary PCA algorithm then follows in Section 9.2. A detailed summary of the properties of PCA is given in Section 9.3. Without attempting to give a complete review of the available research literature, further material concerning PCA may be found in Dunteman (1989); Jackson (2003); Jolliffe (1986); Wold (1978); Chapter 8 in Mardia *et al.* (1979) and Chapter 11 in Anderson (2003).

9.1 The core algorithm

PCA extracts sets of latent variables from a given data matrix $\mathbf{Z}_0 \in \mathbb{R}^{K \times n_z}$, containing K mean-centered and appropriately scaled samples of a variable set $\mathbf{z} \in \mathbb{R}^{n_z}$

$$\mathbf{z} = \mathbf{z}_0 + \bar{\mathbf{z}} \qquad \widetilde{\mathbf{z}}_0 = S\mathbf{z}_0 = S\left(\mathbf{z} - \bar{\mathbf{z}}\right). \tag{9.1}$$

The scaling matrix S is a diagonal matrix and often contains the reciprocal values of the estimated standard deviation of the recorded variables

$$\widehat{\sigma}_j = \sqrt{\frac{1}{K-1}\sum_{k=1}^{K}\left(z_j(k) - \widehat{\bar{z}}_j\right)^2}. \tag{9.2}$$

Statistical Monitoring of Complex Multivariate Processes: With Applications in Industrial Process Control, First Edition. Uwe Kruger and Lei Xie.

Based on $\mathbf{Z}_0^T = \begin{bmatrix} \widetilde{\mathbf{z}}_0(1) & \cdots & \widetilde{\mathbf{z}}_0(k) & \cdots & \widetilde{\mathbf{z}}_0(K) \end{bmatrix}$, PCA determines a series of rank-one matrices

$$\mathbf{Z}_0 = \sum_{j=1}^{n} \widehat{\mathbf{t}}_j \widehat{\mathbf{p}}_j^T + \mathbf{G} \tag{9.3}$$

constructed from the estimated score and loading vectors, $\widehat{\mathbf{t}}_j \in \mathbb{R}^K$ and $\widehat{\mathbf{p}}_j \in \mathbb{R}^{n_z}$, respectively. After extracting a total of n such rank one matrices, the residual matrix is $\mathbf{G} \in \mathbb{R}^{K \times n_z} = \mathbf{Z}_0 - \sum_{j=1}^{n} \widehat{\mathbf{t}}_j \widehat{\mathbf{p}}_j^T$. According to the non-causal data structure in (2.2), for which a detailed discussion is available in Section 6.1, an estimate of the column space of the parameter matrix $\boldsymbol{\Xi}$ is given by the loading matrix, i.e. $\widehat{\mathbf{P}} = \begin{bmatrix} \widehat{\mathbf{p}}_1 & \widehat{\mathbf{p}}_2 & \cdots & \widehat{\mathbf{p}}_n \end{bmatrix}$. As outlined in Section 6.1 the residual matrix is equal to the matrix product of $\mathbf{Z}_0$ and an orthogonal complement of $\widehat{\mathbf{P}}$, which spans the residual subspace. The construction of the model subspace, that is, the estimation of the column vectors of $\widehat{\mathbf{P}}$, and the residual subspace using PCA is now discussed. For simplicity, the subscript j is omitted for outlining the PCA algorithm.

Capturing the maximum amount of information from the data matrix $\mathbf{Z}_0$, each rank one matrix relies on a constraint objective function. Pre-multiplying $\mathbf{Z}_0$ by $\boldsymbol{\tau}^T \in \mathbb{R}^K$ allows extracting information from its column space

$$\boldsymbol{\pi} = \mathbf{Z}_0^T \boldsymbol{\tau} \qquad \boldsymbol{\tau}^T \boldsymbol{\tau} = 1. \tag{9.4}$$

Post-multiplying $\mathbf{Z}_0$ by $\mathbf{p} \in \mathbb{R}^{n_z}$ allows extracting information from its row space

$$\mathbf{t} = \mathbf{Z}_0 \mathbf{p} \qquad \mathbf{p}^T \mathbf{p} = 1. \tag{9.5}$$

The two equations above give rise to the following objective functions

$$J_p = \mathbf{p}^T \mathbf{Z}_0^T \mathbf{Z}_0 \mathbf{p} = \mathbf{p}^T \mathbf{M}_{pp} \mathbf{p} \tag{9.6a}$$

$$J_\tau = \boldsymbol{\tau}^T \mathbf{Z}_0 \mathbf{Z}_0^T \boldsymbol{\tau} = \boldsymbol{\tau}^T \mathbf{M}_{\tau\tau} \boldsymbol{\tau}, \tag{9.6b}$$

which are subject to the constraints

$$C_p = \mathbf{p}^T \mathbf{p} - 1 = 0 \tag{9.7a}$$

$$C_\tau = \boldsymbol{\tau}^T \boldsymbol{\tau} - 1 = 0. \tag{9.7b}$$

In the above equations, $\mathbf{M}_{pp} \in \mathbb{R}^{n_z \times n_z}$ and $\mathbf{M}_{\tau\tau} \in \mathbb{R}^{K \times K}$ are, up to a scalar factor, covariance matrices of the column space and the row space of $\mathbf{Z}_0$, respectively. Both matrices are symmetric and, assuming that $\mathbf{z}$ can be constructed from the data structure in (2.2), $\mathbf{M}_{pp}$ and $\mathbf{M}_{\tau\tau}$ being positive definite and positive semi-definite, respectively, if $K \gg n_z$. Moreover, $\|\mathbf{t}\|^2 = \mathbf{p}^T \mathbf{Z}_0^T \mathbf{Z}_0 \mathbf{p}$ is a scaled variance measure and $\|\boldsymbol{\pi}\|^2 = \boldsymbol{\tau}^T \mathbf{Z}_0 \mathbf{Z}_0^T \boldsymbol{\tau}$ is a sum of squares measure for the column and row space of $\mathbf{Z}_0$, respectively.

The solutions to the constraint objective functions in (9.6) are given by

$$\frac{\partial J_p}{\partial \mathbf{p}} - \mu_p \frac{\partial C_p}{\partial \mathbf{p}} = \mathbf{0} \tag{9.8a}$$

$$\frac{\partial J_\tau}{\partial \boldsymbol{\tau}} - \mu_\tau \frac{\partial C_\tau}{\partial \boldsymbol{\tau}} = \mathbf{0}, \tag{9.8b}$$

where μ_p and μ_τ are Lagrangian multipliers. Substituting the definition of J_p, J_τ, C_p and C_τ in (9.6) and carrying out the above relationships gives rise to

$$\widehat{\mathbf{p}} = \arg\max_{\mathbf{p}} \mathbf{p}^T \mathbf{M}_{pp} \mathbf{p} - \mu_p \left(\mathbf{p}^T \mathbf{p} - 1\right) \Rightarrow 2\mathbf{M}_{pp}\widehat{\mathbf{p}} - 2\mu_p \widehat{\mathbf{p}} = \mathbf{0} \tag{9.9a}$$

$$\widehat{\boldsymbol{\tau}} = \arg\max_{\boldsymbol{\tau}} \boldsymbol{\tau}^T \mathbf{M}_{\tau\tau} \boldsymbol{\tau} - \mu_\tau \left(\boldsymbol{\tau}^T \boldsymbol{\tau} - 1\right) \Rightarrow 2\mathbf{M}_{\tau\tau}\widehat{\boldsymbol{\tau}} - 2\mu_\tau \widehat{\boldsymbol{\tau}} = \mathbf{0}. \tag{9.9b}$$

Equation (9.9) implies that

$$\mu_p = \widehat{\mathbf{p}}^T \left.\frac{\partial J_p}{\partial \mathbf{p}}\right|_{\mathbf{p}=\widehat{\mathbf{p}}} = J_p \tag{9.10a}$$

$$\mu_\tau = \widehat{\boldsymbol{\tau}} \left.\frac{\partial J_\tau}{\partial \boldsymbol{\tau}}\right|_{\boldsymbol{\tau}=\widehat{\boldsymbol{\tau}}} = J_\tau \tag{9.10b}$$

and hence,

$$\left[\mathbf{M}_{pp} - \mu_p \mathbf{I}\right] \widehat{\mathbf{p}} = \mathbf{0} \tag{9.11a}$$

$$\left[\mathbf{M}_{\tau\tau} - \mu_\tau \mathbf{I}\right] \widehat{\boldsymbol{\tau}} = \mathbf{0}. \tag{9.11b}$$

Lemma 9.1.1 *The Lagrangian multipliers μ_p and μ_τ are larger than zero, identical and are equal to the estimated variance of the jth largest principal component, $1 \le j \le n$, if $K \gg n_z$.*

Proof. That μ_p and μ_τ are larger than zero follows from the objective functions in (9.6)

$$J_p = \mu_p = \widehat{\mathbf{t}^T \mathbf{t}} \ge 0 \tag{9.12a}$$

$$J_\tau = \mu_\tau = \widehat{\boldsymbol{\pi}}^T \widehat{\boldsymbol{\pi}} \ge 0. \tag{9.12b}$$

Given that $\|\widehat{\mathbf{p}}\| = \|\widehat{\boldsymbol{\tau}}\| = 1$ and $\mathbf{Z}_0 \ne \mathbf{0}$, it follows that $\widehat{\mathbf{t}^T \mathbf{t}} = \widehat{\mathbf{p}}^T \mathbf{M}_{pp} \widehat{\mathbf{p}} > 0$ and $\widehat{\boldsymbol{\pi}}^T \widehat{\boldsymbol{\pi}} = \widehat{\boldsymbol{\tau}}^T \mathbf{M}_{\tau\tau} \widehat{\boldsymbol{\tau}} > 0$. To proof that $J_p = J_\tau$ and hence, $\mu_p = \mu_\tau$, consider a singular value decomposition[1] of $\mathbf{Z}_0 = \mathcal{U}\mathcal{S}\mathcal{V}^T$, where the matrices $\mathcal{U}$ and $\mathcal{V}$

[1] Note that the column vectors of $\mathcal{U}$ and $\mathcal{V}$ are orthonormal and are the eigenvectors of $\mathbf{M}_{pp}$ and $\mathbf{M}_{\tau\tau}$, respectively (Golub and van Loan 1996).

contain the left and right singular vectors, respectively, and $\mathcal{S}$ is a diagonal matrix storing the singular values in descending order, which yields

$$\mathbf{M}_{pp} = \mathcal{V}\mathcal{S}^2\mathcal{V}^T = \mathcal{V}\boldsymbol{L}\mathcal{V}^T \tag{9.13a}$$

$$\mathbf{M}_{\tau\tau} = \mathcal{U}\mathcal{S}^2\mathcal{U}^T = \mathcal{U}\boldsymbol{L}\mathcal{U}^T, \tag{9.13b}$$

where $\boldsymbol{L} = \mathcal{S}^2$. Since the matrix expressions $\mathcal{V}\boldsymbol{L}\mathcal{V}^T$ and $\mathcal{U}\boldsymbol{L}\mathcal{U}^T$ represent the eigendecomposition of $\mathbf{M}_{pp}$ and $\mathbf{M}_{\tau\tau}$, respectively, it follows that the eigenvalues of $\mathbf{M}_{pp}$ and $\mathbf{M}_{\tau\tau}$ are identical and equal to the diagonal elements of $\boldsymbol{\Lambda}$. Using

- the fact that $\mu_p = \widehat{\mathbf{p}}^T\mathbf{M}_{pp}\widehat{\mathbf{p}}$ and $\mu_\tau = \widehat{\boldsymbol{\tau}}^T\mathbf{M}_{\tau\tau}\widehat{\boldsymbol{\tau}}$, which results from (9.9) to (9.11);
- the fact that $\widehat{\mathbf{p}}$ and $\widehat{\boldsymbol{\tau}}$ are the dominant eigenvectors[2] of $\mathbf{M}_{pp}$ and $\mathbf{M}_{\tau\tau}$, respectively; and
- a reintroduction of the subscript j,

it follows that

$$\begin{aligned}
\mu_{p_j} &= \widehat{\mathbf{p}}_j^T\mathcal{V}\boldsymbol{L}\mathcal{V}^T\widehat{\mathbf{p}}_j \\
\mu_{p_j} &= \widehat{\mathbf{p}}_j^T\begin{bmatrix}\widehat{\mathbf{p}}_1 & \cdots & \widehat{\mathbf{p}}_j & \cdots & \widehat{\mathbf{p}}_n\end{bmatrix}\begin{bmatrix}\mu_1 & \cdots & 0 & \cdots & 0 \\ \vdots & \ddots & & & \vdots \\ 0 & \cdots & \mu_j & \cdots & 0 \\ \vdots & & & \ddots & \vdots \\ 0 & \cdots & 0 & \cdots & \mu_n\end{bmatrix}\begin{bmatrix}\widehat{\mathbf{p}}_1^T \\ \vdots \\ \widehat{\mathbf{p}}_j^T \\ \vdots \\ \widehat{\mathbf{p}}_n^T\end{bmatrix}\widehat{\mathbf{p}}_j \\
\mu_{\tau_j} &= \widehat{\boldsymbol{\tau}}_j^T\mathcal{U}\boldsymbol{L}\mathcal{U}^T\widehat{\boldsymbol{\tau}}_j \\
\mu_{\tau_j} &= \widehat{\boldsymbol{\tau}}_j^T\begin{bmatrix}\widehat{\boldsymbol{\tau}}_1 & \cdots & \widehat{\boldsymbol{\tau}}_j & \cdots & \widehat{\boldsymbol{\tau}}_n\end{bmatrix}\begin{bmatrix}\mu_1 & \cdots & 0 & \cdots & 0 \\ \vdots & \ddots & & & \vdots \\ 0 & \cdots & \mu_j & \cdots & 0 \\ \vdots & & & \ddots & \vdots \\ 0 & \cdots & 0 & \cdots & \mu_n\end{bmatrix}\begin{bmatrix}\widehat{\boldsymbol{\tau}}_1^T \\ \vdots \\ \widehat{\boldsymbol{\tau}}_j^T \\ \vdots \\ \widehat{\boldsymbol{\tau}}_n^T\end{bmatrix}\widehat{\boldsymbol{\tau}}_j \\
\mu_{p_j} &= \mu_{\tau_j} = \mu_j,
\end{aligned} \tag{9.14}$$

since the eigenvectors are mutually orthonormal.

Finally, to prove that μ_j is equal to the contribution of the jth component matrix to the data matrix $\mathbf{Z}_0$ relies on post- and pre-multiplying the relationships in (9.9) by $\widehat{\boldsymbol{\tau}}_j^T$ and $\widehat{\mathbf{p}}_j^T$, which yields

$$\widehat{\boldsymbol{\tau}}_j\widehat{\mathbf{p}}_j^T\mathbf{M}_{pp} = \mu_j\widehat{\boldsymbol{\tau}}_j\widehat{\mathbf{p}}_j^T \qquad \mathbf{M}_{\tau\tau}\widehat{\boldsymbol{\tau}}_j\widehat{\mathbf{p}}_j^T = \mu_j\widehat{\boldsymbol{\tau}}_j\widehat{\mathbf{p}}_j^T. \tag{9.15}$$

[2] A dominant eigenvector is the eigenvector that corresponds to the largest eigenvalue of a given squared matrix.

Equation (9.15) can be further simplified by defining $\mathbf{Z}_0\widehat{\mathbf{p}}_j/\sqrt{\mu_j} = \widehat{\boldsymbol{\tau}}_j$, $\mathbf{Z}_0^T\widehat{\boldsymbol{\tau}}_j/\sqrt{\mu_j} = \widehat{\mathbf{p}}_j$, $\widehat{\mathbf{p}}_j = \widehat{\boldsymbol{\pi}}_j/\sqrt{\mu_j}$ and $\widehat{\boldsymbol{\tau}}_j = \widehat{\mathbf{t}}_j/\sqrt{\mu_j}$, since

$$\widehat{\boldsymbol{\tau}}_j\widehat{\boldsymbol{\tau}}_j^T\mathbf{Z}_0 = \mathbf{Z}_0\widehat{\mathbf{p}}_j\widehat{\mathbf{p}}_j^T = \widehat{\boldsymbol{\tau}}_j\sqrt{\mu_j}\widehat{\mathbf{p}}_j^T = \widehat{\mathbf{t}}_j\widehat{\mathbf{p}}_j^T = \widehat{\boldsymbol{\tau}}_j\widehat{\boldsymbol{\pi}}_j^T. \tag{9.16}$$

The contribution of the jth rank-one component matrix to the data matrix $\mathbf{Z}_0$ is equal to the squared Frobenius norm of $\widehat{\boldsymbol{\tau}}_j\sqrt{\mu_j}\widehat{\mathbf{p}}_j^T$

$$\left\|\widehat{\boldsymbol{\tau}}_j\sqrt{\mu_j}\widehat{\mathbf{p}}_j^T\right\|^2 = \mu_j\sum_{i=1}^{n_z}\sum_{k=1}^{K}\widehat{\tau}_{kj}^2\widehat{p}_{ij}^2 = \mu_j\underbrace{\sum_{i=1}^{n_z}\widehat{p}_{ij}^2}_{=1}\underbrace{\sum_{k=1}^{K}\widehat{\tau}_{kj}^2}_{=1} = \mu_j. \tag{9.17}$$

Theorem 9.1.2 *If the covariance matrix*

$$\mathbf{S}_{z_0z_0} = \lim_{K\to\infty}\frac{1}{K-1}\mathbf{M}_{pp} = \lim_{K\to\infty}\frac{1}{K-1}\sum_{k=1}^{K}\left(\mathbf{z}(k)-\widehat{\bar{\mathbf{z}}}\right)\left(\mathbf{z}(k)-\widehat{\bar{\mathbf{z}}}\right)^T, \tag{9.18}$$

where

$$\widehat{\bar{\mathbf{z}}} = \lim_{K\to\infty}\frac{1}{K}\sum_{k=1}^{K}\mathbf{z}(k) \tag{9.19}$$

is used in the above analysis, the variance of the jth t-score variable is equal to the Lagrangian multiplier λ_j, that is, the jth largest largest eigenvalue of $\mathbf{S}_{z_0z_0}$.

Proof. The variance of the score variable t_j, $E\{t_j^2\}$ is given by

$$E\left\{t_j^2\right\} = E\left\{\mathbf{p}_j^T\mathbf{z}_0\mathbf{z}_0^T\mathbf{p}_j\right\} = \mathbf{p}_j^T\mathbf{S}_{z_0z_0}\mathbf{p}_j = \lambda_j. \tag{9.20}$$

On the basis of the objective functions in (9.6) and the constraints in (9.7), the first rank-one component matrix, $\widehat{\mathbf{t}}_1\widehat{\mathbf{p}}_1^T$, has the most significant contribution to $\mathbf{Z}_0$. This follows from $\widehat{\mathbf{p}}_1$, which, according to (9.9a) and (9.9b), is the most dominant eigenvector of $\mathbf{M}_{pp}$ and the fact that $\widehat{\mathbf{t}}_1 = \mathbf{Z}_0\widehat{\mathbf{p}}_1$, which follows from (9.16). It should also be noted that $\mu_j = (K-1)\lambda_j$ when the estimate of $\mathbf{S}_{z_0z_0}$ is or used.

After subtracting or *deflating* the rank-one matrix $\widehat{\mathbf{t}}_1\widehat{\mathbf{p}}_1^T$ from $\mathbf{Z}_0$, that is $\mathbf{Z}^{(2)} = \mathbf{Z}^{(1)} - \widehat{\mathbf{t}}_1\widehat{\mathbf{p}}_1^T$ with $\mathbf{Z}^{(1)} = \mathbf{Z}_0$, the rank-one matrix $\widehat{\mathbf{t}}_2\widehat{\mathbf{p}}_2^T$ has the most significant contribution to $\mathbf{Z}^{(2)}$. Moreover, $\widehat{\mathbf{t}}_j = \mathbf{Z}^{(j)}\widehat{\mathbf{p}}_j$ and $\widehat{\mathbf{p}}_j$ is the most dominant eigenvector of $\mathbf{M}_{pp}^{(j)} = \mathbf{Z}^{(j)^T}\mathbf{Z}^{(j)}$,

$$\mathbf{M}_{pp}^{(j+1)} = \mathbf{M}_{pp}^{(j)} - \widehat{\mathbf{p}}_j\widehat{\mu}_j\widehat{\mathbf{p}}_j^T. \tag{9.21}$$

Utilizing the deflation procedure, the first n eigenvalues and eigenvectors of $\widehat{\mathbf{S}}_{z_0z_0} = 1/(K-1)\mathbf{M}_{pp}^{(j)}$ can be obtained using the Power method (Geladi and

Kowalski 1986; Golub and van Loan 1996). Alternatively, a PCA model can also be determined by a singular value or eigendecomposition of $\widehat{\mathbf{S}}_{z_0z_0}$, which is computationally more economic (Wold 1978). Based on the covariance matrix, it should also be noted that the objective function for determining the ith loading vector can be formulated as follows

$$\mathbf{p}_i = \arg\max_{\mathbf{p}} \mathbf{p}^T \mathbf{S}_{z_0z_0} \mathbf{p} - \lambda_i \left(\mathbf{p}^T \mathbf{p} - 1 \right). \tag{9.22}$$

9.2 Summary of the PCA algorithm

The above analysis showed that PCA determines linear combinations of the variable set $\mathbf{z}_0$ such that the variance for each linear combination is maximized. The variance contribution from each set of linear combinations is then subtracted from $\mathbf{z}_0$ before determining the next set. This gives rise to a total of n_z combinations that are referred to as *score variables*. Combining the parameters for each of the linear combinations into a vector yields *loading vectors* that are of unit length. If the covariance matrix of the data vector $\mathbf{z}_0$, i.e. $\mathbf{S}_{z_0z_0}$ is available, and the data vector follows the data structure $\mathbf{z}_0 = \mathbf{\Xi}\mathbf{s} + \mathbf{g}$, where $\mathbf{s} \sim \mathcal{N}\left\{\mathbf{0}, \mathbf{S}_{ss}\right\}$, $\mathbf{g} \sim \mathcal{N}\left\{\mathbf{0}, \sigma_g^2 \mathbf{I}\right\}$, $\mathbf{s} \in \mathbb{R}^n$, $\mathbf{g} \in \mathbb{R}^{n_z}$ and $E\left\{\left(\boldsymbol{\xi}_j^T \mathbf{s}\right)\right\} > \sigma_g^2$ for all $1 \leq j \leq n_z$, the following holds true:

Table 9.1 Iterative PCA algorithm.

Step	Description	Equation
1	Initiate iteration	$j = 1$
2	Obtain initial matrix	$\widetilde{\mathbf{M}}_{pp}^{(1)} = \mathbf{S}_{z_0z_0}$
3	Set-up initial loading vector	${}_0\mathbf{p}_j = \widetilde{\mathbf{M}}_{pp}^{(j)}(:,1) / \left\| \widetilde{\mathbf{M}}_{pp}^{(j)}(:,1) \right\|$
4	Calculate matrix-vector product	$\widetilde{\mathbf{p}}_j = \widetilde{\mathbf{M}}_{pp}^{(j)} \left({}_0\mathbf{p}_j \right)$
5	Compute eigenvalue	$\lambda_j = \widetilde{\mathbf{p}}_j^T \widetilde{\mathbf{p}}_j$
6	Scale eigenvector	${}_1\mathbf{p}_j = \mathbf{p}_j / \sqrt{\lambda_j}$
7	Check for convergence	If $\left\| {}_1\mathbf{p}_j - {}_0\mathbf{p}_j \right\| > \varepsilon$, set ${}_0\mathbf{p}_j = {}_1\mathbf{p}_j$ and go to Step 4 else set $\mathbf{p}_j = {}_1\mathbf{p}_j$ and go to Step 8
8	Deflate matrix	$\widetilde{\mathbf{M}}_{pp}^{(j+1)} = \widetilde{\mathbf{M}}_{pp}^{(j)} - \lambda_j \mathbf{p}_k \mathbf{p}_j^T$
9	Check for dimension	If $j < n_z$ set $j = j + 1$ and go to Step 3 else terminate iteration procedure

- the largest n eigenvalues of $\mathbf{S}_{z_0z_0}$ are larger than σ_g^2 and represent the variance of the dominant score variables, that is, the *retained score variables*;
- the remaining $n_z - n$ eigenvalues are equal to σ_g^2 and describe the variance of the residuals, that is, the *discarded score variables*; and
- the first n eigenvectors allow one to extract linear combinations that describe the source signals superimposed by some error vector.

The above properties are relevant and important for process monitoring and are covered in more detail in Subsections 1.2.3, 1.2.4, 3.1.1, 3.1.2 and 6.1.1. In essence, the eigendecomposition of $\mathbf{S}_{z_0z_0}$ contains all relevant information to establish a process monitoring model. Table 9.1 summarizes the steps for iteratively computing the eigendecomposition of $\widehat{\mathbf{S}}_{z_0z_0}$ using the deflation procedure in (9.21). It should be noted, however, that the eigenvalues and eigenvectors of $\mathbf{S}_{z_0z_0}$ can also be obtained simultaneously without a deflation procedure, which is computationally and numerically favorable over the iterative computation. The algorithm in Table 9.1 is based on the covariance matrix $\mathbf{S}_{z_0z_0} = \lim_{K\to\infty} = {}^1/_{(K-1)}\mathbf{M}_{pp} = \widetilde{\mathbf{M}}_{pp}^{(1)}$.

9.3 Properties of a PCA model

The PCA algorithm has the following properties:

1. Each rank-one component matrix produces a maximum variance contribution to the data matrix $\mathbf{Z}_0$ in successive order, that is, $\widehat{\mathbf{t}}_1\widehat{\mathbf{p}}_1^T$ produces the largest, $\widehat{\mathbf{t}}_2\widehat{\mathbf{p}}_2^T$ the second largest and so on.
2. The t-score vectors are mutually orthogonal.
3. The p-loading vectors are mutually orthonormal.
4. Under the assumption that the source variables are statistically independent, the t-score variables follow asymptotically a Gaussian distribution.
5. It is irrelevant whether the score vector $\widehat{\mathbf{t}}_j$ is computed as the matrix vector product of the original data matrix, $\mathbf{Z}_0$, or the deflated data matrix, $\mathbf{Z}^{(j)}$, and the loading vector $\widehat{\mathbf{p}}_j$.
6. If that the rank of $\mathbf{Z}_0$ is $\bar{n}_z \le n_z$, $\mathbf{Z}_0$ is completely exhausted after $\bar{n}_z$ deflation procedures have been carried out.
7. The data covariance matrix $\mathbf{S}_{z_0z_0}$ can be reconstructed by a sum of $\bar{n}_z$ scaled loading vectors $\mathbf{p}_j$.

These properties are now mathematically formulated and proven. Apart from Properties 4 and 7, the proofs rely on the data matrix $\mathbf{Z}_0$ and estimates of the score and loading vectors. Properties 4 and 7 are based on a known, that is, not estimated, data structure $\mathbf{z}_0 = \Xi\mathbf{s} + \mathbf{g}$ and covariance matrix $\mathbf{S}_{z_0z_0}$, respectively.

Property 9.3.1 – Contribution of Score Vectors to $\mathbf{Z}_0$. The contribution of each score vector to the recorded data matrix is expressed by the rank-one component matrices $\widehat{\mathbf{t}}_j\widehat{\mathbf{p}}_j^T$. Theorem 9.3.1 formulates the contribution of each of these component matrices to recorded data matrix.

Theorem 9.3.1 *In subsequent order, the contribution of each rank-one component matrix, $\widehat{\mathbf{t}}_j\widehat{\mathbf{p}}_j^T$, is maximized using the PCA algorithm.*

Proof. Knowing that the process variables are mean-centered, the sum of variances of each process variable, $\sum_{i=1}^{n_z}\widehat{\sigma}_i^2$, is equal to the squared Frobenius norm of $\mathbf{Z}_0$ up to a scaling factor. Moreover, the eigendecomposition of $\widehat{\mathbf{S}}_{z_0z_0}$ describes, in fact, a rotation of the n_z dimensional Euclidian base vectors to be the eigenvectors of $\widehat{\mathbf{S}}_{z_0z_0}$. Under the assumption that $\widehat{\mathbf{S}}_{z_0z_0}$ has full rank n_z, this implies that (9.3) can be rewritten as follows

$$\mathbf{Z}_0 = \sum_{i=1}^{n} \widehat{\mathbf{t}}_i\widehat{\mathbf{p}}_i^T + \sum_{j=n+1}^{n_z} \widehat{\mathbf{t}}_j\widehat{\mathbf{p}}_j^T = \sum_{i=1}^{n} \widehat{\mathbf{t}}_i\widehat{\mathbf{p}}_i^T + \mathbf{G}. \tag{9.23}$$

The next step involves working out the Frobenius norm of (9.23), which gives rise to

$$\left\|\mathbf{Z}_0\right\|^2 = \left\| \sum_{i=1}^{n_z} \widehat{\mathbf{t}}_i\widehat{\mathbf{p}}_i^T \right\|^2. \tag{9.24}$$

Simplifying (9.24) by determining the sum of the squared elements of $\mathbf{Z}_0$ yields

$$\left\|\mathbf{Z}_0\right\|^2 = \left\| \begin{bmatrix} \sum_{i=1}^{n_z} \widehat{t}_i(1)\widehat{p}_{i1} & \sum_{i=1}^{n_z} \widehat{t}_i(1)\widehat{p}_{i2} & \cdots & \sum_{i=1}^{n_z} \widehat{t}_i(1)\widehat{p}_{in_z} \\ \sum_{i=1}^{n_z} \widehat{t}_i(2)\widehat{p}_{i1} & \sum_{i=1}^{n_z} \widehat{t}_i(2)\widehat{p}_{i2} & \cdots & \sum_{i=1}^{n_z} \widehat{t}_i(2)\widehat{p}_{in_z} \\ \vdots & \vdots & \ddots & \vdots \\ \sum_{i=1}^{n_z} \widehat{t}_i(K)\widehat{p}_{i1} & \sum_{i=1}^{n_z} \widehat{t}_i(K)\widehat{p}_{i2} & \cdots & \sum_{i=1}^{n_z} \widehat{t}_i(K)\widehat{p}_{in_z} \end{bmatrix} \right\|^2$$

$$\left\|\mathbf{Z}_0\right\|^2 = \sum_{k=1}^{K}\sum_{j=1}^{n_z}\left(\sum_{i=1}^{n_z} \widehat{t}_j(k)\widehat{p}_{ij}\right)^2$$

$$\left\|\mathbf{Z}_0\right\|^2 = \sum_{k=1}^{K}\sum_{j=1}^{n_z}\left(\sum_{i=1}^{n_z} \widehat{t}_j^2(k)\widehat{p}_{ij}^2 + 2\sum_{i=1}^{n_z-1}\sum_{m=i+1}^{n_z} \widehat{t}_i(k)\widehat{t}_m(k)\widehat{p}_{ij}\widehat{p}_{mj}\right)$$

$$\left\|\mathbf{Z}_0\right\|^2 = \sum_{k=1}^{K}\sum_{j=1}^{n_z}\widehat{t}_j^2(k)\underbrace{\sum_{i=1}^{n_z}\widehat{p}_{ij}^2}_{\|\mathbf{p}_k\|^2=1} + 2\sum_{k=1}^{K}\sum_{i=1}^{n_z-1}\sum_{m=k+1}^{n_z}\widehat{t}_i(k)\widehat{t}_m(k)\underbrace{\sum_{j=1}^{n_z}\widehat{p}_{ij}\widehat{p}_{mj}}_{\widehat{\mathbf{p}}_i^T\widehat{\mathbf{p}}_m=0 \text{ for all } m\neq j}$$

$$\|\mathbf{Z}_0\|^2 = \sum_{i=1}^{n_z} \widehat{\mathbf{t}}_i^T \widehat{\mathbf{t}}_i = \sum_{i=1}^{n_z} \mu_i \underbrace{\widehat{\boldsymbol{\tau}}_i^T \widehat{\boldsymbol{\tau}}_i}_{=1} \tag{9.25}$$

and hence

$$\boxed{\|\mathbf{Z}_0\|^2 = \sum_{i=1}^{n} \mu_i + \|\mathbf{G}\|^2}. \tag{9.26}$$

The Lagrangian multiplier μ_j, however, represents the maximum of the objective functions J_{p_j} and J_{τ_j}. Hence, the variance contribution of $\widetilde{\mathbf{t}}_j \widehat{\mathbf{p}}_j^T$ to $\mathbf{Z}_0$ is the jth largest possible. That $\sum_{j=1}^{n_z} \widehat{p}_{ij} \widehat{p}_{mj} = 0$ for all $i \neq j$ follows from the fact that the eigenvectors of $\mathbf{M}_{pp}$ are mutually orthonormal and hence, $\widehat{\mathbf{P}}^T \widehat{\mathbf{P}} = \widehat{\mathbf{P}}\widehat{\mathbf{P}}^T = \mathbf{I}$. Moreover, Theorem 9.1.2 outlines that the jth largest eigenvalue of the data covariance matrix $\widehat{\mathbf{S}}_{z_0 z_0}$ is equal to the variance of the jth score variable $\widehat{t}_j = \widehat{\mathbf{p}}_j^T \mathbf{z}_0$.

Property 9.3.2 – Orthogonality of the t-Score Vectors. Next, examining the deflation procedure allows showing that the t-score vectors are mutually orthogonal. It should be noted, however, that the orthogonality properties of these vectors also follow from the t-score vectors are dominant eigenvectors of the symmetric and positive semi-definite matrix $\mathbf{M}_{\tau\tau}$, respectively. Theorem 9.3.2 formulates the orthogonality of the t-score vectors.

Theorem 9.3.2 *The deflation procedure produces orthogonal t-score vectors, that is,* $\widehat{\boldsymbol{\tau}}_i^T \widehat{\boldsymbol{\tau}}_j = \delta_{ij}$, $\widehat{\boldsymbol{\tau}}_i = {}^{\mathbf{t}_i}/_{\sqrt{\lambda_i}}$, $\|\widehat{\boldsymbol{\tau}}_i\|^2 = 1$.

Proof. The expression $\widehat{\boldsymbol{\tau}}_i^T \widehat{\boldsymbol{\tau}}_j$ can alternatively be written as ${}^1/_{\sqrt{\lambda_i \lambda_j}} \widehat{\mathbf{p}}_i^T \mathbf{Z}^{(i)^T} \widehat{\mathbf{t}}_j$, which follows from (9.16). Without restriction of generality, assuming that $i > j$ the deflation procedure to reconstruct $\mathbf{Z}^{(i)}$ gives rise to[3]

$$\begin{aligned}
\mathbf{Z}^{(i)} &= \mathbf{Z}^{(i-1)} - \widehat{\mathbf{t}}_{i-1}\widehat{\mathbf{p}}_{i-1}^T = \mathbf{Z}^{(i-1)}\left[\mathbf{I} - \widehat{\mathbf{p}}_{i-1}\widehat{\mathbf{p}}_{i-1}^T\right] \\
&\vdots \\
\mathbf{Z}^{(i)} &= \mathbf{Z}^{(j+1)}\left[\mathbf{I} - \widehat{\mathbf{p}}_{j+1}\widehat{\mathbf{p}}_{j+1}^T\right]\cdots\left[\mathbf{I} - \widehat{\mathbf{p}}_{i-1}\widehat{\mathbf{p}}_{i-1}^T\right] \\
\mathbf{Z}^{(i)} &= \left[\mathbf{I} - \frac{\widehat{\mathbf{t}}_j \widehat{\mathbf{t}}_j^T}{\widehat{\mathbf{t}}_j^T \widehat{\mathbf{t}}_j}\right]\mathbf{Z}^{(j)}\left[\mathbf{I} - \widehat{\mathbf{p}}_{j+1}\widehat{\mathbf{p}}_{j+1}^T\right]\cdots\left[\mathbf{I} - \widehat{\mathbf{p}}_{i-1}\widehat{\mathbf{p}}_{i-1}^T\right].
\end{aligned} \tag{9.27}$$

Substituting (9.27) into ${}^1/_{\sqrt{\lambda_i \lambda_j}} \widehat{\mathbf{p}}_i^T \mathbf{Z}^{(i)^T} \widehat{\mathbf{t}}_j$ produces

$$\frac{1}{\sqrt{\lambda_i \lambda_j}} \widehat{\mathbf{p}}_i^T \left[\mathbf{I} - \widehat{\mathbf{p}}_{i-1}\widehat{\mathbf{p}}_{i-1}^T\right]\cdots\left[\mathbf{I} - \widehat{\mathbf{p}}_{j+1}\widehat{\mathbf{p}}_{j+1}^T\right]\mathbf{Z}^{(j)^T}\left[\mathbf{I} - \frac{\widehat{\mathbf{t}}_j \widehat{\mathbf{t}}_j^T}{\widehat{\mathbf{t}}_j^T \widehat{\mathbf{t}}_j}\right]\widehat{\mathbf{t}}_j \tag{9.28}$$

[3] Note that the relationship below takes advantage of the fact that $\widehat{\mathbf{p}}_i^T = \widehat{\mathbf{t}}_i^T \mathbf{Z}_i / \widehat{\mathbf{t}}_i^T \widehat{\mathbf{t}}_i$, which follows from (9.15) and (9.16)

and hence

$$\frac{1}{\sqrt{\lambda_i\lambda_j}}\widehat{\mathbf{p}}_i\left[\mathbf{I}-\widehat{\mathbf{p}}_{i-1}\widehat{\mathbf{p}}_{i-1}^T\right]\cdots\left[\mathbf{I}-\widehat{\mathbf{p}}_{j+1}\widehat{\mathbf{p}}_{j+1}^T\right]\mathbf{Z}^{(j)^T}\left(\widehat{\mathbf{t}}_j-\frac{\widehat{\mathbf{t}}_j\widehat{\mathbf{t}}_j^T\widehat{\mathbf{t}}_j}{\widehat{\mathbf{t}}_j^T\widehat{\mathbf{t}}_j}\right)=$$
$$\frac{1}{\sqrt{\lambda_i\lambda_j}}\widehat{\mathbf{p}}_i^T\left[\mathbf{I}-\widehat{\mathbf{p}}_{i-1}\widehat{\mathbf{p}}_{i-1}^T\right]\cdots\left[\mathbf{I}-\widehat{\mathbf{p}}_{j+1}\widehat{\mathbf{p}}_{j+1}^T\right]\mathbf{Z}^{(j)^T}\left(\widehat{\mathbf{t}}_j-\widehat{\mathbf{t}}_j\right)=0. \tag{9.29}$$

Consequently,

$$\boxed{\widehat{\boldsymbol{\tau}}_i^T\widehat{\boldsymbol{\tau}}_j=\delta_{ij}\ \text{and}\ \widehat{\mathbf{t}}_i^T\widehat{\mathbf{t}}_j=\lambda_i\delta_{ij}}.$$

Property 9.3.3 – Orthogonality of the p-Loading Vectors. The deflation procedure also allows showing mutual orthogonality of the p-loading vectors, which is discussed in Theorem 9.3.3. It is important to note that the orthogonality of the p-loading vectors also follows from the fact that they are eigenvectors of the symmetric and positive definite matrix $\mathbf{M}_{pp}$.

Theorem 9.3.3 *The deflation procedure produces orthonormal p-loading vectors, i.e.* $\widehat{\mathbf{p}}_i^T\widehat{\mathbf{p}}_j=\delta_{ij}$.

Proof. The proof commences by rewriting $\widehat{\mathbf{p}}_i^T\widehat{\mathbf{p}}_j={}^{\widehat{\mathbf{t}}_i^T\mathbf{Z}^{(i)}}/_{\lambda_i}\widehat{\mathbf{p}}_j$. Reformulating the deflation procedure for $\mathbf{Z}^{(i)}$

$$\begin{aligned}\mathbf{Z}^{(i)}&=\left[\mathbf{I}-\frac{\widehat{\mathbf{t}}_{i-1}\widehat{\mathbf{t}}_{i-1}^T}{\widehat{\mathbf{t}}_{i-1}^T\widehat{\mathbf{t}}_{i-1}}\right]\mathbf{Z}^{(i-1)}\\ &\vdots\\ \mathbf{Z}^{(i)}&=\left[\mathbf{I}-\frac{\widehat{\mathbf{t}}_{i-1}\widehat{\mathbf{t}}_{i-1}^T}{\widehat{\mathbf{t}}_{i-1}^T\widehat{\mathbf{t}}_{i-1}}\right]\cdots\left[\mathbf{I}-\frac{\widehat{\mathbf{t}}_{j+1}\widehat{\mathbf{t}}_{j+1}^T}{\widehat{\mathbf{t}}_{j+1}^T\widehat{\mathbf{t}}_{j+1}}\right]\mathbf{Z}^{(j+1)}\\ \mathbf{Z}^{(i)}&=\left[\mathbf{I}-\frac{\widehat{\mathbf{t}}_{i-1}\widehat{\mathbf{t}}_{i-1}^T}{\widehat{\mathbf{t}}_{i-1}^T\widehat{\mathbf{t}}_{i-1}}\right]\cdots\left[\mathbf{I}-\frac{\widehat{\mathbf{t}}_{j+1}\widehat{\mathbf{t}}_{j+1}^T}{\widehat{\mathbf{t}}_{j+1}^T\widehat{\mathbf{t}}_{j+1}}\right]\mathbf{Z}^{(j)}\left[\mathbf{I}-\widehat{\mathbf{p}}_j\widehat{\mathbf{p}}_j^T\right]\end{aligned} \tag{9.30}$$

and substituting this matrix expression into ${}^{\widehat{\mathbf{t}}_i^T\mathbf{Z}^{(i)}}/_{\lambda_i}\widehat{\mathbf{p}}_j$ yields

$$\begin{aligned}\widehat{\mathbf{p}}_i^T\widehat{\mathbf{p}}_j&=\frac{\widehat{\mathbf{t}}_i^T}{\lambda_i}\left[\mathbf{I}-\frac{\widehat{\mathbf{t}}_{i-1}\widehat{\mathbf{t}}_{i-1}^T}{\widehat{\mathbf{t}}_{i-1}^T\widehat{\mathbf{t}}_{i-1}}\right]\cdots\left[\mathbf{I}-\frac{\widehat{\mathbf{t}}_{j+1}\widehat{\mathbf{t}}_{j+1}^T}{\widehat{\mathbf{t}}_{j+1}^T\widehat{\mathbf{t}}_{j+1}}\right]\mathbf{Z}^{(j)}\left[\mathbf{I}-\widehat{\mathbf{p}}_j\widehat{\mathbf{p}}_j^T\right]\widehat{\mathbf{p}}_j\\ \widehat{\mathbf{p}}_i^T\widehat{\mathbf{p}}_j&=\frac{\widehat{\mathbf{t}}_i^T}{\lambda_i}\left[\mathbf{I}-\frac{\widehat{\mathbf{t}}_{i-1}\widehat{\mathbf{t}}_{i-1}^T}{\widehat{\mathbf{t}}_{i-1}^T\widehat{\mathbf{t}}_{i-1}}\right]\cdots\left[\mathbf{I}-\frac{\widehat{\mathbf{t}}_{j+1}\widehat{\mathbf{t}}_{j+1}^T}{\widehat{\mathbf{t}}_{j+1}^T\widehat{\mathbf{t}}_{j+1}}\right]\mathbf{Z}^{(j)}\left(\widehat{\mathbf{p}}_j-\widehat{\mathbf{p}}_j\right)=0.\end{aligned} \tag{9.31}$$

Thus,

$$\boxed{\widehat{\mathbf{p}}_i^T\widehat{\mathbf{p}}_j=\delta_{ij}}.$$

Property 9.3.4 – Asymptotic Distribution of t-Score Variables. It is expected that for a large numbers of source variables the score variables are approximately Gaussian distributed. A more precise statement to this effect is given by the Liapounoff theorem. A proof of this theorem, in the context of the PCA score variables, relies on the data structure $\mathbf{z}_0 = \Xi\mathbf{s} + \mathbf{g}$ for the error covariance matrix $\mathbf{S}_{\mathbf{gg}} = \sigma_{\mathbf{g}}^2\mathbf{I}$[4]. This data structure guarantees that the covariance matrix $\mathbf{S}_{z_0z_0}$ has full rank n_z. Moreover, $\mathbf{S}_{z_0z_0}$ has the following eigendecomposition $\mathbf{P}\Lambda\mathbf{P}^T$, where the loading vectors $\mathbf{p}_1, \ldots, \mathbf{p}_{n_z}$ are stored as column vectors in $\mathbf{P}$ and the diagonal matrix Λ stores the variances of the score variables $\lambda_1, \ldots, \lambda_n$ and a total of $n_z - n$ times the variance of the error variables $\sigma_{\mathbf{g}}^2$.

Theorem 9.3.4 *For the data structure* $\mathbf{z}_0 = \Xi\mathbf{s} + \mathbf{g}$, $\mathbf{S}_{\mathbf{gg}} = \sigma_{\mathbf{g}}^2\mathbf{I}$, $\Xi = \begin{bmatrix}\boldsymbol{\xi}_1 & \cdots & \boldsymbol{\xi}_n\end{bmatrix}$ *and the assumption that the source variables* $\mathbf{s}$ *are independently but not identically distributed with unknown distribution functions, the score variables are approximately Gaussian distributed if* n *is sufficiently large and none of the individual source variables have a significant influence on the determination of the score variables, which can be expressed by the following condition (Liapounoff 1900, 1901)*

$$\lim_{n\to\infty} \frac{\varrho_{t_m}}{\sigma_{t_m}} \longrightarrow 0 \qquad m = 1, 2, \ldots, n \tag{9.32}$$

where

$$\begin{aligned} \varrho_{t_m}^3 &= \sum_{j=1}^{n} \varrho_{t_{m_j}}^3 \qquad \sigma_{t_m}^2 = \sum_{j=1}^{n} \sigma_{t_{m_j}}^2 = \lambda_m \\ \varrho_{t_{m_j}}^3 &= E\left\{\left|\left(\mathbf{p}_m^T\boldsymbol{\xi}_j\right)s_j + \tfrac{1}{n}\mathbf{p}_m^T\mathbf{g}\right|^3\right\} \\ \sigma_{t_{m_j}}^2 &= E\left\{\left(\left(\mathbf{p}_m^T\boldsymbol{\xi}_j\right)s_j + \tfrac{1}{n}\mathbf{p}_m^T\mathbf{g}\right)^2\right\} = \left(\mathbf{p}_m^T\boldsymbol{\xi}_j\right)^2\sigma_{s_j}^2 + \tfrac{1}{n}\sigma_{\mathbf{g}}^2, \end{aligned} \tag{9.33}$$

and $\sigma_{s_j}^2$ *is the variance of the* j*th source variable.*

Proof. In the context of the t-score variables for PCA, obtained as follows

$$t_m = \mathbf{p}_m^T\mathbf{z}_0 = \sum_{j=1}^{n}\left(\mathbf{p}_m^T\boldsymbol{\xi}_j\right)s_j + \mathbf{p}_m^T\mathbf{g} = \sum_{j=1}^{n}\mathbf{p}_m^T\left(\boldsymbol{\xi}_j s_j + \tfrac{1}{n}\mathbf{g}\right), \tag{9.34}$$

the Liapounoff theorem outlines that if n is sufficiently large and none of the sum elements $\mathbf{p}_m^T\left(\boldsymbol{\xi}_i s_i + \frac{1}{n}\mathbf{g}\right)$ is significantly larger than the remaining ones, the score variables are asymptotically Gaussian distributed as $n \to \infty$. A proof for the case where the distribution function of the n sum elements $\mathbf{p}_m^T\left(\boldsymbol{\xi}_j s_j + \frac{1}{n}\mathbf{g}\right)$ are i.i.d. is given in Subsection 8.7.1 under a simplified version of the Lindeberg-Lévy

[4] The assumption of $\mathbf{S}_{\mathbf{gg}} = \sigma_{\mathbf{g}}^2\mathbf{I}$ is imposed for convenience and does not represent a restriction of generality as the following steps are also applicable for $\mathbf{S}_{\mathbf{gg}} \neq \sigma_{\mathbf{g}}^2\mathbf{I}$.

theorem. The presented proof of Theorem 9.3.4 assumes that these elements are only i.d. and meet the condition outlined in (9.31) and (9.33).

It should be noted, however, that the CLT holds true for more general conditions than that postulated by Liapounoff (1900, 1901). A detailed discussion of more general conditions for the CLT may be found in Billingsley (1995); Bradley (2007); Cramér (1946); Durrett (1996); Feller (1936, 1937); Fisher (2011) and Gut (2005) for example. More precisely, the assumption of statistically independent source signals can be replaced by mixing conditions (Bradley 2007). An exhaustive treatment of mixing conditions, however, is beyond the scope of this book and the utilization of the Liapounoff theorem, imposing statistical independence upon the source signals, serves as an extension to the i.i.d. assumptions associated with the Lindeberg-Lévy theorem. Interested readers can resort to the cited literature above for a more general treatment of the CLT.

Assuming that the random variables $\mathbf{p}_m^T\left(\boldsymbol{\xi}_j s_j + \frac{1}{n}\mathbf{g}\right)$ have the distribution functions $F_1\left(s_1\right), F_2\left(s_2\right), \ldots, F_n\left(s_n\right)$, the proof of Theorem 9.3.4 relies on the characteristic function of $\tau_m = {}^{t_m}/\sqrt{\lambda_m}$, which is defined as

$$\gamma\left(c\right) = E\left\{e^{i\tau_m c}\right\} = \int_{-\infty}^{\infty} e^{\mathrm{i}\tau_m c} \mathrm{d}F\left(\tau_m\right) \tag{9.35}$$

where $\mathrm{i} = \sqrt{-1}$ and the scaled score variables τ_m has zero mean, unity variance and the distribution function $F\left(\mathbf{s}\right)$. As $n \to \infty$ the term $\frac{\mathbf{g}}{n} \to \mathbf{0}$ and can, consequently, be omitted. Next, substituting (9.34) into the definition of the characteristic function of τ_m yields

$$\gamma\left(c\right) = \int_{-\infty}^{\infty} e^{\mathrm{i}\left(\frac{1}{\sqrt{\lambda_m}}\sum_{j=1}^{n}\mathbf{p}_m^T\left(\boldsymbol{\xi}_j s_j\right)\right)c} \mathrm{d}F(\mathbf{s}). \tag{9.36}$$

The remaining elements of the sum, $\sum_{j=1}^{n}\mathbf{p}_m^T\boldsymbol{\xi}_j s_j$, are i.d. according to the assumptions made in Theorem 9.3.4. Hence, (9.36) can be rewritten as follows

$$\gamma\left(c\right) = \prod_{j=1}^{n}\int_{-\infty}^{\infty} e^{\mathrm{i}\left(\frac{1}{\sqrt{\lambda_m}}\mathbf{p}_m^T\boldsymbol{\xi}_j s_j\right)c} \mathrm{d}F_j(s_j). \tag{9.37}$$

If

$$\lim_{n\to\infty}\prod_{j=1}^{n}\gamma_j\left(c\right) \to e^{-c^2/2} \quad \text{for all } c \in \mathbb{R} \tag{9.38}$$

the term $\sum_{j=1}^{n}\mathbf{p}_m^T\left(\boldsymbol{\xi}_j s_j\right)$ is Gaussian distributed, as $e^{-c^2/2}$ is the characteristic function of a zero mean Gaussian distribution of unity variance. Approximating the expression for the product terms $\gamma_j\left(c\right)$ by a Taylor series expansion, developed around $c = 0$, shows that $\prod_{j=1}^{n} \gamma_j(c) \to e^{-c^2/2}$ as $n \to \infty$

$$\gamma_j(c) = \gamma_j(0) + \gamma_j'(0)c + \tfrac{1}{2}\gamma_j''(0)c^2 + R(\tau)c^3. \tag{9.39}$$

Here, $0 \le \tau \le c$ and $R(\tau) = {}^1\!/_6\gamma_j'''(\tau)$ is the remainder in Lagrangian form. The relationships ${}^{d^{\mathrm{m}}\gamma_j(c)}\!/_{dc^{\mathrm{m}}}$, are given by

$$\left.\frac{d^{\mathrm{m}}\gamma_j(c)}{dc^{\mathrm{m}}}\right|_{c=0} = \mathrm{i}^{\mathrm{m}} \int_{-\infty}^{\infty} \left(\frac{\mathbf{p}_m^T \boldsymbol{\xi}_j s_j}{\sqrt{\lambda_m}}\right)^{\mathrm{m}} \mathrm{d}F_j(s_j)$$
$$\left.\frac{d^{\mathrm{m}}\gamma_j(c)}{dc^{\mathrm{m}}}\right|_{c=0} = \mathrm{i}^{\mathrm{m}} E\left\{\left(\frac{\mathbf{p}_m^T \boldsymbol{\xi}_j s_j}{\sqrt{\lambda_m}}\right)^{\mathrm{m}}\right\} \tag{9.40}$$

and yields:

- $\mathrm{m} = 0 \quad : \quad \gamma_j(0) = \mathrm{i}^0 \int_{-\infty}^{\infty} \mathrm{d}F_j(s_j) = 1$;
- $\mathrm{m} = 1 \quad : \quad \gamma_j'(0) = \mathrm{i}E\left\{\frac{\mathbf{p}_m^T \boldsymbol{\xi}_j s_j}{\sqrt{\lambda_m}}\right\} = 0$; and
- $\mathrm{m} = 2 \quad : \quad \gamma_j''(0) = \mathrm{i}^2 E\left\{\left(\frac{\mathbf{p}_m^T \boldsymbol{\xi}_j s_j}{\sqrt{\lambda_m}}\right)^2\right\} = -\frac{\sigma_{t_{m_j}}^2}{\sigma_{t_m}^2}$

and for the Lagrangian remainder

$$\gamma_j'''(\tau) = \mathrm{i}^3 \int_{-\infty}^{\infty} \left(\frac{\mathbf{p}_m^T \boldsymbol{\xi}_j s_j}{\sqrt{\lambda_m}}\right)^3 e^{\mathrm{i}\frac{\mathbf{p}_m^T \boldsymbol{\xi}_j s_j}{\sqrt{\lambda_m}}\tau} \mathrm{d}F(s_j) = \vartheta E\left\{\left|\frac{\mathbf{p}_m^T \boldsymbol{\xi}_j s_j}{\sqrt{\lambda_m}}\right|^3\right\} = \vartheta \frac{\varrho_{t_{m_j}}^3}{\sigma_{t_m}^3}$$

Here, $|\vartheta| < 1$ is a complex constant. Thus, the Taylor series in (9.39) reduces to

$$\gamma_j(c) = 1 - \frac{1}{2}\frac{\sigma_{t_{m_j}}^2}{\sigma_{t_m}^2} c^2 + R(\tau)c^3, \tag{9.41}$$

where $R(\tau) = \frac{1}{6}\vartheta\, \varrho_{t_{m_j}}^3 / \sigma_{t_m}^3$. Recall that (9.41) expresses the jth terms of $\prod_{j=1}^{n} \gamma_j(c)$. Rewriting this equation in logarithmic form allows transforming the product into a sum, since

$$\log\left(\prod_{j=1}^{n} \gamma_j(c)\right) = \sum_{j=1}^{n} \log \gamma_j(c), \tag{9.42}$$

which simplifies the remainder of this proof. Equation (9.43) gives the logarithmic form of (9.41)

$$\log\left(\gamma_j(c)\right) = \log\left(1 - \frac{\sigma_{t_{m_j}}^2}{2\sigma_{t_m}^2} c^2 + R(\tau)c^3\right) = \log(1 + z). \tag{9.43}$$

Here, $z = -{}^{\sigma^2_{t_{m\,j}}}/_{2\sigma^2_{t_m}} c^2 + R(\tau)c^3$. For a sufficiently large number of source signals n, it follows from (9.32) and (9.33) that

$$\frac{\varrho_{t_{m\,j}}}{\sigma_{t_m}} \le \frac{\varrho_{t_m}}{\sigma_{t_m}} < 1. \tag{9.44}$$

Equation (9.45) presents a different way to formulate z

$$z = \widetilde{\vartheta}\frac{\varrho^2_{t_{m\,j}}}{2\sigma^2_{t_m}}c^2 + \tfrac{1}{6}\widetilde{\vartheta}\frac{\varrho^3_{t_{m\,j}}}{\sigma^3_{t_m}}c^3 = \widetilde{\vartheta}\frac{\varrho^2_{t_{m\,j}}}{\sigma^2_{t_m}}\left(\frac{c^2}{2} + \frac{|c|^3}{6}\right). \tag{9.45}$$

Here, $\widetilde{\vartheta}$ is a small correction term. Equations (9.32), (9.44) and (9.45) highlight that

$$\lim_{n\to\infty} z \to 0 \qquad \text{for all } c \in \mathbb{R}, \tag{9.46}$$

as $\lim_{n\to\infty} {}^{\varrho^2_{t_{m\,j}}}/_{\sigma^2_{t_m}} \to 0$. This implies that if the number of source signals n is large enough, $|z| < 0.5$, which, in turn, allows utilizing the Taylor expansion for $\log(1+z)$ to produce

$$\log(1+z) = \sum_{\mathrm{m}=1}^{\infty} (-1)^{\mathrm{m}+1}\frac{z^m}{\mathrm{m}!} = z + \vartheta^\star\frac{z^2}{2}\sum_{\mathrm{m}=0}^{\infty}\frac{1}{2^{\mathrm{m}}} = z + \widetilde{\vartheta}z^2 \tag{9.47}$$

where $\vartheta^\star \in \mathbb{C}$ is, as before, a small correction term. Hence,

$$\begin{aligned}
\log\left(\gamma_j(c)\right) &= \underbrace{-\tfrac{1}{2}\frac{\sigma^2_{t_{m\,j}}}{\sigma_{t_m}}c^2 + \tfrac{1}{6}\vartheta\frac{\varrho^3_{t_{m\,j}}}{\sigma^3_{t_m}}c^3}_{z} + \vartheta^*\underbrace{\frac{\varrho^4_{t_{m\,j}}}{\sigma^4_{t_m}}\left(\frac{c^2}{2} + \frac{|c|^3}{6}\right)^2}_{z^2} \\
&= -\tfrac{1}{2}\frac{\sigma^2_{t_{m\,j}}}{\sigma^2_{t_m}}c^2 + \vartheta^*\frac{\varrho^3_{t_{m\,j}}}{\sigma^3_{t_m}}\underbrace{\left(\frac{|c|^3}{6} + \left(\frac{c^2}{2} + \frac{|c|^3}{6}\right)^2\right)}_{\delta(c)} \\
&= -\tfrac{1}{2}\frac{\sigma^2_{t_{m\,j}}}{\sigma^2_{t_m}}c^2 + \vartheta^*\frac{\varrho^3_{t_{m\,j}}}{\sigma^3_{t_m}}\delta(c)
\end{aligned} \tag{9.48}$$

with ϑ^* being, again, a small correction term. For $\grave{\vartheta}$ being another small correction term, the final step is summing the individual terms, $\gamma_j(c)$, according to (9.42), which yields

$$\log\gamma(c) = \sum_{i=1}^{n}\log\gamma_j(c) = -\tfrac{1}{2}\frac{\sum_{j=1}^{n}\sigma^2_{t_{m\,j}}}{\sigma^2_{t_m}}c^2 + \grave{\vartheta}\frac{\sum_{j=1}^{n}\varrho^3_{t_{m\,j}}}{\sigma^3_{t_m}}\delta(c). \tag{9.49}$$

According to (9.32) and (9.33), (9.49) becomes

$$\lim_{n\to\infty} \log \gamma(c) = \lim_{n\to\infty} -\tfrac{1}{2}c^2 + \tfrac{1}{6}\vartheta \underbrace{\frac{\varrho_{t_m}^3}{\sigma_{t_m}^3}}_{=0} \delta(c) = -\tfrac{1}{2}c^2 \tag{9.50}$$

and hence

$$\boxed{\lim_{n\to\infty} \gamma(c) \to e^{-\frac{1}{2}c^2}}$$

$$\text{with} \tag{9.51}$$

$$\gamma(c) = E\left\{e^{\mathrm{i}\tau_m c}\right\} = \int_{-\infty}^{\infty} e^{\mathrm{i}\left(\sum_{j=1}^{n} \mathbf{p}_m^T \boldsymbol{\xi}_j s_j/\sqrt{\lambda_m}\right)c}\, \mathrm{d}F(s_j).$$

Consequently, the mth t-score variable asymptotically follows a Gaussian distribution under the assumption that the source signals are statistically independent. As stated above, however, the conditions for which the CLT holds true are less restrictive than those formulated in the Liapounoff and the simplified Lindeberg-Lévy theorems used in this book. The interested reader can refer to the references given in this proof for a more comprehensive treatment of conditions under which the CLT holds true.

Property 9.3.5 – Computation of the t-Score Vectors. After proving the asymptotic properties of the t-score variables with respect to the number of source variables, the impact of the deflation procedure upon the computation of the t-score vectors is analyzed next.

Theorem 9.3.5 *It is irrelevant whether the* j*th t-score vector* $\widehat{\mathbf{t}}_j$ *are obtained from the original or the deflated data matrix, that is,* $\widehat{\mathbf{t}}_j = \mathbf{Z}^{(j)}\widehat{\mathbf{p}}_j = \mathbf{Z}_0\widehat{\mathbf{p}}_j$.

Proof. Starting with $\mathbf{Z}^{(j)}\mathbf{p}_j$ and revisiting the deflation of $\mathbf{Z}^{(j)}$

$$\mathbf{Z}^{(j)} = \mathbf{Z}_0 \prod_{i=1}^{j-1} \left[\mathbf{I} - \widehat{\mathbf{p}}_i\widehat{\mathbf{p}}_i^T\right] \tag{9.52}$$

which yields

$$\begin{aligned}
\widehat{\mathbf{t}}_j &= \mathbf{Z}^{(j)}\widehat{\mathbf{p}}_j = \mathbf{Z}_0 \prod_{i=1}^{j-2} \left[\mathbf{I} - \widehat{\mathbf{p}}_i\widehat{\mathbf{p}}_i^T\right]\left[\mathbf{I} - \widehat{\mathbf{p}}_{j-1}\widehat{\mathbf{p}}_{j-1}^T\right]\widehat{\mathbf{p}}_j \\
\widehat{\mathbf{t}}_j &= \mathbf{Z}_0 \prod_{i=1}^{j-2} \left[\mathbf{I} - \widehat{\mathbf{p}}_i\widehat{\mathbf{p}}_i^T\right]\left(\widehat{\mathbf{p}}_j - \widehat{\mathbf{p}}_{j-1}\underbrace{\widehat{\mathbf{p}}_{j-1}^T\widehat{\mathbf{p}}_j}_{=0}\right)
\end{aligned} \tag{9.53}$$

$$\vdots$$

$$\widehat{\mathbf{t}}_j = \mathbf{Z}_0\widehat{\mathbf{p}}_j - \mathbf{Z}_0\widehat{\mathbf{p}}_1 \underbrace{\widehat{\mathbf{p}}_1^T\widehat{\mathbf{p}}_j}_{=0}$$

$$\boxed{\widehat{\mathbf{t}}_j = \mathbf{Z}_0\widehat{\mathbf{p}}_j = \mathbf{Z}^{(j)}\widehat{\mathbf{p}}_j}$$

That the p-loading vectors are mutually orthonormal is subject of Theorem 9.3.3.

Property 9.3.6 – Exhausting the Data Matrix $\mathbf{Z}_0$. The deflation procedure allows exhausting the data matrix if enough latent components are extracted.

Theorem 9.3.6 *For $K \gg n_z$, if the column rank of the data matrix, $\mathbf{Z}_0$, is $\bar{n}_z \leq n_z$ applying a total of $\bar{n}_z$ deflation steps completely exhausts $\mathbf{Z}_0$, i.e. $\mathbf{Z}^{(\bar{n}_z)} = \mathbf{0}$.*

Proof. It is straightforward to prove the case $\bar{n}_z = n_z$, which is shown first. The general case $\bar{n}_z < n_z$ is then analyzed by a geometric reconstruction of $\mathbf{Z}_0$ using the rank-one component matrices shown in (9.3). If $\bar{n}_z = n_z$ the following holds true

$$\begin{aligned} \mathbf{Z}^{(n_z)} &= \mathbf{Z}_0 - \sum_{j=1}^{n_z} \widehat{\mathbf{t}}_j\widehat{\mathbf{p}}_j^T \\ \mathbf{Z}^{(n_z)} &= \mathbf{Z}_0 - \sum_{j=1}^{n_z} \mathbf{Z}_0\widehat{\mathbf{p}}_j\widehat{\mathbf{p}}_j^T \\ \boxed{\mathbf{Z}^{(n_z)} = \mathbf{Z}_0 - \mathbf{Z}_0\widehat{\mathbf{P}}\widehat{\mathbf{P}}^T = \mathbf{Z}_0 - \mathbf{Z}_0 = \mathbf{0}} \end{aligned} \tag{9.54}$$

which follows from the fact that (i) the estimated p-loading vectors are orthonormal and (ii) that $\mathbf{Z}^{(j)}\widehat{\mathbf{p}}_j = \mathbf{Z}_0\widehat{\mathbf{p}}_j$.

In the general case where $\bar{n}_z < n_z$, the observations stored as row vectors in $\mathbf{Z}_0$ lie in a subspace of dimension $\bar{n}_z$. This follows from the fact that any $n_z - \bar{n}_z$ column vectors of $\mathbf{Z}_0$ are linear combinations of the remaining $\bar{n}_z$ columns. We can therefore remove $n_z - \bar{n}_z$ columns from $\mathbf{Z}_0$, which yields a reduced dimensional data matrix $\mathbf{Z}_{0_{red}}$. According to (9.54), $\mathbf{Z}_{0_{red}}$ is exhausted after $\bar{n}_z$ deflation steps have been carried out. Given that the columns that were removed from $\mathbf{Z}_0$ are linear combinations of those that remained in $\mathbf{Z}_{0_{red}}$, deflating the reduced data matrix $\mathbf{Z}_{0_{red}}$ automatically deflates the column vectors not included in $\mathbf{Z}_{0_{red}}$.

Property 9.3.7 – Exhausting the Covariance Matrix $\mathbf{S}_{z_0z_0}$. The final property is concerned with the reconstruction of the given covariance matrix $\mathbf{S}_{z_0z_0}$ using the loading vectors $\mathbf{p}_1, \ldots, \mathbf{p}_{\bar{n}_z}$, where $\bar{n}_z \leq n_z$ is the rank of $\mathbf{S}_{z_0z_0}$. Whilst this is a property that simply follows from the eigendecomposition of a symmetric positive semi-definite matrix, its analysis is useful in providing an insight into the deflation procedure of the PLS algorithm, discussed in Section 10.1.

Lemma 9.3.7 *The covariance matrix $\mathbf{S}_{z_0z_0}$ can be reconstructed by a sum of $\bar{n}_z \le n_z$ scaled loading vectors $\mathbf{p}_j$, where the scaling factor is given by the standard deviation of the score variables.*

Proof. A reformulation of (9.6) leads to

$$J_{p_j} = E\left\{t_j^2\right\} = E\left\{\mathbf{p}_j^T \mathbf{z}_0 \mathbf{z}_0^T \mathbf{p}_j\right\} = \mathbf{p}_j^T \mathbf{S}_{z_0z_0} \mathbf{p}_j - \lambda_j \left(\mathbf{p}_j^T \mathbf{p}_j - 1\right) \tag{9.55}$$

and hence

$$\mathbf{p}_j = \arg\max_{\mathbf{p}} \mathbf{p}^T \mathbf{S}_{z_0z_0} \mathbf{p} - \lambda_j \left(\mathbf{p}^T \mathbf{p} - 1\right). \tag{9.56}$$

Given that the rank of $\mathbf{S}_{z_0z_0} = \bar{n}_z$, there are a total of $\bar{n}_z$ eigenvalues λ_j that are larger than zero. Equation (9.56) can be expanded to allow a simultaneous determination of the $\bar{n}_z$ eigenvectors

$$\mathbf{P} = \arg \max_{\mathfrak{P} \in \mathbb{R}^{n_z \times \bar{n}_z}} \mathfrak{P}^T \mathbf{S}_{z_0z_0} \mathfrak{P} - \begin{bmatrix} \lambda_1 & 0 & \cdots & 0 \\ 0 & \lambda_2 & \cdots & 0 \\ \vdots & \vdots & \ddots & \vdots \\ 0 & 0 & \cdots & \lambda_{\bar{n}_z} \end{bmatrix} \left[\mathfrak{P}^T \mathfrak{P} - \mathbf{I}\right] \tag{9.57}$$

which follows from the fact that the eigenvectors are mutually orthonormal (Theorem 9.3.3) and that $\mathbf{Z}^{(j)} \widehat{\mathbf{p}}_j = \mathbf{Z}_0 \widehat{\mathbf{p}}_j$ (Theorem 9.3.6). The optimum of the individual objective functions, combined in (9.57), is given by

$$\mathbf{P} = \arg \frac{\partial}{\partial \mathfrak{P}} \left(\mathfrak{P}^T \mathbf{S}_{z_0z_0} \mathfrak{P} - \begin{bmatrix} \lambda_1 & 0 & \cdots & 0 \\ 0 & \lambda_2 & \cdots & 0 \\ \vdots & \vdots & \ddots & \vdots \\ 0 & 0 & \cdots & \lambda_{\bar{n}_z} \end{bmatrix} \left[\mathfrak{P}^T \mathfrak{P} - \mathbf{I}\right] \right), \tag{9.58}$$

which gives rise to

$$\mathbf{P} = 2 \arg \left(\mathbf{S}_{z_0z_0} \mathfrak{P} - \mathfrak{P} \begin{bmatrix} \lambda_1 & 0 & \cdots & 0 \\ 0 & \lambda_2 & \cdots & 0 \\ \vdots & \vdots & \ddots & \vdots \\ 0 & 0 & \cdots & \lambda_{\bar{n}_z} \end{bmatrix} \right) = \mathbf{0}. \tag{9.59}$$

Substituting the fact that $\mathfrak{P} = \mathbf{P}$, pre-multiplication of the above equation by $\mathbf{P}^T$ yields

$$\mathbf{P}^T \mathbf{S}_{z_0z_0} \mathbf{P} = \begin{bmatrix} \lambda_1 & 0 & \cdots & 0 \\ 0 & \lambda_2 & \cdots & 0 \\ \vdots & \vdots & \ddots & \vdots \\ 0 & 0 & \cdots & \lambda_{\bar{n}_z} \end{bmatrix} \tag{9.60}$$

which represents the diagonalization of $\mathbf{S}_{z_0 z_0}$. On the other hand, the relationship above also implies that

$$\mathbf{S}_{z_0 z_0} = \mathbf{P} \begin{bmatrix} \lambda_1 & 0 & \cdots & 0 \\ 0 & \lambda_2 & \cdots & 0 \\ \vdots & \vdots & \ddots & \vdots \\ 0 & 0 & \cdots & \lambda_{\bar{n}_z} \end{bmatrix} \mathbf{P}^T \tag{9.61}$$

and hence

$$\boxed{\mathbf{S}_{z_0 z_0} = \sum_{j=1}^{\bar{n}_z} \mathbf{p}_j \lambda_j \mathbf{p}_j^T}. \tag{9.62}$$

It should be noted that (9.61) holds true since $\mathbf{PP}^T$ projects any vector onto the subspace describing the relationship between any $n_z - \bar{n}_z$ variables that are linearly dependent upon the remaining $\bar{n}_z$ ones. This subspace is spanned by the $\bar{n}_z$ eigenvectors stored in $\mathbf{P}$.

10

Partial least squares

This chapter provides a detailed analysis of PLS and its maximum redundancy formulation. The data models including the underlying assumptions for obtaining a PLS and a MRPLS model are outlined in Sections 2.2 and 2.3, respectively.

Section 10.1 presents preliminaries of projecting the recorded samples of the input variables, $\mathbf{x}_0 \in \mathbb{R}^{n_x}$ onto an n-dimensional subspace, $n \leq n_x$, and show how a sequence of rank-one matrices extract variation from the sets of input and output variables $\mathbf{x}_0$ and $\mathbf{y}_0 \in \mathbb{R}^{n_y}$, respectively. Section 10.2 then develops a PLS algorithm and Section 10.3 summarizes the basic steps of this algorithm.

Section 10.4 then analyzes the statistical and geometric properties of PLS and finally, Section 10.5 discusses the properties of MRPLS. Further material covering the development and analysis of PLS may be found in de Jong (1993); Geladi and Kowalski (1986); Höskuldsson (1988, 1996); ter Braak and de Jong (1998); Wold *et al.* (1984) and Young (1994).

10.1 Preliminaries

In a similar fashion to PCA, PLS extracts information from the input and output data matrices, $\mathbf{X}_0 \in \mathbb{R}^{K \times n_x}$ and $\mathbf{Y}_0 \in \mathbb{R}^{K \times n_y}$ by defining a series of rank-one matrices

$$\mathbf{X}_0 = \sum_{j=1}^{n} \widehat{\mathbf{t}}_j \widehat{\mathbf{p}}_j^T + \mathbf{E}^{(n+1)} \tag{10.1a}$$

$$\mathbf{Y}_0 = \sum_{j=1}^{n} \widehat{\mathbf{t}}_j \widehat{\mathbf{q}}_j^T + \mathbf{F}^{(n+1)}. \tag{10.1b}$$

Statistical Monitoring of Complex Multivariate Processes: With Applications in Industrial Process Control, First Edition. Uwe Kruger and Lei Xie.

The data matrices store mean-centered observations of the input and output variable sets, that is $\mathbf{x}_0 = \mathbf{x} - \bar{\mathbf{x}}$ and $\mathbf{y}_0 = \mathbf{y} - \bar{\mathbf{y}}$ with $\bar{\mathbf{x}}$ and $\bar{\mathbf{y}}$ being mean vectors. In the above equation, $\widehat{\mathbf{t}}_j\widehat{\mathbf{p}}_j^T \in \mathbb{R}^{K\times n_x}$ and $\widehat{\mathbf{t}}_j\widehat{\acute{\mathbf{q}}}_j^T \in \mathbb{R}^{K\times n_y}$ are the rank-one matrices for the input and output matrices, respectively, the n vectors $\widehat{\mathbf{t}}_j \in \mathbb{R}^K$ are t-score vectors which are estimated from the input matrix, $\widehat{\mathbf{p}}_j \in \mathbb{R}^{n_x}$ and $\widehat{\acute{\mathbf{q}}}_j \in \mathbb{R}^{n_y}$ are estimated loading vectors for the input and output matrices, respectively, and $\mathbf{E}^{(n+1)} \in \mathbb{R}^{K\times n_x}$ and $\mathbf{F}^{(n+1)} \in \mathbb{R}^{K\times n_y}$ are residual matrices of the input and output matrices, respectively. It should be noted that the residual matrices have a negligible or no contribution to the prediction of the output data matrix.

To establish (10.1), the PLS algorithm determines a sequence of parallel projections, one sequence for the observations stored in the input matrix and a second sequence for the observations stored in the output matrix. Reformulating (10.1)

$$\mathbf{E}^{(j+1)} = \mathbf{X}_0 - \sum_{i=1}^{j-1} \mathbf{t}_i\mathbf{p}_i^T \tag{10.2a}$$

$$\mathbf{F}^{(j+1)} = \mathbf{Y}_0 - \sum_{i=1}^{j} \mathbf{t}_j\acute{\mathbf{q}}_j^T \tag{10.2b}$$

and defining

$$\mathbf{E}^{(1)} = \mathbf{X}_0 \tag{10.3a}$$

$$\mathbf{F}^{(1)} = \mathbf{Y}_0 \tag{10.3b}$$

allows determining the sequence of projections for the input and output variables

$$\mathbf{t}_j = \mathbf{E}^{(j)}\mathbf{w}_j \qquad \left\|\mathbf{w}_j\right\|^2 = 1 \tag{10.4a}$$

$$\mathbf{u}_j = \mathbf{F}^{(j)}\mathbf{q}_j \qquad \left\|\mathbf{q}_j\right\|^2 = 1. \tag{10.4b}$$

Here, $\mathbf{u}_j \in \mathbb{R}^K$ is the u-score vector of the output matrix, and $\mathbf{w}_j \in \mathbb{R}^{n_x}$ and $\mathbf{q}_j \in \mathbb{R}^{n_y}$ are the weight vectors for the input and output variable sets, respectively. Finally, according to (10.4), the score variables, t_k and u_k, are given by

$$\begin{aligned} t_j &= \mathbf{w}_j^T\mathbf{e}^{(j)} & u_j &= \mathbf{q}_j^T\mathbf{f}^{(j)} \\ \mathbf{e}^{(j)} &= \mathbf{x}_0 - \sum_{i=1}^{j-1} t_i\mathbf{p}_i & \mathbf{f}^{(j)} &= \mathbf{y}_0 - \sum_{i=1}^{j-1} t_i\acute{\mathbf{q}}_i \\ \mathbf{e}^{(1)} &= \mathbf{x}_0 & \mathbf{f}^{(1)} &= \mathbf{y}_0. \end{aligned} \tag{10.5}$$

The set of weight and loading vectors, $\mathbf{w}_j$, $\mathbf{q}_j$, $\mathbf{p}_j$ and $\acute{\mathbf{q}}_j$, as well as the set of score variables, t_j and u_j, make up the jth latent variable set.

10.2 The core algorithm

PLS determines the score variables t_j and u_j such that they maximize an objective function describing their covariance, which is subject to the constraint that the projection vectors $\mathbf{w}_j$ and $\mathbf{q}_j$ are of unit length

$$\begin{aligned}\begin{pmatrix}\mathbf{w}_j\\ \mathbf{q}_j\end{pmatrix} &= \arg\max_{\mathbf{w},\mathbf{q}} E\left\{t_j u_j\right\} - \tfrac{1}{2}\lambda_j^{(1)}\left(\mathbf{w}^T\mathbf{w}-1\right) - \tfrac{1}{2}\lambda_j^{(2)}\left(\mathbf{q}^T\mathbf{q}-1\right)\\ \begin{pmatrix}\mathbf{w}_j\\ \mathbf{q}_j\end{pmatrix} &= \arg\max_{\mathbf{w},\mathbf{q}} E\left\{\left(\mathbf{w}^T\mathbf{e}^{(j)}\right)\left(\mathbf{q}^T\mathbf{f}^{(j)}\right)\right\} -\\ &\qquad \tfrac{1}{2}\lambda_j^{(1)}\left(\mathbf{w}^T\mathbf{w}-1\right) - \tfrac{1}{2}\lambda_j^{(2)}\left(\mathbf{q}^T\mathbf{q}-1\right)\\ \begin{pmatrix}\mathbf{w}_j\\ \mathbf{q}_j\end{pmatrix} &= \arg\max_{\mathbf{w},\mathbf{q}} \mathbf{w}^T\mathbf{S}_{ef}^{(j)}\mathbf{q} - \tfrac{1}{2}\lambda_j^{(1)}\left(\mathbf{w}^T\mathbf{w}-1\right) - \tfrac{1}{2}\lambda_j^{(2)}\left(\mathbf{q}^T\mathbf{q}-1\right),\end{aligned} \tag{10.6}$$

where $E\left\{\mathbf{e}^{(j)}\mathbf{f}^{(j)^T}\right\} = \mathbf{S}_{ef}^{(j)}$. The optimal solution for the objective function in (10.6) is given by

$$\begin{aligned}\begin{pmatrix}\mathbf{w}_j\\ \mathbf{q}_j\end{pmatrix} &= \arg\frac{\partial}{\partial\mathbf{w}}\left(\mathbf{w}^T\mathbf{S}_{ef}^{(j)}\mathbf{q} - \tfrac{1}{2}\lambda_j^{(1)}\left(\mathbf{w}^T\mathbf{w}-1\right) - \tfrac{1}{2}\lambda_j^{(2)}\left(\mathbf{q}^T\mathbf{q}-1\right)\right) = \mathbf{0}\\ &\text{and}\\ \begin{pmatrix}\mathbf{w}_j\\ \mathbf{q}_j\end{pmatrix} &= \arg\frac{\partial}{\partial\mathbf{q}}\left(\mathbf{w}^T\mathbf{S}_{ef}^{(j)}\mathbf{q} - \tfrac{1}{2}\lambda_j^{(1)}\left(\mathbf{w}^T\mathbf{w}-1\right) - \tfrac{1}{2}\lambda_j^{(2)}\left(\mathbf{q}^T\mathbf{q}-1\right)\right) = \mathbf{0}\end{aligned} \tag{10.7}$$

which have to be solved simultaneously. This yields

$$\begin{aligned}\mathbf{S}_{ef}^{(j)}\mathbf{q}_j - \lambda_j^{(1)}\mathbf{w}_j &= \mathbf{0}\\ \mathbf{S}_{fe}^{(j)}\mathbf{w}_j - \lambda_j^{(2)}\mathbf{q}_j &= \mathbf{0}.\end{aligned} \tag{10.8}$$

Note that $\mathbf{S}_{ef}^{(j)^T} = \mathbf{S}_{fe}^{(j)}$. Equation (10.8) also confirms that the two Lagrangian multipliers are identical, since

$$\begin{aligned}&\arg\mathbf{w}_j^T\frac{\partial}{\partial\mathbf{w}}\left(\mathbf{w}^T\mathbf{S}_{ef}^{(j)}\mathbf{q} - \tfrac{1}{2}\lambda_j^{(1)}\left(\mathbf{w}^T\mathbf{w}-1\right) - \tfrac{1}{2}\lambda_j^{(2)}\left(\mathbf{q}^T\mathbf{q}-1\right)\right) = \mathbf{0}\\ &\quad\mathbf{w}_j^T\left(\mathbf{S}_{ef}^{(j)}\mathbf{q}_j - \lambda_j^{(1)}\mathbf{w}_j\right) = \mathbf{0} \Rightarrow \lambda_j^{(1)} = \mathbf{w}_j^T\mathbf{S}_{ef}^{(j)}\mathbf{q}_j\\ &\text{and}\\ &\arg\mathbf{q}_j^T\frac{\partial}{\partial\mathbf{q}}\left(\mathbf{w}^T\mathbf{S}_{ef}^{(j)}\mathbf{q} - \tfrac{1}{2}\lambda_j^{(2)}\left(\mathbf{w}^T\mathbf{w}-1\right) - \tfrac{1}{2}\lambda_j^{(2)}\left(\mathbf{q}^T\mathbf{q}-1\right)\right) = \mathbf{0}\\ &\quad\mathbf{q}_j^T\left(\mathbf{S}_{fe}^{(j)}\mathbf{w}_j - \lambda_j^{(2)}\mathbf{q}_j\right) = \mathbf{0} \Rightarrow \lambda_j^{(2)} = \mathbf{q}_j^T\mathbf{S}_{fe}^{(j)}\mathbf{w}_j = \lambda_j^{(1)}.\end{aligned} \tag{10.9}$$

Hence, $\lambda_j^{(1)} = \lambda_j^{(2)} = \lambda_j$. Combining the two expressions in (10.8) gives rise to

$$\left[\mathbf{S}_{ef}^{(j)}\mathbf{S}_{fe}^{(j)} - \lambda_j^2\mathbf{I}\right]\mathbf{w}_j = \mathbf{0} \tag{10.10a}$$

$$\left[\mathbf{S}_{fe}^{(j)}\mathbf{S}_{ef}^{(j)} - \lambda_j^2\mathbf{I}\right]\mathbf{q}_j = \mathbf{0}. \tag{10.10b}$$

The weight vectors $\mathbf{w}_j$ and $\mathbf{q}_j$ are therefore the dominant eigenvectors of the matrix expressions $\mathbf{S}_{ef}^{(j)}\mathbf{S}_{fe}^{(j)}$ and $\mathbf{S}_{fe}^{(j)}\mathbf{S}_{ef}^{(j)}$, respectively. The score vectors for $\mathbf{E}^{(j)}$ and $\mathbf{F}^{(j)}$ can now be computed using (10.4).

After determining the weight and score vectors, the next step involves the calculation of a regression coefficient between the score variables t_j and u_j. It is important to note, however, that the determination of this regression coefficient can be omitted, as this step can be incorporated into the calculation of the $\acute{\mathbf{q}}$-loading vector, which is proven in Section 10.4. For a better understanding of the geometry of the PLS algorithm, however, the introduction of the PLS algorithm here includes this step. Equation (10.11) shows the least squares solution for determining the regression parameter

$$b_j = \frac{\mathbf{w}_j^T\mathbf{S}_{ef}^{(j)}\mathbf{q}_j}{\mathbf{w}_j^T\mathbf{S}_{ee}^{(j)}\mathbf{w}_j} \tag{10.11}$$

where $\mathbf{S}_{ee}^{(j)} = E\{\mathbf{e}^{(j)}\mathbf{e}^{(j)^T}\}$. The final step to complete the determination of the jth set of latent variables requires formulation of objective functions for computing the loading vectors

$$\mathbf{p}_j = \arg\min_{\mathbf{p}} \left\| E\left\{\left(\mathbf{e}^{(j)} - t_j\mathbf{p}\right)\left(\mathbf{e}^{(j)} - t_j\mathbf{p}\right)^T\right\}\right\|^2$$

and (10.12)

$$\acute{\mathbf{q}}_j = \arg\min_{\acute{\mathbf{q}}} \left\| E\left\{\left(\mathbf{f}^{(j)} - u_jb_j\acute{\mathbf{q}}\right)\left(\mathbf{f}^{(j)} - u_jb_j\acute{\mathbf{q}}\right)^T\right\}\right\|^2.$$

The solutions to (10.12) are

$$\mathbf{p}_j = \arg\frac{\partial}{\partial\mathbf{p}}\left\| E\left\{\mathbf{e}^{(j)}\mathbf{e}^{(j)^T}\right\} - 2E\left\{\mathbf{e}^{(j)}\mathbf{e}^{(j)^T}\mathbf{w}_j\mathbf{p}^T\right\} + E\left\{\mathbf{p}\mathbf{w}_j^T\mathbf{e}^{(j)}\mathbf{e}^{(j)^T}\mathbf{w}_j\mathbf{p}^T\right\}\right\|^2 = \mathbf{0}$$

and (10.13)

$$\acute{\mathbf{q}}_j = \arg\frac{\partial}{\partial\acute{\mathbf{q}}}\left\| E\left\{\mathbf{f}^{(j)}\mathbf{f}^{(j)^T}\right\} - 2E\left\{\mathbf{f}^{(j)}\mathbf{e}^{(j)^T}\mathbf{w}_jb_j\acute{\mathbf{q}}^T\right\} + E\left\{\acute{\mathbf{q}}b_j\mathbf{w}_j^T\mathbf{e}^{(j)}\mathbf{e}^{(j)^T}\mathbf{w}_jb_j\acute{\mathbf{q}}^T\right\}\right\|^2 = \mathbf{0}.$$

Working out the relationships yields

$$-2\mathbf{S}_{ee}^{(j)}\mathbf{w}_j + 2\left(\mathbf{w}_j^T\mathbf{S}_{ee}^{(j)}\mathbf{w}_j\right)\mathbf{p}_j = \mathbf{0} \Rightarrow \mathbf{p}_j = \frac{\mathbf{S}_{ee}^{(j)}\mathbf{w}_j}{\mathbf{w}_j^T\mathbf{S}_{ee}^{(j)}\mathbf{w}_j}$$

and (10.14)

$$-2b_j\mathbf{S}_{fe}^{(j)}\mathbf{w}_j + 2b_j^2\left(\mathbf{w}_j^T\mathbf{S}_{ee}^{(j)}\mathbf{w}_j\right)\acute{\mathbf{q}}_j = \mathbf{0} \Rightarrow \acute{\mathbf{q}}_j = \frac{\mathbf{S}_{fe}^{(j)}\mathbf{w}_j}{b_j\mathbf{w}_j^T\mathbf{S}_{ee}^{(j)}\mathbf{w}_j}.$$

Before computing the $(j+1)$th set of LVs, (10.5) highlights that the contribution of the jth set of latent variables must be subtracted from $\mathbf{e}^{(j)}$ and $\mathbf{f}^{(j)}$

$$\mathbf{e}^{(j+1)} = \mathbf{e}^{(j)} - t_j\mathbf{p}_j \quad (10.15a)$$

$$\mathbf{f}^{(j+1)} = \mathbf{f}^{(j)} - t_jb_j\acute{\mathbf{q}}_j. \quad (10.15b)$$

It should be noted that substituting (10.14) into (10.15) gives rise to

$$\mathbf{f}^{(j+1)} = \mathbf{f}^{(j)} - t_jb_j\mathbf{q}_j = \mathbf{f}^{(j)} - t_j\left(\frac{b_j}{b_j}\right)\frac{\mathbf{S}_{fe}^{(j)}\mathbf{w}_j}{\mathbf{w}_j^T\mathbf{S}_{ee}^{(j)}\mathbf{w}_j} \quad (10.16)$$

which, however, requires the $\acute{\mathbf{q}}$-loading vector to be determined as follows

$$\acute{\mathbf{q}}_j = \frac{\mathbf{S}_{fe}^{(j)}\mathbf{w}_j}{\mathbf{w}_j^T\mathbf{S}_{ee}^{(j)}\mathbf{w}_j}. \quad (10.17)$$

It should also be noted that the deflation procedure can be applied directly to the covariance matrix $\mathbf{S}_{ee}^{(j)}$ and the cross-covariance matrix $\mathbf{S}_{ef}^{(j)}$

$$\mathbf{S}_{ee}^{(j+1)} = \mathbf{S}_{ee}^{(j)} - \mathbf{p}_j\sigma_{t_j}^2\mathbf{p}_j^T = \mathbf{S}_{ee}^{(j)} - \mathbf{p}_j\mathbf{w}_j^T\mathbf{S}_{ee}^{(j)}\mathbf{w}_j\mathbf{p}_j^T$$

$$\mathbf{S}_{ee}^{(j+1)} = \mathbf{S}_{ee}^{(j)} - \mathbf{S}_{ee}^{(j)}\mathbf{w}_j\mathbf{p}_j^T = \mathbf{S}_{ee}^{(j)}\left[\mathbf{I} - \mathbf{w}_j\mathbf{p}_j^T\right]$$

and (10.18)

$$\mathbf{S}_{fe}^{(j+1)} = \mathbf{S}_{fe}^{(j)} - \acute{\mathbf{q}}_j\sigma_{t_j}^2\mathbf{p}_j^T = \mathbf{S}_{fe}^{(j)} - \acute{\mathbf{q}}_j\mathbf{w}_j^T\mathbf{S}_{ee}^{(j)}\mathbf{w}_j\mathbf{p}_j^T$$

$$\mathbf{S}_{fe}^{(j+1)} = \mathbf{S}_{fe}^{(j)} - \mathbf{S}_{fe}^{(j)}\mathbf{w}_j\mathbf{p}_j^T = \mathbf{S}_{fe}^{(j)}\left[\mathbf{I} - \mathbf{w}_j\mathbf{p}_j^T\right].$$

The above relationship relies on (10.14), (10.15) and (10.17).

The steps of the PLS algorithm can be carried out using the NIPALS algorithm (Geladi and Kowalski 1986), the SIMPLS algorithm (de Jong 1993) or the computationally more efficient Kernel algorithms (Dayal and MacGregor 1997a; Lindgren *et al.* 1993; Rännar *et al.* 1994). Each of these algorithms are iterative in nature, that is, one pair of latent variables are obtained and the contribution of the t-score vector is deflated from the input and output matrices in one iteration step.

10.3 Summary of the PLS algorithm

The preceding analysis showed that PLS extracts covariance information from the input and output variables, $\mathbf{x}_0$ and $\mathbf{y}_0$, by defining a sequence of score variables which are extracted from the input variable set (10.5). The contribution of each score variable is maximized by the determination of loading vectors, such that the original variable sets are defined by $\mathbf{x}_0 = \sum_{j=1}^{n} t_j \mathbf{p}_j + \mathbf{e}^{(n+1)}$ and $\mathbf{y}_0 = \sum_{j=1}^{n} t_j \acute{\mathbf{q}}_j + \mathbf{f}^{(n+1)}$ (10.15).

The calculation of the n score variables, $t_1, t_2, \ldots, t_n$ relies on an objective function that maximizes a covariance criterion between t_i and a score variable that is extracted from the output variable set u_i, $E\left\{t_i u_i\right\}$ (10.6). In other words,

Table 10.1 PLS algorithm developed from the steps in Section 10.2.

Step	Description	Equation
1	Initiate iteration	$j = 1$
2	Obtain covariance matrix	$\mathbf{S}_{x_0x_0} = \mathbf{S}_{ee}^{(1)}$
3	Determine cross-covariance matrix	$\mathbf{S}_{y_0x_0} = \mathbf{S}_{fe}^{(1)}$
4	Set-up initial q-weight vector	${}_0\mathbf{q}_j = \mathbf{S}_{fe}^{(j)}(:,1) / \left\|\mathbf{S}_{fe}^{(j)}(:,1)\right\|$
5	Calculate w-weight vector	$\mathbf{w}_j = \mathbf{S}_{ef}^{(j)}\, {}_0\mathbf{q}_j$
6	Scale w-weight vector to unit length	$\mathbf{w}_j / \|\mathbf{w}_j\|$
7	Compute q-weight vector	${}_1\mathbf{q}_j = \mathbf{S}_{fe}^{(j)} \mathbf{w}_j$
8	Scale q-weight vector to unit length	${}_1\mathbf{q}_j / \|{}_1\mathbf{q}_j\|$
9	Check for convergence	If $\left\|{}_1\mathbf{q}_j - {}_0\mathbf{q}_j\right\| > \epsilon$, set ${}_0\mathbf{q}_j = {}_1\mathbf{q}_j$ and go to Step 5; else go to Step 10
10	Determine p-loading vector	$\mathbf{p}_j = \dfrac{\mathbf{S}_{ee}^{(j)}\mathbf{w}_j}{\mathbf{w}_j \mathbf{S}_{ee}^{(j)} \mathbf{w}_j}$
11	Calculate q́-loading vector	$\acute{\mathbf{q}}_j = \dfrac{\mathbf{S}_{fe}^{(j)}\mathbf{w}_j}{\mathbf{w}_j \mathbf{S}_{ee}^{(j)} \mathbf{w}_j}$
12	Deflate cross-covariance matrix	$\mathbf{S}_{fe}^{(j+1)} = \mathbf{S}_{fe}^{(j)}\left[\mathbf{I} - \mathbf{w}_j \acute{\mathbf{q}}_j^T\right]$
13	Check whether there is significant variation left in the cross-covariance matrix	If so, go to Step 14 if not, terminate modeling procedure
14	Deflate covariance matrix	$\mathbf{S}_{ee}^{(j+1)} = \mathbf{S}_{ee}^{(j)}\left[\mathbf{I} - \mathbf{w}_j \mathbf{p}_j^T\right]$
15	Check for dimension	If $j < n_x$, set $j = j + 1$ and go to Step 4 if not, terminate modeling procedure

This algorithm is similar to the Kernel PLS algorithm by Lindgren *et al.* (1993).

there are a total of n score variables computed from the input variable set and n score variables calculated from the output variable set. These score variables are obtained in pairs, t_i and u_i and are given by a projection of the input and output variable set onto the weight vectors $\mathbf{w}_i$ and $\mathbf{q}_i$ (10.5). The solution for the pairs of weight vectors gives rise to the determination of dominant eigenvectors of symmetric and positive semi-definite matrices.

Unlike the PCA algorithm, the sets of latent variables can only be determined sequentially using the power method (Geladi and Kowalski 1986). This is an iterative method for determining the dominant eigenvector of a symmetric positive semi-definite matrix (Golub and van Loan 1996). Using the basic steps, developed in the previous subsection, Table 10.1 presents a PLS algorithm for determining the weight and loading vectors from the covariance matrix $\mathbf{S}_{x_0x_0}$ and the cross-covariance matrix $\mathbf{S}_{y_0x_0}$. The next subsection presents a detailed statistical and geometric analysis of the PLS algorithm and introduces a computationally more efficient algorithm to that described in Table 10.1.

10.4 Properties of PLS

The PLS algorithm, developed and summarized in the last two subsections, has the statistical and geometrical properties listed below. For a detailed discussion of these properties, it is important to note that the preceding discussion has assumed the availability of the covariance matrix $\mathbf{S}_{x_0x_0}$ and the cross-covariance matrix $\mathbf{S}_{x_0y_0}$. This has been for the convenience and simplicity of the presentation. Unless stated otherwise, the analysis that follows, however, removes this assumption and relies on the available data matrices $\mathbf{X}_0$ and $\mathbf{Y}_0$, whilst acknowledging that the covariance and cross-covariance matrices can be estimated from these data matrices. Hence, the weight, score and loading vectors become estimates.

1. The weight vectors, $\mathbf{w}_j$ and $\mathbf{q}_j$ are the dominant left and right singular vectors and the maximum of the objective function λ_j is the largest singular value of a singular value decomposition of $\mathbf{S}_{ef}^{(j)}$.
2. The t-score vectors are mutually orthogonal.
3. The matrix vector products $\widehat{\mathbf{t}}_j^T\mathbf{E}^{(j)}$ and $\widehat{\mathbf{t}}_j^T\mathbf{X}_0$ are equivalent.
4. The matrix-vector product $\widehat{\mathbf{t}}_i^T\mathbf{E}^{(j)} = 0$ for all $i < j$.
5. The ith t-score and the jth u-score vectors are orthogonal for all $i < j$.
6. It is sufficient to either deflate the input or the output data matrix.
7. The w-weight vectors are mutually orthonormal.
8. The ith w-weight vector and the jth p-loading vector are orthogonal for all $j > i$ and equal to 1 if $i = j$.
9. The value of the regression coefficient b_j is equal to the length of the $\acute{\mathrm{q}}$-loading vector.

10. The jth q-weight and q́-loading vector point in the same direction.

11. The t-score variables are asymptotically Gaussian distributed.

12. The PLS q-weight and p-loading vectors and the value of the objective function λ_j allow reconstructing $\mathbf{S}_{y_0x_0}$.

13. If the covariance matrix $\mathbf{S}_{x_0x_0}$ has full rank n_x and the maximum number of latent variable sets have been computed, the PLS regression matrix between the input and output variables, $\mathcal{B}_{PLS}$, is equivalent to that calculated by the ordinary least squares solution, $\mathcal{B}_{OLS} = \left[\mathbf{X}_0^T\mathbf{X}_0\right]^{-1}\mathbf{X}_0^T\mathbf{Y}_0$.

14. In contrast to ordinary least squares, PLS does not require a matrix inversion to compute the regression matrix $\mathcal{B}_{PLS}$.

15. Comparing with the algorithm, discussed in the previous subsection, the computation of a PLS model can be considerably simplified leading to a computationally efficient algorithm.

The above properties are now formulated mathematically and proven.

Property 10.4.1 – Singular value decomposition of $\mathbf{S}_{ef}^{(k)}$. If the cross-covariance matrix $\mathbf{S}_{x_0y_0} = \mathbf{S}_{ef}^{(1)}$ is available, there exists the following relationship between the jth pair of weight vectors and the maximum of the objective function for determining these vectors.

Theorem 10.4.1 *The weight vectors $\mathbf{w}_j$ and $\mathbf{q}_j$ and the value of the objective function in (10.6), λ_j, are the left and right singular vector and the largest singular value of the singular value decomposition of $\mathbf{S}_{ef}^{(j)}$, respectively (Kaspar and Ray 1993).*

Proof. Equation (10.10) shows that the weight vectors $\mathbf{w}_j$ and $\mathbf{q}_j$ are the dominant eigenvectors of $\mathbf{S}_{ef}^{(j)}\mathbf{S}_{fe}^{(j)}$ and $\mathbf{S}_{fe}^{(j)}\mathbf{S}_{ef}^{(j)}$, respectively. Moreover, the largest eigenvalue of both matrices is λ_j^2. On the other hand, a singular value decomposition of a matrix $\mathcal{A}$ of arbitrary dimension is equal to $\mathcal{A} = \mathcal{U}\mathcal{S}\mathcal{V}^T$, where the column vectors of $\mathcal{U}$, that is, the left singular vectors, are the eigenvectors of $\mathcal{A}\mathcal{A}^T$, the column vectors of $\mathcal{V}$, that is, the right singular vectors, are the eigenvectors of $\mathcal{A}^T\mathcal{A}$, and the elements of the diagonal matrix $\mathcal{S}$ are the square root of the eigenvalues of $\mathcal{A}^T\mathcal{A}$ or $\mathcal{A}\mathcal{A}^T$ (Golub and van Loan 1996). Note that the eigenvectors of $\mathcal{A}^T\mathcal{A}$ or $\mathcal{A}\mathcal{A}^T$ are scaled to unit length. Now, replacing $\mathcal{A}$ with $\mathbf{S}_{ef}^{(j)}$, it follows that the first column vector of $\mathcal{U}$ is $\mathbf{w}_j$, the first column vector of $\mathcal{U}$ is $\mathbf{q}_j$ and square root of the eigenvalue of $\mathcal{A}^T\mathcal{A}$ or $\mathcal{A}\mathcal{A}^T$ is λ_j, is the first diagonal element of $\mathcal{S}$, that is, the largest singular value of $\mathcal{A}$. This largest singular value, however, is equal to the maximum of the objective function in (10.6), which concludes this proof.

Property 10.4.2 – Orthogonality of the t-score vectors. The pair of t-score vectors $\widehat{\mathbf{t}}_i$ and $\widehat{\mathbf{t}}_j$ has the following geometric property.

Theorem 10.4.2 *The t-score vectors* $\widehat{\mathbf{t}}_i$ *and* $\widehat{\mathbf{t}}_j$, $i \neq j$, *are mutually orthogonal, that is* $\widehat{\mathbf{t}}_i^T \widehat{\mathbf{t}}_j = 0$.

Proof. First, revisiting the determination of the kth pair of loading vectors yields

$$\begin{aligned}\widehat{\mathbf{p}}_j &= \frac{\widehat{\mathbf{S}}_{ee}^{(j)}\widehat{\mathbf{w}}_j}{\widehat{\mathbf{w}}_j^T\widehat{\mathbf{S}}_{ee}^{(j)}\widehat{\mathbf{w}}_j} = \frac{K-1}{K-1}\frac{\mathbf{E}^{(j)^T}\mathbf{E}^{(j)}\widehat{\mathbf{w}}_j}{\widehat{\mathbf{w}}_j^T\mathbf{E}^{(j)^T}\mathbf{E}^{(j)}\widehat{\mathbf{w}}_j} = \frac{\mathbf{E}^{(j)^T}\widehat{\mathbf{t}}_j}{\widehat{\mathbf{t}}_j^T\widehat{\mathbf{t}}_j} \\ &\text{and} \\ \widehat{\mathbf{q}}_j &= \frac{\widehat{\mathbf{S}}_{fe}^{(j)}\widehat{\mathbf{w}}_j}{\widehat{\mathbf{w}}_j^T\widehat{\mathbf{S}}_{ee}^{(j)}\widehat{\mathbf{w}}_j} = \frac{K-1}{K-1}\frac{\mathbf{F}^{(j)^T}\mathbf{E}^{(j)}\widehat{\mathbf{w}}_j}{\widehat{\mathbf{w}}_j^T\mathbf{E}^{(j)^T}\mathbf{E}^{(j)}\widehat{\mathbf{w}}_j} = \frac{\mathbf{F}^{(j)^T}\widehat{\mathbf{t}}_j}{\widehat{\mathbf{t}}_j^T\widehat{\mathbf{t}}_j}.\end{aligned} \tag{10.19}$$

With respect to (10.15), utilizing (10.19) gives rise to the following deflation procedure for $\mathbf{E}^{(j)}$ and $\mathbf{F}^{(j)}$

$$\begin{aligned}\mathbf{E}^{(j+1)} &= \mathbf{E}^{(j)} - \widehat{\mathbf{t}}_j\widehat{\mathbf{p}}_j^T = \mathbf{E}^{(j)} - \widehat{\mathbf{t}}_j\frac{\widehat{\mathbf{t}}_j^T\mathbf{E}^{(j)}}{\widehat{\mathbf{t}}_j^T\widehat{\mathbf{t}}_j} = \left[\mathbf{I} - \frac{\widehat{\mathbf{t}}_j\widehat{\mathbf{t}}_j^T}{\widehat{\mathbf{t}}_j^T\widehat{\mathbf{t}}_j}\right]\mathbf{E}^{(j)} \\ &\text{and} \\ \mathbf{F}^{(j+1)} &= \mathbf{F}^{(j)} - \widehat{\mathbf{t}}_j\widehat{\mathbf{q}}_j^T = \mathbf{F}^{(j)} - \widehat{\mathbf{t}}_j\frac{\widehat{\mathbf{t}}_j^T\mathbf{F}^{(j)}}{\widehat{\mathbf{t}}_j^T\widehat{\mathbf{t}}_j} = \left[\mathbf{I} - \frac{\widehat{\mathbf{t}}_j\widehat{\mathbf{t}}_j^T}{\widehat{\mathbf{t}}_j^T\widehat{\mathbf{t}}_j}\right]\mathbf{F}^{(j)}.\end{aligned} \tag{10.20}$$

The deflation procedure can, alternatively, also be carried out as

$$\begin{aligned}\mathbf{E}^{(j+1)} &= \mathbf{E}^{(j)} - \mathbf{E}^{(j)}\widehat{\mathbf{w}}_j\widehat{\mathbf{p}}_j^T = \mathbf{E}^{(j)}\left[\mathbf{I} - \widehat{\mathbf{w}}_j\widehat{\mathbf{p}}_j^T\right] \\ &\text{and} \\ \mathbf{F}^{(j+1)} &= \mathbf{F}^{(j)} - \mathbf{E}^{(j)}\widehat{\mathbf{w}}_j\widehat{\mathbf{q}}_j^T.\end{aligned} \tag{10.21}$$

Next, applying the above expressions for deflating $\mathbf{E}^{(j)}$ to simplify the expression

$$\widehat{\mathbf{t}}_i^T\widehat{\mathbf{t}}_j = \widehat{\mathbf{t}}_i^T\mathbf{E}^{(j)}\widehat{\mathbf{w}}_j \tag{10.22}$$

by assuming that $i < j$, yields

$$\mathbf{E}^{(j)} = \left[\mathbf{I} - \frac{\widehat{\mathbf{t}}_i\widehat{\mathbf{t}}_i^T}{\widehat{\mathbf{t}}_i^T\widehat{\mathbf{t}}_i}\right]\mathbf{E}^{(i)}\left[\mathbf{I} - \widehat{\mathbf{w}}_{i+1}\widehat{\mathbf{p}}_{i+1}^T\right]\cdots\left[\mathbf{I} - \widehat{\mathbf{w}}_{j-1}\widehat{\mathbf{p}}_{j-1}^T\right]. \tag{10.23}$$

Now, substituting (10.23) into (10.22) gives rise to

$$\begin{aligned}\widehat{\mathbf{t}}_i^T\widehat{\mathbf{t}}_j &= \widehat{\mathbf{t}}_i^T\left[\mathbf{I} - \frac{\widehat{\mathbf{t}}_i\widehat{\mathbf{t}}_i^T}{\widehat{\mathbf{t}}_i^T\widehat{\mathbf{t}}_i}\right]\mathbf{E}^{(i)}\left[\mathbf{I} - \widehat{\mathbf{p}}_{i+1}\widehat{\mathbf{w}}_{i+1}^T\right]\cdots\left[\mathbf{I} - \widehat{\mathbf{p}}_{j+1}\widehat{\mathbf{w}}_{j+1}^T\right]\widehat{\mathbf{w}}_j \\ \widehat{\mathbf{t}}_i^T\widehat{\mathbf{t}}_j &= \left(\widehat{\mathbf{t}}_i^T - \frac{\widehat{\mathbf{t}}_i^T\widehat{\mathbf{t}}_i\widehat{\mathbf{t}}_i^T}{\widehat{\mathbf{t}}_i^T\widehat{\mathbf{t}}_i}\right)\mathbf{E}^{(i)}\left[\mathbf{I} - \widehat{\mathbf{p}}_{i+1}\widehat{\mathbf{w}}_{i+1}^T\right]\cdots\left[\mathbf{I} - \widehat{\mathbf{p}}_{j+1}\widehat{\mathbf{w}}_{j+1}^T\right]\widehat{\mathbf{w}}_j \\ \boxed{\widehat{\mathbf{t}}_i^T\widehat{\mathbf{t}}_j = \left(\widehat{\mathbf{t}}_i^T - \widehat{\mathbf{t}}_i^T\right)\mathbf{E}^{(i)}\left[\mathbf{I} - \widehat{\mathbf{p}}_{i+1}\widehat{\mathbf{w}}_{i+1}^T\right]\cdots\left[\mathbf{I} - \widehat{\mathbf{p}}_{j+1}\widehat{\mathbf{w}}_{j+1}^T\right]\widehat{\mathbf{w}}_j = 0}&.\end{aligned} \tag{10.24}$$

It is interesting to note that the orthogonality property of the t-score vectors implies that the estimated covariance matrix of the score variables is a diagonal matrix

$$\widehat{\mathbf{S}}_{tt} = \frac{1}{K-1} \operatorname{diag} \left\{ \widehat{\mathbf{t}}_1^T\widehat{\mathbf{t}}_1 \quad \widehat{\mathbf{t}}_2^T\widehat{\mathbf{t}}_2 \quad \cdots \quad \widehat{\mathbf{t}}_n^T\widehat{\mathbf{t}}_n \right\}, \tag{10.25}$$

The orthogonality property of the t-score vectors also results in interesting geometric properties in conjunction with the deflated matrix $\mathbf{E}^{(j)}$, which is discussed next.

Property 10.4.3 – Matrix-vector products $\widehat{\mathbf{t}}_j^T\mathbf{E}^{(j)}$ and $\widehat{\mathbf{t}}_j^T\mathbf{X}_0$. The mutual orthogonality of the t-score vectors gives rise to the following relationship for the matrix vector products $\widehat{\mathbf{t}}_j^T\mathbf{E}^{(j)}$ and $\widehat{\mathbf{t}}_j^T\mathbf{X}_0$.

Lemma 10.4.3 *The products $\widehat{\mathbf{t}}_j^T\mathbf{E}^{(j)}$ and $\widehat{\mathbf{t}}_j^T\mathbf{E}^{(1)} = \widehat{\mathbf{t}}_j^T\mathbf{X}_0$ are equivalent*

Proof. Using the deflation procedure to compute $\mathbf{E}^{(j)}$ yields

$$\widehat{\mathbf{t}}_j^T\mathbf{E}^{(j)} = \widehat{\mathbf{t}}_j^T\left[\mathbf{I} - \frac{\widehat{\mathbf{t}}_{j-1}\widehat{\mathbf{t}}_{j-1}^T}{\widehat{\mathbf{t}}_{j-1}^T\widehat{\mathbf{t}}_{j-1}}\right]\left[\mathbf{I} - \frac{\widehat{\mathbf{t}}_{j-2}\widehat{\mathbf{t}}_{j-2}^T}{\widehat{\mathbf{t}}_{j-2}^T\widehat{\mathbf{t}}_{j-2}}\right]\cdots\left[\mathbf{I} - \frac{\widehat{\mathbf{t}}_1\widehat{\mathbf{t}}_1^T}{\widehat{\mathbf{t}}_1^T\widehat{\mathbf{t}}_1}\right]\mathbf{X}_0$$

$$\widehat{\mathbf{t}}_j^T\mathbf{E}^{(j)} = \left(\widehat{\mathbf{t}}_j^T - \frac{\overbrace{\widehat{\mathbf{t}}_j^T\widehat{\mathbf{t}}_{j-1}}^{=0}\widehat{\mathbf{t}}_{j-1}^T}{\widehat{\mathbf{t}}_{j-1}^T\widehat{\mathbf{t}}_{j-1}}\right)\left[\mathbf{I} - \frac{\widehat{\mathbf{t}}_{j-2}\widehat{\mathbf{t}}_{j-2}^T}{\widehat{\mathbf{t}}_{j-2}^T\widehat{\mathbf{t}}_{j-2}}\right]\cdots\left[\mathbf{I} - \frac{\widehat{\mathbf{t}}_1\widehat{\mathbf{t}}_1^T}{\widehat{\mathbf{t}}_1^T\widehat{\mathbf{t}}_1}\right]\mathbf{X}_0$$

$$\widehat{\mathbf{t}}_j^T\mathbf{E}^{(j)} = \left(\widehat{\mathbf{t}}_j^T - \frac{\overbrace{\widehat{\mathbf{t}}_j^T\widehat{\mathbf{t}}_{j-2}}^{=0}\widehat{\mathbf{t}}_{j-2}^T}{\widehat{\mathbf{t}}_{j-2}^T\widehat{\mathbf{t}}_{j-2}}\right)\left[\mathbf{I} - \frac{\widehat{\mathbf{t}}_{j-3}\widehat{\mathbf{t}}_{j-3}^T}{\widehat{\mathbf{t}}_{j-3}^T\widehat{\mathbf{t}}_{j-3}}\right]\cdots\left[\mathbf{I} - \frac{\widehat{\mathbf{t}}_1\widehat{\mathbf{t}}_1^T}{\widehat{\mathbf{t}}_1^T\widehat{\mathbf{t}}_1}\right]\mathbf{X}_0$$

$$\vdots \tag{10.26}$$

$$\widehat{\mathbf{t}}_j^T\mathbf{E}^{(j)} = \left(\widehat{\mathbf{t}}_j^T - \frac{\overbrace{\widehat{\mathbf{t}}_j^T\widehat{\mathbf{t}}_1}^{=0}\widehat{\mathbf{t}}_1^T}{\widehat{\mathbf{t}}_1^T\widehat{\mathbf{t}}_1}\right)\mathbf{X}_0$$

$$\boxed{\widehat{\mathbf{t}}_j^T\mathbf{E}^{(j)} = \widehat{\mathbf{t}}_j^T\mathbf{X}_0}.$$

Property 10.4.4 – Matrix-vector product $\widehat{\mathbf{t}}_i^T\mathbf{E}^{(j)}$. The mutual orthogonality of the t-score variable leads to the following property for the matrix-vector product $\widehat{\mathbf{t}}_i^T\mathbf{E}^{(j)}$ if $i < j$.

Lemma 10.4.4 *The matrix vector product $\widehat{\mathbf{t}}_i^T\mathbf{E}^{(j)} = 0$ for all $i < j$ and $\widehat{\mathbf{p}}_i^T\left(\widehat{\mathbf{t}}_i^T\widehat{\mathbf{t}}_i\right)$ for all $i \geq j$.*

Proof. For $i < j$, the application of the deflation procedure for $\mathbf{E}^{(j)}$ gives rise to

$$\mathbf{E}^{(j)} = \left[\mathbf{I} - \frac{\widehat{\mathbf{t}}_{j-1}\widehat{\mathbf{t}}_{j-1}^T}{\widehat{\mathbf{t}}_{j-1}^T\widehat{\mathbf{t}}_{j-1}}\right]\cdots\left[\mathbf{I} - \frac{\widehat{\mathbf{t}}_i\widehat{\mathbf{t}}_i^T}{\widehat{\mathbf{t}}_i^T\widehat{\mathbf{t}}_i}\right]\mathbf{E}^{(i)}. \tag{10.27}$$

Substituting the above equation into the matrix-vector product $\widehat{\mathbf{t}}_i^T\mathbf{E}^{(j)}$ yields

$$\begin{aligned}
\widehat{\mathbf{t}}_i^T\mathbf{E}^{(j)} &= \widehat{\mathbf{t}}_i^T\left[\mathbf{I} - \frac{\widehat{\mathbf{t}}_{j-1}\widehat{\mathbf{t}}_{j-1}^T}{\widehat{\mathbf{t}}_{j-1}^T\widehat{\mathbf{t}}_{j-1}}\right]\cdots\left[\mathbf{I} - \frac{\widehat{\mathbf{t}}_i\widehat{\mathbf{t}}_i^T}{\widehat{\mathbf{t}}_i^T\widehat{\mathbf{t}}_i}\right]\mathbf{E}^{(i)} \\
\widehat{\mathbf{t}}_i^T\mathbf{E}^{(j)} &= \left(\widehat{\mathbf{t}}_i^T - \frac{\overbrace{\widehat{\mathbf{t}}_i^T\widehat{\mathbf{t}}_{j-1}}^{=0}\widehat{\mathbf{t}}_{j-1}^T}{\widehat{\mathbf{t}}_{j-1}^T\widehat{\mathbf{t}}_{j-1}}\right)\left[\mathbf{I} - \frac{\widehat{\mathbf{t}}_{j-2}\widehat{\mathbf{t}}_{j-2}^T}{\widehat{\mathbf{t}}_{j-2}^T\widehat{\mathbf{t}}_{j-2}}\right]\cdots\left[\mathbf{I} - \frac{\widehat{\mathbf{t}}_i\widehat{\mathbf{t}}_i^T}{\widehat{\mathbf{t}}_i^T\widehat{\mathbf{t}}_i}\right]\mathbf{E}^{(i)} \\
&\vdots \\
&\boxed{\widehat{\mathbf{t}}_i^T\mathbf{E}^{(j)} = \left(\widehat{\mathbf{t}}_i^T - \frac{\widehat{\mathbf{t}}_i^T\widehat{\mathbf{t}}_i\widehat{\mathbf{t}}_i^T}{\widehat{\mathbf{t}}_i^T\widehat{\mathbf{t}}_i}\right)\mathbf{E}^{(i)} = \left(\widehat{\mathbf{t}}_i^T - \widehat{\mathbf{t}}_i^T\right)\mathbf{X}_0 = \mathbf{0}}.
\end{aligned} \tag{10.28}$$

For $i \geq j$, Lemma 10.4.3 highlights that $\widehat{\mathbf{t}}_i^T\mathbf{E}^{(j)} = \widehat{\mathbf{t}}_i^T\mathbf{X}_0$. Equation (10.19) shows that $\widehat{\mathbf{t}}_i^T\mathbf{X}_0 = \widehat{\mathbf{t}}_i^T\mathbf{E}^{(i)}$ forms part of the calculation of the p-loading vector

$$\boxed{\widehat{\mathbf{p}}_i^T = \frac{\widehat{\mathbf{t}}_i^T\mathbf{E}^{(i)}}{\widehat{\mathbf{t}}_i^T\widehat{\mathbf{t}}_i} = \frac{\widehat{\mathbf{t}}_i^T\mathbf{X}_0}{\widehat{\mathbf{t}}_i^T\widehat{\mathbf{t}}_i} \Rightarrow \widehat{\mathbf{p}}_i^T\left(\widehat{\mathbf{t}}_i^T\widehat{\mathbf{t}}_i\right) = \widehat{\mathbf{t}}_i^T\mathbf{X}_0}. \tag{10.29}$$

Property 10.4.5 – Orthogonality of the t- and u-score vectors. The mutual orthogonality of any pair of t-score vectors also implies the following geometric property for the t- and u-score vectors.

Lemma 10.4.5 *The ith t-score vector is orthogonal to the jth u-score vector, that is, $\widehat{\mathbf{t}}_i^T\widehat{\mathbf{u}}_j = 0$ for all $i < j$.*

Proof. With $\widehat{\mathbf{u}}_j = \mathbf{F}^{(j)}\widehat{\mathbf{q}}_i$, the scalar product $\widehat{\mathbf{t}}_i^T\widehat{\mathbf{u}}_j$ becomes

$$\widehat{\mathbf{t}}_i^T\widehat{\mathbf{u}}_j = \widehat{\mathbf{t}}_i^T\mathbf{F}^{(j)}\widehat{\mathbf{q}}_i. \tag{10.30}$$

For $j > i$, tracing the deflated output matrix $\mathbf{F}^{(j)}$ from j back to i gives rise to

$$\widehat{\mathbf{t}}_i^T\widehat{\mathbf{u}}_j = \widehat{\mathbf{t}}_i^T\left[\mathbf{I} - \frac{\widehat{\mathbf{t}}_{j-1}\widehat{\mathbf{t}}_{j-1}^T}{\widehat{\mathbf{t}}_{j-1}^T\widehat{\mathbf{t}}_{j-1}}\right]\mathbf{F}^{(j-1)}\widehat{\mathbf{q}}_j$$

$$\widehat{\mathbf{t}}_i^T\widehat{\mathbf{u}}_j = \widehat{\mathbf{t}}_i^T\left[\mathbf{I} - \frac{\widehat{\mathbf{t}}_{j-1}\widehat{\mathbf{t}}_{j-1}^T}{\widehat{\mathbf{t}}_{j-1}^T\widehat{\mathbf{t}}_{j-1}}\right]\cdots\left[\mathbf{I} - \frac{\widehat{\mathbf{t}}_i\widehat{\mathbf{t}}_i^T}{\widehat{\mathbf{t}}_i^T\widehat{\mathbf{t}}_i}\right]\mathbf{F}^{(i)}\widehat{\mathbf{q}}_j$$

$$\widehat{\mathbf{t}}_i^T\widehat{\mathbf{u}}_j = \left(\widehat{\mathbf{t}}_i^T - \frac{\overbrace{\widehat{\mathbf{t}}_i^T\widehat{\mathbf{t}}_{j+1}}^{=0}\widehat{\mathbf{t}}_{j+1}^T}{\widehat{\mathbf{t}}_{j+1}^T\widehat{\mathbf{t}}_{j+1}}\right)\left[\mathbf{I} - \frac{\widehat{\mathbf{t}}_{j-2}\widehat{\mathbf{t}}_{j-2}^T}{\widehat{\mathbf{t}}_{j-2}^T\widehat{\mathbf{t}}_{j-2}}\right]\cdots\left[\mathbf{I} - \frac{\widehat{\mathbf{t}}_i\widehat{\mathbf{t}}_i^T}{\widehat{\mathbf{t}}_i^T\widehat{\mathbf{t}}_i}\right]\mathbf{F}^{(i)}\widehat{\mathbf{q}}_j \tag{10.31}$$

$$\vdots$$

$$\boxed{\widehat{\mathbf{t}}_i^T\widehat{\mathbf{u}}_j = \left(\widehat{\mathbf{t}}_i^T - \frac{\widehat{\mathbf{t}}_i^T\widehat{\mathbf{t}}_i\widehat{\mathbf{t}}_i^T}{\widehat{\mathbf{t}}_i^T\widehat{\mathbf{t}}_i}\right)\mathbf{F}^{(i)}\widehat{\mathbf{q}}_j = \left(\widehat{\mathbf{t}}_i^T - \widehat{\mathbf{t}}_i^T\right)\mathbf{F}^{(i)}\widehat{\mathbf{q}}_j = 0}.$$

Property 10.4.6 – Deflation of the data matrices. The analysis focuses now on the deflation procedure, which yields that only one of the output variable sets needs to be deflated and not both simultaneously. Therefore, the following holds true for the deflation of the data matrices.

Theorem 10.4.6 *The deflation procedure requires the deflation of the output data matrix or the input data matrix only.*

Proof. First, we examine the deflation of the output data matrix. This analysis also yields the necessary condition to show that it is sufficient to deflate the input data matrix only, which culminates in Corollary 10.4.8. Examining the deflation procedure of the PLS algorithm in Table 10.1 highlights that the deflation procedure is applied to the covariance and cross-covariance matrices. These matrices can be replaced by the matrix products $\mathbf{E}^{(j)^T}\mathbf{E}^{(j)}$ and $\mathbf{E}^{(j)^T}\mathbf{F}^{(j)}$, respectively. The deflation of these matrix products leads to

$$\mathbf{E}^{(j+1)^T}\mathbf{E}^{(j+1)} = \mathbf{E}^{(j)^T}\left[\mathbf{I} - \frac{\widehat{\mathbf{t}}_j\widehat{\mathbf{t}}_j^T}{\widehat{\mathbf{t}}_j^T\widehat{\mathbf{t}}_j}\right]\left[\mathbf{I} - \frac{\widehat{\mathbf{t}}_j\widehat{\mathbf{t}}_j^T}{\widehat{\mathbf{t}}_j^T\widehat{\mathbf{t}}_j}\right]\mathbf{E}^{(j)}$$

$$\mathbf{E}^{(j+1)^T}\mathbf{E}^{(j+1)} = \mathbf{E}^{(j)^T}\left[\mathbf{I} - \frac{\widehat{\mathbf{t}}_j\widehat{\mathbf{t}}_j^T}{\widehat{\mathbf{t}}_j^T\widehat{\mathbf{t}}_j} - \frac{\widehat{\mathbf{t}}_j\widehat{\mathbf{t}}_j^T}{\widehat{\mathbf{t}}_j^T\widehat{\mathbf{t}}_j} + \frac{\widehat{\mathbf{t}}_j\widehat{\mathbf{t}}_j^T}{\widehat{\mathbf{t}}_j^T\widehat{\mathbf{t}}_j}\right]\mathbf{E}^{(j)} \tag{10.32}$$

$$\mathbf{E}^{(j+1)^T}\mathbf{E}^{(j+1)} = \mathbf{E}^{(j)^T}\left[\mathbf{I} - \frac{\widehat{\mathbf{t}}_j\widehat{\mathbf{t}}_j^T}{\widehat{\mathbf{t}}_j^T\widehat{\mathbf{t}}_j}\right]\mathbf{E}^{(j)} = \mathbf{E}^{(j+1)^T}\mathbf{E}^{(j)^T} = \mathbf{E}^{(j)^T}\mathbf{E}^{(j+1)}.$$

If $j = 1$, it follows from (10.32) that $\mathbf{E}^{(2)^T}\mathbf{E}^{(2)} = \mathbf{E}^{(1)^T}\mathbf{E}^{(2)} = \mathbf{X}_0^T\mathbf{E}^{(2)}$. To prove the general case, the deflation procedure allows computing $\mathbf{E}^{(j)}$ from the output data matrix $\mathbf{X}_0$ using the score vectors $\widehat{\mathbf{t}}_1, \ldots, \widehat{\mathbf{t}}_{j-1}$.

$$\mathbf{E}^{(j+1)^T}\mathbf{E}^{(j+1)} = \mathbf{E}^{(j)^T}\left[\mathbf{I} - \frac{\widehat{\mathbf{t}}_j\widehat{\mathbf{t}}_j^T}{\widehat{\mathbf{t}}_j^T\widehat{\mathbf{t}}_j}\right]\mathbf{E}^{(j)}$$

$$\mathbf{E}^{(j+1)^T}\mathbf{E}^{(j+1)} = \mathbf{E}^{(j-1)^T}\left[\mathbf{I} - \frac{\widehat{\mathbf{t}}_{j-1}\widehat{\mathbf{t}}_{j-1}^T}{\widehat{\mathbf{t}}_{j-1}^T\widehat{\mathbf{t}}_{j-1}}\right]\left[\mathbf{I} - \frac{\widehat{\mathbf{t}}_j\widehat{\mathbf{t}}_j^T}{\widehat{\mathbf{t}}_j^T\widehat{\mathbf{t}}_j}\right]\mathbf{E}^{(j)}$$

$$\vdots \tag{10.33}$$

$$\mathbf{E}^{(j+1)^T}\mathbf{E}^{(j+1)} = \mathbf{X}_0^T\prod_{i=1}^{j}\left[\mathbf{I} - \frac{\widehat{\mathbf{t}}_i\widehat{\mathbf{t}}_i^T}{\widehat{\mathbf{t}}_i^T\widehat{\mathbf{t}}_i}\right]\mathbf{E}^{(j)}$$

$$\mathbf{E}^{(j+1)^T}\mathbf{E}^{(j+1)} = \mathbf{X}_0^T\left[\mathbf{I} - \sum_{i=1}^{j}\frac{\widehat{\mathbf{t}}_i\widehat{\mathbf{t}}_i^T}{\widehat{\mathbf{t}}_i^T\widehat{\mathbf{t}}_i}\right]\mathbf{E}^{(j)}$$

$$\boxed{\mathbf{E}^{(j+1)^T}\mathbf{E}^{(j+1)} = \mathbf{X}_0^T\left[\mathbf{I} - \frac{\widehat{\mathbf{t}}_j\widehat{\mathbf{t}}_j^T}{\widehat{\mathbf{t}}_j^T\widehat{\mathbf{t}}_j}\right]\mathbf{E}^{(j)} = \mathbf{X}_0^T\mathbf{E}^{(j+1)} = \mathbf{E}^{(j+1)^T}\mathbf{X}_0}.$$

The above relationship relies on the fact that the t-score vectors are mutually orthogonal, as described in Theorem 10.4.2, and that $\widehat{\mathbf{t}}_i^T\mathbf{E}^{(j)} = 0$ for all $i < j$, outlined in Lemma 10.4.4. Applying the same steps yields

$$\boxed{\mathbf{E}^{(j+1)^T}\mathbf{F}^{(j+1)} = \mathbf{X}_0^T\mathbf{F}^{(j+1)} = \mathbf{E}^{(j+1)^T}\mathbf{Y}_0}. \tag{10.34}$$

The conclusion of this proof requires showing that the calculation of the t-score vectors can be carried out directly from the input data matrix, since

$$\begin{aligned} \widehat{\mathbf{p}}_j &= \frac{\mathbf{X}_0^T\widehat{\mathbf{t}}_j}{\widehat{\mathbf{t}}_j^T\widehat{\mathbf{t}}_j} & \widehat{\mathbf{q}}_j &= \frac{\mathbf{F}^{(j)^T}\widehat{\mathbf{t}}_j}{\widehat{\mathbf{t}}_j^T\widehat{\mathbf{t}}_j} \\ \widehat{\mathbf{q}}_j &\propto \widehat{\mathbf{F}}^{(j)^T}\mathbf{X}_0\mathbf{w}_j & \widehat{\mathbf{w}}_j &\propto \mathbf{X}_0\mathbf{F}^{(j)}\widehat{\mathbf{q}}_j \end{aligned} \tag{10.35}$$

$$\widehat{\mathbf{t}}_j \neq \mathbf{X}_0\widehat{\mathbf{w}}_j,$$

which is formulated below.

Lemma 10.4.7 *The definition of the r-weight vectors*

$$\widehat{\mathbf{r}}_1 = \widehat{\mathbf{w}}_1 \qquad \widehat{\mathbf{r}}_j = \widehat{\mathbf{w}}_j - \sum_{i=1}^{j-1}\widehat{\mathbf{p}}_i^T\widehat{\mathbf{w}}_j\widehat{\mathbf{r}}_i \tag{10.36}$$

enables calculation of the t-score vectors directly from the input data matrix, that is, $\widehat{\mathbf{t}}_j = \mathbf{X}_0\widehat{\mathbf{r}}_j$, $1 \le j \le n$.

Proof. Revisiting the calculation of the kth t-score vector yields

$$\begin{aligned}
\widehat{\mathbf{t}}_j &= \mathbf{E}^{(j)}\widehat{\mathbf{w}}_j = \left[\mathbf{E}^{(j-1)} - \widehat{\mathbf{t}}_{j-1}\widehat{\mathbf{p}}_{j-1}^T\right]\widehat{\mathbf{w}}_j \\
\widehat{\mathbf{t}}_j &= \left[\mathbf{E}^{(j-2)} - \widehat{\mathbf{t}}_{j-2}\widehat{\mathbf{p}}_{j-2}^T - \widehat{\mathbf{t}}_{j-1}\widehat{\mathbf{p}}_{j-1}^T\right]\widehat{\mathbf{w}}_j \\
\widehat{\mathbf{t}}_j &= \left[\mathbf{X}_0 - \widehat{\mathbf{t}}_1\widehat{\mathbf{p}}_1^T - \widehat{\mathbf{t}}_2\widehat{\mathbf{p}}_2^T - \cdots - \widehat{\mathbf{t}}_{j-2}\widehat{\mathbf{p}}_{j-2}^T - \widehat{\mathbf{t}}_{j-1}\widehat{\mathbf{p}}_{j-1}^T\right]\widehat{\mathbf{w}}_j \qquad (10.37) \\
\widehat{\mathbf{t}}_j &= \left[\mathbf{X}_0 - \mathbf{X}_0\widehat{\mathbf{r}}_1\widehat{\mathbf{p}}_1^T - \mathbf{X}_0\widehat{\mathbf{r}}_2\widehat{\mathbf{p}}_2^T - \cdots - \mathbf{X}_0\widehat{\mathbf{r}}_{j-2}\widehat{\mathbf{p}}_{j-2}^T - \mathbf{X}_0\widehat{\mathbf{r}}_{j-1}\widehat{\mathbf{p}}_{j-1}^T\right]\widehat{\mathbf{w}}_j \\
\widehat{\mathbf{t}}_j &= \mathbf{X}_0\left[\mathbf{I} - \widehat{\mathbf{r}}_1\widehat{\mathbf{p}}_1^T - \widehat{\mathbf{r}}_2\widehat{\mathbf{p}}_2^T - \cdots - \widehat{\mathbf{r}}_{j-2}\widehat{\mathbf{p}}_{j-2}^T - \widehat{\mathbf{r}}_{j-1}\widehat{\mathbf{p}}_{j-1}^T\right]\widehat{\mathbf{w}}_j
\end{aligned}$$

which gives rise to the following iterative calculation of the r-weight vectors

$$\boxed{\widehat{\mathbf{r}}_1 = \widehat{\mathbf{w}}_1 \qquad \widehat{\mathbf{r}}_j = \widehat{\mathbf{w}}_j - \sum_{i=1}^{j-1}\widehat{\mathbf{r}}_i\widehat{\mathbf{p}}_i^T\widehat{\mathbf{w}}_j = \widehat{\mathbf{w}}_j - \sum_{i=1}^{j-1}\widehat{\mathbf{p}}_i^T\widehat{\mathbf{w}}_j\widehat{\mathbf{r}}_i}. \qquad (10.38)$$

Equation (10.35) highlights that only the output matrix $\mathbf{Y}_0 \rightarrow \mathbf{F}^{(k)}$ needs to be deflated, given that the r-weight vectors allow the computation of the t-score vectors directly, which concludes the proof of Theorem 10.4.6. Moreover, it is also important to note the following.

Corollary 10.4.8 *It is also sufficient to deflate* $\mathbf{X}_0$ *instead of* $\mathbf{Y}_0$.

Corollary 10.4.8 follows from the fact that $\mathbf{E}^{(j)^T}\mathbf{F}^{(j)} = \mathbf{X}_0\mathbf{F}^{(j)} = \mathbf{E}^{(j)^T}\mathbf{Y}_0$, discussed in (10.34). Whilst this does not require the introduction of the r-weight vectors in Lemma 10.4.7, it requires the deflation of two matrix products, that is, $\mathbf{X}_0^T\mathbf{E}^{(j)} \rightarrow \mathbf{X}_0^T\mathbf{E}^{(j+1)}$ and $\mathbf{X}_0^T\mathbf{F}^{(j)} \rightarrow \mathbf{X}_0^T\mathbf{F}^{(j+1)}$, for computing the pairs of weight and loading vectors. It is, however, computationally more expensive to deflate both matrix products. The following rank-one modification presents a numerically expedient way to deflate the matrix product $\mathbf{E}^{(j)^T}\mathbf{F}^{(j)}$

$$\begin{aligned}
\mathbf{E}^{(j+1)^T}\mathbf{F}^{(j+1)} &= \mathbf{E}^{(j)^T}\left[\mathbf{I} - \frac{\widehat{\mathbf{t}}_j\widehat{\mathbf{t}}_j^T}{\widehat{\mathbf{t}}_j^T\widehat{\mathbf{t}}_j}\right]\mathbf{F}^{(j)} \\
\mathbf{E}^{(j+1)^T}\mathbf{F}^{(j+1)} &= \mathbf{E}^{(j)^T}\mathbf{F}^{(j)} - \underbrace{\mathbf{E}^{(j)^T}\widehat{\mathbf{t}}_j}_{=\widehat{\mathbf{p}}_j\left(\widehat{\mathbf{t}}_j^T\widehat{\mathbf{t}}_j\right)}\underbrace{\frac{\widehat{\mathbf{t}}_j^T\mathbf{F}^{(j)}}{\widehat{\mathbf{t}}_j^T\widehat{\mathbf{t}}_j}}_{=\widehat{\mathbf{q}}_j^T} \qquad (10.39) \\
&\boxed{\mathbf{E}^{(j+1)^T}\mathbf{F}^{(j+1)} = \mathbf{E}^{(j)^T}\mathbf{F}^{(j)} - \widehat{\mathbf{p}}_j\left(\widehat{\mathbf{t}}_j^T\widehat{\mathbf{t}}_j\right)\widehat{\mathbf{q}}_j^T}.
\end{aligned}$$

It should be noted that the scalar product $\widehat{\mathbf{t}}_j^T\widehat{\mathbf{t}}_j$ is required for the calculation of the loading vectors and hence available for the deflation of $\mathbf{E}^{(j)^T}\mathbf{F}^{(j)}$. The relationship of (10.39) relies on (10.19), (10.29) and (10.32).

Property 10.4.7 – Orthogonality of the w-weight vectors. We now focus on orthogonality properties of the w-weight vectors and start with the geometry property of any pair, $\widehat{\mathbf{w}}_i$ and $\widehat{\mathbf{w}}_j$, which has the following property.

Theorem 10.4.9 *The w-weight vectors $\widehat{\mathbf{w}}_i$ and $\widehat{\mathbf{w}}_j$, $i \neq j$, are mutually orthonormal, that is $\widehat{\mathbf{w}}_i^T \widehat{\mathbf{w}}_j = 0$.*

Proof. Assuming that $i > j$, the scalar product $\widehat{\mathbf{w}}_i^T \widehat{\mathbf{w}}_j$ can be rewritten as

$$\frac{\widehat{\mathbf{w}}_i^T \mathbf{E}^{(i)^T} \mathbf{Y}_0 \mathbf{Y}_0^T \mathbf{E}^{(i)^T} \widehat{\mathbf{w}}_j}{\left\| \mathbf{E}^{(i)^T} \mathbf{Y}_0 \mathbf{Y}_0^T \mathbf{E}^{(i)^T} \widehat{\mathbf{w}}_i \right\|}, \tag{10.40}$$

which follows from (10.10). Next, analyzing the term $\mathbf{E}^{(i)} \widehat{\mathbf{w}}_j$ reveals that it is equal to zero

$$\begin{aligned}
\mathbf{E}^{(i)} \widehat{\mathbf{w}}_j &= \left[\mathbf{I} - \frac{\widehat{\mathbf{t}}_{i-1} \widehat{\mathbf{t}}_{i-1}^T}{\widehat{\mathbf{t}}_{i-1}^T \widehat{\mathbf{t}}_{i-1}} \right] \cdots \left[\mathbf{I} - \frac{\widehat{\mathbf{t}}_j \widehat{\mathbf{t}}_j^T}{\widehat{\mathbf{t}}_j^T \widehat{\mathbf{t}}_j} \right] \mathbf{E}^{(j)} \widehat{\mathbf{w}}_j \\
\mathbf{E}^{(i)} \widehat{\mathbf{w}}_j &= \left[\mathbf{I} - \frac{\widehat{\mathbf{t}}_{i-1} \widehat{\mathbf{t}}_{i-1}^T}{\widehat{\mathbf{t}}_{i-1}^T \widehat{\mathbf{t}}_{i-1}} \right] \cdots \left[\mathbf{I} - \frac{\widehat{\mathbf{t}}_j \widehat{\mathbf{t}}_j^T}{\widehat{\mathbf{t}}_j^T \widehat{\mathbf{t}}_j} \right] \widehat{\mathbf{t}}_j \\
\mathbf{E}^{(i)} \widehat{\mathbf{w}}_j &= \left[\mathbf{I} - \frac{\widehat{\mathbf{t}}_{i-1} \widehat{\mathbf{t}}_{i-1}^T}{\widehat{\mathbf{t}}_{i-1}^T \widehat{\mathbf{t}}_{i-1}} \right] \cdots \left[\mathbf{I} - \frac{\widehat{\mathbf{t}}_{j+1} \widehat{\mathbf{t}}_{j+1}^T}{\widehat{\mathbf{t}}_{j+1}^T \widehat{\mathbf{t}}_{j+1}} \right] \left(\widehat{\mathbf{t}}_j - \widehat{\mathbf{t}}_j \right) = \mathbf{0}
\end{aligned} \tag{10.41}$$

which implies that

$$\boxed{\widehat{\mathbf{w}}_i^T \widehat{\mathbf{w}}_j = 0}.$$

Property 10.4.8 – Orthogonality of the w-weight and p-loading vectors. The following holds true for the scalar product $\widehat{\mathbf{p}}_i^T \widehat{\mathbf{w}}_j$.

Lemma 10.4.10 *The ith p-loading and the jth w-weight vector are orthogonal if $i > j$ and equal to 1 for $i = j$.*

Proof. According to (10.19), the scalar product $\widehat{\mathbf{p}}_i^T \widehat{\mathbf{w}}_j$ is given by

$$\widehat{\mathbf{p}}_i^T \widehat{\mathbf{w}}_j = \frac{\widehat{\mathbf{t}}_i^T \mathbf{E}^{(i)} \widehat{\mathbf{w}}_j}{\widehat{\mathbf{t}}_i^T \widehat{\mathbf{t}}_i}. \tag{10.42}$$

For $i > j$, tracing the deflation procedure for $\mathbf{E}^{(i)}$ from i back to j yields

$$\frac{\widehat{\mathbf{t}}_i^T \mathbf{E}^{(i)} \widehat{\mathbf{w}}_j}{\widehat{\mathbf{t}}_i^T \widehat{\mathbf{t}}_i} = \frac{\widehat{\mathbf{t}}_i^T \left[\mathbf{I} - \frac{\mathbf{t}_{i-1} \mathbf{t}_{i-1}^T}{\mathbf{t}_{i-1}^T \mathbf{t}_{i-1}} \right] \mathbf{E}^{(i-1)} \widehat{\mathbf{w}}_j}{\widehat{\mathbf{t}}_i^T \widehat{\mathbf{t}}_i}$$

$$\frac{\widehat{\mathbf{t}}_i^T \mathbf{E}^{(i)} \widehat{\mathbf{w}}_j}{\widehat{\mathbf{t}}_i^T \widehat{\mathbf{t}}_i} = \frac{\widehat{\mathbf{t}}_i^T \left[\mathbf{I} - \frac{\mathbf{t}_{i-1}\mathbf{t}_{i-1}^T}{\mathbf{t}_{i-1}^T \mathbf{t}_{i-1}}\right] \cdots \left[\mathbf{I} - \frac{\mathbf{t}_j \mathbf{t}_j^T}{\mathbf{t}_j^T \mathbf{t}_j}\right] \overbrace{\mathbf{E}^{(j)} \widehat{\mathbf{w}}_j}^{=\widehat{\mathbf{t}}_j}}{\widehat{\mathbf{t}}_i^T \widehat{\mathbf{t}}_i} \tag{10.43}$$

$$\boxed{\frac{\widehat{\mathbf{t}}_i^T \mathbf{E}^{(i)} \widehat{\mathbf{w}}_j}{\widehat{\mathbf{t}}_i^T \widehat{\mathbf{t}}_i} = \frac{\widehat{\mathbf{t}}_i^T \left[\mathbf{I} - \frac{\widehat{\mathbf{t}}_{i-1}\widehat{\mathbf{t}}_{i-1}^T}{\widehat{\mathbf{t}}_{i-1}^T \widehat{\mathbf{t}}_{i-1}}\right] \cdots \left[\mathbf{I} - \frac{\widehat{\mathbf{t}}_{j+1}\widehat{\mathbf{t}}_{j+1}^T}{\widehat{\mathbf{t}}_{j+1}^T \widehat{\mathbf{t}}_{j+1}}\right]}{\widehat{\mathbf{t}}_i^T \widehat{\mathbf{t}}_i} \left(\widehat{\mathbf{t}}_j - \widehat{\mathbf{t}}_j\right) = 0}.$$

That $\widehat{\mathbf{p}}_i^T \widehat{\mathbf{w}}_j = 1$ for $i = j$ follows from the computation of the p-loading vector

$$\boxed{\widehat{\mathbf{p}}_i^T \widehat{\mathbf{w}}_j = \frac{\widehat{\mathbf{t}}_i^T \overbrace{\mathbf{E}^{(i)} \widehat{\mathbf{w}}_j}^{=\widehat{\mathbf{t}}_j}}{\widehat{\mathbf{t}}_i^T \widehat{\mathbf{t}}_i} = \frac{\widehat{\mathbf{t}}_j^T \widehat{\mathbf{t}}_j}{\widehat{\mathbf{t}}_j^T \widehat{\mathbf{t}}_j} = 1}. \tag{10.44}$$

Property 10.4.9 – Calculation of the regression coefficient $\widehat{b}_j$. There is the following relationship between the estimated regression coefficient of the jth pair of score variables, t_j and u_j, and the length of the jth q́-loading vector

Theorem 10.4.11 *The estimated regression coefficient $\widehat{b}_j$ is equal to the norm of the q́-loading vector $\acute{\mathbf{q}}_j^T = \widehat{\mathbf{t}}_j \mathbf{F}^{(j)} / \widehat{\mathbf{t}}_j^T \widehat{\mathbf{t}}_j$, that is $\widehat{b}_j = \|\acute{\mathbf{q}}_j\|$.*

Proof. Determining the length of $\acute{\mathbf{q}}_j$ yields

$$\|\acute{\mathbf{q}}_j\| = \sqrt{\frac{\widehat{\mathbf{t}}_j^T \mathbf{F}^{(j)} \mathbf{F}^{(j)^T} \widehat{\mathbf{t}}_j}{\left(\widehat{\mathbf{t}}_j^T \widehat{\mathbf{t}}_j\right)^2}} = \frac{\sqrt{\widehat{\mathbf{w}}_j^T \mathbf{E}^{(j)^T} \mathbf{F}^{(j)} \mathbf{F}^{(j)^T} \mathbf{E}^{(j)} \widehat{\mathbf{w}}_j}}{\widehat{\mathbf{t}}_j^T \widehat{\mathbf{t}}_j}. \tag{10.45}$$

However, since $\widehat{\mathbf{w}}_j$ is the dominant eigenvector of $\mathbf{E}^{(j)^T} \mathbf{F}^{(j)} \mathbf{F}^{(j)^T} \mathbf{E}^{(j)}$, the expression $\widehat{\mathbf{w}}_j^T \mathbf{E}^{(j)^T} \mathbf{F}^{(j)} \mathbf{F}^{(j)^T} \mathbf{E}^{(j)} \widehat{\mathbf{w}}_j$ is equal to the largest eigenvalue. According to (10.6) and (10.10) this eigenvalue is equal to the square of the Lagrangian multiplier of the objective function for computing the jth pair of weight vectors. Moreover, the eigenvalue of $\mathbf{E}^{(j)^T} \mathbf{F}^{(j)} \mathbf{F}^{(j)^T} \mathbf{E}^{(j)}$ is $(K-1)^2$ times the eigenvalue of $\widehat{\mathbf{S}}_{ef}^{(j)} \widehat{\mathbf{S}}_{fe}^{(j)}$, $\widehat{\lambda}_j^2$, and hence, equal to $\left((K-1)\widehat{\lambda}_j\right)^2$. On the other hand, $\widehat{\mathbf{t}}_j^T \widehat{\mathbf{u}}_k$ is the estimate for $K-1$ times the covariance between the t- and u-score variables. Consequently, (10.45) becomes

$$\boxed{\|\acute{\mathbf{q}}_j\| = \frac{\sqrt{\widehat{\lambda}_j^2}}{\widehat{\mathbf{t}}_j^T \widehat{\mathbf{t}}_j} = \frac{\sqrt{\left(\widehat{\mathbf{t}}_j^T \widehat{\mathbf{u}}_j\right)^2}}{\widehat{\mathbf{t}}_j^T \widehat{\mathbf{t}}_j} = \frac{\widehat{\mathbf{t}}_j^T \widehat{\mathbf{u}}_j}{\widehat{\mathbf{t}}_j^T \widehat{\mathbf{t}}_j}} \tag{10.46}$$

which is, according to (10.11), equal to the estimate $\widehat{b}_k$.

Property 10.4.10 – Relationship between the q-weight and q́-loading vectors. The following relationship between the jth pair of q-weight and q́-loading vectors exists.

Theorem 10.4.12 *The q-weight vector* $\widehat{\mathbf{q}}_j$ *and the q́-loading vector* $\widehat{\acute{\mathbf{q}}}_j$ *have the same direction and the scaling factor between these vectors is the regression coefficient* $\widehat{b}_j$.

Proof. According to (10.8), the q-weight vector can be written as

$$\widehat{\mathbf{q}}_j = \frac{\mathbf{F}^{(j)^T}\mathbf{E}^{(j)}\widehat{\mathbf{w}}_j}{\left\|\mathbf{F}^{(j)^T}\mathbf{E}^{(j)}\widehat{\mathbf{w}}_j\right\|} = \frac{\mathbf{F}^{(j)^T}\widehat{\mathbf{t}}_j}{\left\|\mathbf{F}^{(j)^T}\widehat{\mathbf{t}}_j\right\|} \tag{10.47}$$

whilst the q́-loading vector is given by

$$\widehat{\acute{\mathbf{q}}}_j = \frac{\mathbf{F}^{(j)^T}\widehat{\mathbf{t}}_j}{\widehat{\mathbf{t}}_j^T\widehat{\mathbf{t}}_j} = \frac{\widehat{\mathbf{q}}_j\left\|\mathbf{F}^{(j)^T}\widehat{\mathbf{t}}_j\right\|}{\widehat{\mathbf{t}}_j^T\widehat{\mathbf{t}}_j}. \tag{10.48}$$

Since

$$\left\|\mathbf{F}^{(j)^T}\widehat{\mathbf{t}}_j\right\| = \sqrt{\widehat{\mathbf{t}}_j^T\mathbf{F}^{(j)}\mathbf{F}^{(j)^T}\widehat{\mathbf{t}}_j} = \sqrt{\widehat{\mathbf{w}}_j^T\mathbf{E}^{(j)^T}\mathbf{F}^{(j)}\mathbf{F}^{(j)^T}\mathbf{E}^{(j)}\widehat{\mathbf{w}}_j} \tag{10.49}$$

is, according to Theorem 10.4.11, equal to $\widehat{\mathbf{t}}_k^T\widehat{\mathbf{u}}_k$. Equation (10.48) therefore becomes

$$\boxed{\widehat{\acute{\mathbf{q}}}_j = \widehat{\mathbf{q}}_j\frac{\widehat{\mathbf{t}}_j^T\widehat{\mathbf{u}}_j}{\widehat{\mathbf{t}}_j^T\widehat{\mathbf{t}}_j} = \widehat{\mathbf{q}}_j\widehat{b}_j}. \tag{10.50}$$

This, however, implies that $\widehat{\acute{\mathbf{q}}}_j \propto \widehat{\mathbf{q}}_j$, where the scaling factor between both vectors is the regression coefficient $\widehat{b}_j$.

Property 10.4.11 – Asymptotic distribution of t-score variables. Equations (2.23) and (2.24) describe the data structure for PLS models, which gives rise to the following asymptotic distribution of the t-score variables.

Theorem 10.4.13 *Under the assumption that the source variables have zero mean and are statistically independent, the t-score variables asymptotically follow a Gaussian distribution under the Liapounoff theorem, detailed in (9.31), since*

$$\begin{aligned}
\varrho_{t_j}^3 &= \sum_{i=1}^{n}\varrho_{t_{j_i}}^3 \qquad \sigma_{t_j}^2 = \sum_{i=1}^{n}\sigma_{t_{j_i}}^2 \\
\varrho_{t_{j_i}}^3 &= E\left\{\left|\left(\mathbf{r}_j^T\mathbf{p}_i\right)s_i + \tfrac{1}{n}\mathbf{r}_j^T\mathbf{e}\right|^3\right\} \\
\sigma_{t_{j_i}}^2 &= E\left\{\left(\left(\mathbf{r}_j^T\mathbf{p}_i\right)s_i + \tfrac{1}{n}\mathbf{r}_j^T\mathbf{e}\right)^2\right\} = \left(\mathbf{r}_j^T\mathbf{p}_i\right)^2\sigma_{s_i}^2 + \tfrac{1}{n}\sum_{i=1}^{n_x}r_{ji}^2\sigma_{e_i}^2.
\end{aligned} \tag{10.51}$$

Proof. The calculation of the t-score variables

$$t_j = \mathbf{r}_j^T \mathbf{x}_0 = \sum_{i=1}^{n_x} r_{ji} x_{0_i} = \sum_{i=1}^{n} \mathbf{r}_j^T \left(\mathbf{p}_i s_i + \tfrac{1}{n} \mathbf{e} \right) \qquad (10.52)$$

becomes asymptotically

$$\lim_{n \to \infty} t_j = \lim_{n \to \infty} \sum_{i=1}^{n} \mathbf{r}_j^T \mathbf{p}_i s_i . \qquad (10.53)$$

Replacing $\mathbf{r}_j$, $\mathbf{b}_i$ and $\mathbf{e}$ by $\mathbf{p}_j$, $\boldsymbol{\xi}_i$ and $\mathbf{g}$, respectively, (10.52) shows the same formulation as that for computing the t-score variables using PCA. Consequently, the proof of Theorem 9.3.4 is also applicable to the proof of Theorem 10.4.13.

Property 10.4.12 – Reconstruction of the Cross-covariance matrix $\mathbf{S}_{x_0 y_0}$. The focus now shifts on the reconstruction of the cross-covariance matrix $\mathbf{S}_{x_0 y_0}$ using the sets of LVs computed by the PLS algorithm.

Theorem 10.4.14 *If the covariance matrix $\mathbf{S}_{x_0 x_0}$ has full rank n_x, the n_x sets of LVs allow a complete reconstruction of the cross-covariance matrix $\mathbf{S}_{x_0 y_0}$ using the n_x p-loading vectors, the n_x value of the objective functions function and the n_x q-weight vectors.*

Proof. The reconstruction of the covariance matrix $\mathbf{S}_{x_0 y_0}$ follows from

$$\begin{aligned}
\mathbf{S}_{x_0 y_0} &= \lim_{K \to \infty} \tfrac{1}{K-1} \mathbf{X}_0^T \mathbf{Y}_0 \\
\mathbf{S}_{x_0 y_0} &= \lim_{K \to \infty} \tfrac{1}{K-1} \mathbf{X}_0^T \left[\mathbf{T} \acute{\mathbf{Q}}^T + \mathbf{F} \right] = \lim_{K \to \infty} \tfrac{1}{K-1} \mathbf{X}_0^T \mathbf{T} \acute{\mathbf{Q}}^T \\
\mathbf{S}_{x_0 y_0} &= \lim_{K \to \infty} \tfrac{1}{K-1} \left[\sum_{j=1}^{n_x} \mathbf{p}_j \mathbf{t}_j^T \right] \left[\sum_{j=1}^{n_x} \mathbf{t}_j \acute{\mathbf{q}}_j^T \right] \\
\mathbf{S}_{x_0 y_0} &= \lim_{K \to \infty} \tfrac{1}{K-1} \left[\sum_{j=1}^{n_x} \mathbf{p}_j \mathbf{t}_j^T \right] \left[\sum_{j=1}^{n_x} \mathbf{t}_j b_j \mathbf{q}_j^T \right] \qquad (10.54) \\
\mathbf{S}_{x_0 y_0} &= \lim_{K \to \infty} \tfrac{1}{K-1} \sum_{j=1}^{n_x} \mathbf{p}_j \mathbf{t}_j^T \mathbf{t}_j b_j \mathbf{q}_j^T \\
\mathbf{S}_{x_0 y_0} &= \lim_{K \to \infty} = \tfrac{1}{K-1} \sum_{j=1}^{n_x} \mathbf{p}_j \mathbf{t}_j^T \mathbf{u}_j \mathbf{q}_j = \sum_{j=1}^{n_x} \mathbf{p}_j \left(\mathbf{w}_j^T \mathbf{S}_{ef}^{(j)} \mathbf{q}_j \right) \mathbf{q}_j^T
\end{aligned}$$

$$\boxed{\mathbf{S}_{x_0 y_0} = \sum_{j=1}^{n_x} \mathbf{p}_j \lambda_j \mathbf{q}_j^T} .$$

The above holds true, since:

- $\lim_{K\to\infty} \frac{1}{K-1}\mathbf{X}_0^T\mathbf{F} = \mathbf{P} \lim_{K\to\infty} \frac{1}{K-1}\mathbf{T}^T\mathbf{F}$
  $\lim_{K\to\infty} \frac{1}{K-1}\mathbf{X}_0^T\mathbf{F} = \mathbf{P} \lim_{K\to\infty} \frac{1}{K-1}\mathbf{T}^T\left[\mathbf{Y}_0 - \mathbf{T}\acute{\mathbf{Q}}^T\right]$
  $\lim_{K\to\infty} \frac{1}{K-1}\mathbf{X}_0^T\mathbf{F} = \mathbf{P} \lim_{K\to\infty} \frac{1}{K-1}\mathbf{T}^T\left[\mathbf{I} - \mathbf{T}\left[\mathbf{T}^T\mathbf{T}\right]^{-1}\mathbf{T}^T\right]\mathbf{Y}_0$
  $\lim_{K\to\infty} \frac{1}{K-1}\mathbf{X}_0^T\mathbf{F} = \mathbf{P}\left[\mathbf{S}_{\mathrm{t}y_0} - \mathbf{S}_{\mathrm{tt}}\mathbf{S}_{\mathrm{tt}}^{-1}\mathbf{S}_{\mathrm{t}y_0}\right]$
  $\lim_{K\to\infty} \frac{1}{K-1}\mathbf{X}_0^T\mathbf{F} = \mathbf{0}$;
- the t-score vectors are mutually orthogonal, which Theorem 10.4.2 outlines;
- $\acute{\mathbf{q}}_k = b_k\mathbf{q}_k$, which Theorem 10.4.11 confirms;
- $\lim_{K\to\infty} \frac{1}{K-1}\mathbf{t}_k^T\mathbf{u}_k = b_k \lim_{K\to\infty} \frac{1}{K-1}\mathbf{t}_k^T\mathbf{t}_k$, which follows from (10.11); and
- $\lambda_j = \mathbf{w}_j^T\mathbf{S}_{ef}^{(j)}\mathbf{q}_j$, which follows from (10.9).

Property 10.4.13 – Accuracy of PLS regression model. The following relationship between the PLS regression model and the regression model obtained by the ordinary least squares solution exists.

Theorem 10.4.15 *Under the assumption that the rank of the covariance matrix* $\mathbf{S}_{x_0x_0}$ *is* n_x *the PLS regression model is identical to that obtained by an ordinary least squares solution, that is* $\mathcal{B} = \mathcal{B}_{OLS} = \mathbf{S}_{x_0x_0}^{-1}\mathbf{S}_{x_0y_0}$.

Proof. Starting by revisiting the data structure in (2.23)

$$\mathbf{y}_0 = \mathcal{B}^T\mathbf{x}_0 + \mathfrak{f} \qquad E\left\{\mathbf{x}_0\mathfrak{f}^T\right\} = \mathbf{0} \qquad \mathcal{B} = \mathbf{S}_{x_0x_0}^{-1}\mathbf{S}_{x_0y_0}. \tag{10.55}$$

Using PLS, the prediction of output vector $\mathbf{y}_0$ becomes

$$\mathbf{y}_0 = \acute{\mathbf{Q}}\mathbf{t} + \mathbf{f} = \mathbf{S}_{y_0\mathrm{t}}\mathbf{S}_{\mathrm{tt}}^{-1}\mathbf{t} + \mathbf{f}. \tag{10.56}$$

Next, analyzing the relationship between $\mathbf{S}_{\mathrm{tt}}$ and $\mathbf{S}_{x_0x_0}$ as well as between $\mathbf{S}_{y_0\mathrm{t}}$ and $\mathbf{S}_{y_0x_0}$ concludes this proof, since

$$\begin{aligned}\mathbf{S}_{\mathrm{tt}} &= E\left\{\mathbf{t}\mathbf{t}^T\right\} = E\left\{\mathbf{R}^T\mathbf{x}_0\mathbf{x}_0^T\mathbf{R}\right\} = \mathbf{R}^T\mathbf{S}_{x_0x_0}\mathbf{R}\\ \mathbf{S}_{y_0\mathrm{t}} &= E\left\{\mathbf{y}_0\mathbf{t}^T\right\} = E\left\{\mathbf{y}_0\mathbf{x}_0^T\mathbf{R}\right\} = \mathbf{S}_{y_0x_0}\mathbf{R}\end{aligned} \tag{10.57}$$

which gives rise to

$$\mathbf{y}_0 = \mathbf{S}_{y_0x_0}\mathbf{R}\overbrace{\mathbf{R}^{-1}\mathbf{S}_{x_0x_0}^{-1}\mathbf{R}^{-T}}^{=\mathbf{S}_{\mathrm{tt}}^{-1}}\overbrace{\mathbf{R}^T\mathbf{x}_0}^{=\mathbf{t}} + \mathbf{f}. \tag{10.58}$$

With $\mathbf{R}\mathbf{R}^{-1} = \mathbf{R}^{-T}\mathbf{R}^T$ reducing to the identity matrix, (10.58) becomes

$$\boxed{\mathbf{y}_0 = \acute{\mathbf{Q}}\mathbf{t} + \mathfrak{f} = \mathbf{S}_{y_0x_0}\mathbf{S}_{x_0x_0}^{-1}\mathbf{x}_0 + \mathbf{f} = \mathcal{B}^T\mathbf{x}_0 + \mathfrak{f}}. \tag{10.59}$$

Property 10.4.14 – Computing the estimate of $\mathcal{B}$. Using the n_x sets of LVs, computed from the PLS algorithm, the following holds true for estimating the parameter matrix $\mathcal{B}$, $\mathbf{y}_0 = \mathcal{B}^T \mathbf{x}_0 + \mathfrak{f}$.

Lemma 10.4.16 *If the covariance matrix $\mathbf{S}_{x_0 x_0}$ has full rank n_x, the n_x sets of LVs allow one computation of an estimate of the parameter matrix $\mathcal{B}$ without requiring the inversion of any squared matrix.*

Proof. The prediction of the output vector $\mathbf{y}_0$ using the n_x sets of LVs is

$$\boxed{\mathbf{y}_0 = \acute{\mathbf{Q}}\mathbf{t} + \mathfrak{f} = \acute{\mathbf{Q}}\mathbf{R}^T \mathbf{x}_0 + \mathfrak{f} = \mathcal{B}^T \mathbf{x}_0 + \mathfrak{f}}. \tag{10.60}$$

The column vectors of the matrices $\acute{\mathbf{Q}}$ and $\mathbf{R}$, however, can be computed iteratively

$$\begin{aligned} &\mathbf{S}_{ef}^{(j)}\mathbf{S}_{fe}^{(j)}\mathbf{w}_j = \lambda_j^2 \mathbf{w}_j \\ &\boxed{\mathbf{r}_j = \mathbf{w}_j - \sum_{i=1}^{j-1} \mathbf{p}_i^T \mathbf{w}_j \mathbf{r}_i \quad \acute{\mathbf{q}}_j = \frac{\mathbf{S}_{fe}^{(j)}\mathbf{r}_j}{\mathbf{r}_j^T \mathbf{S}_{x_0 x_0}\mathbf{r}_j}} \\ &\mathbf{p}_j = \frac{\mathbf{S}_{x_0 x_0}\mathbf{r}_j}{\mathbf{r}_j^T \mathbf{S}_{x_0 x_0}\mathbf{r}_j} \\ &\mathbf{S}_{ef}^{(j+1)} = \mathbf{S}_{ef}^{(j)} - \acute{\mathbf{q}}_j \left(\mathbf{r}_j^T \mathbf{S}_{x_0 x_0}\mathbf{r}_j\right)\mathbf{p}_j = \mathbf{S}_{ef}^{(j)}\left[\mathbf{I} - \mathbf{w}_j \mathbf{p}_j^T\right]. \end{aligned} \tag{10.61}$$

The expression for determining the q-loading vector follows from (10.35) and (10.37). Hence, unlike the OLS solution, PLS does not require any matrix inversion to iteratively estimate $\mathcal{B}$. Subsection 6.2.2 presents an excellent example to demonstrate the benefit of the iterative PLS procedure over OLS.

Property 10.4.15 – Computationally efficient PLS algorithm. The preceding analysis into the properties of PLS algorithm has shown that the deflation procedure only requires the deflation of the input or the output data matrix and that introducing the r-weight vectors allows the t-score vectors to be be directly computed from the input data matrix. This gives rise to the development of a computationally efficient PLS algorithm. Table 10.2 shows the steps of the revised PLS algorithm. To cover any possible combination in terms of the number of input and output variables n_x and n_y, the revised algorithm includes the case of $n_y = 1$ and obtains the w-weight or the q-loading vector using the iterative power method, depending on whether $n_x < n_y$ or $n_x \geq n_y$, respectively. More precisely, the dimension of the symmetric and positive semi-definite matrix products $\mathbf{X}_0^T \mathbf{F}^{(j)} \mathbf{F}^{(j)^T} \mathbf{X}_0$ and $\mathbf{F}^{(j)^T} \mathbf{X}_0 \mathbf{X}_0^T \mathbf{F}^{(j)}$ are $n_x \times n_x$ and $n_y \times n_y$, respectively. Given that there is the following linear relationship between the weight vectors

$$\mathbf{S}_{ef}^{(j)}\mathbf{q}_j = \lambda_j \mathbf{w}_j \qquad \mathbf{S}_{fe}^{(j)}\mathbf{w}_j = \lambda_j \mathbf{q}_j \tag{10.62}$$

Table 10.2 Computationally efficient PLS algorithm.

Step	Description	Equation
1	Initiate iteration	$j = 1$, $\mathbf{S}_{ef}^{(1)} = \mathbf{S}_{x_0 y_0}$
2	Set up matrix product	$\mathbf{M}_w = \mathbf{S}_{ef}^{(j)} \mathbf{S}_{fe}^{(j)}$ if $n_x < n_y$ else $\mathbf{M}_q = \mathbf{S}_{fe}^{(j)} \mathbf{S}_{ef}^{(j)}$
3	Check dimension of $\mathbf{y}_0$	if $n_y = 1$, $\mathbf{w}_j = \mathbf{S}_{ef}^{(j)} / \left\lVert \mathbf{S}_{ef}^{(j)} \right\rVert$ and go to Step 9, if not go to Step 4
4	Initiate power method	${}_0\mathbf{w}_j = \mathbf{M}_w(:,1) / \lVert \mathbf{M}_w(:,1) \rVert$ if $n_x < n_y$ else ${}_0\mathbf{q}_j = \mathbf{M}_q(:,1) / \lVert \mathbf{M}_q(:,1) \rVert$
5	Compute matrix-vector product	${}_1\mathbf{w}_j = \mathbf{M}_w \left({}_0\mathbf{w}_j \right)$ if $n_x < n_y$ else ${}_1\mathbf{q}_j = \mathbf{M}_q \left({}_0\mathbf{q}_k \right)$
6	Scale weight vector	${}_1\mathbf{w}_j = {}_1\mathbf{w}_j / \lVert {}_1\mathbf{w}_j \rVert$ if $n_x < n_y$ else ${}_1\mathbf{q}_j = {}_1\mathbf{q}_j / \lVert {}_1\mathbf{q}_j \rVert$
7	Check for convergence	if $\left\lVert {}_1\mathbf{w}_j - {}_0\mathbf{w}_j \right\rVert > \epsilon$ or $\left\lVert {}_1\mathbf{q}_j - {}_0\mathbf{q}_j \right\rVert > \epsilon$ ${}_0\mathbf{w}_j = {}_1\mathbf{w}_j$ or ${}_0\mathbf{q}_j = {}_1\mathbf{q}_j$ and go to Step 5 else set $\mathbf{w}_j = {}_0\mathbf{w}_j$ or $\mathbf{q}_j = {}_0\mathbf{q}_j$ and go to Step 8
8	Calculate 2nd weight vector	$\mathbf{q}_j = \mathbf{S}_{fe}^{(j)} \mathbf{w}_j / \left\lVert \mathbf{S}_{fe}^{(j)} \mathbf{w}_j \right\rVert$ if $n_x < n_y$ else $\mathbf{w}_j = \mathbf{S}_{ef}^{(j)} \mathbf{q}_j / \left\lVert \mathbf{S}_{ef}^{(j)} \mathbf{q}_j \right\rVert$
9	Compute r-weight vector	$\mathbf{r}_j = \mathbf{w}_j - \sum_{i=1}^{j-1} \mathbf{p}_i^T \mathbf{w}_j \mathbf{r}_i$
10	Determine scalar	$\tau_j^2 = \mathbf{r}_j^T \mathbf{S}_{x_0 x_0} \mathbf{r}_j$
11	Calculate p-loading vector	$\mathbf{p}_j = \mathbf{S}_{x_0 x_0} \mathbf{r}_j / \tau_j^2$
12	Obtain q́-loading vector	$\acute{\mathbf{q}}_j = \mathbf{S}_{fe}^{(j)} \mathbf{r}_j / \tau_j^2$
13	Deflate cross-covariance matrix	$\mathbf{S}_{ef}^{(j+1)} = \mathbf{S}_{ef}^{(j)} - \mathbf{p}_j \tau_j^2 \acute{\mathbf{q}}_j^T$
14	Check whether there is significant variation left in $\mathbf{S}_{ef}^{(j+1)}$	If so, go to Step 15 if not go to Step 16
15	Check for dimension	If $j < n_x$, set $j = j + 1$ and go to Step 2, if not go to Step 16
16	Compute regression matrix	$\mathcal{B} = \mathbf{R}\acute{\mathbf{Q}}^T$

only one of the dominant eigenvectors needs to be computed. It is therefore expedient to apply the power method to the smaller matrix product if $n_x \neq n_y$. If $n_y = 1$, the covariance matrix $\mathbf{S}_{x_0y_0}$ reduces to a vector of dimension n_x. In this case, the w-weight vector is proportional to $\mathbf{S}_{ef}^{(j)}$. It should be noted that the algorithm in Table 10.2 assumes the availability of the covariance and cross-covariance matrices $\mathbf{S}_{x_0x_0}$ and $\mathbf{S}_{x_0y_0}$. As they are not available in most practical cases, they need to be estimated from the recorded samples stored in $\mathbf{X}_0$ and $\mathbf{Y}_0$ and the computed weight and loading vectors become, accordingly, estimates. It should also be noted that the PLS algorithm in Table 10.2 is similar to that reported in Dayal and MacGregor (1997a).

10.5 Properties of maximum redundancy PLS

Section 2.3 introduces MRPLS as a required extension to PLS to model the data structure in (2.51). This section offers a detailed examination of MRPLS in term of its geometric properties and develops a numerically more efficient algorithm. Readers who are predominantly interested in the application of the methods discussed in this book can note the computationally efficient MRPLS algorithm in Table 10.3 or the batch algorithm for simultaneously computing the n q-loading and w-weight vectors in Table 10.4.

The analysis of the properties of the MRPLS algorithm concentrates on the geometric properties of score, loading and weight vectors first. The results enable a further analysis regarding the deflation procedure and contribute to the numerically and computationally efficient algorithm that is summarized in Table 10.3.

It should be noted that the proposed MRPLS algorithm in Table 2.3 incorporates the fact that only one of the matrices needs to be deflated and that the length-constraint for the w-weight vectors $\mathbf{w}_i^T \mathbf{E}^{(i)^T} \mathbf{E}^{(i)} \mathbf{w}_i - 1$ is equal to $\mathbf{w}_i^T \mathbf{X}_0^T \mathbf{X}_0 \mathbf{w}_i - 1$. This is also proven in this section as part of the analysis of the deflation procedure.

The properties of the MRPLS algorithm are as follows.

1. The t-score vectors are mutually orthonormal.
2. The t- and u-score vectors are mutually orthogonal.
3. The products $\mathbf{E}^{(j)^T} \mathbf{E}^{(j)}$ and $\mathbf{E}^{(j)^T} \mathbf{F}^{(j)}$ are equal to $\mathbf{X}_0^T \mathbf{E}^{(j)}$ and $\mathbf{X}_0^T \mathbf{F}^{(j)}$, respectively.
4. The w-weight, the auxiliary w-weight and p-loading vectors are mutually orthogonal.
5. The q-weight vectors are mutually orthonormal and point in the same direction as the q́-loading vectors.
6. The constraint of the MRPLS objective function $\mathbf{w}_j^T \mathbf{E}^{(j)^T} \mathbf{E}^{(j)} \mathbf{w}_j - 1 = 0$ is equal to $\mathbf{w}_j^T \mathbf{X}_0^T \mathbf{X}_0 \mathbf{w}_j - 1 = 0$.

7. The MRPLS objective function $\mathbf{w}_j^T \mathbf{E}^{(j)^T} \mathbf{F}^{(j)} \mathbf{q}_j$ is equal to $\mathbf{w}_j^T \mathbf{X}_0^T \mathbf{F}^{(j)} \mathbf{q}_j$.
8. The w-weight and auxiliary w-weight vector are the left and right eigenvectors of $[\mathbf{X}_0^T \mathbf{X}_0]^{-1} \mathbf{X}_0^T \mathbf{F}^{(j)} \mathbf{Y}_0^T \mathbf{X}_0$.
9. The q-weight vectors are right eigenvectors of the $\mathbf{Y}_0^T \mathbf{X}_0 [\mathbf{X}_0 \mathbf{X}_0^T]^{-1} \mathbf{X}_0^T \mathbf{F}^{(j)}$.
10. The w- and q-weight vectors can be simultaneously and independently computed as the left eigenvectors of the matrix products $[\mathbf{X}_0^T \mathbf{X}_0]^{-1} \mathbf{X}_0^T \mathbf{Y}_0 \mathbf{Y}_0^T \mathbf{X}_0$ and $\mathbf{Y}_0^T \mathbf{X}_0 [\mathbf{X}_0 \mathbf{X}_0^T]^{-1} \mathbf{X}_0^T \mathbf{Y}_0$, respectively.

Property 10.5.1 – Orthogonality of the t-score vectors. The first property relates to the geometry of the t-score vectors, which Theorem 10.5.1 describes.

Theorem 10.5.1 *The t-score vectors are mutually orthonormal, that is,* $\widehat{\mathbf{t}}_i^T \widehat{\mathbf{t}}_j = \delta_{ij}$.

Proof. The proof for Theorem 10.5.1 requires a detailed analysis of the deflation procedure and starts by reformulating $\widehat{\mathbf{t}}_i^T \widehat{\mathbf{t}}_j$

$$\widehat{\mathbf{t}}_i^T \widehat{\mathbf{t}}_j = \widehat{\mathbf{w}}_i^T \mathbf{E}^{(i)^T} \mathbf{E}^{(j)} \widehat{\mathbf{w}}_j. \tag{10.63}$$

Now, incorporating the deflation procedure for the data matrix $\mathbf{E}^{(j)} = \mathbf{E}^{(j-1)} - \widehat{\mathbf{t}}_{j-1} \widehat{\mathbf{p}}_{j-1} = \left[\mathbf{I} - \widehat{\mathbf{t}}_{j-1} \widehat{\mathbf{t}}_{j-1}^T\right] \mathbf{E}^{(j-1)} = \mathbf{E}^{(j-1)} \left[\mathbf{I} - \widehat{\mathbf{w}}_{j-1} \widehat{\mathbf{p}}_{j-1}^T\right]$ yields

$$\widehat{\mathbf{t}}_i^T \mathbf{E}^{(j-1)} \left[\mathbf{I} - \widehat{\mathbf{w}}_{j-1} \widehat{\mathbf{p}}_{j-1}^T\right] \widehat{\mathbf{w}}_j \tag{10.64}$$

under the assumption that $i < j$. Applying the deflation procedure a total of $j - i - 1$ times yields

$$\boxed{\widehat{\mathbf{t}}_i^T \left[\mathbf{I} - \widehat{\mathbf{t}}_i \widehat{\mathbf{t}}_i^T\right] \mathbf{E}^{(i)} \left[\prod_{m=i+1}^{j-1} \left[\mathbf{I} - \widehat{\mathbf{w}}_m \widehat{\mathbf{p}}_m^T\right]\right] \widehat{\mathbf{w}}_j = 0}. \tag{10.65}$$

The vector-matrix product to the left, however, reduces to $\widehat{\mathbf{t}}_i^T - \widehat{\mathbf{t}}_i^T$, which implies that $\widehat{\mathbf{t}}_i^T \widehat{\mathbf{t}}_j = 0$ if $i \neq j$ and 1 if $i = j$.

Property 10.5.2 – Orthogonality of the t- and u-score vectors. The t- and u-score vectors have the following geometric property.

Theorem 10.5.2 *The t- and u-score vectors are mutually orthogonal, that is,* $\widehat{\mathbf{t}}_i^T \widehat{\mathbf{u}}_j = \delta_{ij} \widehat{\lambda}_i$

Proof. The proof of $\widehat{\mathbf{u}}_i^T \widehat{\mathbf{t}}_j^T = \delta_{ij} \widehat{\lambda}_i$ commences with

$$\widehat{\mathbf{u}}_i^T \widehat{\mathbf{t}}_j = \widehat{\mathbf{q}}_i^T \mathbf{F}^{(i)^T} \mathbf{E}^{(j)} \widehat{\mathbf{w}}_j. \tag{10.66}$$

Table 10.3 Computationally efficient MRPLS algorithm.

Step	Description	Equation
1	Initiate iteration	$n = 1,\ j = 1$
2	Set up matrix products	$\mathbf{M}_{x_0 x_0} = \mathbf{X}_0^T \mathbf{X}_0$, $\mathbf{M}_{x_0 y_0}^{(1)} = \mathbf{X}_0^T \mathbf{Y}_0$ and $\mathbf{M}_{x_0 x_0}^{-1}$
3	Set up initial q-weight vector	${}_0\widehat{\mathbf{q}}_j^T = \mathbf{M}_{x_0 y_0}^{(j)}(1,:) / \left\Vert \mathbf{M}_{x_0 y_0}^{(j)}(1,:) \right\Vert$
4	Compute auxiliary weight vector	$\widehat{\mathfrak{w}}_j = \mathbf{M}_{x_0 y_0}^{(j)} \left({}_0\widehat{\mathbf{q}}_j \right)$
5	Calculate w-weight vector	if $j = n$ $\widehat{\mathbf{w}}_j = \mathbf{M}_{x_0 x_0}^{-1} \widehat{\mathfrak{w}}_j / \sqrt{\widehat{\mathbf{w}}_j^T \mathbf{M}_{x_0 x_0} \widehat{\mathbf{w}}_k}$ else : $\widehat{\mathbf{w}}_j = \widehat{\mathfrak{w}}_j$
6	Determine q-weight vector	${}_1\widehat{\mathbf{q}}_j = \mathbf{M}_{x_0 y_0}^{(j)^T} \widehat{\mathbf{w}}_j / \left\Vert \mathbf{M}_{x_0 y_0}^{(j)^T} \widehat{\mathbf{w}}_j \right\Vert$
7	Check for convergence	if $\left\Vert {}_1\widehat{\mathbf{q}}_j - {}_0\widehat{\mathbf{q}}_j \right\Vert > \epsilon$ set ${}_0\widehat{\mathbf{q}}_j = {}_1\widehat{\mathbf{q}}_j$ and go to Step 4 else set $\widehat{\mathbf{q}}_j = {}_1\widehat{\mathbf{q}}_j$ and go to Step 8
8	Compute r-weight vector	if $j = n : \widehat{\mathbf{r}}_j = \widehat{\mathbf{w}}_j$ else : $\widehat{\mathbf{r}}_j = \widehat{\mathbf{w}}_j - \sum_{i=1}^{j-1} \widehat{\mathbf{p}}_i^T \widehat{\mathbf{w}}_j \widehat{\mathbf{r}}_i$
9	Determine p-loading vector	if $j = n : \widehat{\mathbf{p}}_j = \mathbf{M}_{x_0 x_0} \widehat{\mathbf{w}}_j$ else compute $\tau_j = \widehat{\mathbf{r}}_j^T \mathbf{M}_{x_0 x_0} \widehat{\mathbf{r}}_j$ and $\widehat{\mathbf{p}}_j = \mathbf{M}_{x_0 x_0} \widehat{\mathbf{w}}_j / \tau_j$
10	Determine q́-loading vector	if $j = n$ $\acute{\mathbf{q}}_j = \mathbf{M}_{x_0 y_0}^{(j)^T} \widehat{\mathbf{w}}_j$ else $\acute{\mathbf{q}}_j = \mathbf{M}_{x_0 y_0}^{(j)^T} \widehat{\mathbf{w}}_j / \tau_j$
11	Deflate cross-product matrix	if $j = n$ $\mathbf{M}_{x_0 y_0}^{(j+1)} = \mathbf{M}_{x_0 y_0}^{(j)} - \widehat{\mathbf{p}}_j \widehat{\acute{\mathbf{q}}}_j^T$ else $\mathbf{M}_{x_0 y_0}^{(j+1)} = \mathbf{M}_{x_0 y_0}^{(j)} - \widehat{\mathbf{p}}_j \tau_j \widehat{\acute{\mathbf{q}}}_j^T$
12	Check whether there is still significant variation remaining in $\mathbf{M}_{x_0 y_0}^{(j+1)}$	if so $j = j + 1$, $n = n + 1$ and go to Step 4 if not $j = j + 1$, go to Step 13
13	Check whether $j = n_x$	if so then terminate else go to Step 3

Table 10.4 Simultaneous MRPLS algorithm for LV sets.

Step	Description	Equation
1	Form matrix products	$\mathbf{M}_{x_0x_0} = \mathbf{X}_0^T\mathbf{X}_0$ and $\mathbf{M}_{x_0y_0} = \mathbf{X}_0^T\mathbf{Y}_0$
2	Compute SVD of $\mathbf{M}_{x_0x_0}$	$\mathbf{M}_{x_0x_0} = \mathcal{U}_v\mathcal{S}_v\mathcal{U}_v^T$
3	Form matrix product	$\mathbf{M}_{x_0x_0}^{-1/2} = \mathcal{U}_v\mathcal{S}_v^{-1/2}\mathcal{U}_v^T$
4	Form matrix product	$\mathbf{S} = \mathbf{M}_{x_0x_0}^{-1/2}\mathbf{M}_{x_0y_0}$
5	Calculate SVD of $\mathbf{S}$	$\mathbf{S} = \mathcal{U}\ \mathrm{diag}\{\boldsymbol{\lambda}\}\,\mathcal{V}$
6	Determine w-weight matrix	$\widehat{\mathbf{W}} = \mathbf{M}_{x_0x_0}^{-1/2}\mathcal{U}$
7	Compute q-weight matrix	$\widehat{\mathbf{Q}} = \mathcal{V}$
8	Calculate w-loading matrix	$\widehat{\mathbf{P}} = \mathbf{M}_{x_0x_0}\widehat{\mathbf{W}}$
9	Obtain q́-loading matrix	$\widehat{\acute{\mathbf{Q}}} = \mathbf{M}_{x_0y_0}^T\widehat{\mathbf{W}}$

Assuming that $i < j$ and applying the deflation procedure for $\mathbf{E}^{(j)}$ a total of $j - i - 1$ times gives rise to

$$\widehat{\mathbf{q}}_i^T\mathbf{F}^{(i)}\mathbf{E}^{(i)}\left[\prod_{m=i}^{j-1}\left[\mathbf{I} - \widehat{\mathbf{w}}_m\widehat{\mathbf{p}}_m^T\right]\right]\widehat{\mathbf{w}}_j. \tag{10.67}$$

Equation (2.70) yields that $\widehat{\mathbf{q}}_i^T\mathbf{F}^{(i)^T}\mathbf{E}^{(i)} = \widehat{\lambda}_i\widehat{\mathbf{w}}_i^T\mathbf{E}^{(i)^T}\mathbf{E}^{(i)}$, which yields

$$\widehat{\lambda}_i\widehat{\mathbf{w}}_i^T\mathbf{E}^{(i)^T}\mathbf{E}^{(i)}\left[\prod_{m=i}^{j-1}\left[\mathbf{I} - \widehat{\mathbf{w}}_m\widehat{\mathbf{p}}_m^T\right]\right]\widehat{\mathbf{w}}_j \tag{10.68}$$

and hence

$$\widehat{\mathbf{u}}_i^T\widehat{\mathbf{t}}_j = \widehat{\lambda}_i\widehat{\mathbf{w}}_i^T\mathbf{E}^{(i)^T}\mathbf{E}^{(j)}\widehat{\mathbf{w}}_j = \widehat{\lambda}_i\widehat{\mathbf{t}}_i^T\widehat{\mathbf{t}}_j = 0. \tag{10.69}$$

The above conclusion, however, is only valid for $i < j$. For the case of $i > j$, (10.66) can be rewritten as follows

$$\widehat{\mathbf{u}}_i^T\widehat{\mathbf{t}}_j = \widehat{\mathbf{q}}_i^T\mathbf{F}^{(i)^T}\widehat{\mathbf{t}}_j = \widehat{\mathbf{q}}_i^T\mathbf{F}^{(j)^T}\left[\prod_{m=j}^{i-1}\left[\mathbf{I} - \widehat{\mathbf{t}}_m\widehat{\mathbf{t}}_m^T\right]\right]\widehat{\mathbf{t}}_j. \tag{10.70}$$

Given that the t-score vectors are mutually orthonormal, the matrix-vector product on the right hand side of (10.70) reduces to

$$\boxed{\widehat{\mathbf{u}}_i^T\widehat{\mathbf{t}}_j = \widehat{\mathbf{q}}_i^T\mathbf{F}^{(j)^T}\left(\widehat{\mathbf{t}}_j - \widehat{\mathbf{t}}_j\right) = 0}. \tag{10.71}$$

Finally, for $i = j$, (10.69) yields $\widehat{\mathbf{u}}_j^T\widehat{\mathbf{t}}_j = \widehat{\lambda}_i$. Hence, $\widehat{\mathbf{u}}_j^T\widehat{\mathbf{t}}_j = \widehat{\lambda}_i\delta_{ij}$.

Property 10.5.3 – Matrix products $\mathbf{E}^{(j)^T}\mathbf{E}^{(j)}$ and $\mathbf{E}^{(j)^T}\mathbf{F}^{(j)}$. The analysis of the geometric properties of the t- and u-score variables is now followed by examining the effect that mutually orthonormal t-score variables have upon the deflation procedure. Lemma 10.5.3 describes this in detail.

Lemma 10.5.3 *The mutually orthonormal t-score vectors simplify the deflation step to guarantee that only one of the cross product matrices needs to be deflated, that is,* $\mathbf{E}^{(j)^T}\mathbf{E}^{(j)} = \mathbf{X}_0^T\mathbf{E}^{(j)} = \mathbf{E}^{(j)^T}\mathbf{X}_0$ *and* $\mathbf{E}^{(j)^T}\mathbf{F}^{(j)} = \mathbf{X}_0^T\mathbf{F}^{(j)} = \mathbf{E}^{(j)^T}\mathbf{Y}_0$.

Proof. Starting with the deflation of the input data matrix, which is given by

$$\begin{aligned}
\mathbf{E}^{(j)^T} &= \mathbf{E}^{(j-1)^T}\left[\mathbf{I} - \widehat{\mathbf{t}}_{j-1}\widehat{\mathbf{t}}_{j-1}^T\right] \\
\mathbf{E}^{(j)^T} &= \mathbf{E}^{(j-2)^T}\left[\mathbf{I} - \widehat{\mathbf{t}}_{j-2}\widehat{\mathbf{t}}_{j-2}^T\right]\left[\mathbf{I} - \widehat{\mathbf{t}}_{j-1}\widehat{\mathbf{t}}_{j-1}^T\right] \\
\mathbf{E}^{(k)^T} &= \mathbf{X}_0^T\prod_{i=1}^{j-1}\left[\mathbf{I} - \widehat{\mathbf{t}}_i\widehat{\mathbf{t}}_i^T\right] = \mathbf{X}_0^T\left[\mathbf{I} - \widehat{\mathbf{T}}_{j-1}\widehat{\mathbf{T}}_{j-1}^T\right]
\end{aligned} \tag{10.72}$$

where $\widehat{\mathbf{T}}_{j-1} = \left[\begin{array}{ccc}\widehat{\mathbf{t}}_1 & \cdots & \widehat{\mathbf{t}}_{j-1}\end{array}\right]$. Similarly, the deflation of $\mathbf{F}^{(j)}$ is given by

$$\begin{aligned}
\mathbf{F}^{(j)^T} &= \mathbf{F}^{(j-1)^T}\left[\mathbf{I} - \widehat{\mathbf{t}}_{j-1}\widehat{\mathbf{t}}_{j-1}^T\right] \\
\mathbf{F}^{(j)^T} &= \mathbf{F}^{(j-2)^T}\left[\mathbf{I} - \widehat{\mathbf{t}}_{j-2}\widehat{\mathbf{t}}_{j-2}^T\right]\left[\mathbf{I} - \widehat{\mathbf{t}}_{j-1}\widehat{\mathbf{t}}_{j-1}^T\right] \\
\mathbf{F}^{(j)^T} &= \mathbf{Y}_0^T\prod_{i=1}^{j-1}\left[\mathbf{I} - \widehat{\mathbf{t}}_i\widehat{\mathbf{t}}_i^T\right] = \mathbf{Y}_0^T\left[\mathbf{I} - \widehat{\mathbf{T}}_{j-1}\widehat{\mathbf{T}}_{j-1}^T\right].
\end{aligned} \tag{10.73}$$

Next, incorporating the above deflation procedures gives rise to

$$\begin{aligned}
\mathbf{E}^{(j)^T}\mathbf{E}^{(j)} &= \mathbf{X}_0^T\left[\mathbf{I} - \widehat{\mathbf{T}}_{j-1}\widehat{\mathbf{T}}_{j-1}^T\right]\left[\mathbf{I} - \widehat{\mathbf{T}}_{j-1}\widehat{\mathbf{T}}_{j-1}^T\right]\mathbf{X}_0 \\
\mathbf{E}^{(j)^T}\mathbf{E}^{(j)} &= \mathbf{X}_0^T\left[\mathbf{I} - \widehat{\mathbf{T}}_{j-1}\widehat{\mathbf{T}}_{j-1}^T - \widehat{\mathbf{T}}_{j-1}\widehat{\mathbf{T}}_{j-1}^T + \widehat{\mathbf{T}}_{j-1}\widehat{\mathbf{T}}_{j-1}^T\right]\mathbf{X}_0 \\
&\boxed{\mathbf{E}^{(j)^T}\mathbf{E}^{(j)} = \mathbf{X}_0^T\left[\mathbf{I} - \widehat{\mathbf{T}}_{j-1}\widehat{\mathbf{T}}_{j-1}^T\right]\mathbf{X}_0 = \mathbf{X}_0^T\mathbf{E}^{(j)} = \mathbf{E}^{(j)^T}\mathbf{X}_0}
\end{aligned} \tag{10.74}$$

and

$$\boxed{\mathbf{E}^{(j)^T}\mathbf{F}^{(j)} = \mathbf{X}_0^T\left[\mathbf{I} - \widehat{\mathbf{T}}_{j-1}\widehat{\mathbf{T}}_{j-1}^T\right]\mathbf{Y}_0 = \mathbf{X}_0^T\mathbf{F}^{(j)} = \mathbf{E}^{(j)^T}\mathbf{Y}_0},$$

respectively. It follows from Lemma 10.5.3 that

$$\widehat{\mathbf{w}}_i^T \mathbf{E}^{(i)^T} \mathbf{E}^{(j)} \widehat{\mathbf{w}}_j = \widehat{\mathbf{w}}_i^T \mathbf{X}_0^T \widehat{\mathbf{t}}_j \qquad \text{if} \quad j > i$$

and (10.75)

$$\widehat{\mathbf{w}}_i^T \mathbf{E}^{(i)^T} \mathbf{E}^{(j)} \widehat{\mathbf{w}}_j = \widehat{\mathbf{t}}_i^T \mathbf{X}_0^T \widehat{\mathbf{w}}_j \qquad \text{if} \quad i > j.$$

According to Theorem 10.5.1, $\widehat{\mathbf{w}}_i^T \mathbf{X}_0^T \widehat{\mathbf{t}}_j = \widehat{\mathbf{t}}_i^T \mathbf{X}_0^T \widehat{\mathbf{w}}_j = \delta_{ij}$.

Property 10.5.4 – Orthogonality of the weight and vectors of the input variables. Starting with the orthogonality properties of the weight and loading vectors associated with the input variables, Theorem 10.5.4 highlights the geometric relationships between the weight and loading vectors.

Theorem 10.5.4 *The w-weight vectors are mutually orthonormal to the p-loading vectors and mutually orthogonal to the auxiliary w-weight vector, that is,* $\widehat{\mathbf{w}}_i^T \widehat{\mathbf{p}}_j = 0$ *and* $\widehat{\mathbf{w}}_i^T \widehat{\mathfrak{w}}_j = 0$ *if* $i \neq j$, *and the vectors* $\widehat{\mathbf{p}}_i$ *and* $\widehat{\mathfrak{w}}_i$ *are equal up to a scaling factor.*

Proof. The first step is to show that $\widehat{\mathbf{p}}_i \propto \widehat{\mathfrak{w}}_i$, which follows from

$$\widehat{\mathbf{w}}_i \propto \left[\mathbf{X}_0^T \mathbf{E}^{(i)}\right]^{\dagger} \widehat{\mathfrak{w}}_i \quad \Rightarrow \quad \widehat{\mathfrak{w}}_i \propto \mathbf{X}_0^T \mathbf{E}^{(i)} \widehat{\mathbf{w}}_i = \widehat{\mathbf{p}}_i. \tag{10.76}$$

It is therefore sufficient to prove $\widehat{\mathbf{w}}_i^T \widehat{\mathbf{p}}_j = 0$, as this includes the case $\widehat{\mathbf{w}}_i^T \widehat{\mathfrak{w}}_j = 0$ for all $i \neq j$. Given that $\widehat{\mathbf{p}}_j = \mathbf{E}^{(j)^T} \mathbf{E}^{(j)} \widehat{\mathbf{w}}_j$, $\widehat{\mathbf{w}}_i^T \widehat{\mathbf{p}}_j$ can be written as

$$\boxed{\widehat{\mathbf{w}}_i^T \mathbf{X}_0^T \mathbf{E}^{(j)} \widehat{\mathbf{w}}_j = \widehat{\mathbf{w}}_i^T \mathbf{X}_0^T \widehat{\mathbf{t}}_j = \delta_{ij}}. \tag{10.77}$$

Theorem 10.5.1 confirms that (10.77) is δ_{ij}.

Property 10.5.5 – Orthogonality of the q-weight and q́-loading vectors. With regards to the weight and loading vectors of output variables, Theorem 10.5.5 summarizes the geometric relationships between and among them.

Theorem 10.5.5 *The q-weight vectors are mutually orthonormal and the q-weight and q́-loading vectors are mutually orthogonal, i.e.* $\widehat{\mathbf{q}}_i^T \widehat{\mathbf{q}}_j = \mathbf{q}_i^T \acute{\mathbf{q}}_j = \acute{\mathbf{q}}_i^T \acute{\mathbf{q}}_j = 0$ *for all* $i \neq j$.

Proof. Substituting the relationship between the w- and q-weight vectors, that is, $\widehat{\lambda}_j \widehat{\mathbf{q}}_j = \mathbf{F}^{(j)^T} \mathbf{E}^{(j)} \widehat{\mathbf{w}}_j$, into $\widehat{\mathbf{q}}_i^T \widehat{\mathbf{q}}_j$ under the assumption that $i < j$, gives rise to

$$\boxed{\widehat{\mathbf{q}}_i^T \widehat{\mathbf{q}}_j = \frac{\widehat{\mathbf{q}}_i^T \mathbf{F}^{(j)^T} \mathbf{E}^{(j)} \widehat{\mathbf{w}}_j}{\widehat{\lambda}_j} = \frac{\widehat{\mathbf{u}}_i^T \prod_{m=i}^{j-1} \left[\mathbf{I} - \widehat{\mathbf{t}}_m \widehat{\mathbf{t}}_m^T\right] \widehat{\mathbf{t}}_j}{\widehat{\lambda}_j} = 0}. \tag{10.78}$$

Given that $\widehat{\mathbf{u}}_i^T\widehat{\mathbf{t}}_j = 0$ and $\widehat{\mathbf{t}}_i^T\widehat{\mathbf{t}}_j = 0$, for all $i \neq j$, (10.78) reduces to $\widehat{\mathbf{q}}_i^T\widehat{\mathbf{q}}_j = \delta_{ij}$. Next, that $\widehat{\mathbf{q}}_i^T\acute{\mathbf{q}}_j = 0$ for all $i \neq j$ follows from

$$\acute{\mathbf{q}}_j = \mathbf{F}^{(j)^T}\mathbf{E}^{(j)}\widehat{\mathbf{w}}_j \qquad \widehat{\lambda}_j\widehat{\mathbf{q}}_j = \mathbf{F}^{(j)^T}\mathbf{E}^{(j)}\widehat{\mathbf{w}}_j. \tag{10.79}$$

Hence,

$$\boxed{\widehat{\mathbf{q}}_i^T\acute{\mathbf{q}}_j = \widehat{\mathbf{q}}_i^T\widehat{\mathbf{q}}_j\widehat{\lambda}_j = 0} \tag{10.80}$$

and consequently

$$\boxed{\acute{\mathbf{q}}_i^T\acute{\mathbf{q}}_j = \widehat{\lambda}_i\widehat{\lambda}_j\widehat{\mathbf{q}}_i^T\widehat{\mathbf{q}}_j = \widehat{\lambda}_j^2\delta_{ij}}, \tag{10.81}$$

which completes the proof of Theorem 10.5.5.

Property 10.5.6 – Simplification of constraint $\mathbf{w}_j^T\mathbf{E}^{(j)^T}\mathbf{E}^{(j)}\mathbf{w}_j - 1 = 0$. Theorem 10.5.4 can be taken advantage of to simplify the constraint for the w-weight vector $\widehat{\mathbf{w}}_j^T\mathbf{E}^{(j)^T}\mathbf{E}^{(j)}\widehat{\mathbf{w}}_j - 1 = 0$, which is discussed in Lemma 10.5.6.

Lemma 10.5.6 *The constraint* $\widehat{\mathbf{w}}_j^T\mathbf{E}^{(j)^T}\mathbf{E}^{(j)}\widehat{\mathbf{w}}_j = 1$ *is equal to* $\widehat{\mathbf{w}}_j^T\mathbf{X}_0^T\mathbf{X}_0\widehat{\mathbf{w}}_j = 1$.

Proof. Lemma 10.5.3 highlights that $\mathbf{E}^{(j)^T}\mathbf{E}^{(j)}$ is equal to $\mathbf{X}_0^T\mathbf{E}^{(k)} = \mathbf{E}^{(j)^T}\mathbf{X}_0$. Next, incorporating the fact that the w-weight and the p-loading vectors are mutually orthonormal (Theorem 10.5.4) gives rise to

$$\boxed{\widehat{\mathbf{w}}_j^T\mathbf{X}_0^T\mathbf{E}^{(j)}\widehat{\mathbf{w}}_j = \widehat{\mathbf{w}}_j^T\mathbf{X}_0^T\mathbf{X}_0\prod_{i=1}^{j-1}\left[\mathbf{I} - \widehat{\mathbf{w}}_i\widehat{\mathbf{p}}_i^T\right]\widehat{\mathbf{w}}_j = \widehat{\mathbf{w}}_j^T\mathbf{X}_0^T\mathbf{X}_0\widehat{\mathbf{w}}_j}. \tag{10.82}$$

Property 10.5.7 – Simplification of the MRPLS objective function. Theorem 10.5.4 and Lemmas 10.5.3 and 10.5.6 yield a simplification for solving the MRPLS objective function, which is described in Theorem 10.5.7.

Theorem 10.5.7 *The relationships of the MRPLS objective function,*

$$\begin{pmatrix}\widehat{\mathbf{w}}_j\\ \widehat{\mathbf{q}}_j\end{pmatrix} = \arg\max_{\mathbf{w},\mathbf{q}} \mathbf{w}^T\mathbf{E}^{(j)^T}\mathbf{F}^{(j)}\mathbf{q} - \tfrac{1}{2}\lambda\left(\mathbf{q}^T\mathbf{q} - 1\right) - \tfrac{1}{2}\lambda\left(\mathbf{w}^T\mathbf{E}^{(j)^T}\mathbf{E}^{(j)}\mathbf{w} - 1\right),$$

with respect to $\mathbf{w}$ *and* $\mathbf{q}$,

$$\mathbf{E}^{(j)^T}\mathbf{F}^{(j)}\widehat{\mathbf{q}}_j - \widehat{\lambda}_j\mathbf{F}^{(j)^T}\mathbf{F}^{(j)}\widehat{\mathbf{w}}_j = \mathbf{0} \text{ and } \mathbf{F}^{(j)^T}\mathbf{E}^{(j)}\widehat{\mathbf{w}}_j - \widehat{\lambda}_j\widehat{\mathbf{q}}_j = \mathbf{0},$$

are equal to

$$\mathbf{X}_0^T\mathbf{F}^{(j)}\widehat{\mathbf{q}}_j - \widehat{\lambda}_j\mathbf{X}_0^T\mathbf{X}_0\widehat{\mathbf{w}}_j \text{ and } \mathbf{Y}_0^T\mathbf{X}_0\widehat{\mathbf{w}}_j - \widehat{\lambda}_j\widehat{\mathbf{q}}_j = \mathbf{0},$$

respectively.

Proof. Directly applying Lemmas 10.5.3 and 10.5.6 to the solution of the MRPLS objective function yields

$$\boxed{\mathbf{X}_0^T\mathbf{F}^{(j)}\widehat{\mathbf{q}}_j - \widehat{\lambda}_j\mathbf{X}_0^T\mathbf{X}_0\widehat{\mathbf{w}}_j \text{ and } \mathbf{Y}_0^T\mathbf{E}^{(j)}\widehat{\mathbf{w}}_j - \widehat{\lambda}_j\widehat{\mathbf{q}}_j = \mathbf{0}}. \tag{10.83}$$

Next, incorporating the results described in Theorem 10.5.4 to the matrix-vector product $\mathbf{E}^{(j)}\widehat{\mathbf{w}}_j$ gives rise to

$$\mathbf{E}^{(j)}\widehat{\mathbf{w}}_j = \mathbf{X}_0\prod_{i=1}^{j-1}\left[\mathbf{I} - \widehat{\mathbf{w}}_i\widehat{\mathbf{p}}_i^T\right]\widehat{\mathbf{w}}_j = \mathbf{X}_0\widehat{\mathbf{w}}_j. \tag{10.84}$$

Consequently,

$$\boxed{\mathbf{Y}_0^T\mathbf{E}^{(j)}\widehat{\mathbf{w}}_j - \widehat{\lambda}_j\widehat{\mathbf{q}}_j = \mathbf{0} \text{ becomes } \mathbf{Y}_0^T\mathbf{X}_0\widehat{\mathbf{w}}_j - \widehat{\lambda}_j\widehat{\mathbf{q}}_j = \mathbf{0}}.$$

Property 10.5.8 – Relationship between weight vectors for input variables. Theorem 10.5.8 describes the relationship between the jth w-weight vector, $\widehat{\mathbf{w}}_j$, and the jth auxiliary weight vector $\widehat{\mathfrak{w}}_j$.

Theorem 10.5.8 *The jth w-weight and auxiliary weight vectors are the left and right eigenvectors of the matrix product $\left[\mathbf{X}_0^T\mathbf{X}_0\right]^{-1}\mathbf{X}_0^T\mathbf{F}^{(j)}\mathbf{Y}_0^T\mathbf{X}_0$, respectively.*

Proof. That $\widehat{\mathbf{w}}_j$ is the left eigenvector of $\left[\mathbf{X}_0^T\mathbf{X}_0\right]^{-1}\mathbf{X}_0^T\mathbf{F}^{(j)}\mathbf{Y}_0^T\mathbf{X}_0$, associated with the largest eigenvalue, can be confirmed by solving the relationships of the MRPLS objective function

$$\left[\mathbf{X}_0^T\mathbf{X}_0\right]^{-1}\mathbf{X}_0^T\mathbf{F}^{(j)}\widehat{\mathbf{q}}_j = \widehat{\lambda}_j\widehat{\mathbf{w}}_j \qquad \widehat{\mathbf{q}}_j = \frac{\mathbf{Y}_0^T\mathbf{X}_0\widehat{\mathbf{w}}_j}{\widehat{\lambda}_j}, \tag{10.85}$$

which yields

$$\boxed{\left[\mathbf{X}_0^T\mathbf{X}_0\right]^{-1}\mathbf{X}_0^T\mathbf{F}^{(j)}\mathbf{Y}_0^T\mathbf{X}_0\widehat{\mathbf{w}}_j = \widehat{\lambda}_j^2\widehat{\mathbf{w}}_j}. \tag{10.86}$$

According to the MRPLS algorithm in Table 2.3, the auxiliary weight vector, $\mathfrak{w}_j$, is initially determined as the matrix-vector product $\mathbf{E}^{(j)^T}\widehat{\mathbf{u}}_j$. By substituting Steps 4 to 8 into Step 3 yields

$$\widehat{\mathfrak{w}}_j \propto \mathbf{E}^{(j)^T}\widehat{\mathbf{u}}_j$$

$$\widehat{\mathfrak{w}}_j \propto \mathbf{X}_0^T \mathbf{F}^{(j)} \widehat{\mathbf{q}}_j$$

$$\widehat{\mathfrak{w}}_j \propto \mathbf{X}_0^T \mathbf{F}^{(j)} \mathbf{Y}_0^T \mathbf{X}_0 \widehat{\mathbf{w}}_j \tag{10.87}$$

$$\widehat{\mathfrak{w}}_j \propto \mathbf{X}_0^T \mathbf{F}^{(j)} \mathbf{Y}_0^T \mathbf{X}_0 \left[\mathbf{X}_0^T \mathbf{X}_0\right]^{-1} \widehat{\mathfrak{w}}_j$$

$$\boxed{\widehat{\lambda}_j^2 \widehat{\mathfrak{w}}_j^T = \widehat{\mathfrak{w}}_j^T \left[\mathbf{X}_0^T \mathbf{X}_0\right]^{-1} \mathbf{X}_0^T \mathbf{F}^{(j)} \mathbf{Y}_0^T \mathbf{X}_0}.$$

Therefore, $\widehat{\mathbf{w}}_j$ is the dominant right eigenvector and $\widehat{\mathfrak{w}}_j$ is the dominant left eigenvector of $\left[\mathbf{X}_0^T \mathbf{X}_0\right]^{-1} \mathbf{X}_0^T \mathbf{F}^{(j)} \mathbf{Y}_0^T \mathbf{X}_0$.

Property 10.5.9 – Calculation of the *k*th q-weight vector. Before introducing a computationally efficient MRPLS algorithm, Lemma 10.5.9 shows that q-weight and q́-loading vectors are also eigenvectors of a specific matrix product.

Lemma 10.5.9 *The jth q-weight and q́-loading vector,* $\widehat{\mathbf{q}}_j$ *and* $\widehat{\acute{\mathbf{q}}}_j$*, are the dominant eigenvectors of the matrix product* $\mathbf{X}_0^T \mathbf{Y}_0 \left[\mathbf{X}_0^T \mathbf{X}_0\right]^{-1} \mathbf{X}_0^T \mathbf{F}^{(j)}$.

Proof. Lemma 10.5.9 directly follows from the relationships of the MRPLS objective function in (10.85)

$$\mathbf{w}_j = \frac{\left[\mathbf{X}_0^T \mathbf{X}_0\right]^{-1} \mathbf{X}_0^T \mathbf{F}^{(j)} \mathbf{q}_j}{\lambda_j} \qquad \mathbf{Y}_0^T \mathbf{X}_0 \mathbf{w}_j = \lambda_j \mathbf{q}_j. \tag{10.88}$$

Substituting the equation on the left hand side into that of the right hand side yields

$$\boxed{\mathbf{Y}_0^T \mathbf{X}_0 \left[\mathbf{X}_0^T \mathbf{X}_0\right]^{-1} \mathbf{X}_0^T \mathbf{F}^{(j)} \mathbf{q}_j = \lambda_j^2 \mathbf{q}_j}. \tag{10.89}$$

Property 10.5.10 – Computationally efficient MRPLS algorithm. After discussing the geometric orthogonality properties of the weight, score and loading vectors as well as their impact upon the deflation procedure, a computationally efficient MRPLS algorithm can now be introduced. Table 10.3 summarizes the steps of the implementation of the revised MRPLS algorithm. Computational savings are made by removing the calculation of the score vectors and reducing the deflation procedure to the rank-one modification $\mathbf{X}_0^T \mathbf{F}^{(j+1)} = \mathbf{X}_0^T \mathbf{F}^{(j)} - \mathbf{p}_j \acute{\mathbf{q}}_j^T$. Finally, this section concludes with the development of a batch algorithm for simultaneously computing the n q- and w-weight vectors.

Simultaneous computation of weight vectors. Recall that the derivation of the MRPLS algorithm in Subsection 2.3.3 consists of two steps. The first step involves the computation of the q-loading vectors by solving (2.63). The second step subsequently determines the w-weight vector by solving (2.66). It has then been shown that both steps can be combined. More precisely, the solution of the combined objective function in (2.68) is equivalent to the individual solutions of (2.63) and (2.66).

Coming back to the facts (i) that the q-weight vectors can be determined independently from the w-weight vectors and (ii) that the jth q-weight vector is the dominant eigenvector of the matrix $\mathbf{Y}_0^T\mathbf{X}_0\left[\mathbf{X}_0^T\mathbf{X}_0\right]^{-1}\mathbf{X}_0^T\mathbf{F}^{(j)}$ gives rise to the following theorem.

Theorem 10.5.10 *The kth q-weight vector is the eigenvector associated with the kth largest eigenvalue of* $\mathbf{Y}_0^T\mathbf{X}_0\left[\mathbf{X}_0^T\mathbf{X}_0\right]^{-1}\mathbf{X}_0^T\mathbf{Y}_0 \in \mathbb{R}^{n_y \times n_y}$*, which is a symmetric and positive semi-definite matrix of rank* $n \leq n_y$*.*

Proof. The proof of Theorem 10.5.10 commences by showing that the matrix product $\mathbf{Y}_0^T\mathbf{X}_0\left[\mathbf{X}_0^T\mathbf{X}_0\right]^{-1}\mathbf{X}_0^T\mathbf{Y}_0$ is of rank $n \leq n_y$. With regards to the data structure in (2.51), the covariance and cross-covariance matrices $E\left\{\mathbf{x}_0\mathbf{x}_0^T\right\}$ and $E\left\{\mathbf{x}_0\mathbf{y}_0^T\right\}$ are equal to

$$\mathbf{S}_{x_0y_0} = \mathfrak{P}\mathbf{S}_{ss}\mathfrak{Q}^T \in \mathbb{R}^{n_x \times n_y} \tag{10.90}$$

and

$$\mathbf{S}_{x_0x_0} = \mathfrak{P}\mathbf{S}_{ss}\mathfrak{P}^T + \mathfrak{P}'\mathbf{S}_{s's'}\mathfrak{P}'^T \in \mathbb{R}^{n_x \times n_x}, \tag{10.91}$$

respectively. The matrix products $\mathbf{Y}_0^T\mathbf{X}_0\left[\mathbf{X}_0^T\mathbf{X}_0\right]^{-1}\mathbf{X}_0^T\mathbf{Y}_0$ and $\widehat{\mathbf{S}}_{y_0x_0}\widehat{\mathbf{S}}_{x_0x_0}^{-1}\widehat{\mathbf{S}}_{x_0y_0}$ are equal up to the scaling factor $K-1$. As the rank of the true cross-covariance matrix is n, given that $\mathfrak{P} \in \mathbb{R}^{n_x \times n}$, $\mathfrak{Q} \in \mathbb{R}^{n_y \times n}$ and $\mathbf{S}_{ss} \in \mathbb{R}^{n \times n}$, the rank of $\widehat{\mathbf{S}}_{y_0x_0}\widehat{\mathbf{S}}_{x_0x_0}^{-1}\widehat{\mathbf{S}}_{x_0y_0}$ is asymptotically n, under the assumption that $\mathbf{S}_{ss}$ has full rank rank n, that is, the source signals are not linearly dependent. The jth q-weight vector is an eigenvector associated with the eigenvalue $\widehat{\mathbf{q}}_j^T\mathbf{Y}_0^T\mathbf{X}_0$ $\left[\mathbf{X}_0^T\mathbf{X}_0\right]^{-1}\mathbf{X}_0^T\mathbf{F}^{(j)}\widehat{\mathbf{q}}_j$, which suggests a different deflation for $\mathbf{Y}_0^T\mathbf{X}_0\left[\mathbf{X}_0^T\mathbf{X}_0\right]^{-1}\mathbf{X}_0^T$ $\mathbf{F}^{(j)}$. Abbreviating this matrix expression by $\mathbf{M}_q^{(j)}$ and defining $\mathbf{M}_q^{(1)} = \mathbf{Y}_0^T\mathbf{X}_0\left[\mathbf{X}_0^T\mathbf{X}_0\right]^{-1}\mathbf{X}_0^T\mathbf{Y}_0$, the first q-weight vector satisfies

$$\widehat{\lambda}_1 = \widehat{\mathbf{q}}_1^T\mathbf{M}_q^{(1)}\widehat{\mathbf{q}}_1. \tag{10.92}$$

Given that $\mathbf{M}_q^{(1)}$ is symmetric and positive definite, the deflation procedure for determining the second q-weight vector that is orthogonal to the first one is as follows

$$\mathbf{M}_q^{(2)} = \mathbf{M}_q^{(1)} - \widehat{\mathbf{q}}_1\widehat{\lambda}_1\widehat{\mathbf{q}}_1^T \tag{10.93}$$

which is the principle of the power method to determine subsequent eigenpairs of symmetric and positive definite matrices. After determining the jth eigenpair, the deflation procedure becomes

$$\boxed{\mathbf{M}_q^{(j+1)} = \mathbf{M}_q^{(j)} - \widehat{\mathbf{q}}_j\widehat{\lambda}_j\widehat{\mathbf{q}}_j^T = \mathbf{M}_q^{(1)} - \sum_{i=1}^{j}\widehat{\lambda}_i\widehat{\mathbf{q}}_i\widehat{\mathbf{q}}_i^T}, \tag{10.94}$$

accordingly. Given that the rank of $\mathbf{M}_q^{(1)}$ is n, a total of n eigenpairs can be determined by the iterative power method with deflation. On the other hand, the n eigenvectors and eigenvalues can also be determined simultaneously in a batch mode, for example discussed in Chapter 8 in Golub and van Loan (1996). Once the n q-weight vectors are available, the w-weight vectors can be computed. Before demonstrating that this can also be done in a batch mode, Lemma 10.5.11 shows how to compute the u-scores directly from the output data matrix.

Lemma 10.5.11 *If the q-weight vectors are available, the u-score variables can be directly computed from the output data matrix, that is,* $\widehat{u}_j = \mathbf{y}_0^T\widehat{\mathbf{q}}_j$.

Proof. In the preceding discussion, the u-score variables have been computed from the deflated output data, that is, $\widehat{u}_j = \mathbf{f}^{(j)^T}\widehat{\mathbf{q}}_j$. However, the fact that the q-weight vectors are mutually orthonormal yields

$$\boxed{\widehat{u}_j = \left(\mathbf{y}_0^T - \sum_{i=1}^{j-1}\widehat{t}_i\widehat{\acute{\mathbf{q}}}_i^T\right)\widehat{\mathbf{q}}_i = \left(\mathbf{y}_0^T - \sum_{i=1}^{j-1}\widehat{t}_i\widehat{\lambda}_i\widehat{\mathbf{q}}_i^T\right)\widehat{\mathbf{q}}_i = \mathbf{y}_0^T\widehat{\mathbf{q}}_j}. \tag{10.95}$$

The above relationship incorporates $\widehat{\lambda}_j\widehat{\mathbf{q}}_j = \widehat{\acute{\mathbf{q}}}_j$, which (10.79) highlights.

Next, incorporating $\widehat{\mathbf{u}}_j = \mathbf{Y}_0\widehat{\mathbf{q}}_j$ into the objective function for determining the w-weight vector in (2.66) gives rise to

$$\widehat{\mathbf{w}}_j = \arg\max_{\mathbf{w}} \widehat{\mathbf{u}}_j^T\mathbf{X}_0\mathbf{w} - \tfrac{1}{2}\lambda_j\left(\mathbf{w}^T\mathbf{X}_0^T\mathbf{X}_0\mathbf{w} - 1\right). \tag{10.96}$$

Taking advantage of the fact that $\widehat{\mathbf{u}}_i^T\widehat{\mathbf{t}}_j = 0$ and $\widehat{\mathbf{t}}_i^T\widehat{\mathbf{t}}_j = 0$ for all $i \neq j$, (10.96) can be expanded upon

$$\begin{aligned}\begin{bmatrix}\widehat{\mathbf{w}}_1 & \widehat{\mathbf{w}}_1 & \cdots & \widehat{\mathbf{w}}_n\end{bmatrix} = \arg\max_{\mathbf{W}} \begin{bmatrix}\widehat{\mathbf{u}}_1^T \\ \widehat{\mathbf{u}}_2^T \\ \vdots \\ \widehat{\mathbf{u}}_n^T\end{bmatrix}\mathbf{X}_0\begin{bmatrix}\mathbf{w}_1 & \mathbf{w}_2 & \cdots & \mathbf{w}_n\end{bmatrix} - \\ \tfrac{1}{2}\mathrm{diag}\begin{bmatrix}\lambda_1 & \lambda_2 & \cdots & \lambda_n\end{bmatrix} \times\end{aligned} \tag{10.97}$$

$$\begin{bmatrix} \mathbf{w}_1^T \\ \mathbf{w}_2^T \\ \vdots \\ \mathbf{w}_n \end{bmatrix} \mathbf{X}_0^T \mathbf{X}_0 \begin{bmatrix} \mathbf{w}_1 & \mathbf{w}_2 & \cdots & \mathbf{w}_n \end{bmatrix},$$

where $\widehat{\mathbf{W}} = \begin{bmatrix} \widehat{\mathbf{w}}_1 & \widehat{\mathbf{w}}_1 & \cdots & \widehat{\mathbf{w}}_n \end{bmatrix}$. Equation (10.97) can also be written as

$$\begin{aligned}
&\widehat{\mathbf{U}}^T \mathbf{X}_0 \widehat{\mathbf{W}} - \tfrac{1}{2}\mathrm{diag}\begin{bmatrix} \widehat{\lambda}_1 & \widehat{\lambda}_2 & \cdots & \widehat{\lambda}_n \end{bmatrix} \widehat{\mathbf{W}}^T \mathbf{X}_0^T \mathbf{X}_0 \widehat{\mathbf{W}} = \\
&\widehat{\mathbf{U}}^T \widehat{\mathbf{T}} - \tfrac{1}{2}\mathrm{diag}\begin{bmatrix} \widehat{\lambda}_1 & \widehat{\lambda}_2 & \cdots & \widehat{\lambda}_n \end{bmatrix} = \\
&\mathrm{diag}\begin{bmatrix} \widehat{\lambda}_1 & \widehat{\lambda}_2 & \cdots & \widehat{\lambda}_n \end{bmatrix} - \tfrac{1}{2}\mathrm{diag}\begin{bmatrix} \widehat{\lambda}_1 & \widehat{\lambda}_2 & \cdots & \widehat{\lambda}_n \end{bmatrix} = \\
&\tfrac{1}{2}\mathrm{diag}\begin{bmatrix} \widehat{\lambda}_1 & \widehat{\lambda}_2 & \cdots & \widehat{\lambda}_n \end{bmatrix},
\end{aligned} \qquad (10.98)$$

which follows from Theorem 10.5.1 and yields in this batch formulation the same solution as those obtained individually. More precisely, defining the maximum of the combined objective by $\sum_{i=1}^{n} \widehat{\lambda}_i$, it can only be a maximum if each of the sum elements are a maximum, according to the Bellman principle of optimality (Bellman 1957). As shown in (10.97), storing the n w-weight vectors to form the matrix $\widehat{\mathbf{W}}$, the solution of the objective function in (10.97) is given by

$$\begin{aligned}
\widehat{\mathbf{W}} = \arg \frac{\partial}{\partial \mathbf{W}} \mathrm{trace} \Bigg\{ &\begin{bmatrix} \widehat{\mathbf{u}}_1 \\ \widehat{\mathbf{u}}_2 \\ \vdots \\ \widehat{\mathbf{u}}_n \end{bmatrix} \mathbf{X}_0 \begin{bmatrix} \mathbf{w}_1 & \mathbf{w}_2 & \cdots & \mathbf{w}_n \end{bmatrix} \\
&- \tfrac{1}{2}\mathrm{diag}\begin{bmatrix} \lambda_1 & \lambda_2 & \cdots & \lambda_n \end{bmatrix} \\
&\times \begin{bmatrix} \mathbf{w}_1^T \\ \mathbf{w}_2^T \\ \vdots \\ \mathbf{w}_n \end{bmatrix} \mathbf{X}_0^T \mathbf{X}_0 \begin{bmatrix} \mathbf{w}_1 & \mathbf{w}_2 & \cdots & \mathbf{w}_n \end{bmatrix} \Bigg\} = \mathbf{0}.
\end{aligned} \qquad (10.99)$$

Working out the partial relationships, (10.99) becomes

$$\mathbf{X}_0^T \widehat{\mathbf{U}} - \mathbf{X}_0^T \mathbf{X}_0 \mathrm{diag}\left\{\widehat{\boldsymbol{\lambda}}\right\} \widehat{\mathbf{W}} = \mathbf{0}, \qquad (10.100)$$

where $\widehat{\mathbf{U}} = \begin{bmatrix} \widehat{\mathbf{u}}_1 & \widehat{\mathbf{u}}_2 & \cdots & \widehat{\mathbf{u}}_n \end{bmatrix}$ and $\mathrm{diag}\left\{\widehat{\boldsymbol{\lambda}}\right\} = \mathrm{diag}\begin{bmatrix} \widehat{\lambda}_1 & \widehat{\lambda}_2 & \cdots & \widehat{\lambda}_n \end{bmatrix}$.

To simultaneously calculate the n w-weight vectors in batch form, the solution to the objective function in (10.97) is therefore

$$\mathrm{diag}\left\{\widehat{\boldsymbol{\lambda}}\right\} \widehat{\mathbf{W}} = \left[\mathbf{X}_0^T \mathbf{X}_0\right]^{-1} \mathbf{X}_0^T \mathbf{Y}_0 \widehat{\mathbf{Q}}. \qquad (10.101)$$

Equations (10.86) and (10.89) outline that the n diagonal elements of the matrix $\text{diag}\{\widehat{\boldsymbol{\lambda}}\}$ are equal to the square root of the eigenvalues of $\mathbf{Y}_0^T\mathbf{X}_0\left[\mathbf{X}_0^T\mathbf{X}_0\right]^{-1}\mathbf{X}_0^T\mathbf{Y}_0$ and $\left[\mathbf{X}_0^T\mathbf{X}_0\right]^{-1}\mathbf{X}_0\mathbf{Y}_0^T\mathbf{Y}_0^T\mathbf{X}_0$. The conclusion of this proof requires to show that

$$\widehat{\mathbf{Q}} = \mathbf{Y}_0^T\mathbf{X}_0\widehat{\mathbf{W}}\text{diag}\left\{\widehat{\boldsymbol{\lambda}}\right\}^{-1}. \tag{10.102}$$

To start with, Theorem 10.5.5 shows that $\widehat{\mathbf{Q}}^T\widehat{\mathbf{Q}} = \mathbf{I}$, i.e. the column vectors or $\widehat{\mathbf{Q}}$ are mutually orthonormal. Next, (10.83) and (10.84) highlight that

$$\widehat{\lambda}_j\widehat{\mathbf{q}}_j = \mathbf{Y}_0^T\mathbf{X}_0\widehat{\mathbf{w}}_j \tag{10.103}$$

which confirms (10.102).

In summary, Theorem 10.5.10 outlined that the q-weight vectors can be simultaneously computed as eigenvectors of the matrix product $\mathbf{Y}_0^T\mathbf{X}_0\left[\mathbf{X}_0^T\mathbf{X}_0\right]^{-1}\mathbf{X}_0^T\mathbf{Y}_0$, which is positive semi-definite, whose rank asymptotically converges to $n \le n_y$. On the other hand, it is also possible to simultaneously compute the w-weight vectors as eigenvectors of the matrix product $\left[\mathbf{X}_0^T\mathbf{X}_0\right]^{-1}\mathbf{X}_0^T\mathbf{Y}_0\mathbf{Y}_0^T\mathbf{X}_0$, which is a positive semi-definite matrix with an asymptotic rank of $n \le n_y$. Furthermore, the computation of the w-weight vectors is independent of the determination of the q-weight vectors. Equations (10.105) and (10.106) finally summarize the eigendecomposition of these matrices

$$\left[\mathbf{X}_0^T\mathbf{X}_0\right]^{-1}\mathbf{X}_0^T\mathbf{Y}_0\mathbf{Y}_0^T\mathbf{X}_0\widehat{\mathbf{W}} = \text{diag}\left\{\widehat{\boldsymbol{\lambda}}\right\}^2\widehat{\mathbf{W}} \tag{10.104}$$

and

$$\mathbf{Y}_0^T\mathbf{X}_0\left[\mathbf{X}_0^T\mathbf{X}_0\right]^{-1}\mathbf{X}_0^T\mathbf{Y}_0\widehat{\mathbf{Q}} = \text{diag}\left\{\widehat{\boldsymbol{\lambda}}\right\}^2\widehat{\mathbf{Q}}. \tag{10.105}$$

For canonical correlation analysis (CCA), a multivariate statistical method developed by Harold Hotelling in the 1930s (Hotelling 1935, 1936) that determines weight vectors to produce score variables that have a maximum correlation, a batch algorithm has been proposed to simultaneously determine $\mathbf{Q}$ and $\mathbf{W}$. This solution for simultaneously computing the n q- and w-weight vectors can be taken advantage of by simultaneously computing the weight vectors of the MRPLS algorithm. Prior to that, the next paragraph discusses the similarities between the objective functions for CCA and maximum redundancy.

As outlined in Stewart and Love (1968), ten Berge (1985) and van den Wollenberg (1977), the CCA objective function does not consider the predictability of the output variables in the same way as maximum redundancy does. More precisely, the CCA objective function is given by

$$\begin{aligned}\begin{pmatrix}\mathbf{w}_j\\ \mathbf{q}_j\end{pmatrix} &= \arg\max_{\mathbf{w},\mathbf{q}} E\left\{\mathbf{w}^T\mathbf{x}_0\mathbf{y}_0^T\mathbf{q}\right\} - \\ &\quad \tfrac{1}{2}\lambda_j E\left\{\left(\mathbf{w}^T\mathbf{x}_0\mathbf{x}_0^T\mathbf{w}\right)\right\} - \tfrac{1}{2}\lambda_j E\left\{\left(\mathbf{q}^T\mathbf{y}_0\mathbf{y}_0^T\mathbf{q}\right)\right\}.\end{aligned} \tag{10.106}$$

Similar to the maximum redundancy formulation, the Lagrangian multipliers are identical and the solution of the CCA objective function for the jth pair of weight vectors, or canonical variates, is given by (Anderson 2003)

$$\left[\mathbf{X}_0^T\mathbf{X}_0\right]^{-1}\mathbf{X}_0^T\mathbf{Y}_0\left[\mathbf{Y}_0^T\mathbf{Y}_0\right]^{-1}\mathbf{Y}_0^T\mathbf{X}_0\widehat{\mathbf{w}}_j=\widehat{\lambda}_j^2\widehat{\mathbf{w}}_j \tag{10.107}$$

and

$$\left[\mathbf{Y}_0^T\mathbf{Y}_0\right]^{-1}\mathbf{Y}_0^T\mathbf{X}_0\left[\mathbf{X}_0^T\mathbf{X}_0\right]^{-1}\mathbf{X}_0^T\mathbf{Y}_0\widehat{\mathbf{q}}_k=\widehat{\lambda}_k^2\widehat{\mathbf{q}}_k. \tag{10.108}$$

It is interesting to note that the difference in the objective function between the CCA and the maximum redundancy is the presence of the matrix product $\left[\mathbf{Y}_0^T\mathbf{Y}_0\right]^{-1}$, which results from the different constraint for the q-weight vector. For the simultaneous computation of the n weight vectors, a batch algorithm that relies on a series of singular value decompositions (SVDs) has been developed (Anderson 2003). Table 10.4 summarizes the steps of the resultant batch algorithm for simultaneously determining the n q- and w-weight vectors.

References

Abbott, T.S. and Person, L.H. (1991) Method and system for monitoring and displaying engine performance parameters. United States Patent, patent number: 5050081.

Akaike, H. (1974) A new look at statistical model identification. *IEEE Transactions on Automatic Control*, 19(6):716–723.

Akbari, H., Levinson, R., and Rainer, L. (2005) Monitoring the energy-use effects of cool roofs on California commercial buildings. *Energy and Buildings*, 37(10):1007–1016.

Aldrich, J. (1997) R.A. Fisher and the making of maximum likelihood 1912–1922. *Statistical Science*, 12(3):162–176.

Al-Ghazzawi, A. and Lennox, B. (2008) Monitoring a complex refining process using multivariate statistics. *Control Engineering Practice*, 16(3):294–307.

Allen, D.M. (1974) The relationship between variable selection and data augmentation and a method for prediction. *Technometrics*, 16(1):125–127.

Anderson, T.W. (1951) Estimating linear restrictions on regression coefficients for multivariate normal distributions. *The Annals of Mathematical Statistics*, 22(3):327–351.

Anderson, T.W. (1963) Asymptotic theory for principal component analysis. *The Annals of Mathematical Statistics*, 34(1):122–148.

Anderson, T.W. (2003) *An Introduction into Multivariate Statistical Analysis*. John Wiley & Sons, New York, USA, 3rd. edition.

Aparisi, F. (1997–1998) Sampling plans for the multivariate T^2 control chart. *Quality Engineering*, 10(1):141–147.

Aravena, J.L. (1990) Recursive moving window DFT algorithm. *IEEE Transactions on Computers*, 39(1):145–148.

Arteaga, F. and Ferrer, A. (2002) Dealing with missing data in MSPC: several methods, different interpretations, some examples. *Journal of Chemometrics*, 16(8-10):408–418.

Bardou, O. and Sidahmed, M. (1994) Early detection of leakages in the exhaust and discharge systems of reciprocating machines by vibration analysis. *Mechanical Systems & Signal Processing*, 8(5):551–570.

Bartelmus, W. and Zimroz, R. (2009) Vibration condition monitoring of planetary gearbox under varying external load. *Mechanical Systems & Signal Processing*, 23(1):246–257.

Basseville, M. (1988) Detecting changes in signals and systems – a survey. *Automatica*, 24(3):309–326.

Statistical Monitoring of Complex Multivariate Processes: With Applications in Industrial Process Control, First Edition. Uwe Kruger and Lei Xie.

Baydar, N. and Ball, A.D. (2001) A comparative study of acoustics and vibration signals in detection of gear failures using Wigner-Ville distribution. *Mechanical Systems & Signal Processing*, 15(6):1091–1107.

Baydar, N., Ball, A.D., and Kruger, U. (1999) Detection of incipient tooth defect in helical gears using principal components. In *1st International Conference on the Integrating of Dynamics, Monitoring and Control*, pp.93–100, Manchester, UK.

Baydar, N., Chen, Q., Ball, A.D., and Kruger, U. (2001) Detection of incipient tooth defect in helical gears using multivariate statistics. *Mechanical Systems & Signal Processing*, 15(2):303–321.

Bellman, R. (1957) *Dynamic Programming*. Princeton University Press, Priceton, NJ, USA.

Billingsley, P. (1995) *Probability and Measures*. John Wiley & Sons, New York, 3rd edition.

Bissessur, Y., Martin, E.B., and Morris, A.J. (1999) Monitoring the performance of the paper making process. *Control Engineering Practice*, 7(11):1357–1368.

Bissessur, Y., Martin, E.B., Morris, A.J., and Kitson, P. (2000) Fault detection in hot steel rolling using neural networks and multivariate statistics. *IEE Proceedings, Part D – on Control Theory and Applications*, 147(6):633–640.

Björck, Å. (1996) *Numerical Methods for Least Squares Problems*. SIAM Publishing, Philadelphia, PA, USA.

Boller, C. (2000) Next generation structural health monitoring and its integration into aircraft design. *International Journal of Systems Science*, 31(11):1333–1349.

Box, G.E.P. (1954) Some theorems on quadratic forms applied in teh study of analysis of variance problems: Effect of inequality of variance in one-way classification. *Annals of Mathematical Statistics*, 25(2):290–302.

Box, G.E.P., Hunter, W.G., MacGregor, J.F., and Erjavec, J. (1973) Some problems associated with the analysis of multiresponse data. *Technometrics*, 15(1):33–51.

Bradley, R.C. (2007) *Introduction to Strong Mixing Conditions, Volume 1–3*. Kendrick Press, Heber City, Utah, USA.

Brussee, W. (2004) *Statistics for Six Sigma Made Easy*. McGraw-Hill, New York, USA.

Buckley, P.S., Luyben, W.L., and Shunta, J.P. (1985) *Design of distillation column control systems*. Edward Arnold, London.

Bunch, J.R., Nielsen, C.P., and Sorensen, D.C. (1978) Rank-one modifcation of the symmetric eigenproblem. *Numerische Mathematik*, 31:31–48.

Burr, J.T. (2005) *Elementary Statistical Quality Control*. Marcel Dekker, New York, USA.

Byrd, R.H., Lu, P., Nocedal, J., and Zhu, C. (1995) A limited memory algorithm for bound constrained optimization. *SIAM Journal on Scientific Computing*, 16(5):1190–1208.

Cattell, R.B. (1966) The scree test for the number of factors. *Multivariate Behavioral Research*, 1(2):245–276.

Cattell, R.B. and Vogelmann, S. (1977) A comprehensive trial of the scree and KG criteria for determining the number of factors. *Multivariate Behavioral Research*, 12(3):289–325.

Champagne, B. (1994) Adaptive eigendecomposition of data covariance matrices based on first order perturbations. *IEEE Transactions on Signal Processing*, 42(10):2758–2770.

Chatterjee, C., Kang, Z., and Roychowdhury, V.P. (2000) Algorithms for accelerated convergence of adaptive pca. *IEEE Transactions on Neural Networks*, 11(2):338–355.

Chen, J. and Liu, K.C. (2004) On-line batch process monitoring using dynamic pca and dynamic PLS models. *Chemical Engineering Science*, 57(1):63–75.

Chen, Q. and Kruger, U. (2006) Analysis of extended partial least squares for monitoring large-scale processes. *IEEE Transactions on Control Systems Technology*, 15(5):807–813.

Chen, Y.D., Du, R., and Qu, L.S. (1995) Fault features of large rotating machinery and diagnosis using sensor fusion. *Journal of Sound & Vibration*, 188(2):227–242.

Chiang, L.H., Russel, E.L., and Braatz, R.D. (2001) *Fault Detection and Diagnosis in Industrial Systems*. Springer-Verlag, London.

Coello, C.A.C., Pulido, G.T., and Lechuga, M.S. (2004) Handling multiple objectives with particle swarm optimization. *IEEE Transactions on Evolutionary Computation*, 8(3):256–279.

Cramér, H. (1946) *Mathematical Methods in Statistics*. Princeton University Press, Princeton, NJ, USA.

Crosby, P.B. (1979) *Quality is Free*. McGraw Hill, New York.

Cullum, J.K. and Willoughby, R.A. (2002) *Lanczos Algorithms for Large Symmetric Eigenvalue Computations: Theory*. SIAM Publishing, Philadelphia, PA.

Daszykowski, M. (2007) From projection pursuit to other unsupervised chemometric techniques. *Journal of Chemometrics*, 21(7-9):270–279.

Davies, P.L. (1992) The asymptitics of Rousseeuw's minimum volume ellipsoid estimator. *The Annals of Statistics*, 20(4):1828–1843.

Dayal, B.S. and MacGregor, J.F. (1996) Identification of finite impulse response models: Methods and robustness issues. *Industrial & Engineering Chemistry Research*, 35(11):4078–4090.

Dayal, B.S. and MacGregor, J.F. (1997a) Improved PLS algorithms. *Journal of Chemometrics*, 11(1):73–85.

Dayal, B.S. and MacGregor, J.F. (1997b) Multi-output process identification. *Journal of Process Control*, 7(4):269–282.

Dayal, B.S. and MacGregor, J.F. (1997c) Recursive exponentially weighted PLS and its applications to adaptive control and prediction. *Journal of Process Control*, 7(3):169–179.

de Jong, S. (1993) Simpls, an alternative approach to partial least squares regression. *Chemometrics and Intelligent Laboratory Systems*, 18:251–263.

de Jong, S., Wise, B.M., and Ricker, N.L. (2001) Canonical partial least squares and continuum power regression. *Journal of Chemometrics*, 15(2):85–100.

Deming, W.E. and Birge, R.T. (1934) On the statistical theory of errors. *Review of Modern Physics*, 6(3):119–161.

Ding, S.X. (2008) *Model-based Fault Diagnosis Techniques: Design Schemes, Algorithms and Tools*. Springer, Berlin.

Dionisio, R.M.A. and Mendes, D.A. (2006) Entropy-based independence test. *Nonlinear Dynamics*, 44:351–357.

Doebling, S.W., Farrar, C.R., Prime, M.B., and Shevitz, D.W. (1996) Damage Identification and Health Monitoring of Structural and Mechanical Systems from Changes in Their Vibration Characteristics: A Literature Review. Technical report, Los Alamos National Laboratory, USA.

Donoho, D.L. (1982) Breakdown properties of multivariate location estimators. Qualifying Paper. Harvard University, Boston, MA, USA.

Doukopoulos, X.G. and Moustakides, G.V. (2008) Fast and stable subspace tracking. *IEEE Transactions on Signal Processing*, 56(4):1452–1465.

Duchesne, C., Kourti, T., and MacGregor, J.F. (2002) Multivariate SPC for startups and grade transitions. *AIChE Journal*, 48(12):2890–2901.

Duchesne, C. and MacGregor, J.F. (2001) Jackknife and bootstrap methods in the identification of dynamic models. *Journal of Process Control*, 11(5):553–564.

Duda, R.O. and Hart, P.E. (1973) *Pattern classification and scene analysis*. John Wiley & Sons, New York.

Dunia, R. and Qin, S.J. (1998) A unified geometric approach to process and sensor fault identification and reconstruction: the unidimensional fault case. *Computers & Chemical Engineering*, 22(7-8):927–943.

Dunia, R., Qin, S.J., Edgar, T.F., and McAvoy, T.J. (1996) Identification of faulty sensors using principal component analysis. *AIChE Journal*, 42(10):2797–2812.

Dunteman, G.H. (1989) *Principal Components Analysis*. Quantitative Applications in the Social Sciences. Sage, London.

Durrett, R. (1996) *Probability: Theory and Examples*. Duxbury Press, Wadsworth Publishing Company, Belmont, CA, USA, 2nd edition.

Eastment, H.T. and Krzanowski, W.J. (1982) Cross-validatory choice of the number of components from a principal component analysis. *Technometrics*, 24(1):73–77.

Efron, B. and Tibshirani, R.J. (1993) *An Introduction to the Bootstrap*. Chapman and Hall, New York.

Farmer, S.A. (1971) An investigation into the results of principal component analysis of data derived from random numbers. *The Statistician*, 20(4):63–72.

Feigenbaum, D. (1951) *Quality Control: Principles, Practice, and Administration*. McGraw-Hill, New York.

Feital, T., Kruger, U., Xie, L., Schubert, U., Lima, E.L., and Pinto, J.C. (2010) A unified statistical framework for monitoring multivariate systems with unknown source and error signals. *Chemometrics & Intelligent Laboratory Systems*, 104(2):223–232.

Feller, W. (1936) Über den zentralen grenzwertsatz der wahrscheinlichkeitsrechnung. *Mathematische Zeitschrift*, 40(1):521–559.

Feller, W. (1937) Über den zentralen grenzwertsatz der wahrscheinlichkeitsrechnung. ii. *Mathematische Zeitschrift*, 42(1):301–312.

Fisher, H. (2011) *A History of the Central Limit Theorem: From Classical to Modern Probability Theory*. Springer Verlag, New York.

Fortier, J.J. (1966) Simultaneous linear prediction. *Psychometrika*, 31(3):369–381.

Fortuna, L., Graziani, S.G. Xibilia, M., and Barbalace, N. (2005) Fuzzy activated neural models for product quality monitoring in refineries. In Zítek, P., editor, *Proceedings of the 16th IFAC World Congress*, volume 16. Elsevier.

Frank, P.M., Ding, S.X., and Marcu, T. (2000) Model-based fault diagnosis in technical processes. *Transactions of the Institute of Measurement and Control*, 22(1): 57–101.

Fuchs, E. and Donner, K. (1997) Fast least-squares polynomial approximation in moving time windows. In *Proceedings of the 1997 IEEE International Conference on Acoustics, Speech, and Signal Processing, ICASSP*, pages 1965–1968, Munich, Germany.

Fugate, M.L., Sohn, H., and Farrar, C.R. (2001) Vibration-based damage detection using statistical process control. *Mechanical System & Signal Processing*, 15(4): 707–721.

Fuller, W.A. (1987) *Measurement Error Models*. John Wiley & Sons, New York, USA.

Gallagher, N.B., Wise, B.M., Butler, S.W., White, D.D., and Barna, G.G. (1997) Development of benchmarking of multivariate statistical process control tools for a semiconductor etch process: Improving robustness through model updating. In *Proceedings of ADCHEM 97*, pages 78–83, Banff, Canada. International Federation of Automatic Control.

Ge, Z., Kruger, U., Lamont, L., Xie, L., and Song, Z. (2010) Fault detection in non-Gaussian vibration systems using dynamic statistical-based approaches. *Mechanical Systems & Signal Processing*, 24(8):2972–2984.

Ge, Z., Xie, L., Kruger, U., and Song, Z. (2011) Local ICA for multivariate statistical fault diagnosis in systems with unknown signal and error distributions. *AIChE Journal*, in print.

Geisser, S. (1974) A predictive approach to the random effect model. *Biometrika*, 61(1):101–107.

Geladi, P. and Kowalski, B.R. (1986) Partial least squares regression: A tutorial. *Analytica Chimica Acta*, 185:231–246.

Gérard, Y., Holdsworth, R.J., and Martin, P.A. (2007) Multispecies *in situ* monitoring of a static internal combustion engine by near-infrared diode laser sensors. *Applied Optics*, 46(19):3937–3945.

Gnanadesikan, R. and Kettenring, J.R. (1972) Robust estimates, residuals, and outliers detection with multiresponse data. *Biometrics*, 28:81–124.

Golub, G.H. (1973) Some modifed matrix eigenvalue problems. *SIAM Review*, 15(2):318–334.

Golub, G.H. and van Loan, C.F. (1996) *Matrix Computation*. John Hopkins, Baltimore, USA 3 edition.

Gosselin, C. and Ruel, M. (2007) Advantages of monitoring the performance of industrial processes. In *ISA EXPO 2007*, Reliant Center, Houston, TX, USA.

Granger, C.W., Maasoumi, E., and Racine, J. (2004) A dependence metric for possibly nonlinear processes. *Journal of Time Series Analysis*, 25(5):649–669.

Graybill, F. (1958) Determining sample size for a specified width confidence interval. *The Annals of Mathematical Statistics*, 29(1):282–287.

Graybill, F.A. and Connell, T.L. (1964) Sample size required for estimating the variance within *d* units of the true value. *The Annals of Mathematical Statistics*, 35(1):438–440.

Graybill, F.A. and Morrison, R.D. (1960) Sample size for a specified width confidence interval on the variance of a normal distribution. *Biometrics*, 16(4):636–641.

Greenwood, J.A. and Sandomire, M.M. (1950) Sample size required for estimating the standard deviation as a percent of its true value. *Journal of the American Statistical Association*, 45(250):257–260.

Gupta, P.L. and Gupta, R.D. (1987) Sample size determination in estimating a covariance matrix. *Computational Statistics & Data Analysis*, 5(3):185–192.

Gut, A. (2005) *Probability: A Graduate Course*. Springer Verlag, New York.

Hall, P., Marshall, D., and Martin, R. (1998) Incrementally computing eigenspace models. In *Proceeding of British Machine Vision Conference*, pages 286–295, Southampton.

Hall, P., Marshall, D., and Martin, R. (2000) Merging and splitting eigenspace models. *IEEE Transactions on Pattern Analysis and Machine Intelligence*, 22(9): 1042–1049.

Hall, P., Marshall, D., and Martin, R. (2002) Adding and subtracting eigenspaces with eigenvalue decomposition and singular value decomposition. *Image and Vision Computing*, 20(13-14):1009–1016.

Hampel, F.R. (1974) The influence curve and its role in robust estimation. *Journal of the American Statistical Association*, 69:383–393.

Hawkins, D.M. (1974) The detection of errors in multivariate data using principal components. *Journal of the American Statistical Association*, 69(346):340–344.

Hawkins, D.M. (1993) Cumulative sum control charting: An underutilized SPC tool. *Quality Engineering*, 5(3):463–477.

Hawkins, D.M. and Olwell, D.H. (1998) *Cumulative Sum Charts and Charting for Quality Improvements*. Springer Verlag, New York, NY, USA.

He, Q.B., Kong, F.R., and Yan, R.Q. (2007) Subspace-based gearbox condition monitoring by kernel principal component analysis. *Mechanical Systems & Signal Processing*, 21(4):1755–1772.

He, Q.B., Yan, R.Q., Kong, F.R., and Du, R.X. (2009) Machine condition monitoring using principal component representations. *Mechanical Systems & Signal Processing*, 23(4):446–466.

Helland, K., Berntsen, H., Borgen, O., and Martens, H. (1991) Recursive algorithm for partial least squares regression. *Chemometrics & Intelligent Laboratory Systems*, 14(1-3):129–137.

Henderson, C.R. (1975) Unbiased estimation and prediction under a selection model. *Biometrics*, 31(2):423–444.

Horn, J.L. (1965) A rationale and test for the number of factors in factor analysis. *Psychometrica*, 30(2):73–77.

Höskuldsson, A. (1988) PLS regression models. *Journal of Chemometrics*, 2(3): 211–228.

Höskuldsson, A. (1994) The H-principle: new ideas, algorithms and methods in applied mathematics and statistics. *Chemometrics & Intellingent Laboratory Systems*, 23(1):1–28.

Höskuldsson, A. (1995) A combined theory for PCA and PLS. *Journal of Chemometrics*, 9(2):91–123.

Höskuldsson, A. (1996) *Prediction Methods in Schience and Technology*. Thor Publishing, Copenhagen, Denmark.

Höskuldsson, A. (2008) H-methods in applied sciences. *Journal of Chemometrics*, 22(3-4):150–177.

Hotelling, H. (1935) The most predictable criterion. *Journal of Educational Psychology*, 26(2):139–142.

Hotelling, H. (1936) Relations between two sets of variates. *Biometrica*, 28(3-4):321–377.

Hotelling, H. (1947) Multivariate quality control, illustrated by the air testing of sample bombsights. In Eisenhart, C., Hastay, M.W., and Wallis, W.A., editors, *Selected Techniques of Statistical Analysis*, pages 111–184. McGraw-Hill, New York.

Howlett, R.J., de Zoysa, M.M., Walters, S.D., and Howson, P.A. (1999) Neural network techniques for monitoring and control of internal combustion engines. In *Proceedings of the International Symposium on Intelligent Industrial Automation*, Genova, Italy.

Hu, N., Chen, M., and Wen, X. (2003) The application of stochastic resonance theory for early detecting rub-impact fault of rotor system. *Mechanical Systems and Signal Processing*, 17(4):883–895.

Hu, Q., He, Z., Zhang, Z., and Zi, Y. (2007) Fault diagnosis of rotating machinery based on improved wavelet package transform and SVMS ensemble. *Mechanical Systems & Signal Processing*, 21(2):688–705.

Hunter, J.S. (1986) The exponentially weighted moving average. *Journal of Quality Technology*, 18(4):203–210.

Hyvarinen, A. (1999) Gaussian moments for noisy independent component analysis. *IEEE Signal Processing Letters*, 6:145–147.

Hyvärinen, A., Karhunen, J., and Oja, E. (2001) *Independent Component Analysis*. John Wiley & Sons, New York.

Iserman, R. (1993) Fault diagnosis of machines via parameter estimation and knowledge processing: Tutorial paper: Fault detection, supervision and safety for technical processes. *Automatica*, 29(4):815–835.

Isermann, R. (2006) *Fault-Diagnosis Systems – An Introduction from Fault Detection to Fault Tolerance*, volume XVIII of *Robotics*. Springer Verlag GmbH.

Isermann, R. and Ballé, P. (1997) Trends in the application of model-based fault detection and diagnosis of technical processes. *Control Engineering Practice*, 5(5):709–719.

Ishikawa, K. (1985) *What is Total Quality Control?* Prentice Hall, Englewood Cliffs, NJ, USA.

Isserlis, L. (1918) On a formula for the product-moment coefficient of any order of a normal frequency distribution in any number of variables. *Biometrika*, 12(1/2):134–139.

Jackson, J.E. (1959) Quality control methods for several related variables. *Technometrics*, 1(4):359–377.

Jackson, J.E. (1980) Principal components and factor analysis: Part I: Principal components. *Journal of Quality Control*, 12(4):201–213.

Jackson, J.E. (2003) *A Users Guide to Principal Components*. Wiley Series in Probability and Mathematical Statistics. John Wiley, New York.

Jackson, J.E. and Morris, R.H. (1956) Quality control methods for two related variables. *Industrial Quality Control*, 12(7):2–6.

Jackson, J.E. and Morris, R.H. (1957) An application of multivariate quality control to photographic processing. *Journal of the American Statistical Association*, 52(278):186–199.

Jackson, J.E. and Mudholkar, G.S. (1979) Control procedures for residuals associated with principal component analysis. *Technometrics*, 21:341–349.

Jaw, L.C. (2005) Recent advancements in aircraft engine health management (EHM) technologies and recommendations for the next step. In *Proceedings of the ASME Turbo Expo 2005: Power for Land, Sea, and Air*, pages 683–695, Reno, USA.

Jaw, L.C. and Mattingly, J.D. (2008) *Aircraft Engine Controls: Design, System Analysis, And Health Monitoring*. American Institute of Aeronautics and Astronautics Education Series.

Jolliffe, I.T. (1972) Discarding variables in principal component analysis. I: Artificial data. *Applied Statistics*, 21(2):160–173.

Jolliffe, I.T. (1973) Discarding variables in a principal component analysis. II: Real data. *Applied Statistics*, 22(1):21–31.

Jolliffe, I.T. (1986) *Principal Component Analysis*. Springer, New York.

Juran, J.M. and Godfrey, A.B. (2000) *Juran's Quality Handbook*. McGraw Hill, New York, 5th edition.

Kaiser, H.F. (1960) The application of electronic computers to factor analysis. *Educational & Psychological Measurement*, 20(1):141–151.

Kaspar, M.H. and Ray, W.H. (1992) Chemometric method for process monitoring. *AIChE Journal*, 38(10):1593–1608.

Kaspar, M.H. and Ray, W.H. (1993) Partial least squares modelling as successive singular value decompositions. *Computers & Chemical Engineering*, 17(10):985–989.

Kenney, J., Linda, Y., and Leon, J. (2002) Statistical process control integration systems and methods for monitoring manufacturing processes. United States Patent, patent number 6445969.

Kiencke, U. and Nielsen, L. (2000) *Automotive Control Systems*. Springer-Verlag, Berlin, Heidelberg.

Kim, K. and Parlos, A.G. (2003) Reducing the impact of false alarms in induction motor fault diagnosis. *Journal of Dynamic Systems, Measurement and Control, Transactions of the ASME*, 125(1):80–95.

Knutson, J.W. (1988) Techniques for determining the state of control of a multivariate process. In *Proceedings of the Seventh Annual Control Engineering Conference*, pages 1–14, Rosemount, IL, USA. Tower Conference Management.

Ko, J.M. and Ni, Y.Q. (2005) Technology developments in structural health monitoring of large-scale bridges. *Engineering Structures*, 27(12):1715–1725.

Kosanovich, K.A. and Piovoso, M.J. (1991) Process data analysis using multivariate statistical methods. In *Proceedings of the American Control Conference*, Chicago, IL, USA.

Kourti, T. (2005) Application of latent variable methods to process control and multivariate statistical process control in industry. *International Journal of Adaptive Control and Signal Processing*, 19(4):213–246.

Kourti, T. and MacGregor, J.F. (1995) Process analysis, monitoring and diagnosis using multivariate projection methods. *Chemometrics & Intelligent Laboratory Systems.*, 28:3–21.

Kourti, T. and MacGregor, J.F. (1996) Multivariate SPC methods for process and product management. *Journal of Quality Technology*, 28:409–428.

Kramer, M.A. and Palowitch, B.L. (1987) A rule-based approach to fault diagnosis using the signed directed graph. *AIChE Journal*, 33:1067–1078.

Kresta, J.V., MacGregor, J.F., and Marlin, T.E. (1989) Multivariate statistical monitoring of process performance. In *AIChE Annual Meeting*, San Francisco, CA, USA.

Kresta, J.V., MacGregor, J.F., and Marlin, T.E. (1991) Multivariate statistical monitoring of process operating performance. *The Canadian Journal of Chemical Engineering*, 69:35–47.

Kruger, U., Chen, Q., Sandoz, D.J., and McFarlane, R.C. (2001) Extended PLS approach for enhanced condition monitoring of industrial processes. *AIChE Journal*, 47(9):2076–2091.

Kruger, U. and Dimitriadis, G. (2008) Diagnosis of process faults in chemical systems using a local partial least squares approach. *AIChE Journal*, 54(10):2581–2596.

Kruger, U., Kumar, S., and Littler, T. (2007) Improved principal component modelling using the local approach. *Automatica*, 43(9):1532–1542.

Kruger, U., Zhou, Y., Wang, X., Rooney, D., and Thompson, J. (2008a) Robust partial least squares regression – part III, outlier analysis and application studies. *Journal of Chemometrics*, 22(5):323–334.

Kruger, U., Zhou, Y., Wang, X., Rooney, D., and Thompson, J. (2008b) Robust partial least squares regression: Part I, algorithmic developments. *Journal of Chemometrics*, 22(1):1–13.

Kruger, U., Zhou, Y., Wang, X., Rooney, D., and Thompson, J. (2008c) Robust partial least squares regression: Part II, new algorithm and benchmark studies. *Journal of Chemometrics*, 22(1):14–22.

Kumar, S., Martin, E.B., and Morris, A.J. (2002) Detection of process model changes in pca based performance monitoring. In *Proceedings of the American Control Confernece*, pages 2719–2724, Anchorage, AK.

Kwon, O.K., Kong, H.S., Kim, C.H., and Oh, P.K. (1987) Condition monitoring techniques for an internal combustion engine. *Tribology International*, 20(3): 153–159.

Lane, S., Martin, E.B., Morris, A.J., and Glower, P. (2003) Application of exponentially weighted principal component analysis for the monitoring of a polymer film manufacturing process. *Transactions of the Institute of Measurement and Control*, 25(1):17–35.

Lee, D.S. and Vanrolleghem, P.A. (2003) Monitoring of a sequencing batch reactor using adaptive multiblock principal component analysis. *Biotechnology and Bioengineering*, 82(4):489–497.

Lehane, M., Dube, F., Halasz, M., Orchard, R., Wylie, R., and Zaluski, M. (1998) Integrated diagnosis system (IDS) for aircraft fleet maintenance. In Mostow, J. and Rich, C., editors, *Proceedings of the Fifteenth National Conference on Artificial Intelligence (AAAI '98)*, Menlo Park, CA, USA. AAAI Press.

Lennox, B., Montague, G.A., Hiden, H.G., Kornfeld, G., and Goulding, P.R. (2001) Process monitoring of an industrial fed-batch fermentation. *Biotechnology and Bioengineering*, 74(2):125–135.

Leone, F.C., Rutenberg, Y.H., and Topp, C.W. (1950) The use of sample quasi-ranges in setting confidence intervals for the population standard deviation. *Journal of the American Statistical Association*, 56(294):260–272.

Li, W., Yue, H., Valle-Vervantes, S., and Qin, S.J. (2000) Recursive PCA for adaptive process monitoring. *Journal of Process Control*, 10(5):471–486.

Liang, Y.Z. and Kvalheim, O.M. (1996) Robust methods for multivaraite analysis – a tutorial review. *Chemometrics & Intelligent Laboratory Systems*, 32(1):1–10.

Liapounoff, A.M. (1900) Sur une proposition de la théorie des probabilités. *Bulletin de l'Academie Impériale des Sciences de St. Pétersbourg*, 5(13):359–386.

Liapounoff, A.M. (1901) Nouvelle forme du théoréme sur la limite de probabilités, Mémoires de l'Academie Impériale des Sciences de St. Pétersbourg. *Classe Physicomathématique*, 8(12):1–24.

Lieftucht, D., Kruger, U., and Irwin, G.W. (2004) Improved diagnosis of sensor faults using multivariate statistics. In *Proceedings of the American Control Conference*, pages 4403–4407, Boston.

Lieftucht, D., Kruger, U., and Irwin, G.W. (2006a) Improved reliability in diagnosing faults using multivariate statistics. *Computers & Chemical Engineering*, 30(5):901–912.

Lieftucht, D., Kruger, U., Irwin, G.W., and Treasure, R.J. (2006b) Fault reconstruction in linear dynamic systems using multivariate statistics. *IEE Proceedings, Part D – On Control Theory and Applications*, 153(4):437–446.

Lieftucht, D., Völker, M., Sonntag, C., *et al.* (2009) Improved fault diagnosis in multivariate systems using regression-based reconstruction. *Control Engineering Practice*, 17(4):478–493.

Lindberg, W., Persson, J.-A., and Wold, S. (1983) Partial least-squares method for spectrofluorimetric analysis of mixtures of humic acid and ligninsulfonate. *Analytical Chemistry*, 55(4):643–648.

Lindgren, F., Geladi, P., and Wold, S. (1993) The kernal algorithm for PLS. *Journal of Chemometrics*, 7(1):45–59.

Ljung, L. (1999) *System Identification: Theory for the User*. Prentice Hall, Upper Saddle River, NJ, USA, 2nd edition.

Lohmoeller, J.B. (1989) *Latent Variable Path Modelling with Partial Least Squares*. Physica-Verlag, Heidelberg, Germany.

Lucas, J.M. and Saccucci, M.S. (1990) Expnentially weighted moving average schemes: Properties and enhancements. *Technometrics*, 32(1):1–12.

MacGregor, J.F. (1997) Using on-line process data to improve quality: challenges for statisticians. *International Statistical Review*, 65(3):309–323.

MacGregor, J.F., Jaeckle, C., Kiparissides, C., and Koutoudi, M. (1994) Process monitoring andd iagnosis by multiblock plsmethods. *AIChE Journal*, 40(5):826–838.

MacGregor, J.F. and Kourti, T. (1995) Statistical process control of multivariate processes. *Control Engineering Practice*, 3(3):403–414.

MacGregor, J.F., Marlin, T.E., Kresta, J.V., and Skagerberg, B. (1991) Multivariate statistical methods in process analysis and control. In *AIChE Symposium Proceedings of the 4th International Conference on Chemical Process Control*, pages 79–99, New York. AIChE Publication, No. P-67.

Malhi, A. and Gao, R. (2004) PCA-based feature selection scheme for machine defect classification. *IEEE Transaction on Instrumentation and Measurement*, 53(6):1517–1525.

Malinowski, E.R. (1977) Theory of error in factor analysis. *Analytical Chemistry*, 49(4):606–612.

Marcon, M., Dixon, T.W., and Paul, A. (2005) Multivariate SPC applications in the calcining business. In *134th Annual Meeting & Exhibition (TMS 2005)*, Moscone West Convention Center, San Francisco, CA, USA 13–17 February.

Mardia, K.V., Kent, J.T., and Bibby, J.M. (1979) *Multivariate Analysis*. Probability and Mathematical Statistics. Academic Press, London.

Maronna, R.A. (1976) Robust M-estimator of multivariate location and scatter. *The Annals of Statistics*, 4(1):51–67.

Martin, E.B., Morris, A.J., and Lane, S. (2002) Monitoring process manufacturing performance. *IEEE Control Systems Magazine*, 22(5):26–39.

Mason, R.L. and Young, J.C. (2001) *Multivariate Statistical Process Control with Industrial Applications*. ASA-SIAM series on statistical and applied probability. ASA-SIAM, Philadelphia, PA, USA.

Mastronardi, N., van Camp, E., and van Barel, M. (2005) Divide and conquer algorithms for computing the eigendecomposition of symmetric diagonal-plus-semiseparable matrices. *Numerical Algorithms*, 39(4):379–398.

McDowell, N., McCullough, G., Wang, X., Kruger, U., and W., I.G. (2008) Fault diagnostics for internal combustion engines – current and future techniques. In *Proceedings of the 8th International Conference on Engines for Automobiles*, Capri (Naples), Italy.

McFarlane, R.C., Reineman, R.C., Bartee, J.F., and Georgakis, C. (1993) Dynamic simulator of a model IV Fluid catalytic cracking unit. *Computers and Chemical Engineering*, 17(3):275–300.

Meronk, M. (2001) The application of model based predictive control and multivariate statistical process control in industry. Master's thesis, The University of Manchester, Manchester, UK.

Miletic, I., Quinn, S., Dudzic, M., Vaculik, V., and Champagne, M. (2004) An industrial perspective on implementing on-line applications of multivariate statstics. *Journal of Process Control*, 14(8):821–836.

Miller, P., Swanson, R.E., and Heckler, C.F. (1998) Contribution plots: A missing link in multivariate quality control. *Applied Mathematics and Computer Science*, 8(4):775–792.

Ming, R., Haibin, Y., and Heming, Y. (1998) Integrated distribution intelligent system architecture for incidents monitoring and diagnosis. *Computers in Industry*, 37(2):143–151.

Møller, S.F., von Frese, J., and Bro, R. (2005) Robust methods for multivariate data analysis. *Journal of Chemometrics*, 19(10):549–563.

Monostori, L. and Prohaszka, J. (1993) A step towards intelligent manufacturing: modelling and monitoring of manufacturing processes through artificial neural networks. *CIRP Annals – Manufacturing Technology*, 42(1):485–488.

Montgomery, D.C. (2005) *Introduction to Statistical Quality Control*. John Wiley & Sons, Hoboken, NJ, USA 5th edition.

Morud, T. (1996) Multivariate statistical process control; example from the chemical process industry. *Journal of Chemometrics*, 10:669–675.

Mosteller, F. and Wallace, D.L. (1963) Inference in an authorship problem. *Journal of the American Statistical Association*, 58(302):275–309.

Muirhead, R.J. (1982) *Aspects of Multivariate Statistical Theory*. John Wiley & Sons, New York, NY, USA.

Narasimhan, S. and Shah, S.L. (2008) Model identification and error covariance estimation from noisy data using PCA. *Control Engineering Practice*, 16(1):146–155.

Nelson, P.R.C., MacGregor, J.F., and Taylor, P.A. (2006) The impact of missing measurements on PCA and PLS prediction and monitoring applications. *Chemometrics & Intelligent Laboratory Systems*, 80(1):1–12.

Nelson, P.R.C., Taylor, P.A., and MacGregor, J.F. (1996) Missing data methods for PCA and PLS: score calculations with incomplete observations. *Chemometrics & Intelligent Laboratory Systems*, 35(1):45–65.

Nimmo, I. (1995) Adequate address abnormal situation operations. *Chemical Engineering Progress*, 91(1):36–45.

Nomikos, P. and MacGregor, J.F. (1994) Monitoring of batch processes using multiway principal component analysis. *AIChE Journal*, 40:1361–1375.

Nomikos, P. and MacGregor, J.F. (1995) Multivariate SPC charts for monitoring batch processes. *Technometrics*, 37(1):41–59.

Oakland, J.S. (2008) *Statistical Process Control*. Butterworth-Heinemann, Oxford, UK, 6th edition.

Paige, C.C. (1980) Accuracy and effectiveness of the Lanczos algorithm for the symmetric eigenproblem. *Linear Algebra and its Applications*, 34:235–258.

Parlett, B.N. (1980) *The Symmetric Eigenvalue Problem*. Prentice Hall, Englewood Cliffs, NJ, USA.

Pearson, C. (1901) On lines and planes of closest fit to systems of points in space. *Phil. Mag., Series B*., 2(11):559–572.

Pfafferott, J., Herkela, S., and Wambsganß, M. (2004) Design, monitoring and evaluation of a low energy office building with passive cooling by night ventilation. *Energy and Buildings*, 36(5):455–465.

Phillips, G.R. and Eyring, M.E. (1983) Comparison of conventional and robust regression analysis of chemical data. *Analytical Chemistry*, 55(7):1134–1138.

Piovoso, M.J. and Kosanovich, K.A. (1992) Process data chemometric. *IEEE Transactions on Instrumentation and Measurement*, 41(2):262–268.

Piovoso, M.J., Kosanovich, K.A., and Pearson, P.K. (1991) Monitoring process performance in real-time. In *Proceedings of the American Control Conference*, Chicago, IL, USA.

Powers, W.F. and Nicastri, P.R. (1999) Automotive vehicle control challenges in the twenty-first century. In *Proceedings of the 14th IFAC World Congress*, pages 11–29, Beijing, P.R. China.

Pranatyasto, T.N. and Qin, S.J. (2001) Sensor validation and process fault diagnosis for FCC units under MPC feedback. *Control Engineering Practice*, 9(8):877–888.

Pujadó, P.R. and Moser, M. (2006) Catalytic reforming. In *Handbook of Petroleum Processing*, pages 217–237. Springer, The Netherlands.

Qin, S.J. (1998) Recursive PLS algorithms for adaptive data modelling. *Computers and Chemical Engineering*, 22(4-5):503–514.

Qin, S.J., Cherry, G., Good, R., Wang, J., and Harrison, C.A. (2006) Semiconductor manufacturing process control and monitoring: a fab-wide framework. *Journal of Process Control*, 16(3):179–191.

Qin, S.J. and Dunia, R. (2000) Determining the number of principal components for best reconstruction. *Journal of Process Control*, 10(2-3):245–250.

Qin, S.J., Valle, S., and Piovoso, M.J. (2001) On unifying multiblock analysis with application to decentralized process monitoring. *Journal of Chemometrics*, 15(9):715–742.

Qing, Z. and Zhihan, X. (2004) Design of a novel knowledge-based fault detection and isolation scheme. *IEEE Transactions on Systems, Man and Cybernetics, Part B (Cybernetics)*, 34(2):1089–1095.

Raich, A. and Çinar, A. (1996) Statistical process monitoring and disturbance diagnosis in multivariable continuous processes. *AIChE Journal*, 42(4):995–1009.

Ramaker, H.-J., van Sprang, E.N., Westerhuis, J.A., and Smilde, A.K. (2004) The effect of the size of the training set and number of principal components on the false alarm rate in statistical process monitoring. *Chemometrics & Intelligent Laboratory Systems*, 73(2):181–187.

Rännar, S., Geladi, P., Lindgren, F., and Wold, S. (1995) A PLS kernel algorithm for data sets with many variables and fewer objects. part II: Cross-validation, missing data and examples. *Journal of Chemometrics*, 9(6):459–470.

Rännar, S., Lindgren, F., Geladi, P., and Wold, S. (1994) A PLS kernel algorithm for data sets with many variables and fewer objects. part I: Theory and algorithm. *Journal of Chemometrics*, 8(2):111–125.

Rissanen, J. (1978) Modelling by shortest data description. *Automatica*, 14(2):465–471.

Rocke, D.M. and Woodruff, D.L. (1996) Identification of outliers in multivariate data. *Journal of the American Statistical Association*, 91:1047–1061.

Rousseeuw, P.J. (1984) Least median of squares regression. *Journal of the American Statistical Association*, 79:871–880.

Rousseeuw, P.J. and Croux, C. (1993) Alternatives to median absolute deviation. *Journal of the American Statistical Association*, 88:1273–1283.

Rousseeuw, P.J. and Driessen, K. (1999) A fast algorithm for the minimum covariance determinant estimator. *Technometrics*, 41(3):212–223.

Rousseeuw, P.J. and Hubert, M. (2011) Robust statistics for outlier detection. *Wiley Interdisciplinary Reviews: Data Mining and Knowledge Discovery*, 1(1):73–79.

Rozett, R.W. and Petersen, E.M. (1975) Methods of factor analysis of mass spectra. *Analytical Chemistry*, 47(8):1301–1308.

Russell, N.S., Farrier, D.R., and Howell, J. (1985) Evaluation of multinormal probabilities using Fourier series expansions. *Journal of the Royal Statistical Society. Series C (Applied Statistics)*, 34(1):49–53.

Satterthwaite, F.E. (1941) Synthesis of variance. *Psychometrica*, 6:309–316.

Schmidt-Traub, H. and Górak, A. (2006) *Integrated Reaction and Separation Operations: Modelling and Experimental Validation*. Chemische Technik/Verfahrenstechnik. Springer-Verlag, Berlin.

Schubert, U., Kruger, U., Arellano-Garcia, H., Feital, T., and Wozny, G. (2011) Unified model-based fault diagnosis for three industrial application studies. *Control Engineering Practice*, 19(5):479–490.

Schuler, H. (2006) Automation in chemical industry. *ATP Automatisierungstechnische Praxis*, 54(8):363–371.

Schwarz, G. (1978) Estimating the dimension of a model. *Annals of Statistics*, 6(2):461–464.

Sharma, S.K. and Irwin, G.W. (2003) Fuzzy coding of genetic algorithms. *IEEE Transactions on Evolutionary Computation*, 7(4):344–355.

Shewhart, W.A. (1931) *Economic Control of Quality of Manufactured Product*. Van Nostrand Reinhold, Princeton, NJ, USA.

Shewhart, W.A. (1939) *Statistical Method from the Viewpoint of Quality Control*. The Graduate School of the Department of Agriculture, Washington DC, USA. Reprinted by Dover: Toronto, 1986.

Shin, R. and Lee, L.S. (1995) Use of fuzzy cause-effect digraph for resolution fault diagnosis of process plants I. fuzzy cause-effect digraph. *Industrial & Engineering Chemistry Research*, 34:1688–1702.

Shing, C.T. and Chee, P.L. (2004) Application of an adaptive neural network with symbolic rule extraction to fault detection and diagnosis in a power generation plant. *IEEE Transactions on Energy Conversion*, 19(2):369–377.

Simani, S., Patton, R.J., and Fantuzzi, C. (2002) *Model-Based Fault Diagnosis in Dynamic Systems Using Identification Techniques*. Springer-Verlag New York.

Simoglou, A., Martin, E.B., and Morris, A.J. (2000) Multivariate statistical process control of an industrial fluidised-bed reactor. *Control Engineering Practice*, 8(8):893–909.

Skogestad, S. (2007) The do's and dont's of distillation column control. *Chemical Engineering Research and Design*, 85(1):13–23.

Smith, G.M. (2003) *Statistical Process Control and Quality Improvement*. Prentice Hall, Upper Saddel River, NJ, USA, 5th edition.

Söderström, T. (2007) Error-in-variable methods in system identification. *Automatica*, 43(7):939–958.

Söderström, T. and Stoica, P. (1994) *System Identification*. Prentice Hall, Upper Saddle River, NJ, USA.

Sohn, H., Allen, D.W., Worden, K., and Farrar, C.R. (2005) Structural damage classification using extreme value statistics. *Transactions of the ASME. Journal of Dynamic Systems, Measurement and Control*, 127(1):125–132.

Stahel, W.A. (1981) *Robust Estimation, Infinitisimal Optimality and Covariance Matrix Estimator*. PhD thesis, ETH Zurich, Zurich, Switzerland.

Stander, C.J., Heyns, P.S., and Schoombie, W. (2002) Using vibration monitoring for local fault detection on gears operating under fluctuating load conditions. *Mechanical Systems & Signal Processing*, 16(6):1005–1024.

Staszewski, W.J. and Tomlinson, G.R. (1997) Time-frequency analysis in gearbox fault detection using the Wigner-Ville distribution and pattern recognition. *Mechanical Systems & Signal Processing*, 11(5):673–692.

Stewart, D. and Love, W.A. (1968) A general canonical correlation index. *Psychological Bulletin*, 70(3):160–163.

Stewart, G.W. and Sun, J.-G. (1990) *Matrix Perturbation Theory*. Academic Press, San Diego, CA, USA.

Stone, M. (1974) Cross-validatory choice and assessment of statistical prediction (with discussion). *Journal of the Royal Statistical Society (Series B)*, 36:111–133.

Taguchi, G. (1986) *Introduction to Quality Engineering and Redesigning Quality Into Products and Processes*. Asian Productivity Organisation, Tokyo.

Tan, C. and Mba, D. (2005) Limitation of acoustic emission for identifying seeded defects in gearboxes. *Nondestructive Evaluation*, 24(1):11–28.

Tate, R.F. and Klett, G.W. (1959) Optimal confidence intervals for the variance of a normal distribution. *Journal of the American Statistical Association*, 54(287):674–682.

Tates, A.A., Louwerse, D.J., Smilde, A.K., Koot, G.L.M., and Berndt, H. (1999) Monitoring a PVC batch process with multivariate statistical process control charts. *Industrial & Engineering Chemistry Research*, 38(12):4769–4776.

ten Berge, J.M.F. (1985) On the relationship between Fortier's simultaneous linear prediction and van den Wollenberg's redundancy analysis. *Psychometrika*, 50(1):121–122.

ter Braak, C.J.F. and de Jong, S. (1998) The objective function of partial least squares regression. *Journal of Chemometrics*, 12(1):41–54.

Thompson, J.R. and Koronacki, J. (2002) *Statistical Process Control – The Deming Paradigm and Beyond*. Chapman & Hall/CRC Press, Boca Raton, 2nd edition.

Thompson, W.A. and Endriss, J. (1961) The required sample size when estimating variances. *The American Statistician*, 15(3):22–23.

Tracey, N.D., Young, J.C., and Mason, R.L. (1992) Multivariate control charts for individual observations. *Journal of Quality Technology*, 24(2):88–95.

Tumer, I.Y. and Bajwa, A. (1999) A survey of aircraft engine health monitoring systems. In *Proceedings of the 35th AIAA/ASME/SAE/ASEE Joint Propulsion Conference and Exhibit*, Los Angeles, U.S.A.

Upadhyaya, B.R., Zhao, K., and Lu, B. (2003) Fault monitoring of nuclear power plant sensors and field devices. *Progress in Nuclear Energy*, 43(1-4):337–342.

Valle, S., Li, W., and Qin, S.J. (1999) Selection of the number of principal components: The variance of the reconstruction error criterion compared to other methods. *Industrial & Engineering Chemistry Research*, 38:4389–4401.

van den Wollenberg, A.L. (1977) Redundancy analysis and alternative for canonical correlation analysis. *Psychometrika*, 42(2):207–219.

van Huffel, S. and Vandewalle, J. (1991) *The Total Least Squares Problem*. Society for Industrial and Applied Mathematics, Philadelphia PA, UAS.

van Sprang, E.N.M., Ramaker, H.J., Westerhuis, J.A., Gurden, S.P., and Smilde, A.K. (2002) Critical evaluation of approaches for on-line batch process monitoring. *Chemical Engineering Science*, 57(18):3979–3991.

Vedam, H. and Venkatasubramanian, V. (1999) PCA-SDG based process monitoring and fault detection. *Control Engineering Practice*, 7(7):903–917.

Velicer, W.F. (1976) Determining the number of components from the matrix of partial correlations. *Psychometrika*, 41(3):321–327.

Veltkamp, D.J. (1993) Multivariate monitoring and modeling of chemical processes using chemometrics. In *International Forum on Process Analytical Chemistry IFPAC*, volume 5 of *Process Control & Quality*, pages 205–217, Galveston, TX, USA.

Venkatasubramanian, V., Rengaswamy, R., Yin, K., and Kavuri, S.N. (2003) A review of process fault detection and diagnosis: Part I: Quantitative model-based methods. *Computers & Chemical Engineering*, 27(3):293–311.

Walczak, B. and Massart, D.L. (1996) The radial basis functions – partial least squares approach as a flexible non-linear regression technique. *Analytica Chimica Acta*, 331:177–185.

Wang, W. (2008) Autoregressive model-based diagnostics for gears and bearings. *Insight-Non-Destructive Testing and Condition Monitoring*, 50(8):414–418.

Wang, X., Kruger, U., and Irwin, G.W. (2005) Process monitoring approach using fast moving window PCA. *Industrial & Engineering Chemistry Research*, 44(15):5691–5702.

Wang, X., Kruger, U., Irwin, G.W., McCullough, G., and McDowell, N. (2008) Nonlinear PCA with the local approach for diesel engine fault detection and diagnosis. *IEEE Transactions on Control Systems Technology*, 16(1):122–129.

Wang, X., Kruger, U., and Lennox, B. (2003) Recursive partial least squares algorithms for monitoring complex industrial processes. *Control Engineering Practice*, 11(6):613–632.

Wangen, L.E. and Kowalski, B.R. (1989) A multiblock partial least squares algorithm for investigating complex chemical systems. *Journal of Chemometrics*, 3(1):3–20.

Wax, M. and Kailath, T. (1985) Detection of signals by information theoretic criteria. *IEEE Transactions on Acoustics, Speech, and Signal Processing*, 33(2):387–392.

Wentzell, P.D., Andrews, D.T., Hamilton, D.C., Faber, K., and Kowalski, B.R. (1997) Maximum likelihood principal component analysis. *Journal of Chemometrics*, 11(4):339–366.

Westergren, K.E., Hans Höberg, H., and Norlén, U. (1999) Monitoring energy consumption in single-family houses. *Energy and Buildings*, 29(3):247–257.

Westerhuis, J.A., Kourti, T., and F., M.J. (1998) Analysis of multiblock and hierarchical PCA and PLS models. *Journal of Chemometrics*, 12(5):301–321.

Wikström, C., Albano, C., Eriksson, L., *et al.* (1998) Multivariate process and quality monitoring applied to an electrolysis process; part I. process supervision with multivariate control charts. *Chemometrics & Intelligent Laboratory Systems*, 42:221–231.

Willink, T. (2008) Efficient adaptive adaptive SVD algorithm for MIMO applications. *IEEE Transactions on Signal Processing*, 56(2):615–622.

Wilson, D.J.W. (2001) Plant-wide multivariate SPC in fiber manufacturing. *Chemical Fibers International*, 51(1):72–73.

Wise, B.M. and Gallagher, N.B. (1996) The process chemometrics approach to process monitoring and fault detection. *Journal of Process Control*, 6(6):329–348.

Wise, B.M., Ricker, N.L., and Veltkamp, D.J. (1989a) Upset and sensor failure detection in multivariate processes. Technical report, Eigenvector Research, Inc., Wenatchee, WA, USA.

Wise, B.M., Ricker, N.L., and Veltkamp, D.J. (1989b) Upset and sensor detection in a multivariate process. In *AIChE Annual Meeting*, San Francisco, CA, USA.

Wise, B.M., Veltkamp, D.J., Davis, B., Ricker N.L. and Kowalski, B.R. (1988) Principal Components Analysis for Monitoring the West Valley Liquid Fed Ceramic Melter. Waste Management '88 Proceedings, pp, 811–818, Tucson AZ USA.

Wold, H. (1966a) Estimation of principal components and related models by iterative least squares. In Krishnaiah, P.R., editor, *Multivariate Analysis*, pages 391–420. Academic Press, NY USA.

Wold, H. (1966b) Non-linear estimation by iterative least squares procedures. In David, F., editor, *Research Papers in Statistics*. Wiley, NY USA.

Wold, S. (1978) Cross validatory estimation of the number of principal components in factor and principal component models. *Technometrics*, 20(4):397–406.

Wold, S., Esbensen, K., and Geladi, P. (1987) Principal component analysis. *Chemometrics and Intelligent Laboratory Systems*, 2:37–52.

Wold, S., Ruhe, A., Wold, H., and Dunn, W.J. (1984) The collinearity problem in linear regression. The partial least squares (PLS) approach to generalised inverses. *SIAM Journal on Scientific and Statistical Computing*, 5(3):735–743.

Wu, E.H.C., Yu, P.L.H., and Li, W.K. (2009) A smooth bootstrap test for independence based on mutual information. *Computational Statistics and Data Analysis*, 53:2524–2536.

Wu, J., Hsu, C., and Wu, G. (2008) Fault gear identification and classification using discrete wavelet transform and adaptive neuro-fuzzy inference. *Expert Systems with Applications*, 36(3):6244–6255.

Yang, Y.M. and Guo, C.H. (2008) Gaussian moments for noisy unifying model. *Neurocomputing*, 71:3656–3659.

Yoon, S. and MacGregor, J.F. (2000) Statistical causal model-based approaches to fault detection and isolation. *AIChE Journal*, 46(9):1813–1824.

Yoon, S. and MacGregor, J.F. (2001) Fault diagnosis with multivariate statistical models part I: Using steady state fault signatures. *Journal of Process Control*, 11(4):287–400.

Young, P.J. (1994) A reformulation of the partial least squares regression algorithm. *SIAM Journal on Scientific Computing*, 15(1):225–230.

Yue, H.H. and Qin, S.J. (2001) Reconstruction-based fault identification using a combined index. *Industrial & Engineering Chemistry Research*, 40(20):4403–4414.

Index

References to figures are given in italic type. References to tables are given in bold type.

Statistical Monitoring of Complex Multivariate Processes: With Applications in Industrial Process Control, First Edition. Uwe Kruger and Lei Xie.

Statistics in Practice

Human and Biological Sciences

Berger – Selection Bias and Covariate Imbalances in Randomized Clinical Trials
Berger and Wong – An Introduction to Optimal Designs for Social and Biomedical Research
Brown and Prescott – Applied Mixed Models in Medicine, Second Edition
Carstensen – Comparing Clinical Measurement Methods
Chevret (Ed) – Statistical Methods for Dose-Finding Experiments
Ellenberg, Fleming and DeMets – Data Monitoring Committees in Clinical Trials: A Practical Perspective
Hauschke, Steinijans & Pigeot – Bioequivalence Studies in Drug Development: Methods and Applications
Källén – Understanding Biostatistics
Lawson, Browne and Vidal Rodeiro – Disease Mapping with Win-BUGS and MLwiN
Lesaffre, Feine, Leroux & Declerck – Statistical and Methodological Aspects of Oral Health Research
Lui – Statistical Estimation of Epidemiological Risk
Marubini and Valsecchi – Analysing Survival Data from Clinical Trials and Observation Studies
Millar – Maximum Likelihood Estimation and Inference: With Examples in R, SAS and ADMB
Molenberghs and Kenward – Missing Data in Clinical Studies
O'Hagan, Buck, Daneshkhah, Eiser, Garthwaite, Jenkinson, Oakley & Rakow – Uncertain Judgements: Eliciting Expert's Probabilities
Parmigiani – Modeling in Medical Decision Making: A Bayesian Approach
Pintilie – Competing Risks: A Practical Perspective
Senn – Cross-over Trials in Clinical Research, Second Edition
Senn – Statistical Issues in Drug Development, Second Edition
Spiegelhalter, Abrams and Myles – Bayesian Approaches to Clinical Trials and Health-Care Evaluation
Walters – Quality of Life Outcomes in Clinical Trials and Health-Care Evaluation

Welton, Sutton, Cooper and Ades – Evidence Synthesis for Decision Making in Healthcare
Whitehead – Design and Analysis of Sequential Clinical Trials, Revised Second Edition
Whitehead – Meta-Analysis of Controlled Clinical Trials
Willan and Briggs – Statistical Analysis of Cost Effectiveness Data
Winkel and Zhang – Statistical Development of Quality in Medicine

Earth and Environmental Sciences

Buck, Cavanagh and Litton – Bayesian Approach to Interpreting Archaeological Data
Chandler and Scott – Statistical Methods for Trend Detection and Analysis in the Environmental Statistics
Glasbey and Horgan – Image Analysis in the Biological Sciences
Haas – Improving Natural Resource Management: Ecological and Political Models
Helsel – Nondetects and Data Analysis: Statistics for Censored Environmental Data
Illian, Penttinen, Stoyan, H and Stoyan D-Statistical Analysis and Modelling of Spatial Point Patterns
McBride – Using Statistical Methods for Water Quality Management
Webster and Oliver – Geostatistics for Environmental Scientists, Second Edition
Wymer (Ed) – Statistical Framework for Recreational Water Quality Criteria and Monitoring

Industry, Commerce and Finance

Aitken – Statistics and the Evaluation of Evidence for Forensic Scientists, Second Edition
Balding – Weight-of-evidence for Forensic DNA Profiles
Brandimarte – Numerical Methods in Finance and Economics: A MATLAB-Based Introduction, Second Edition
Brandimarte and Zotteri – Introduction to Distribution Logistics
Chan – Simulation Techniques in Financial Risk Management
Coleman, Greenfield, Stewardson and Montgomery (Eds) – Statistical Practice in Business and Industry
Frisen (Ed) – Financial Surveillance

Fung and Hu – Statistical DNA Forensics
Gusti Ngurah Agung – Time Series Data Analysis Using EViews
Kenett (Eds) – Operational Risk Management: A Practical Approach to Intelligent Data Analysis
Kenett (Eds) – Modern Analysis of Customer Surveys: With Applications using R
Kruger and Xie – Statistical Monitoring of Complex Multivariate Processes: With Applications in Industrial Process Control
Jank and Shmueli (Ed.) – Statistical Methods in e-Commerce Research
Lehtonen and Pahkinen – Practical Methods for Design and Analysis of Complex Surveys, Second Edition
Ohser and Mücklich – Statistical Analysis of Microstructures in Materials Science
Pourret, Naim & Marcot (Eds) – Bayesian Networks: A Practical Guide to Applications
Taroni, Aitken, Garbolino and Biedermann – Bayesian Networks and Probabilistic Inference in Forensic Science
Taroni, Bozza, Biedermann, Garbolino and Aitken – Data Analysis in Forensic Science